T0399232

# TOMATO PROCESSING BY-PRODUCTS

# Tomato processing BY-PRODUCTS

## Sustainable Applications

Edited by

**MEJDI JEGUIRIM**

*The institute of Materials Science of Mulhouse (IS2M), University of Haute Alsace, University of Strasbourg, CNRS, UMR, Mulhouse, France*

**ANTONIS A. ZORPAS**

*Open University of Cyprus, Faculty of Pure and Applied Sciences, Environmental Conservation and Management, Laboratory of Chemical Engineering and Engineering Sustainability, Latsia, Nicosia, Cyprus*

Academic Press is an imprint of Elsevier
125 London Wall, London EC2Y 5AS, United Kingdom
525 B Street, Suite 1650, San Diego, CA 92101, United States
50 Hampshire Street, 5th Floor, Cambridge, MA 02139, United States
The Boulevard, Langford Lane, Kidlington, Oxford OX5 1GB, United Kingdom



**British Library Cataloguing-in-Publication Data**
A catalogue record for this book is available from the British Library

**Library of Congress Cataloging-in-Publication Data**
A catalog record for this book is available from the Library of Congress

ISBN: 978-0-12-822866-1

For Information on all Academic Press publications visit our website at https://www.elsevier.com/books-and-journals

*Publisher:* Charlotte Cockle
*Acquisitions Editor:* Nina Rosa de Araujo Bandeira
*Editorial Project Manager:* Allison Hill
*Production Project Manager:* Debasish Ghosh
*Cover Designer:* Vicky Pearson Esser

Typeset by Aptara, New Delhi, India

# Contents

# Contributors

**Agapiou Agapios**
Department of Chemistry, University of Cyprus, Nicosia, Cyprus

**Miguel Carmona-Cabello**
Department of Physical Chemistry and Applied Thermodynamics, EPS, Ed. Leonardo da Vinci, Campus of Rabanales, University of Cordoba, Campus of International Agrifood Excellence ceiA3, Spain

**Sofia Chanioti**
Laboratory of Food Chemistry and Technology, School of Chemical Engineering, National Technical University of Athens, Zografou, Greece

**M. Pilar Dorado**
Department of Physical Chemistry and Applied Thermodynamics, EPS, Ed. Leonardo da Vinci, Campus de Rabanales, Universidad de Córdoba, Campus de Excelencia Internacional Agroalimentario ceiA3, Spain

**Trinidad de Evan**
Department of Agricultural Production, Technical University of Madrid (UPM), Madrid, Spain

**Georgia Frakolaki**
Laboratory of Food Chemistry and Technology, School of Chemical Engineering, National Technical University of Athens, Zografou, Greece

**Virginia Giannou**
Laboratory of Food Chemistry and Technology, School of Chemical Engineering, National Technical University of Athens, Zografou, Greece

**Helmi Hamdi**
Center for Sustainable Development, College of Arts and Sciences, Qatar University, Doha, Qatar

**Noureddine Hamdi**
Higher Institute of Water Sciences and Techniques of Gabes, Zrig, Gabes, Tunisia; National Center of Research in Materials Sciences (CNRSM), Borj Cedria Technopole, Tunisia

**Maya Ibrahim**
Lebanese University, Faculty of Sciences, Laboratory of Physical Chemistry of Materials LCPM/PR2N, Fanar, Lebanon

**Selsabil Elghazel Jeguirim**
Textile Research Unit of ISET of Ksar-Hellal, Ksar Hellal, Tunisia

**Mejdi Jeguirim**
The institute of Materials Science of Mulhouse (IS2M), University of Haute Alsace, University of Strasbourg, CNRS, UMR, Mulhouse, France

**Salah Jellali**
PEIE Research Chair for the Development of Industrial Estates and Free Zones, Center for Environmental Studies and Research, Sultan Qaboos University, Al-Khoud, Muscat, Oman

**Maria Katsouli**
Laboratory of Food Chemistry and Technology, School of Chemical Engineering, National Technical University of Athens, Zografou, Greece

**Tryfon Kekes**
Laboratory of Food Chemistry and Technology, School of Chemical Engineering, National Technical University of Athens, Zografou, Greece

**Besma Khiari**
Laboratory of Wastewaters and Environment, Centre of Water Researches and Technologies (CERTE), Technopark Borj Cedria, Tunisia

**Madona Labaki**
Lebanese University, Faculty of Sciences, Laboratory of Physical Chemistry of Materials LCPM/PR2N, Fanar, Lebanon

**George Liadakis**
Laboratory of Food Chemistry and Technology, School of Chemical Engineering, National Technical University of Athens, Zografou, Greece

**Ignacio Gómez Lucas**
Agrochemical and Environment Department, Miguel Hernandez University of Elche, Avd. de la Universidad sn, Elche, Alicante, Spain

**Carlos Navarro Marcos**
Department of Agricultural Production, Technical University of Madrid (UPM), Madrid, Spain

**Stylianou Marinos**
Department of Chemistry, University of Cyprus, Nicosia, Cyprus; Open University of Cyprus, Faculty of Pure and Applied Sciences, Environmental Conservation and Management, Laboratory of Chemical Engineering and Engineering Sustainability, Latsia, Nicosia, Cyprus

**Tsangas Michail**
Open University of Cyprus, Faculty of Pure and Applied Sciences, Environmental Conservation and Management, Laboratory of Chemical Engineering and Engineering Sustainability, Latsia, Nicosia, Cyprus

**Eduarda Molina-Alcaide**
Zaidin Experimental Station (EEZ), Spanish National Research Council, Granada, Spain

**Jose Navarro Pedreño**
Agrochemical and environment department, Miguel Hernandez University of Elche, Avd. de la Universidad sn, Elche, Alicante, Spain

**Khalifa Riahi**
High School of Engineers of Medjez El Bab (ESIM), UR-GDRES, University of Jendouba, Tunisia

**María Desamparados Soriano Soto**
School of Agricultural Engineering and Environment, Polytechnic University of Valencia, Camino de Vera, Valencia, Spain

**Andrius Tamošiūnas**
Lithuanian Energy Institute, Plasma Processing Laboratory, Kaunas, Lithuania

**Psichoula Terpsithea**
Open University of Cyprus, Faculty of Pure and Applied Sciences, Environmental Conservation and Management, Laboratory of Chemical Engineering and Engineering Sustainability, Latsia, Nicosia, Cyprus

**María Dolores Carro Travieso**
Department of Agricultural Production, Technical University of Madrid (UPM), Madrid, Spain

**Constantina Tzia**
Laboratory of Food Chemistry and Technology, School of Chemical Engineering, National Technical University of Athens, Zografou, Greece

**Antonis A. Zorpas**
Open University of Cyprus, Faculty of Pure and Applied Sciences, Environmental Conservation and Management, Laboratory of Chemical Engineering and Engineering Sustainability, Latsia, Nicosia, Cyprus

CHAPTER ONE

# Identification, quantification, and characterization of tomato processing by-products

**George Liadakis, Maria Katsouli, Sofia Chanioti, Virginia Giannou, Constantina Tzia**
Laboratory of Food Chemistry and Technology, School of Chemical Engineering, National Technical University of Athens, Zografou, Greece

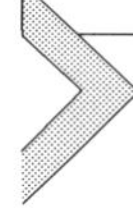

## 1.1 Introduction (quantitative data for raw materials, industries, final products)

Tomato is the edible fruit of the plant *Lycopersicum esculentum*, which belongs to the nightshade family of Solanides (*Solanaceae*) along with potato, eggplant, and pepper. Although it is customarily considered a vegetable, it is actually a fruit, and more specifically a berry, based on its plant parts.

The history of tomato dates back to 700 AD, in the tropical regions of South America (Peru, Bolivia, Ecuador) or Mexico, where even today its wild plants can still be found. The name "tomato" originates from the word "tomatl" used in the native Náhuatl dialect of the Aztecs that gave rise to the Spanish word "tomate." In the literature it has also been referred to as Pomid'oro (golden apple) or pomme d'amour (apple of love). Tomato was first introduced in Europe from Spanish explorers in the early 16$^{th}$ century and was initially used as a food by the Italians. In the northern Europe, tomato was originally cultivated as an ornamental plant and was considered poisonous. However, this is partly true, since all the green parts of the plant contain the neurotoxin solanine. Its cultivation in the USA began two centuries later when it was brought there by European immigrants.

Tomato is an annual or short-lived perennial, self-pollinated plant. It can be grown on a wide range of soils from sandy to heavy clay, with well-drained, sandy or red loam and slightly acidic (pH 5.8 to 6.8) soils being considered ideal. It requires low to medium rainfall, and ideal growth temperatures range between 21 and 24°C. Tomato is one of the most highly cultivated plants around the world, for both commercial or domestic purposes, and its global production reached 180 million tons in 2018 according to the Food and Agriculture Organization (FAO). Its main production

*Tomato Processing by-Products*
DOI: https://doi.org/10.1016/B978-0-12-822866-1.00004-1

areas are located in temperate zones, close to the 40th parallels, mainly in the Northern hemisphere. China is by far the largest producer, accounting for more than one third of the global production, followed by India, USA, and Turkey [41,65].

On domestic level tomatoes are eaten raw, cooked, sun-dried or pickled. However more than 40 million tons of tomatoes are processed each year in the food industry around the world, while their consumption is expected to exceed 50 million tons by 2025. Commercial tomato products include canned tomatoes, tomato juice, puree and paste, ketchup, tomato soup, dried tomatoes or dehydrated pulp.

Industrial tomato processing, however, leads to the production of a significant volume of by-products, which represent about 5% w/w of the processed tomatoes and generate up to 4 million tons of organic matter. This is known as "tomato pomace" and contains tomato peels and seeds and a small fraction of pulp residues. The environmental management of tomato pomace is considered a problem for the tomato industry, since it has no commercial value and, in most cases, remains underutilized, is disposed as waste or used to a limited extent as animal feed. Nevertheless, tomato pomace, if properly treated, can be a rich source of nutrients, bioactive compounds and valuable phytochemicals, such as carotenoids, phenolic compounds and lycopene [15,23,73].

Aim of this chapter is to introduce all the steps involved in the tomato processing industry and the respective generation of the different by-products. In addition, state of the art and advisable practices for the sustainable management of tomato processing by-products will be addressed. Analysis of the legislative frameworks will be referred, whereas the current situation and policy recommendations will be addressed.

## 1.2 Tomato production processing and technology

Tomato (*Solanum lycopersicum*) is the second most consumed vegetable/fruit, ranks third in terms of production volumes and first in processing volumes. Fresh or processed tomatoes are commonly used in households, food processing and restaurants. The global tomato processing industry processes around 40 million tons of tomatoes. The most consumed processed tomatoes are paste, sauce, juice, ketchup, and whole peeled and diced canned products. The majority of the processing tomatoes is manufactured into bulk paste, while the rest is processed as whole peeled or diced peeled tomato products [44,77].

### 1.2.1 Harvest - preharvest factor affecting the tomato quality

The optimum time for tomato harvesting, is based on the maximum percentage of red ripe usable fruit. The time of harvest depends on the variations among different cultivars, maximum and/or minimum temperatures, and rainfall, followed by actual fruit count of a given size. Moreover, the type of soil, the irrigation practices, the date of planting as well as the stress conditions may affect the time of harvest. Specifically, for mechanical harvesting the fruit should be mature and ripe at the same time, however it should be firm and crack-resistant and in a good condition during transit. The plants should not have excessive foliage because vegetative growth interferes with the harvest processing. The stem should be jointless so that the fruits are not punctured during handling [9,65]. The main components of the mechanical tomato harvesters are the following:

- Pick up mechanism
- Fruit and vine separating area
- Hand sorting area
- Discharge or container-loading mechanism

During mechanical harvesting several problems may occur such as:

- Contamination with soil: Soil is present on the surface of the fruit as a result of the method of severing or removing the plant from the ground as it enters the machine.
- Microbiological contamination: Soil in intimate contact with tomatoes during the harvesting operation can create a serious problem for the processor because soil contains many common spoilage microorganisms (*Clostridium botulinum*).
- Loss of product quality: During harvesting all the ripe, overripe, and cull fruits are collected, thus the quality of tomatoes is varied. The final quality (color and defects) of the harvested fruit will be determined during sorting. The amount of damaged and bruised fruit depends on the handling system; the size and type (wet or dry) of container [58,83].

### 1.2.2 Postharvest technologies to maintain tomato fruit quality

Tomato fruit has a relatively reduced postharvest life since many processes generating quality loss take place soon after harvesting. The major limiting factors in the storage of tomato fruit are transpiration, fungal infection, acceleration of the ripening process and senescence. The ripening process of tomato is controlled by the hormone ethylene, involving various biochemical, chemical, physical, and physiological changes that occur during plants' growth and quickly after harvest. These changes can render the fruit at an over-ripe state or even unmarketable. Since the production of ethylene accelerates ripening and leads to senescence, post-harvest storage technologies focus on controlling and altering the actions of this hormone in order to delay the quality losses and to extend the post-harvest shelf life of tomato [13].

Tomato ripens rapidly and becomes unmarketable shortly under ambient temperature, so the use of low temperatures is recommended for delaying and reducing ethylene production. However, only temperatures over 11°C are advisable for postharvest storage as tomatoes are sensitive to chill injury. Moreover, heat treatments are also proposed to maintain tomato quality parameters after harvesting. Generally, hot water dips or hot air treatments (27 to 48°C and 15 to 90 min) can be used to protect the membrane components of the tomato as heat treatments can alleviate chill injuries by reducing decay, loss of firmness, and electrolyte leakage. Apart from the above-mentioned techniques, other postharvest technologies are also proposed such as modified atmosphere storage, 1-methylcyclopropene (1-MCP) application and the use of edible coatings [9,13,65].

### 1.2.3 Tomato transit from field to processing factory

After harvesting, tomatoes are transported using trucks and are primarily delivered to the processing factory in 10 to 25-ton loads. Loads are graded at the fruit inspection station, which are located next to the weight-scale where tomato trailers are weighed. The tomatoes are unloaded by flushing the loads with water. The majority of industries use recycled water for unloading and it is used as long as possible. Flumes should have sufficient size (depth and width with water volume and pressure) to allow fruits to move without clogging the system under maximum capacity. Preferably, the flume should have both air and steam pipes at the bottom to permit agitation of the water. The steam inlets can raise the water temperature to 54°C. A properly sized flume will allow fruits to be held for a period up to 3 min, in order to loosen the soil from the fruit prior to the final wash. Such treatment can be effective in removing drosophila eggs and larvae. The tomatoes are washed into holding flumes (up to 50 tons). In the first flume system considerable amounts of sand, dirt, mud, may accumulate during a given period of time depending of course, on the harvesting conditions [66,83].

Most of tomato industries use water handling because water offers certain advantages over dry handling:

- It prevents cracking and bruising of tomatoes, as the water serves as a cushion for depositing the tomatoes from the harvester into the containers, as well as during transportation from the field to the factory.

- It causes limited drosophila problem, as chlorine-washed tomatoes are less attractive to fruit flies. This problem is nonexistent except for the fruit floating on the top of the water in the tanks.
- It provides decreased mold count values compared to dry bulk-handling, since antimycotic or other chemical agents can be added to the water to control mold growth.
- It enhances tomatoes' quality depending on the variety, maturity of the fruit, and the number of microorganisms in the tanks.

## 1.2.4 Raw material pretreatments

### *1.2.4.1 Sorting*

Loads are subsequently evaluated based on fruit color; soluble solids content; pH; unripe fruit content; defects, such as worm damage, mechanical damage, mold, blemishes, and decay; as well as on the presence of materials other than tomatoes (stems, leaves, vines, roots, dirt, stones, and trash). Probably the most important concern is manual fruit-sorting or the operation of the electronic color sorter. Over-sorting or elimination of good fruit may result in profit loss. Under-sorting may also be costly as the loads will probably have to be re-sorted to maintain quality standards [64,79].

#### 1.2.4.1.1 Dry sorting

The first unit operation in preparing tomatoes for processing is dry sorting. The purpose of dry sorting is to remove material other than tomatoes and defective fruit (green, decomposed, or unfit) which would otherwise contaminate wash waters. A conveyor or roller belt allows loose materials to be separated from the tomatoes. The speed of the belt varies from 15 to 30 rpm. The unit should have a sufficient sorting area to permit adequate sorting, and an installed artificial daylight illumination to allow 24-h operation [64,79].

#### 1.2.4.1.2 Size grading

A vast number of sorting equipment has been developed, because of the variability in the size of tomato varieties. The most commonly used equipment are the diverging-belt graders or clover-leaf drums with varying hole sizes. These machines result in a uniform fill for whole pack portion control. Nowadays, the new varieties are smaller with more uniform size, thus the need for size grading as a unit operation may be lessened in the future [64,79].

#### 1.2.4.1.3 Final sorting and trimming

During the final sorting and trimming, the off-colored and defective fruits or parts such as rotten areas, mold portions, insect, etc. are removed. Specifically, the washed tomatoes are sorted and only the large perfect ones are sent up to the peeler or scalder, the rotten ones to the dump, and the misshapen and small ones to the pulping line. Therefore, sorting can be characterized as a profitable investment. Depending on the condition of the fruit, they are used for different processed tomato products, for instance, only the best tomatoes are used for canning, the overripe and unevenly ripened ones are used only for pulping and the ones with a small amount of rot or green areas are trimmed and canned as extra or standard grade [64,79].

### *1.2.4.2 Peeling*

Peeling is an important unit operation as it does not only affect the palatability, quality, and nutrient value of the final products, but also the subsequent wastewater treatments, as well as the energy and water consumption. Steam or lye peeling are commonly used to loosen the peel from the flesh before peels are mechanically removed. However, both methods are quite water- and energy-intensive. Commonly, hot lye peeling is used by many tomato processing industries due to its higher product yields and superior product quality, even though their discharged wastewater shave excessively high pH values and chemical oxygen demand. Steam peeling in comparison to lye peeling is chemical-free and environmentally friendlier, but it exhibits higher peeling loss and lower quality of the peeled fruits due to the inconsistency of heating during the process. It should be highlighted that for both steam and lye peeling, under-peeling may increase the difficulty in peel removal. However, over-peeling may degrade the texture and flavor of the fruits and may create undesirable color, leading to soft and mushy peeled tomato products.

The procedure applied in each case is briefly described below.

#### 1.2.4.2.1 Steam peeling

Tomatoes are conveyed to the live steam chamber in order to be scalded. However, they should remain in contact with the steam only long enough to completely loosen the peel and not to well heat the fruits or to soften the pulp and the flesh. Live steam conditions vary from 30 to 98 s and 100 to 149°C depending on fruit size and variety, and the stage of maturity or ripeness. Sometimes closed steam coils are added to the scalder above the tomatoes to increase heat inside the chamber. After the scalder, tomatoes are sprayed with cold water in order to crack their skin. The light scalding requires the use of a peeling knife in order to completely remove the peel. However, this added step increases not only the labor cost, but also adds to production time losses, gives undesirable appearance to the peeled tomatoes, and generally leaves some peel on the finished product. The over-scalding should be avoided, because over-softened tomatoes result into excessive product losses and poor quality for the industry.

In another steam peeling system, the tomatoes are immersed in a hot water tank, agitated by steam for a few seconds and then discharge into a hopper where the tomato are sprayed by cold-water. The sudden drop in temperature makes the peel loosen without loss of other tissues [38,58,79].

#### 1.2.4.2.2 Lye peeling

For many years chemical peeling of fruits and vegetables has been used in the food processing industry mainly for economic reasons and due to its simplicity and labor-saving advantages. The active agent in lye peeling, also known as lye, is a water solution of sodium hydroxide (NaOH) at a concentration of 16% to 20% wt depending on temperature and the addition of wetting agent. Suitable wetting agents added to the lye peeling solutions provide more uniform peeling, either in less time or with lower concentrations of lye, and decrease the amount of water required to remove residual lye. Two wetting agents approved by FDA are the anionic surfactants: sodium 2-ethylhexyl sulfate and sodium mono- and dimethyl naphthalene sulfonates. During lye peeling, tomatoes are immersed in a hot lye bath or are sprayed with lye for a definite period of time. Tomatoes are then drained to remove the excess lye and washed. Basic equipment includes the draper-type scalder, Fox lye-spray scalder, mill-wheel scalder, and the combination of lye-filming and pressure steam peeler [38,58,79].

Other peeling methods (e.g., infrared peeling) have also been proposed. After peeling, defective or unpeeled tomatoes are removed using two sequential manual and optical sorting instruments [64].

Peeled tomatoes are diced and blended with tomato juice, while intact peeled tomatoes are used to produce whole peeled tomatoes.

### 1.2.5 Processing - processed tomato products

Following grading, common processing steps include washing, sorting, and sizing of tomatoes. The tomatoes are then segregated into separate streams for further processing based on the target final products. Tomato processing, for bulk paste and diced products, is a seasonal operation. However, throughout the processing season, plants run continuously with intermittent cleaning-in-place. It is important for processing plants to operate on a constant flow of fresh tomatoes in order to achieve their maximum capacity. Otherwise, experiencing downtime outside of cleaning or running below threshold capacity may lead to wastage, both of the fresh tomatoes waiting in trucks, and of the tomatoes at various stages of processing [41,58].

The basic processing steps for the various tomato products are illustrated in Fig. 1.1 and described in more detail below.

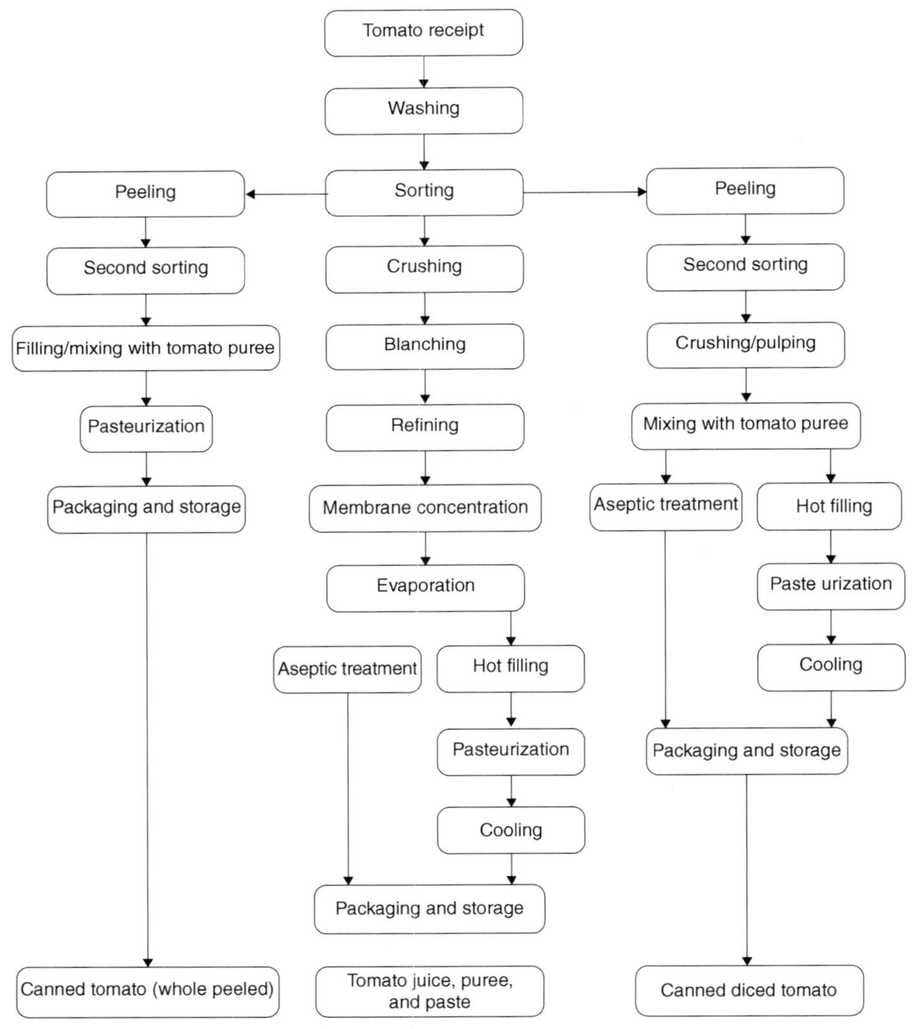


**Fig. 1.1** ***Flow diagram for the processing of tomato products.***

### 1.2.5.1 *Canned tomatoes*

Generally, for canned products, a thorough washing is an essential step. A combination of soaking (3 to 5 min at 54°C) and spraying process that reaches every part of the tomato surface is recommended. Whether soaking is required, it depends largely on the kind of soil and the amount of rain during growth.

After soaking, tomatoes are sorted to dissociate off-color and defective fruit or fruit parts. For efficiency in sorting, tomatoes are turned on roller conveyors, and the sorters separate tomatoes intended for trimming from those to be discarded. Based on the quality standards, tomatoes with stem scars greater than 6 mm in diameter require coring, however, for many new varieties coring is not necessary. Different coring methods exist, such as hand or water-powered machine coring, but machine coring is generally more cost-effective and less wasteful. Following the scalder, tomatoes are subjected to cold water spraying or dry vacuum to crack their skin. Tomatoes are then subjected to peeling using lye, steam or hot water as described above. Peeling along with coring, count approximately 60% of the total labor cost of the canned tomatoes [24,41]. Canned products can be further categorized as follows:

### 1.2.5.2 *Whole peeled tomatoes*

Cans made from differential electrolytic tinplate with lacquered inside bodies and ends and an internal side seam stripe of enamel are commonly used for the canning of whole tomatoes. Filling of tomatoes is typically performed mechanically, and the juice is added by means of a volumetric filler. Sodium chloride–calcium chloride solutions are usually dissolved into the tomato juice and a small amount of a permitted acid is also added in order to lower the pH of the products. The peeled whole tomatoes are then filled into the cans. The presence of any substances (salts, acids, sweeteners, firming agents) apart from tomato must be declared on the label. The tomatoes are generally filled into cans cold. The filled cans must be exhausted so that sufficient vacuum may be obtained to keep the cans from spoiling during storage. The temperature of the exhaust box varies from 88 to 93°C [24].

### 1.2.5.3 *Diced tomatoes*

The selection of an appropriate variety is crucial to produce dice pieces with the highest textural integrity and intense red color. Tomatoes designed for diced products are dry sorted, washed, color sorted, and conveyed into the peeler. The choice of dice size depends upon the characteristics of the final product as well as on the adequate heat penetration into the dice pieces during thermal processing. Following dicing, the products are transferred over a shaker belt to remove slivers of tomato juice and pieces, and are sorted once again. Dice pieces are passed through calcium chloride baths or sprays to improve their textural integrity. Different methods of cooking are employed for diced tomatoes, depending on the availability of equipment, final products' characteristics and dice size. Diced tomatoes are essentially blanched at 100°C for up to 5 min on a wire mesh belt. Otherwise, hot water or hot topping juice or tomato puree as a cooking medium can also be employed. During this step, the cooking medium temperature is close to boiling, and the residence time is usually less than 2 min. At the final step, hot or cold filling into cans is employed. In the case of bulk aseptic dice processing, where dice are filled hot into 200 L drums or bag-in-box containers, dice pieces are usually cooked by blanching through a water or topping juice cooking medium and are rapidly cooled thereinafter. Many operations use rotating drums which spray water at room or cold temperature over the final products. The retail size cans are typically conveyed through a water bath until their internal temperature is reduced to at least 38°C [24].

#### 1.2.5.4 Tomato purée (pulp)

Two general types of tomato pulp exist, one made from whole tomatoes and the other made from tomato by-products; peels and cores from canning tomatoes or partially extracted tomatoes from tomato juice. Tomato pulp or puree must contain a minimum of 8% tomato solids, but less than 24% of salt-free tomato solids. Whole tomatoes are washed, sorted, and trimmed to remove all visible defects. The tomatoes either pass on to a cyclone for a "cold break" process or to a chopper or a crusher followed by a preheater and a cyclone for the "hot break" process. Hot break temperatures are typically above 77°C, whereas cold break temperatures are below 66°C. The cyclone juice from either process goes to the pulp tank for boiling right down to the required concentration. Tomato by-products are put through a pulper either before or after being heated. Evaporation is carried on in tanks with coils or in vacuum pans. Evaporation efficiency depends on pressurized steam supply (>700 kPa) and steam traps. To secure a satisfactory color and flavor, evaporation should be as rapid as possible. The suggested evaporation time of each batch is 30 min. It is preferable that the storage tank is equipped with a steam coil so that the contents are kept hot at all times. Some manufacturers also install a mechanical stirrer to ensure thorough mixing. After evaporation, tomato pulp is run through a finishing machine to remove any rough particles that were driven through by the pulper. It is then filled into cans. Filled containers are passed through a warm water spray to remove any pulp on the outside [24].

#### 1.2.5.5 Tomato paste

Tomato paste must contain a minimum of 24% tomato solids on a salt-free basis while artificial colors are not permitted under any conditions. Only whole tomato paste is called "paste." Commercially, this product nearly always contains salt and basil. The preliminary processing steps of tomato paste are the same as with tomato pulp. Only, the packing of tomato paste differs in the degree to which the concentration is carried out. Depending on the size of the plant, concentration of tomato paste is performed in single- or multiple-effect evaporators. The paste is filled into cans at a temperature of at least 90°C to prevent the survival of microorganisms. No further heating is applied to the cans after closing; therefore, hot filling and immediate seaming are imperative [24].

#### 1.2.5.6 Tomato juice

Tomatoes are dry sorted, washed, sorted, and trimmed as described earlier for whole peeled tomatoes. Extraction of tomato juice can be done either hot or cold. Hot extracting produces juice of a better-quality with respect to flavor, color, and body. The temperature should be raised to at least 82°C as quickly as possible in a matter of seconds. The fruit can be crushed or chopped before heating, otherwise, the whole tomatoes can be heated in large tanks prior to juice extraction. Heating is employed into large tanks equipped with an agitator designed to provide maximum agitation without the incorporation of air. These tanks are steam jacketed and typically contain steam coils to provide rapid heating. Tomatoes are carried through and discharged into the extractor. At cold extraction, tomatoes are initially scalded to loosen the skins so that no tomato flesh will cling to the peels during extracting. Tomatoes are conveyed directly from the scalder, on an inspection belt, to a chopper and then to the extractor. The yield of cold break juice is lower compared to the one from hot break. There are several different types of commercially available extractors, including either the screw type or paddle type extractors. If shaker screens are used, a higher temperature hot break and a tighter setting of the extractor are suggested as stems and other material which cause off-flavors have already been removed [24,77].

After extraction, deaeration is sometimes applied. Dissolved and occluded air incorporated during pulping and extracting is removed by passing through an efficient deaerator. Color, flavor, and vitamin C content are improved by the complete removal of air. Vacuum deaeration is typically performed immediately after extraction of the juice using a 10° flash to remove the dissolved air. Tomato juice can be homogenized to produce a thicker-bodied product and to prevent settling of the solids. However, this is not necessary for "hot break" juice.

Cans made from differential electrolytic double lacquered tinplate bodies and double-lacquered tinplate ends are used. Prior to filling, the cans are spray washed with a relatively large volume of water at a minimum temperature of 82°C so as to remove dust or other foreign material. Tomato juice should be heated at temperatures higher than 100°C to kill Bacillus coagulans (*Bacillus thermoacidurans*). A heat exchanger is used and either steam or water (121 to 135°C) are used as the heating medium. Standard juice fillers are used for filling tomato juice. The juice may be held prior to filling in supply tanks if its temperature is kept over at least 98°C. Vacuum within the cans is typically obtained by filling the product at a minimum temperature of 88°C and immediately closing the cans using an atmospheric sealer. The filled cans should pass through a hot water spray washer after closure to remove any adhering product. Cold water should not be used to avoid cooling the product [24].

## 1.3 Tomato processing by-products: solid wastes

Tomato processing by-products, including solid and liquid wastes from tomato processing, are a serious problem for the tomato industry. Solid wastes include the tomato pomace (peel, seeds, and pulp) that remains after tomato processing, while liquid wastes include wash waters, chemical peeling solutions and peel water, clean-up water and cooling water [53].

Tomato seeds account for approximately 10% of the fruit representing 50% to 55% of the pomace. The amount of tomato pomace produced annually, depends on the characteristics of the tomatoes and the processing conditions and ranges between 5% and 13% of the initial weight of the tomatoes counting from 600 thousand to 2 million tons of disposed organic waste [22,47]. Other tomato by-products include tomato leaves, culled tomatoes, and tomato seed cake. Culled tomatoes are the fruits that do not meet customers' standards due to appearance defects (shape, size, color, etc.). Approximately 2% of the crop is rejected as culled tomato. Tomato seed cake is the residue obtained after the extraction of oil from tomato seeds and is considered a valuable source of proteins, magnesium, phosphorus, calcium, and riboflavin [75].

The current valorization techniques of tomato pomace include its use as an ingredient in animal feeds or as fertilizer as well as its dumping in the landfills [78,88]. A small fraction of the seeds is used by the oil industry. The disposal of this residue has significant economic and environmental concerns for the tomato processing industry. Tomato industry tries to overcome this problem by developing a suitable waste management system in terms of recycling and reutilization of tomato by-products since they consist of valuable bioactive compounds including proteins, carbohydrates, fatty acids, dietary fibers, and phenolic compounds promoting human health [41,65]. Among the minor tomato pomace components, lycopene attracts the greatest attention for its potential health benefits [72].

### 1.3.1 Treatment of tomato pomace

The effectively recovery and utilization of the individual nutrients from tomato pomace requires the separation of the peel and seeds as the main step. At present, air separation is used in the tomato industry in order to isolate the desirable target products from unwanted materials [71]. The main principle of air separators is based on the differences in the terminal velocity for the separation of mixtures of materials into their individual components. The terminal velocity is defined as the velocity of a rising airstream at which the given individual component remains suspended in the air. For example, if a mixture of two different components is fed into a rising airstream with a velocity that is between the terminal velocities of the individual components, one component will rise and the other will drop. Based on the location of the moving air unit, air separators can be classified into pneumatic separators and aspirators. In pneumatic separators, the fan is placed at the intake end of the air column while in aspirators it locates at the discharge end of the air column. The main parameters affecting the efficiency of an air separation system are the air velocity as well as the feeding rate and moisture content of the tomato pomace [74]. Shao et al. [70] showed the feasibility of the separation of tomato peel and seeds by air separation based on their density and terminal velocity by controlling air velocity, feeding rate and moisture content of the tomato pomace. The optimal conditions were 8.0% moisture content, 6.4 m/s air velocity, and 40 kg/h feeding rate, reaching a separation efficiency of 69%.

A pilot-scale flotation-cum-sedimentation system was also developed by Kaur et al. [45] for the separation of peel and seeds. The researchers separated the peels from the seeds using water and achieved separation yields ranging between 58% to 71% for peels and 42% to 65% for seeds. However, serious concerns arose due to the high volume of water required for the separation. Additionally, this process can cause nutrients' loss since they may dissolve and leach into the water [71].

Tomato pomace has a high moisture content which must be removed prior to the separation of valuable nutrients. Additionally, its high moisture content makes its storage difficult. Unfortunately, in many cases tomato pomace is just dumped and allowed to decay. Upon its spoilage a very foul odor is emerged and it can become a breeding place for various pests. Therefore, wet tomato pomace should be dried before storage and further utilization in order to inhibit spoilage microorganisms as well as enzymatic or non-enzymatic reaction in the target material and extend shelf life [4,17].

Numerous drying processes are commonly used for agricultural products and their by-products including spray-drying, freeze-drying, drum drying, hot air-drying, and microwave-vacuum drying. Hot air-drying results in dehydrated products with extended shelf-life as most of the free water is removed by evaporation but unfortunately with reduced quality characteristics mainly in terms of color [37]. The drying rate efficiency is influenced by many factors including air speed and temperature rate [18]. Ibrahim et al. [43] applied mechanical air-drying at 100°C on tomato pomace and produced tomato powder with minimum microbial load, minimum color change and high total carotenoids' content. Freeze-drying is a gentle dehydration method recommended for the recovery of perishable nutrients, especially lycopene, from tomato pomace. Freeze-dried tomato pomace was found to have better retention of color, lycopene (95.25 mg/100 g) and total phenolic compounds (85.30 mg/100 g) compared to the air-dried pomace (80.30 and 65.30 mg/100 g, respectively) [11]. However, it is an expensive method and requires long drying times (12 to 24 h) [43,84]. Concerning spray-drying, its main indices are the efficiency of the dryer and product recovery. Process variables, such as the inlet air temperature, drying air and compressed airflow rates, influence the moisture, solubility and particle density of the recovered powder.

A promising alternative method used for the recovery of heat-sensitive nutrients from agricultural products and their by-products is continuous vacuum belt drying [43]. During this process, the product is continuously fed under vacuum onto a moving belt passing over a series of conduction heaters. Drying times are typically under 1 to 2 h and temperatures are relatively low producing powders with color and nutrients characteristics similar to the freeze-dried products [84]. Microwave assisted drying is also an efficient method in terms of high production rates and low labor requirements offering reduced drying times and improved quality of the dehydrated product [40]. Al-Harahsheh et al. [3] found that microwave drying can be effectively used for the drying of tomato pomace, further reducing processing time.

At present, the typical treatment of wet tomato pomace in the tomato industry includes the following steps. Firstly, the pomace is left in a heap on a slant until the excess water is drained out. The drained tomato pomace is then mechanically pressed by using a filter press and dried by thermal drying through blowing hot air or using a solar drier. If such equipment is not available, the tomato pomace may be spread in a 5 to 7 cm thick layer on a concrete floor under direct sunlight for sun drying. The material should be turned upside down twice a day, till the dry matter reaches around 90%. Finally, the dried tomato pomace is ground in a mill using 1 to 2 mm screens and stored until further use. The drying process can be applied to the tomato peel, seeds, and defective or culled tomatoes. Peels stored under sanitary condition and chemically acidified, if necessary, can be useful as a fiber supplement.

Liquid wastes also pose a serious problem in the tomato industry. They can be digested and then be recycled or reused in the plant following filtration and chlorination. The liquid wastes from chemical peeling are difficult to digest without neutralizing their alkalinity. In some cases, the endogenous acids of the tomato may assist neutralization. Tomato waste waters are acidic (pH 5 to 6) and low in alkalinity (400 to 600 mg/L). The generation of by-products in tomato processing is inevitable, however they can be reduced by careful management of the equipment, water usage and starting material.

## 1.4 Characterization/composition of tomato pomace

Tomato pomace consists mainly of peels and seeds. The proximate composition of tomato pomace is presented in Table 1.1. Dietary fibers are the major compounds of tomato pomace on a dry matter basis (25.0% to 59.0%). Total proteins range between 15.0% and 33%, total carbohydrates between 3% and 43%, total fat between 2% and 20%, and ash content between 3% and 6%. In Table 1.2 the proximate composition of tomato

**Table 1.1** Proximate composition of tomato pomace (peels and seeds) (d.b.).

| Protein | Carbohydrates | Fibers | Oil | Ash | References |
|---|---|---|---|---|---|
| 22.2 | — | 29.6 | 14.2 | 3.3 | [26] |
| 22.0 | — | 28.1 | 13.1 | 5.1 | [10] |
| 19.5 | 5.1 | 11.3 | 15.9 | 3.4 | [25] |
| 21.2 | — | — | — | 7.7 | [48] |
| 22.1 | — | — | 20.1 | 6.0 | [51] |
| 15.0 | 3.1 | 34.1 | 5.1 | 4.0 | [14] |
| 20.1 | — | 28.0 | 13.0 | 6.1 | [42] |
| 23.1 | — | 25.0 | 12.0 | — | [1] |
| 18.0 | — | 50.0 | 10.0 | — | [7] |
| 19.3 | 25.7 | 59.3 | 5.85 | — | [22] |
| 22.70 | — | 58.53 | 16.24 | 4.40 | [71] |
| 17.6 | — | 52.4 | 2.2 | 4.2 | [59] |
| 20.9 | — | 50.7 | 14.1 | 3.5 | [73] |
| 32.7 | 43.1 | 29.4 | 15.4 | 3.6 | [60] |
| 17.38 | 3.50 | 59.17 | 7.33 | 3.56 | [5] |

**Table 1.2** Proximate composition of tomato peels and seeds (% d.b.).

| Protein | Carbohydrates | Fibers | Oil | Ash | References |
|---|---|---|---|---|---|
| | | **Tomato peels** | | | |
| 10.0 | 7.8 | 55.9 | 3.6 | 2.7 | [78] |
| 5.7 | 8.2 | 65.6 | 3.8 | 2.8 | [50] |
| 20.0 | | 46.1 | 1.7 | 5.6 | [49] |
| 13.6 | 1.1 | | 2.5 | 2.9 | [52] |
| 10.1 | | 29.9 | 3.2 | 25.6 | [47] |
| | | **Tomato seeds** | | | |
| 24.5 | 2.9 | 19.1 | 28.1 | 5.4 | [78] |
| 16.6 | 26.0 | 19.4 | 28.5 | 5.6 | [50] |
| 32.6 | | 14.8 | 22.4 | 4.8 | [49] |
| 25.2 | 2.9 | | 15.0 | 3.7 | [52] |
| 20.2 | | 53.8 | 6.4 | 5.2 | [47] |

peels and tomato seeds is presented, respectively. The differences between the results of various studies could be attributed to the composition of the starting material and the processing conditions applied.

### 1.4.1 Amino acids

The amino acids content of tomato pomace (peels and seeds) or tomato seeds and peels is presented in Table 1.3. The main essential amino acids

Table 1.3 Amino acids content of tomato pomace.

| Amino acids | Tomato pomace [78] | Tomato seeds [62] | Tomato peels [28] | Tomato pomace [59] |
|---|---|---|---|---|
| Aspartic acid | 20.2 | 2.8 | 0.7 | 15.7 |
| Glutamic acid | 33.6 | 5.1 | 14.6 | 72.1 |
| Serine | 10.9 | 1.3 | – | 1.7 |
| Glycine | 12.2 | – | 4.3 | 6.3 |
| Threonine | 7.7 | 0.9 | – | 5.5 |
| Arginine | 12.7 | 2.5 | 4.3 | 14.6 |
| Alanine | 7.6 | 1.1 | 5.0 | 7.1 |
| Tyrosine | 6.0 | 1.0 | 3.4 | 6.9 |
| Valine | 8.7 | 1.2 | 4.6 | 5.4 |
| Phenylalanine | 6.7 | 1.3 | 3.1 | 6.1 |
| Isoleucine | 6.3 | 1.1 | 3.9 | 6.9 |
| Leucine | 10.9 | 1.7 | 5.1 | 10.7 |
| Lysine | 9.3 | 1.5 | 4.4 | 8.8 |
| Cystine | 1.1 | 0.4 | 0.4 | 2.3 |
| Methionine | 1.5 | 0.4 | 1.0 | 2.7 |

of tomato pomace are glutamic acid, aspartic acid, serine, glucine, alanine, tyrosine, cysteine, leucine, isoleucine, lysine, methionine, phenylalanine, threonine, arginine, and valine. Glutamic acid is the most abundant amino acid in tomato pomace and in the protein fraction of the tomato peel as well as reported by Elbadrawy and Sello [28]. Persia et al. [62] reported a total amount of glutamic acid and arginine of 5.1% and 2.5%, respectively, in the protein fraction of the tomato seeds.

### 1.4.2 Fatty acids

The concentrations of fatty acids in tomato pomace (peels and seeds), or tomato seeds based on different studies, are presented in Table 1.4. The major fatty acid in tomato pomace is linoleic acid (50% to 52%), followed by oleic acid (18% to 28%) and palmitic acid (12% to 16%). The saturated fatty acids represent 18-23% of the total fatty acids while the monounsaturated and polyunsaturated fatty acids count for 19% to 29% and 53% to 57%, respectively, revealing the predominance of the unsaturated fatty acids over saturated fatty acids in tomato pomace.

### 1.4.3 Total phenolics and carotenoids

The total phenolic content, total flavonoids, carotenoids and antioxidant activity of tomato pomace, according to two different studies, are

**Table 1.4** Concentrations of fatty acids of tomato pomace.

| Fatty acids | | Tomato seeds [16] | Tomato pomace [39] | Tomato pomace [59] | Tomato seeds [46] | Tomato pomace [60] |
|---|---|---|---|---|---|---|
| Myristic | 14:0 | 0.12 | 0.13 | 0.41 | | - |
| Pentadecanoic | 15:0 | | - | 0.09 | | - |
| Pentadecenoic | 15:1 | | - | 0.09 | | - |
| Palmitic | 16:0 | 13.38 | 12.97 | 16.32 | 14 | 12.17 |
| Palmitoleic | 16:1 | 0.25 | 0.34 | 0.64 | | 0.26 |
| Heptadecanoic | 17:0 | | 0.40 | 0.19 | | 0.11 |
| Heptadecenoic | 17:1 | 0.53 | 0.08 | 0.52 | | |
| Stearic | 18:0 | 4.48 | 5.74 | 5.43 | 5 | 5.26 |
| cis-Oleic | 18:1 | 19.65 | 25.71 | 18.50 | 21 | 28.02 |
| cis-Linoleic | 18:2 | 55.45 | 51.90 | 51.91 | 57 | 50.77 |
| γ-Linolenic | 18:3 (n-6) | | - | nd | | - |
| α-Linolenic | 18:3 (n-3) | 2.21 | 1.99 | 3.35 | 1 | 2.32 |
| Octadecatetraenoic | 18:4 (n-3) | | - | 0.48 | | - |
| Eicosadienoic | 20:2 (n-6) | 0.34 | 0.10 | 0.15 | | - |
| Eicosatrienoic | 20:3 (n-6) | | - | 0.01 | | - |
| Docosadienoic | 22:2 (n-6) | | - | 0.02 | | - |
| Docosatrienoic | 22:3 (n-6) | | - | 0.55 | | - |
| Docosatrienoic | 22:3 (n-3) | | 0.14 | 0.13 | | - |
| Eicosapentaenoic | 20:5 (n-3) | | - | 0.26 | | - |
| Lignoceric | 24:0 | | - | 0.29 | | - |
| Other | | 2.63 | - | 0.22 | 2 | - |
| Fatty acids profile | | | | | | |
| Saturated (SFA) | | | 19.82 | 22.72 | | 18.06 |
| Monounsaturated (MUFA) | | | 26.29 | 19.75 | | 28.84 |
| Polyunsaturated (PUFA) | | | 53.89 | 57.29 | | 53.10 |

presented in Table 1.5. Total phenolic content of tomato pomace was reported between 1095 to1538 mg GAE/kg [19]. Tomato pomace contains considerable amounts of flavonoids (415-688 mg QE/kg) and in particular flavonols such as quercetin and kaempferol [19]. Carotenoids are also bioactive compounds of tomato pomace counting for about 272 to 554 mg/kg [57]. Main carotenoids of tomato pomace are lycopene, β–carotene, lutein, cis–β–carotene and zeaxanthin. Additionally, tomato pomace exhibits strong antioxidant activity (0.68 to 2.2 mM Trolox/100 g). Tomato peels are a rich source of carotenoids, including lycopene [47].

### 1.4.4 Minerals

The minerals' content of dried tomato pomace (peels and seeds) is shown in Table 1.6. Among minerals, potassium presents the highest concentration

**Table 1.5** Total phenolic content, total flavonoids, carotenoids, and antioxidant activity of tomato pomace.

| Compounds | Tomato pomace [59] | Tomato pomace [60] |
|---|---|---|
| Total phenolics (mg GAE/kg) | 1229.5 | 1095.7 |
| Total flavonoids (mg QE/kg) | 415.3 | 688.2 |
| Total carotenoids (mg/kg) | - | 272.0 |
| Lycopene (mg/kg) | 510.6 | - |
| β-Carotene (mg/kg) | 95.6 | - |
| Antioxidant activity (mM Trolox/100g) | 0.68 | 2.2 |

**Table 1.6** Minerals content of tomato pomace.

| Mineral (g/kg) | Tomato pomace [6] | Tomato pomace [59] |
|---|---|---|
| Ascorbic acid | 0.278 | - |
| Potassium | 14.57 | 30.30 |
| Phosphorous | 3.33 | - |
| Magnesium | 2.67 | 2.11 |
| Sodium | - | 0.67 |
| Calcium | 1.89 | 1.32 |
| Zinc | 0.19 | 0.063 |
| Manganese | 0.020 | 0.014 |
| Copper | 0.014 | 0.005 |
| Iron | 0.28 | 0.06 |

ranging between 14.6 to 30.3 g/kg. The levels obtained for calcium, magnesium, sodium, iron, manganese, and copper are low, while the sodium content is quite high, thus restricting the inclusion of tomato pomace in animal feed diets [47].

## 1.5 Sustainable management for tomato processing by-products and wastes

Food waste throughout the food chain has become a global concern because of its adverse impacts on natural resources, food security, environment, and human health. The FAO has estimated that every year, nearly 1.3 billion tons of food worth 750 million USD is lost or wasted [54]. Similarly, European Union (EU) in 2010 revealed that 89 million tons of food waste is generated every year, which is equivalent to 170 Mt of $CO_2$ [69]. It was revealed that during food production, up to 70% of waste or by-products are generated. These data make the need for the valorization of these waste materials or by-products even more urgent. Besides, only through the effective reprocessing of these by-products and by adopting a circular economy approach, it would be possible to reduce their environmental impact [55,80,82,85].

Globally, around 180 million tons of tomatoes are produced each year, of which about 40 million tons/year are eventually processed to obtain different tomato products. In Europe, 16.6 million tons of tomato fruits are produced each year [61]. Statistically, recent FAO data show that the area harvested and the global production of tomato are rising. Tomato processing industries produce large quantities of tomato pomace which sometimes exceed 5% wt of the raw material used, while the proximate annual estimate of the tomato pomace produced around the world is 5.4 to 9.0 million tons. It consists of nearly 33% seeds, 27% peels, and 40% pulp, whereas its dried form contains approximately 44% seeds and 56% peels [69].

Another concern for the tomato processing industry is the significant moisture content of tomato pomace waste and its high transport cost which hinder its exploitation. In order to overcome such problems, a thermal drying process of the industrial tomato residues is highly suggested, so that they can be used either for livestock feed production or for boiler fuel as pellets. The most widely used commercial drying plants for numerous industrial agri-food by-products are convective type dryers, in which heat is transferred to the product by means of hot gases. More recently, the infrared drying technique is also proposed which presents several advantages compared to conventional drying: decreased drying time, high energy efficiency and lower air flow through the sample product [68].

Moreover, in terms of sustainable management, another problem engaged in the tomato processing industry is the massive water consumption associated mainly with lye and steam peeling and the subsequent salinity issue in wastewater treatment. Peeling is an important unit operation in tomato processing; approximately 70% of the tomatoes processed are steam peeled and the remaining 30% are lye peeled. Although the environmental impact caused by steam peeling is significantly lower than that generated by lye peeling, it still requires a lot of water, high pressure, and energy. Several novel peeling methods, such as infrared peeling, pulsed electric fields, ohmic peeling, and power ultrasound peeling have been proposed as alternative peeling processes to address these issues. For these reasons, current research is focusing on new sustainable peeling alternatives that can effectively peel tomatoes with minimum losses and a higher-quality end product, while also causing fewer environmental problems and reducing water and energy consumption. However, many of these processes (e.g., ohmic and pulse ultrasound) still involve the use of water, which remains a concern. Moreover, there is lack of confirmed data on these approaches regarding energy saving compared to the conventional peeling methods [67,81].

## 1.5.1 Current practices

### *1.5.1.1 Current practices for tomato pomace*

Tomato pomace, due to its nutrients' composition, seeds content and especially its high levels of protein, is regarded as a potential auxiliary to pasturing or as a nutrient supplement for mixed feed. It is used as a ration additive for cows, sheep, goats, pigs, rabbits, hamsters, chickens, and carp. A significant increase in the animal weight as well as in the yield and nutritional quality of the products derived from pomace-fed animals, including milk, meat, and eggs has been observed [56,81]. However, the high moisture content of the pomace makes it prone to spoilage given that it will be stored on a farm for more than a few days under ambient conditions. As a result, fresh tomato pomace requires either ensiling to preserve it as silage for pastures or drying to extend its shelf life. In the case of drying, most tomato pomace for animal feed is dried to about 8% moisture content [81].

### *1.5.1.2 Current practices for water in tomato processing plants*

During tomato paste processing, the processing facility uses fresh water, condensate generated from evaporators and recycled wash water for a variety of activities, including transporting tomatoes, steam generation, cooling, and condensing vapors from evaporators, make-up water in cooling towers, water softening, and plant cleanup. Three distinct process water streams are produced: (1) condensed steam from the evaporator stream (process condensate); discharged into a cooling pond or cooling towers, (2) wash water stream from transporting tomatoes; discharged to the settling pond, and (3) facility wash-water stream from plant and machinery cleanup; discharged as a commingled stream to the final wash water stream exiting the plant [81].

## 1.5.2 Sustainable management for water in tomato processing plants

Process condensate has various applications in industrial tomato processing. Currently, process condensate is used for several purposes, including make-up water for boilers, cooling towers and ponds by routing it through the feedwater purification and conditioning systems. Furthermore, the relatively clean nature of process condensate makes it a candidate for reuse elsewhere in the processing facility with or without cooling, depending on the process requirement.

More specifically, condensate is commonly used for transporting tomatoes, clean-in-place of processing equipment, and as rinse water for cleaning. Thus, it serves not only as a source of make-up water in the tomato conveying flume, but also prevents tomatoes from cooling, contracting, and sucking in the dirty flume water. In addition, the thermal energy of the condensate water also slightly raises the temperature of tomatoes, working as a pre-heater for tomatoes, and consequently saving energy during hot or cold break. However, those energy recovery strategies depend on the size and properties of the heat exchangers, and the temperature difference between the process condensate generated from the evaporators and the disrupted tomatoes. Moreover, hot process condensate is often routed to cooling towers to bring it down to ambient temperature ahead of discharge.

The clean-in-place of the processing equipment, such as evaporators, hot or cold break-systems, and peelers requires relatively hotter and softer water, and process condensate is suitable in both regards. Other potential uses of process condensate include evaporative coolers, surface and equipment rinsing, and toiletry and laundry use after cooling.

Summarizing, tomato processors and regulators should develop and explore a sustainable and cost-effective wastewater treatment of the discharged wash water from the facilities. Future research should focus on identifying wastewater treatments, including biological, enzyme, chemical, and mechanical treatments that can reduce energy consumption, increase treatment efficiency, minimize the adverse effects of chemicals on ecology, and apply natural processes [81].

### 1.5.3 Sustainable applications of tomato processing by-products (tomato pomace)

Different approaches for the sustainable utilization of tomato pomace, in the concept of bio-economy, as introduced by the European Commission, have been proposed (Fig. 1.2) and are further described in the following paragraphs. According to this approach the economic growth of modern societies should be based on biological resources from the land and sea as inputs to food and feed, industrial and energy production. Additionally, industry should rely on the use of renewable raw materials for the generation of sustainably created products [2,12]. Thus, the sustainable use of plant derived wastes for the recovery of added-value compounds with potential

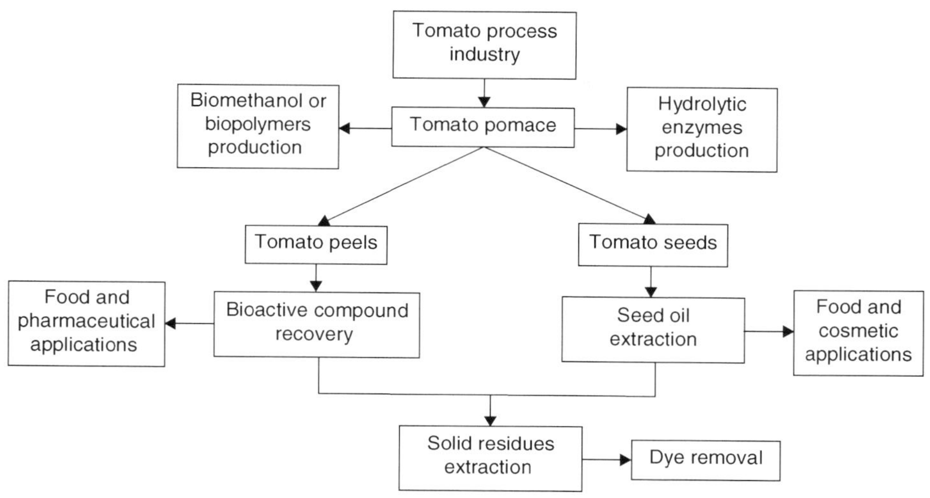


**Fig. 1.2** ***Sustainable valorization of tomato processing by-products.***

application in the food, feed, biotechnological, and pharmaceutical industries may help to tackle the societal challenges of the 21st century [63].

### *1.5.3.1 Food industry*

The valorization of the tomato processing by-products (seeds, peels, and pomace) has gained great interest during the last few years as they can be an excellent source for the recovery of valuable bioactive compounds. Moreover, tomato pomace components can be directly incorporated in different food products (e.g. pasta, bread and cookies, dairy and meat products) as an ingredient thus reducing production cost and providing nutritional and functional benefits as well (Table 1.7).

Wheat flour-based foods offer the greatest potential for the incorporation of tomato pomace and its components. All relevant studies report an increase in the protein, dietary fiber and lycopene content of the resulting products as well as enhanced antistaling properties. Moreover, the addition of tomato pomace powder, up to 10%, did not generate any adverse effects on the acceptability of bread. However, as with most non-wheat flours, a decrease in samples' volume is expected due to the lower gluten levels. Other results showed that the incorporation of 5% wt tomato pomace powder did not have a significant effect on the overall desirability scores of cookies, while a disadvantageous effect was found with tomato pomace concentrations up to 25% wt. Also, crackers with tomato pomace powder presented increased bitterness, which was caused by the presence

**Table 1.7** Food formulations using tomato processing by-products [20].

| By-product | Food | Improved characteristics |
|---|---|---|
| Dry tomato peel | Dry fermented sausages | Lycopene content, sensory, and textural characteristics |
| | Beef hamburgers | Lycopene content |
| | Pasta | Carotenoids and fiber content |
| Tomato pomace | Flat bread | Moisture content, textural characteristics, delayed staling |
| | Cookies | Crude protein and ash content |
| | Bread, crackers, and muffin | Fiber and mineral content, antioxidant activity |
| | Low calorie jams | Total carbohydrates, fibers |
| By-product extracts | Butter | Antioxidant activity |
| | Olive oil | Carotenoids content |
| | Ice cream | Sensory characteristics |
| Seed oil | Tomato slices | Moisture and color preservation, antioxidant activity, germ reduction |
| Seed powder | Noodles | Protein, fiber and lysine content |

of furostanol saponin in tomato pomace. The above findings show that the incorporation of tomato by-products in flour-based foods is feasible, but their effects are dependent on their content and the overall composition of the selected products [56].

Moreover, tomato pomace has also been incorporated into meat products such as hamburgers, sausages and ham. The resulting products not only were rich in dietary fiber and antioxidants but exhibited a more appealing color as tomato pomace can be a natural colorant. Additionally, the presence of tomato pomace fibers can modify all textural properties, pectin in tomato peel can increase springiness, while cellulose and lignin influence hardness and cohesiveness, resulting in increased chewiness [56].

Finally, the utilization of tomato pomace as an ingredient in tomato paste is strongly recommended as well. The lyophilized and powdered tomato pomace can be supplemented into tomato paste up to 2% wt. According to relevant studies, the taste and appearance of the resulting tomato paste is not compromised or negatively altered and remains comparable to that of the commercial tomato paste. Final products can also exhibit higher β-carotene and lycopene content [56].

Despite their high nutritional content and satisfactory overall characteristics, tomato by-products are underutilized as food ingredients due to the additional unit operations (e.g., drying, grinding) required for their preparation. Thus, even though the preliminary results are highly encouraging, more work is needed towards new products' development or the modification of the existing ones, as well as in order to validate process economics and consumer acceptance [9,27,56].

#### *1.5.3.2 Biotechnology and pharmaceutical industry*

Future strategies and endeavors for the utilization of tomato by-products also include research and development focusing on producing novel and high value-added products, especially of nutraceuticals, and pharmaceuticals. Moreover, compost derived from tomato waste is considered a promising option for organic waste management both due to its high abundance and its ability to improve soil characteristics and enhance plant growth. Another potential includes the production of renewable biofuels in liquid or gaseous forms, such as biogas and bioethanol, through the bioconversion of tomato by-products. In fact, bioenergy is a promising, inexhaustible, sustainable source capable to address the rising energy demands and depleting fossil fuels. Finally, hydrolytic enzymes (e.g. xylanase) are also produced from pomace with fungi in solid state fermentations [20,56].

### 1.5.4 Legislation for sustainable management of food waste and valorization of food by-products

Agri-food sector is one of the main priority areas contained in the European Sustainable Production and Consumption policies, aiming to create more sustainable consumption and production patterns, not only for supporting environmental protection, but also for increasing businesses' competitiveness as well as economic development and social wellbeing [87,89,90]. In January 2013, the European Commission announced a Retail Action Plan which would support actions to reduce food waste, and work on developing a long-term policy on food waste.

Furthermore, the European Roadmap to a Resource Efficiency defines some important goals for agriculture in the next future, such as a 50% reduction in waste production, the preservation of biodiversity and ecosystem services, the reduction of land use, the improvement in soil quality and the independence from fossil fuels [29].

Food waste prevention is also an integral part of the Commission's new Circular Economy Package which sets out a concrete program of action, with measures covering the whole cycle, from production and consumption to waste management and the market, for secondary raw materials, thus closing the loop [8,36,76,86].

In 2015, in the framework of the 2030 Sustainable Development Goals, the United Nations General Assembly adopted a target of halving per capita food waste at the retail and consumer level, and reducing food losses along production and supply chains ("United Nations", 2015). The European Common Agricultural Policy (CAP) has identified three priority areas for intervention, namely biodiversity and the preservation and development of "natural" farming and forestry systems, water management and use, and dealing with climate change [34]. At international level, it is well recognized that an integrated sustainability assessment must be based on a life cycle approach, that allows to deeply analyze the whole system or supply chain, thus avoiding burden shifting. Indeed, life cycle thinking has a strategic position in the policy-making process, and it has been embedded by the European Commission in the Integrated Product Policy (IPP) [30], in the European Sustainable and Consumption Production Action Plan [32], and in several framework directives and thematic strategies, such as the Waste Framework Directive, the Thematic Strategies on The Sustainable Use of Natural Resources [31] and on the Prevention and Recycling of Waste [33], and the Communication about Building a single market for green products [35].

In this respect, the European Commission pledged specifically to:

1. Introduce a common methodology across EU countries to measure food waste.
2. Create a new platform (EU Platform on Food Losses and Food Waste) in order to help define measures needed to prevent food waste, share best practice, and evaluate progress.
3. Clarify legislation related to waste, food and feed and facilitate food donation and the use of foodstuffs and by-products for feed production, without compromising food and feed safety.
4. Improve the use of date marking and its understanding by consumers, in particular "best before" labeling [21].

## 1.6 Conclusions - future trends

The management of tomato by-products represents a worldwide problem for both environmental and economical reasons and their disposal or valorization remains a significant issue for the food industry. Moreover, the protection of natural resources, such as water and energy, is also a major priority of the modern industrial ecology.

The large volumes and high-water content of tomato pomace make it even more difficult to manage. However, it contains multiple bioactive compounds, antioxidants and colorants, which if properly utilized may lead to the development of new or improved products (food, nutraceuticals or pharmaceuticals) of high commercial and nutritional value. As an alternative, tomato by-products can be exploited in biofuels' production generating a valuable source of renewable energy.

In addition to the aforementioned strategies, several other ambitious solutions for the successful management of tomato wastes and by-products have been proposed, as well, in the context of environmental protection and circular economy. Nevertheless, their implementation requires not only the creativity and effort of the scientific and research community but also the will and determination of the world's leadership to implement the necessary directives and regulations that will lead towards a green and sustainable waste valorization.

## References

[1] M. Abaza, A. Nour, B. Borhami, K. El-Shazly, Nutritive value of vegetable and fruit waste as animal feeds, Alex. J. Agric. Res. 32 (2) (1987) 75–83.
[2] A. Agapiou, A. Vasileiou, M. Stylianou, K. Mikedi, A.A. Zorpas, Waste aroma profile in the framework of food waste management through household composting, J. Clean. Prod. 257 (2020) 120340. https://doi.org/10.1016/j.jclepro.2020.120340.
[3] M. Al-Harahsheh, A.H. Al-Muhtaseb, T.R.A Magee, Microwave drying kinetics of tomato pomace: effect of osmotic dehydration, Chem. Eng. Process. 48 (1) (2009) 524–531. https://doi.org/10.1016/j.cep.2008.06.010.
[4] A.H. Al-Muhtaseb, M. Al-Harahsheh, M. Hararah, T.R.A Magee, Drying characteristics and quality change of unutilized-protein rich-tomato pomace with and without osmotic pre-treatment, Ind. Crop. Prod. 31 (1) (2010) 171–177. https://doi.org/10.1016/j.indcrop.2009.10.002.
[5] N.K. Alqahtani, A. Helal, T.M. Alnemr, O. Marquez, Influence of tomato pomace inclusion on the chemical, physical and microbiological properties of stirred yoghurt, Int. J. Dairy Sci. 15 (3) (2020) 152–160. https://doi.org/10.3923/ijds.2020.152.160.
[6] A. Alvarado, E. Pacheco-Delahaye, P. Hevia, Value of a tomato byproduct as a source of dietary fiber in rats, Plant Foods Hum. Nutr. 56 (2001) 335–348. https://doi.org/10.1023/A.
[7] M. Alvarado, E. Pacheco-Delahaye, M. Schnell, P. Hevia, Fibradietéticaen el resíduo industrial del tomate y suefectosobre la respuestaglicémica y el colesterolséricoenratas, Arch. Latinoam Nutr. 49 (2) (1999) 138–142.
[8] N. Antoniou, A.A. Zorpas, Quality protocol development to define end-of-waste criteria for tire pyrolysis oil in the framework of circular economy strategy, Waste Manag. 95 (2019) 161–170. https://doi.org/10.1016/j.wasman.2019.05.035.
[9] N. Bertin, M. Génard, Tomato quality as influenced by preharvest factors, Sci. Hortic. 233 (2018) 264–276. https://doi.org/10.1016/j.scienta.2018.01.056.
[10] B. Bhargava, S. Talapatra, Studies on fruit and vegetable factory wastes, Indian Vet. J. 45 (11) (1968) 948–952.
[11] M.A. Bhat, Quality characteristics of freeze and cabinet dried tomato pomace, Int. J. Pure Appl. Biosci. 6 (2) (2018) 891–897. https://doi.org/10.18782/2320-7051.4044.
[12] F. Boccia, P. Di Donato, D. Covino, A. Poli, Food waste and bio-economy: a scenario for the Italian tomato market, J. Clean. Prod. 227 (1 August) (2019) 424–433.

[13] S. Brodt, K.J. Kramer, A. Kendall, G. Feenstra, Comparing environmental impacts of regional and national-scale food supply chains: A case study of processed tomatoes, Food Policy 42 (2013) 106–114. https://doi.org/10.1016/j.foodpol.2013.07.004.
[14] L. Cabrera, I. Berdan, A. Ramírez, Caracterizacion de los desechos de la industria de conservas de tomate. I. Composicionquimica y mineral, Rev CienQuím 15 (1984) 291–294.
[15] D.A. Campos, R. Gómez-García, A.A. Vilas-Boas, A.R. Madureira, M.M. Pintado, Management of fruit industrial by-products - a case study on circular economy approach, Molecules 25 (2020) 320 10.3390/molecules25020320.
[16] C.A. Cassinerio, J.G. Fadel, J. Asmus, J.M. Heguy, S.J. Taylor, E.J. DePeters, Tomato seeds as a novel by-product feed for lactating dairy cows, J. Dairy Sci. 98 (7) (2015) 4811–4828. https://doi.org/10.3168/jds.2014-9121.
[17] M. Castoldi, M.F. Zotarelli, A. Durigon, B.A.M. Carciofi, J.B Laurindo, Production of tomato powder by refractance window drying, Dry. Technol. 33 (12) (2015) 1463–1473. https://doi.org/10.1080/07373937.2014.989327.
[18] Y. Chen, A. Martynenko, Computer vision for real-time measurements of shrinkage and color changes in blueberry convective drying, Dry. Technol. 31 (10) (2013) 1114–1123. https://doi.org/10.1080/07373937.2013.775587.
[19] J.K. Chérif, S. Jémai, N.B. Rahal, A. Jrad, M. Trabelsi-Ayadi, Study of antioxidant and antiradical capacity of fresh and industrial waste of Tunisia tomato. Valorization of Tunisian tomato in bioactive molecules, Tunis. J. Med. Plant. Nat. Prod. 4 (1) (2010) 116–125.
[20] V. Coman, B.E. Teleky, L. Mitrea, G.A. Martău, K. Szabo, L.F. Călinoiu, D.C. Vodnar, Bioactive potential of fruit and vegetable wastes, Adv. Food Nutr. Res. 91 (2020) 157–225. https://doi.org/10.1016/bs.afnr.2019.07.001.
[21] A. Del Borghi, M. Gallo, C. Strazza, M. Del Borghi, An evaluation of environmental sustainability in the food industry through life cycle assessment: the case study of tomato products supply chain, J. Clean. Prod. 78 (2014) 121–130. https://doi.org/10.1016/j.jclepro.2014.04.083.
[22] M. Del Valle, M. Cámara, M.E. Torija, Chemical characterization of tomato pomace, J. Sci. Food Agric. 86 (8) (2006) 1232–1236. https://doi.org/10.1002/jsfa.2474.
[23] R.G Domínguez, P. P., P.E.S. M., Munekata, W. Zhang, G.M Lorenzo, Tomato as potential source of natural additives for meat industry. A review., Antioxidants 9 (2020) 73. 10.3390/antiox9010073.
[24] D.L. Downing, Canned tomato products. In: A Complete Course in Canning and Related Processes, 4th ed., Woodhead Publishing, Cambridge, UK, 2016, pp. 199–228. https://doi.org/10.1016/b978-0-85709-679-1.00005-2.
[25] N.J. Drouliscos, Nutritional evaluation of the protein of dried tomato pomace in the rat, Br. J. Nutr. 36 (1976) 449–456.
[26] P. Edwards, K. Roderick, A. Hoersh, N. Redfield, Recovery of tomato processing wastes, Food Technol. 6 (October) (1952) 383–386.
[27] H.M. El Mashad, L. Zhao, R. Zhang, Z. Pan, Tomato. In: Integrated Processing Technologies for Food and Agricultural By-Products, Academic Press, Cambridge, MA, USA, 2019, pp. 107–131. https://doi.org/10.1016/B978-0-12-814138-0.00005-8.
[28] E. Elbadrawy, A. Sello, Evaluation of nutritional value and antioxidant activity of tomato peel extracts, Arab. J. Chem. 9 (2016) 1010–1018. https://doi.org/10.1016/j.arabjc.2011.11.011.
[29] European Commission (2001). Resource efficiency, the roadmap to a resource efficient europe. COM/2011/571.
[30] European Commission (2003). Policy, integrated product thinking, building on environmental life-cycle. COM/2003/302.
[31] European Commission, Communication from the Commission to the Council, the European Parliament, the European Economic and Social Committee and the

Committee of the Regions. In: Thematic Strategy on the Sustainable Use of Natural Resources, 2005 COM/2005/670.
[32] European Commission (2008). Sustainable consumption and production and sustainable industrial policy action plan. COM/2008/397.
[33] European Commission (2011). Report from the Commission to the European Parliament, the Council, the European Economic and Social Committee and the Committee of the Regions on the Thematic Strategy on the Prevention and Recycling of Waste. COM/2011/13.
[34] European Commission, The Common Agricultural Policy - Partnership Between Europe and Farmers, Publications Office of the European Union, Luxembourg, 2012.
[35] European Commission, Communication from the Commission to the European Parliament and the Council. Building the Single Market for Green Products, Facilitating Better Information on the Environmental Performance of Products and Organizations, 2013. COM/2013/196.
[36] European Commission (2015). Circular economy action plan, a new circular economy action plan for a cleaner and more competitive Europe.
[37] J. Famurewa, A. Raji, Physicochemical characteristics of osmotically dehydrated tomato (Lycopersicon esculentum) under different common drying methods, Int. J. Biol. Chem. Sci. 5 (3) (2011) 1304–1309. https://doi.org/10.4314/ijbcs.v5i3.72284.
[38] E. Garcia, D.M. Barrett, Peelability and yield of processing tomatoes by steam or lye, J. Food Process. Preserv. 30 (1) (2006) 3–14. https://doi.org/10.1111/j.1745-4549.2005.00042.x.
[39] A.M. Giuffrè, M. Capocasale, Physicochemical composition of tomato seed oil for an edible use: the effect of cultivar, Int. Food Res. J. 23 (2) (2016) 583–591.
[40] F. Gogus, M. Maskan, Drying of olive pomace by a combined microwave-fan assisted convection oven, Nahrung/Food 45 (2) (2001) 129–132.
[41] W.A. Gould, Tomato Production, Processing and Technology, CTI Publications, Inc., Baltimore, Maryland, 1992. https://doi.org/10.1533/9781845696146.1.1.
[42] B. Gupta, R. Gupta, P. Maheshwari, A. Sinha, Nutritive potential of some vegetable waste products, Indian J. Anim. Health 24 (1) (1985) 25–28.
[43] A. Ibrahim, M. EL-Iraqi, T. Osman, Y. Hendawey, Bio-engineering studies for tomato pomace powder production as a nutritional valuable material, J. Soil Sci. Agric. Eng. 8 (12) (2017) 671–680. https://doi.org/10.21608/jssae.2017.38212.
[44] W. Jongen, Fruit and Vegetable Processing, CRC Press, Woodhead Publishing Limited, Cambridge, UK, 2002.
[45] D. Kaur, D.S. Sogi, S.K. Garg, A.S. Bawa, Flotation-cum-sedimentation system for skin and seed separation from tomato pomace, J. Food Eng. 71 (4) (2005) 341–344. https://doi.org/10.1016/j.jfoodeng.2004.10.038.
[46] B. Khiari, M. Moussaoui, M. Jeguirim, Tomato-processing by-product combustion: thermal and kinetic analyses, Materials 12 (2) (2019) 1–11. https://doi.org/10.3390/ma12040553.
[47] M. Knoblich, B. Anderson, D. Latshaw, Analyses of tomato peel and seed byproducts and their use as a source of carotenoids, J. Sci. Food Agric. 85 (2005) 1166–1170.
[48] A. Kramer, W.H. Kwee, Utilization of tomato processing wastes, J. Food Sci. 42 (1) (1977) 212–215.
[49] E.S. Lazos, P. Kalathenos, Technical note: Composition of tomato processing wastes, Int. J. Food Sci. Technol. 23 (6) (1988) 649–652. https://doi.org/10.1111/j.1365-2621.1988.tb01052.x.
[50] R. Lasztity, M. Abd El-Samei, M. El-Shafei, Biochemical studies on some non-conventional sources of protein. Part I. Tomato seeds and peels, Nahrung 30 (6) (1986) 615–620.
[51] S. Latlief, D. Knorr, Tomato seeds protein concentrates: effects of methods of recovery upon yield and compositional characteristics, J. Food Sci. 48 (1983) 1583–1586.

[52] G.N. Liadakis, Utilization of Tomato-Processing by-Products, National Technical University of Athens, Greece, 1999 PhD Thesis.
[53] X. Li, Z. Pan, G.G. Atungulu, X. Zheng, D. Wood, M. Delwiche, T.H. McHugh, Peeling of tomatoes using novel infrared radiation heating technology, Innov. Food Sci. Emerg. Technol. 21 (2014) 123–130. https://doi.org/10.1016/j.ifset.2013.10.011.
[54] P. Loizia, N. Neofytou, A.A. Zorpas, The concept of circular economy in food waste management for the optimization of energy production through UASB reactor, Environ. Sci. Pollut. Res. (2018). https://doi.org/10.1007/s11356-018-3519-4.
[55] P. Loizia, I. Voukkali, A.A. Zorpas, J. Navarro Pedreno, G. Chatziparaskeva, V.J. Inglezakis, I. Vardopoulos, Measuring environmental performance in the framework of waste strategy development, Sci. Total Environ. 753 (2021) 141974. https://doi.org/10.1016/j.scitotenv.2020.141974.
[56] Z. Lu, J. Wang, R. Gao, F. Ye, G. Zhao, Sustainable valorisation of tomato pomace: a comprehensive review, Trends Food Sci. Technol. 86 (2019) 172–187. https://doi.org/10.1016/j.tifs.2019.02.020.
[57] E. Luengo, I. Alvarez, J. Raso, Improving carotenoids extraction from tomato waste by pulsed electric fields, Frontier. Nutr. 1 (August) (2014) 1–10.
[58] P.M. Muchinsky, Tomato Production, Processing & Technology, 53, CTI Publications, Baltimore, MD, USA, 1992. https://doi.org/10.1017/CBO9781107415324.004.
[59] V. Nour, T.D. Panaite, M. Ropota, R. Turcu, A.R. Corbu, V. Nour, I. Trandafir, Nutritional and bioactive compounds in dried tomato processing waste, CyTA J. Food 16 (1) (2018) 222–229. https://doi.org/10.1080/19476337.2017.1383514.
[60] Z. Ozbek, K. Celik, P. Ergonul, A. Hepcimen, A promising food waste for food fortification: characterization of dried tomato pomace and its cold pressed oil, J. Food Chem. Nanotechnol. 6 (1) (2020) 9–17.
[61] G. Pataro, G. Ferrari, V.J. Ferreira, A.M. López-Sabirón, G.A. Ferreira, Implementation of PEF treatment at real-scale tomatoes processing considering LCA methodology as an innovation strategy in the agri-food sector, Sustainability (Switzerland) 10 (2018). https://doi.org/10.3390/su10040979.
[62] M. Persia, C. Parsons, M. Schang, J. Azcona, Nutritional evaluation of dried tomato seeds, Poult. Sci. 82 (1) (2003) 141–146.
[63] J. Pinela, M.A. Prieto, M.F. Barreiro, A.M. Carvalho, M. Beatriz, P.P. Oliveira, T.P. Curran, I.C.F.R Ferreira, Valorisation of tomato wastes for development of nutrient-rich antioxidant ingredients: a sustainable approach towards the needs of the today's society, Innov. Food Sci. Emerg. Technol. 41 (June) (2017) 160–171.
[64] S. Porretta, Tomato chemistry, industrial processing and product development, The Royal Society of Chemistry, London, UK, 2019.
[65] V.R. Preedy, R.R. Watson, Tomatoes and Tomato Products, Nutritional, Medicinal and Therapeutic Properties, Science Publishers, Enfield, New Hampshire, USA, 2008, pp. 3–8. https://doi.org/10.1201/9781439843390.
[66] V.R. Preedy, R.R. Watson, Lycopene Nutritional, Medicinal and Therapeutic Properties, CRC Press, Cambridge, UK, 2019.
[67] C. Rock, W. Yang, R. Goodrich-Schneider, H. Feng, Conventional and alternative methods for tomato peeling, Food Eng. Rev. 4 (1) (2012) 1–15. https://doi.org/10.1007/s12393-011-9047-3.
[68] A. Ruiz Celma, F. Cuadros, F. López-Rodríguez, Characterisation of industrial tomato by-products from infrared drying process, Food Bioprod. Process. 87 (4) (2009) 282–291. https://doi.org/10.1016/j.fbp.2008.12.003.
[69] R.K. Saini, S.H. Moon, Y.S. Keum, An updated review on use of tomato pomace and crustacean processing waste to recover commercially vital carotenoids, Food Res. Int. 108 (2018) 516–529. https://doi.org/10.1016/j.foodres.2018.04.003.

[70] D. Shao, C. Venkitasamy, J. Shi, X. Li, W. Yokoyama, Z. Pan, Optimization of tomato pomace separation using air aspirator system by response surface methodology, Trans. ASABE 58 (6) (2015) 1885–1894. https://doi.org/10.13031/trans.58.11256.
[71] D. Shao, G.G. Atungulu, Z. Pan, T. Yue, A. Zhang, X. Chen, Separation methods and chemical and nutritional characteristics of tomato pomace, Trans. ASABE 56 (1) (2013) 261–268. https://doi.org/10.13031/2013.42577.
[72] R. Sharma, H. Oberoi, G. Dhillon, Fruit and vegetable processing waste: renewable feed stocks for enzyme production, in: Dhillon G. Singh, S. Kaur (Eds.), Agro-Industrial Wastes As Feedstock For Enzyme Production, Academic Press, London, U.K., 2016, pp. 23–58.
[73] Y.P.A. Silva, B.C. Borba, V.A. Pereira, M.G. Reis, M. Caliari, M.S.L. Brooks, T.A.P.C Ferreira, Characterization of tomato processing by-product for use as a potential functional food ingredient: nutritional composition, antioxidant activity and bioactive compounds, Int. J. Food Sci. Nutr. 70 (2) (2019) 150–160. https://doi.org/10.1080/09637486.2018.1489530.
[74] S.P. Sonawane, G.P. Sharma, N.J. Thakor, R.C. Verma, Moisture-dependent physical properties of kokum seed (Garcinia indica Choisy), Res. Agric. Eng. 60 (2) (2014) 75–82. https://doi.org/10.17221/59/2011-rae.
[75] J. Stanley, S. Narayanan, R. Ashish, S. Kumar, Other uses of tomato by-products, in: S. Porretta (Ed.), Tomato Chemistry, Industrial Processing and Product Development, The Royal Society of Chemistry, London, UK, 2019, pp. 259–279.
[76] D. Symeonides, P. Loizia, A.A. Zorpas, Tires waste management system in cyprus in the framework of circular economy strategy, J. Environ. Sci. Pollut. Res. (2019). https://doi.org/10.1007/s11356-019-05131-z.
[77] B.R. Thakur, R.K. Singh, P.E. Nelson, Quality attributes of processed tomato products: a review, Food Rev. Int. 12 (3) (1996) 375–401. https://doi.org/10.1080/87559129609541085.
[78] G. Tsatsaronis, D. Boskou, Amino acid and mineral salt content of tomato seed and skin waste, J. Sci. Food Agric. 26 (4) (1975) 421–423.
[79] C. Tzia, P. Sfakianakis, V. Giannou, Raw materials of foods: handling and management, Ch. 1, in: T. Varzakas, C. Tzia (Eds.), Handbook of Food Processing: Food Safety, Quality, and Manufacturing Processes, 2, CRC Press, Boca Raton, FL, USA, 2015, pp. 1–40.
[80] I. Vardopoulos, I. Konstantopoulos, A.A. Zorpas, L. Limousy, S. Bennici, V. Inglezakis, I. Voukkali, Sustainable metropolitan areas perspectives though the assessment of the existing waste management strategy, Environ. Sci. Pollut. Res. (2020). https://doi.org/10.1007/s11356-020-07930-1.
[81] S.K. Vidyarthi, C.W. Simmons, Characterization and management strategies for process discharge streams in California industrial tomato processing, Sci. Total Environ. 723 (2020) 137976. https://doi.org/10.1016/j.scitotenv.2020.137976.
[82] I. Voukali, P. Loizia, J. Navarro Pedreno, A.A. Zorpas, Urban strategies evaluation for waste management in coastal areas in the framework of area metabolism, Waste Manag. Res. (2021), doi:10.1177/0734242X20972773.
[83] K. Waheed, H. Nawaz, M.A. Hanif, R. Rehman, Tomato, Medicinal Plants of South Asia (2020) 631–644. https://doi.org/10.1016/B978-0-08-102659-5.00046-X.
[84] S. Xu, R.B. Pegg, W.L. Kerr, Physical and chemical properties of vacuum belt dried tomato powders, Food Bioproc. Technol. 9 (1) (2016) 91–100. https://doi.org/10.1007/s11947-015-1608-7.
[85] A.A. Zorpas, Sustainable waste management through end-of-waste criteria development, Environ. Sci. Pollut. Res. 23 (8) (2015) 7376–7389. 10.1007/s11356-015-5990-5.
[86] A.A. Zorpas, Strategy development in the framework of waste management, Sci. Total Environ. 716 (2020) 137088. https://doi.org/10.1016/j.scitotenv.2020.137088.

[87] A.A Zorpas, K. Lasaridi, K. Abeliotis, I. Voukkali, P. Loizia, L. Fitiri, C. Chroni, N. Bikaki, Waste prevention campaign regarding the Waste Framework Directive, Fresenius Environ. Bull. 23 (11) (2014) 2876–2883.
[88] A.A. Zorpas, K. Lasaridi, D.-M. Pociovălişteanu, P. Loizia, Household compost monitoring and evaluation in the framework of waste prevention strategy, J. Clean. Prod. 172 (2018) 3567–3577. http://dx.doi.org/10.1016/j.jclepro.2017.03.155.
[89] A.A. Zorpas, K. Lasaridi, I.Voukkali, P. Loizia, C. Chroni, Promoting sustainable waste prevention activities and plan in relation to the waste framework directive in insular communities, Environ. Proc. 2 (2015) 159–173, doi:10.1007/s40710-015-0093-3.
[90] A.A. Zorpas, I. Voukkali, P. Loizia, A prevention strategy plan concerning the waste framework directive in Cyprus, Fresenius Environ. Bull. 26 (2) (2017) 1310–1317.
[91] United Nations, Resolution 70/1, Transforming our World: the 2030 Agenda for Sustainable Development, 25th September 2015.

CHAPTER TWO

# Tomato by-products as animal feed

**María Dolores Carro Travieso[a], Trinidad de Evan[a], Carlos Navarro Marcos[a], Eduarda Molina-Alcaide[b]**
[a]Department of Agricultural Production, Technical University of Madrid (UPM), Madrid, Spain
[b]Zaidin Experimental Station (EEZ), Spanish National Research Council, Granada, Spain

## 2.1 Introduction

According to the statistics of the Food and Agriculture Organization of the United Nations [1], tomato is the second most important vegetable crop worldwide after potato and its production has increased in the last years from $109 \times 10^6$ tonnes in 1999 to $181 \times 10^6$ tonnes of tomato fruits produced on $5.03 \times 10^6$ ha in 2019. Potatoes production was estimated to reach $462 \times 10^6$ tonnes in 2019 [1]. The main tomato producers are China, India, Turkey, United States of America, and Egypt, accounting for 34.8%, 10.5%, 7.1%, 6.0%, and 3.7% of total world production in 2019, respectively [1]. Tomato fruits also rank second after potato as the most consumed vegetable worldwide, and its consumption has been related to considerable health benefits such as decreasing the risk of cancer, cardiovascular diseases, osteoporosis, ultraviolet light-induced skin erythema, and cognitive dysfunction [2,3]. Tomato fruits are rich in carotenoids (lycopene, lutein, and α- and β-carotene), vitamin E, vitamin C, fiber, and potassium. Tomatoes are usually the main source of lycopene in the human diet [4], a powerful antioxidant. Although tomato fruits are widely consumed fresh, an important part of total production is processed to produce a wide range of products (sauces, juice, paste, ketchup, puree, canned and dried tomatoes, soup, …), and it has been shown that lycopene bioavailability in processed tomato products is greater than in fresh tomatoes [5,6].

Cultivation and processing of tomatoes into tomato products generates waste fruits and a large amount of solid and liquid by-products that are usually discarded [7]. A great part of the tomatoes processed is manufactured as bulk paste, which is later remanufactured into different final products, whereas the rest is used to prepare other tomato foods such as whole peeled and diced tomatoes [8]. The first step in processing plants is the grading of

*Tomato Processing by-Products*
DOI: https://doi.org/10.1016/B978-0-12-822866-1.00001-6

tomatoes, during which damaged or defected fruits and those that do not reach the required standards are culled. After grading, tomatoes are generally washed, sorted, sized and separated for further processing depending on the target tomato product [7]. The presence of peels and seeds in most tomato products is undesirable, and therefore tomatoes are peeled and seeds are separated during processing for producing paste, juice, canned tomatoes, and other products. The main by-product of tomato processing for paste and juice production is tomato pomace (TP), which is composed of a mixture of peels, seeds, and small amounts of pulp, and it can represent between 1.5% and 10% of the weight of the original fruits [9–11]. Peels and seeds are by-products of the tomato cannery industry, and they can be removed by different methods [12]. In addition, fresh water is used during processing for different functions such as helping tomatoes transport through the equipment, producing steam, cooling, cleaning,..., and generating process water that should also be disposed of [7].

By-products of tomato processing contain multiple nutrients and are rich in bioactive compounds, such as lycopene, phenolic compounds, and vitamins [11,13], and therefore they can be further valorized providing an extra income for processing plants, reducing the environmental problems caused by their accumulation, and transforming a waste into a usable resource. Tomato by-products are currently used as a source of functional ingredients for the food industry and of raw materials for nutraceuticals and cosmetics production, but also as feed ingredients for animal feeding [14]. The aim of this chapter is to review the literature concerning the use of tomato by-products in animal feeding, with special emphasis on their effects on livestock production and quality of animal products.

## 2.2 Chemical composition and nutritive value of tomato by-products

### 2.2.1 Tomato pomace

TP is quantitatively the most important tomato by-product, and consequently there are abundant data on its chemical composition. Table 2.1 shows chemical composition values for TP, tomato peels, seeds and cull tomatoes reported in different animal feeding trials. As TP is composed of a mixture of peels, seeds and residual pulp, its chemical composition is highly variable depending on the relative proportion of each fraction. TP is characterized by a high-moisture content, usually greater than 70% [7,9], due to the water added at the last steps of the industrial process to help the flow

**Table 2.1** Chemical composition (g/100 g dry matter (DM), except DM content) of samples of different tomato by-products used in animal feeding trials.[a]

| | Tomato pomace | Tomato peels | Tomato seeds | Cull tomatoes |
|---|---|---|---|---|
| Dry matter (DM; g/100 g) | 24.3 ± 4.93 (14.3–38.1); n = 28 | 7.8 ± 2.42 (5.2–10.0); n = 3 | 9.0 ± 1.40 (7.4–10.0); n = 3 | 7.2 ± 1.05 (6.5–8.7); n = 4 |
| Ash | 5.42 ± 2.60 (2.5–14.4); n = 54 | 6.4 ± 7.89 (1.9–25.6); n = 8 | 4.5 ± 1.45 (3.0–7.6); n = 10 | 8.0 ± 2.89 (3.1–10.1); n = 5 |
| Crude protein | 19.0 ± 3.00 (12.4–25.4); n = 54 | 10.5 ± 7.90 (1.0–23.3); n = 8 | 30.3 ± 7.42 (16.5–40.9); n = 10 | 18.5 ± 2.31 (15.3–21.0); n = 5 |
| Crude fibre | 33.6 ± 6.75 (17.8–47.3); n = 19 | 66.6 ± 16.3 (37.8–88.0); n = 7 | 25.2 ± 9.82 (15.1–36.3); n = 5 | 17.8 ± 2.06 (15.6–19.7); n = 3 |
| Neutral detergent fiber | 57.5 ± 4.40 (47.8–68.0); n = 44 | | | 26.5 ± 7.71 (19.1–34.5); n = 3 |
| Acid detergent fiber | 44.5 ± 4.62 (32.6–58.7); n = 35 | | | 20.7 ± 6.40 (13.9–26.6); n = 3 |
| Lignin | 21.8 ± 6.55 (6.4–31.8); n = 25 | | | 12.2 ± 10.3 (4.9–19.5); n = 3 |
| Fat | 9.4 ± 6.75 (4.1–16.6); n = 49 | 4.3 ± 0.95 (2.3–5.5); n = 6 | 22.2 ± 2.93 (17.8–27.1) n = 8 | 4.8 ± 1.87 (2.8–7.2); n = 4 |
| Sugars | 17.2 ± 6.88 (8.2–28.3); n = 22 | | | |
| Ca | 0.51 ± 0.416 (0.19–1.68); n = 11 | 0.02 n = 1 | 0.11 (0.11–0.11) n = 2 | 1.5 ± 1.54 (0.5–3.3); n = 2 |
| P | 0.46 ± 0.064 (0.33–0.56); n = 11 | 0.37 n = 1 | 0.58 (0.58–0.58) n = 2 | 0.43 ± 0.01 (0.42–0.44); n = 2 |

[a]mean ± standard deviation; minimum and maximum values in brackets; n indicates the number of samples. All values were obtained from published papers cited in the reference list.

of tomato residues through the ducts of the equipment [7,15]. This high-moisture content makes TP highly perishable, which represents an important limitation for its use in animal feeding, and therefore fresh TP is used only in livestock farms located near to the processing plants [16]. TP can be dried, but artificial drying increases markedly its costs; alternatively, TP can be either sun dried in areas where climatic conditions are favourable, ensiled [17] or included in multinutrient blocks for livestock feeding [18,19].

TP is a fibrous material, containing as average (dry matter (DM) basis) 33.6%, 57.5%, and 44.5% of crude fiber (CF), neutral detergent fiber (NDF) and acid detergent fiber (ADF), respectively, according to reported values in the literature (Table 2.1). Lignin content in TP can vary from 6.4% [20] to 31.8% [21], mainly depending on peels proportion, and the high lignification reduces fiber digestibility. TP contains relatively high levels of crude protein (CP; 19.0% as average) and fat (9.4% as average), but values for both fractions present great variability. The fat contains a high proportion of unsaturated fatty acids (UFA; ≈76%) [22] and linoleic, oleic, and palmitic acids have been identified as the predominant fatty acids (FA) [22].

Although sugars content of TP has been analyzed in a limited number of studies, available data also indicate high variability, with levels ranging from 8.2 [15] to 28.3% of DM [9]. Fructose has been identified as the most predominant sugar [9]. Average concentrations of Ca and P for 11 samples of TP used in different feeding trials were 0.51% and 0.46% (DM basis), respectively. TP is also a good source of lycopene, with concentrations reaching nearly 300 mg/100 g DM [23]. Del Valle et al. [9] analyzed pectin's concentrations of 21 samples of TP obtained at different steps during tomato processing for paste and reported an average value of 7.55 g/100 g DM. Recently, Marcos et al. [15] analysed total polyphenols content in 12 samples of TP from two different processing plants and obtained values ranging from 0.25 to 0.53 g/100 g DM.

Differences in chemical composition of TP reported in the literature can be due to multiple reasons, including the chemical composition of the processed tomato fruits, which is highly variable depending on factors such as variety, growth conditions, ripeness and time of harvesting, among others [9,14]. The characteristics of the industrial processing of tomato fruits and the relative proportions of seeds, peels and pulp in the final product can also markedly influence the chemical composition of TP [9,11, 15], as peels are rich in fiber, lycopene and phenols, and seeds are rich in CP and fat [14]. Relative proportions of the different fractions can be highly variable. Thus, Silva et al. [11] analyzed six different batches of TP and reported average proportions of 61.5% for peels and 38.5% for seeds, with no pulp being detected. In contrast, Sogi and Bawa [24] reported average proportion of 40%, 27%, and 33% for pulp, peels and seeds, respectively. Finally, differences in the analytical methods used in the laboratories cannot be discarded as an additional contributing to variability in TP composition [9].

The use of TP in animal diets can reduce feeding costs for livestock farmers and contribute to save processing industries the costs of disposal. Fresh TP can be directly fed to animals, but it spoils rapidly and therefore only a small proportion of total production is used fresh in farms close to the processing plants [15]. In most cases, TP is dried (DTP) to increase its shelf life and facilitate storage, although it can also be included in multinutrient blocks or ensiled. However, TP cannot be ensiled alone due to its high moisture content, and it is usually mixed with fibrous materials (cereal or rice straw, corn stovers, hay, ...) before ensiling [17].

Because of its high fibre content, TP can be better utilized by ruminants compared with non-ruminant species. The nutritive value of TP for ruminants is moderate, and its net energy content for dairy animals has been reported to range from 1.23 to 1.53 Mcal/kg DM [25,26]. The high CP and EE content of TP makes it a valuable by-product for ruminant feeding. The reported values for *in situ* DM ruminal degradability of TP range from less than 40% [20] to nearly 80% [27]. This high variability can be attributed to differences in the chemical composition of the TP samples tested in the different studies, but also to variations in the *in situ* procedure that is affected by many factors such as ruminant species and their feeding, incubation procedure, bags washing, etc. [28]. In several *in vivo* studies with ruminants, the average digestibility of TP organic matter (OM) was 56.0% [16,29,30]. The ruminal degradability of TP fiber is low (< 40%) due to high proportions of lignin and polyphenols [15,30].

Although TP contains appreciable levels of CP, its ruminal degradability is low, with values for 16 to 24 h *in situ* measurements ranging from 41% [20] to 65% [29]. This low CP rumen degradability could be partly related to the formation of indigestible compounds during processing, due to Maillard reactions between amino acids and sugars produced as consequence of the high temperatures applied [15]. This resulted in high acid detergent insoluble nitrogen (N) concentrations in TP, which can reach up to 26.0% of total N [31,32]. Intestinal digestibility of the rumen undegraded fraction of CP is also lower than 50% [15], resulting in CP total tract digestibility values lower than those observed for protein concentrates such as soybean meal [33].

The nutritive value of TP for nonruminant species is lower than that for ruminants, but its inclusion at low levels in the diet can improve the quality of animal products, as described later in this chapter. Average energy contents of 1.76 Mcal metabolizable energy/kg DM of DTP have been reported for broilers [34,35] and 0.932 Mcal of net energy/kg DM for pigs [16].

### 2.2.2 Tomato peels

Tomato peels have low DM content (Table 2.1; 5.210.0%) [12,36], and fiber is the most abundant component in the DM, although reported contents in the literature are highly variable (37.8% to 88.0%) [37,38]. The soluble/insoluble fiber ratio in tomato peels (1:5) [39] is much lower than that observed in orange peels (1:1) [40], indicating a low fiber digestibility. The CP content of tomato peels is usually low, but also exhibits high variability as reported values range from 1.0% to 23.3% [36,41], being glutamic and aspartic acids the most abundant amino acids [12]. However, tomato peels have high concentrations of carotenoids (mainly lycopene) and flavanols [11,23], and it has been shown that the intake of tomato paste enriched with peels increases the amount of lycopene absorbed in humans compared with tomato paste without peels [3]. Knoblich et al. [12] reported that the amount of total carotenoids and lycopene was 5.0 and 5.6 times greater, respectively, in tomato peels compared with tomato seeds.

The method used for tomatoes peeling affects chemical composition of tomato peels, especially their content in CP, fiber, and lycopene [42,43]. In addition, when NaOH is added during the peeling process, tomatoes peels can contain high Na levels (>8% of DM) [12], and this should be taken into account in the formulation of the diets because an excess of Na can have negative effects on animal health and performance. Tomato peels also contain high levels of potassium and iron [12].

### 2.2.3 Tomato seeds

Tomato seeds are mainly produced in the tomato cannery industry and contain low DM (Table 2.1; 7.4%–10.0%) [12,44], but this fraction is rich in CP (16.5% to 40.9%) [12,45], fiber (15.1% to 36.3% of CF) [44,45], and fat (17.8% to 27.1%) [46,47]. Fat of tomato seeds has an UFA profile, with linoleic (37.6% to 72.7% of total FA) and oleic (15.5% to 29.7%) being the most abundant FA [22,47–49].

Johns and Gersdorff [50] were the first in reporting that protein of tomato seeds is formed of α- and β-globulins, and that β-globulins contained more lysine (3.8% vs 1.2% of total protein, respectively) and histidine (6.4% vs 4.9%) but less arginine (10.7% vs 14.0%) compared with α-globulins. Anwar et al. [51] observed that chemical composition of tomato seeds meal was similar to that of cotton seed meal, but tomato seeds contained three times more fiber, which is an important limitation in poultry feeding as it decreases the metabolizable energy content of feeds. These authors also observed that the amino acid profile of both seed meals was similar, with the exception of lysine that was 1.4 times greater in tomato seeds meal, and concluded that methionine was the first limiting amino acid in tomato seeds meal. Protein of tomato seeds also contain greater levels of lysine than cereals proteins [52], but the most abundant amino acids are glutamic (19.4% to 24.4% of total amino acids) and aspartic (8.82% to 10.3%) acids [53].

Tomato seeds contain low amount of lycopene (<0.02%) [12], but relatively high amounts of phenols, flavonoids (quercetin-3-O-β- glucoside, naringenin, rutin, kaempferol, *iso*-rhamnetin, etc.) and different acids, such as caffeic, vanillic, gallic, ferulic, and p-coumaric among others [54,55].

Dried tomato seeds has more energy than DTP due to their high fat content, and their practical use in animal feeding has been assessed in several studies, either as partial replacers of corn grains and soybean meal in poultry diets [44] or of whole cottonseed in dairy cows [56].

### 2.2.4 Cull tomatoes

Tomato fruits that do not meet the required quality standards either for sale fresh (damaged, broken, misshapen, mushy fruits, ...) or for processing are discarded. In addition, in situations of excessive production, part of the production can also be discarded to maintain prices [17]. The information on chemical composition and nutritive value for animal feeding of culled tomatoes is more limited than that for TP. Fresh tomatoes usually contain lower DM (7.2% as average) [10,37] than TP due to water elimination by pressing the tomatoes during processing [10], but reported CP levels are similar (average 19.0 and 18.5% of DM for TP and cull tomatoes; Table 2.1). Partial solubilization of minerals during processing can explain the greater ash content in cull tomatoes compared with TP (average 8.0% and 5.42% of DM, respectively) [37,57]. Fresh tomatoes have lower contents of fat and fiber (DM basis) than TP due to the presence of pulp, and average contents of 4.8%, 26.5%, and 20.7% of DM have been reported for fat, NDF and ADF, respectively [10,37,58]. Lignin content in culled tomatoes (12.2%) [10,58] is also lower than in TP (21.8%) due to the lower content of peels and seeds in the fresh fruits. Unripe tomatoes contain α-tomatine, an alkaloid that gives a bitter taste to green tomatoes and reduces their palatability [7]. However, concentrations of α-tomatine are reduced as tomato fruits mature, and there is no evidence of any adverse effects of this compound on animals fed either tomatoes or their by-products [59].

Cull tomatoes have greater digestibility and energy content than TP because they contain the whole pulp. Ventura et al. [10] reported that

DM ruminal degradability (estimated for a rumen passage rate of 6%/h in Canarian goats) of fresh tomatoes was 57.9%, but degradability of CP was much lower (19.4%) due to the high proportion of CP linked to ADF. Ammerman et al. [60] reported that 28.6 and 11.6% of total ingested N was retained when lambs were fed diets containing either soybean meal or cull tomatoes, respectively, which indicates that N in dried culled tomatoes was less digestible and had lower biological value than that in soybean meal. In a feeding trial with steers, DM and CP digestibility values were 63.1% and 54.6%, respectively [60]. These values agrees well with the 63.2% *in vitro* OM digestibility of culled tomatoes reported by Ventura et al. [10] using ruminal fluid from goats. Culled tomatoes have an estimated ME content of 1.694 Mcal of net energy/kg DM for dairy ruminants, and their total digestible nutrient content (TDN; 70.7%) is similar to that in citrus pulp [61].

The high moisture content of cull tomatoes limits their intake by the animals and also makes transport unprofitable. Therefore, tomatoes are usually dried or ensiled (mixed with a fibrous material) before being fed to livestock [62]. Another feasible alternative is including tomatoes as an ingredient of multinutrient blocks [18,19,58]. The influence of dried and ensiled culled tomatoes on livestock performance and quality of animal products has been assessed in several species, including cattle, sheep, goat, and poultry [60,62,63]. Interestingly, the dietary inclusion of tomato fruits has decreased methane production in goats in both *in vivo* [19,58,63,64] and *in vitro* [57,65,66] studies, but no antimethanogenic effect has been reported for TP [15]. Methane is a potent greenhouse gas and represents an energy loss for ruminants that can reach up to 12% of the gross energy intake [67], and therefore reducing methane emissions is currently an important goal for ruminant nutritionists.

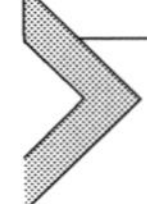

## 2.3 Use of tomato by-products in animal feeding

### 2.3.1 Cattle and buffalo

The potential use of tomato by-products in cattle feeding has being less studied than in small ruminants, possible due to their lower capacity of to utilize low-quality feeds compared with sheep and goats [68]. The results of different studies assessing the influence of feeding tomato by-products to cattle are summarized in Table 2.2, and their results have been sometimes contrasting. In most cases, TP was ensiled and only in a couple of studies TP was dried before feeding.

Table 2.2 Effects of feeding tomato by-products to cattle on animal performance and quality of products.

| Reference | Animals | Tomato by-product | Diet characteristics | Inclusion level (% total diet) | Comments |
|---|---|---|---|---|---|
| [69] | Lactating Holstein cows (100 DIM) | TP silage | TMR (corn silage as forage) | 7.2% | No effects on DMI, nutrient digestibility and milk yield and composition |
| [70] | Lactating Holstein cows (70 DIM) | TP silage or DTP | NR | 10% | No effect on DMI and milk yield and composition |
| [71] | Pregnant and lactating Holstein cows | TP silage | TMR (corn silage as forage) | 10% | Improved DMI and DM<br>Increased vitamin content in milk<br>Improved antioxidant status and immune response |
| [72] | Dry Holstein cows | TP silage | TMR (corn silage as forage) | 13% | Improved DMI<br>No effect on nutrient digestibility |
| [73,74] | Multiparous Holstein cows (91 DIM) | TP silage mixed with apple pomace (1:1) | TMR (alfalfa hay as forage) | 15% and 30% | Increased milk yield; tended to increased fat milk yield<br>Increased DM and OM digestibility<br>Increased ruminal VFA concentrations<br>Reduced chewing and ruminating time<br>Decreased feed efficiency at 30% |
| [56] | Lactating Holsten cows (60 to 100 DIM) | Whole tomato seeds | TMR (alfalfa hay as forage) | 1.1%, 2.4%, and 4.0% | No effect on DMI and milk yield<br>Lineal decrease of milk fat; increased C18:3 in milk<br>Reduced CP and FA digestibility<br>Lineal decrease of milk and blood urea N |
| [33] | Rumen-fistulated beef steers (370 kg BW) | DTP | TMR (rice straw and rice bran as fiber sources) | 3.2%, 8.0%, and 11.2% | No effect on DMI, retained N and nutrient digestibility, excepting decreased CP digestibility<br>Reduced ruminal total VFA concentrations<br>Increased rumen ammonia N and blood urea N concentrations |
| [79] | Crossbred Brahman x Native steers (258 kg BW) | DTP | TMR (rice straw and rice bran as fiber sources) | 3.2%, 8.0%, and 11.2% | Lineal decrease in DMI and final BW<br>No differences in ADG<br>Lineal increase in blood urea N |

ADG, average daily gain; BW, body weight; CP, crude protein; DM, dry matter; DIM, days in milk; DMI, dry matter intake; DTP, dried tomato pomace; FA, fatty acid; FCR, feed conversion rate; N, nitrogen; NR, not reported; OM, organic matter; TMR, total mixed ration; TP, tomato pomace; VFA, volatile fatty acids.

Weiss et al. [69] analyzed the effects of replacing corn silage by ensiled TP (7.2%) in a total mixed ration (TMR) on performance of multiparous Holstein cows at 100 days in milk (DIM), and observed no effects on dry matter intake (DMI), nutrient digestibility, or milk yield and composition. Similarly, Safari et al. [70] did not observe any effect on DMI and milk yield when diets including 10% of either ensiled TP or DTP were fed to Holstein cows (70 DIM), but milk protein content was slightly improved.

More recently, Tuoxunjiang et al. [71] fed a TMR, in which 10% of maize silage was replaced with TP silage, to lactating Holstein cows and observed no changes in milk yield. The cows fed TP silage had greater DMI (1.7%) and DM digestibility, but NDF digestibility was decreased, possibly as consequence of the low fiber degradability of TP. Feeding TP silage also resulted in enhanced concentrations of vitamins A, E, and C in milk, which was attributed to the increase in dietary levels of vitamins when TP silage was included in the diet. Furthermore, the antioxidant status of the cows and their immune response during the transition period were improved by feeding TP silage and the incidence of mastitis was reduced. In other study by the same authors [72], no changes in nutrient digestibility were observed when corn silage was replaced with TP silage (13% of total diet) in the diet of dry Holstein cows, but DMI was increased.

Abdollahzadeh et al. [73] assessed the influence of replacing alfalfa hay with a 1:1 mixture of ensiled TP and apple pomace at 15% or 30% of the TMR of multiparous mid-lactating dairy cows (91 DIM) and did not detect changes in milk composition, but milk yield was increased by 6.3%. In addition, DMI was linearly raised and digestibility of both DM and OM was increased, which may be related to the inclusion of apple pomace in the TMR [74,75] and to the greater pectin and other rapidly-fermented carbohydrates content of TP and apple pomace [9,15,76] compared with alfalfa hay [25]. The greater ruminal concentrations of total volatile fatty acids (VFA), lower ruminal pH values, and enhanced blood concentrations of glucose and total protein reported in the cows fed the diets containing the TP silage and apple pomace mixture compared with control-fed cows [77] indicate greater ruminal degradation in the cows fed the by-products. Alternatively, the lower pH values could be related to the reduction in chewing and ruminating time observed when TP silage and apple pomace was fed to cows [77], as a reduction in these processes decreases saliva secretion. On the contrary, Yuangklang et al. [78] reported that the replacement of soybean meal with DTP in dairy cows diets decreased protein content in milk, which could be related to the low digestibility of CP ofTP compared with soybean meal CP.

The use of other tomato by-products in dairy cows has been little studied. Cassinerio et al. [56] fed three levels of whole tomato seeds (1.1%, 2.4%, and 4.0%) as replacement of whole cottonseed to Holstein dairy cows and observed no negative effect on feed intake (DM, fat, and CP) or milk yield, but milk fat and urea concentrations were linearly decreased as the dietary amount of tomato seeds increased. In addition, the whole-tract digestibility of CP was decreased, as consequence of the low ruminal degradability and digestibility of CP in TP [15,20]. A decrease in FA whole tract digestibility was reported in the same study, which contrasts with the increases in fat digestibility observed in other studies when TP was included in the diet [29,79]. Cassinerio et al. [56] observed the presence of whole tomato seeds in the feces, which supports the lower fat digestibility observed in this study for the cows fed the diets containing tomato seeds. These authors also analyzed the changes in milk FA profile, and observed that concentration of C18:3 increased and C18:2 tended to increase by feeding tomato seeds, but total milk yield of these FA was unaffected, suggesting that only a small fraction of FA in tomato seeds might have by-passed the rumen. Consequently, Cassinerio et al. [56] indicated that tomato seeds are an interesting feed ingredient for dairy cows, but more research is needed and diets with higher proportions should be evaluated.

The potential ofTP as feed ingredient has also been evaluated in dairy buffalos. Choubey et al. [80] studied the influence of including 10% of TP in multinutrient blocks for dairy buffaloes and reported no changes in nutrient digestibility, excepting a reduction in CP digestibility. Feeding the multinutrient blocks containing TP also caused a reduction in the ruminal non-protein N and blood urea-N concentrations, which is in agreement with results reported in Holstein cows [56]. Additionally, the buffaloes fed TP had lower urinary N excretion, but the amount of N retained was unaffected. In a similar study, Wadhwa and Bakshi [81] observed that feeding dairy buffaloes multinutrient blocks, including TP (10%) and other by-products (waste bread and spent sugar syrup) resulted in no negative effects on either nutrient digestibility or animal health status compared with buffaloes fed a control diet.

More recently, Gawad et al. [82] analyzed the effects of replacing wheat bran with DTP (10%) and dried beet pulp (10%) in the diet of Egyptian dairy buffaloes and observed no changes in milk yield and milk fat production. However, milk FA profile was shifted to greater polyunsaturated FA

**Table 2.3** Effects of feeding tomato by-products to sheep and goats on animal performance and products quality.

| Reference | Animals | Tomato by-product | Diet characteristics | Inclusion level (% total diet) | Comments |
|---|---|---|---|---|---|
| [87] | Lactating Awassi ewes (67 DIM) | DTP | Forage:concentrate (30:70) | 30% | No effects on DMI and milk yield<br>Increased milk concentrations of lactose and MUFA<br>Decreased milk SFA |
| [88] | Lactating Comisana sheep | DTP | TMR (barley straw as forage) | 20.4% | No effect on milk yield<br>Increased concentrations of milk fat, PUFA and CLA |
| [89] | Lactating Saanen goats (90 DIM) | DTP | TMR (corn silage and alfalfa hay as forage) | 24% | No effect on DMI and milk yield<br>Increased milk concentrations of MUFA and CLA |
| [90] | Lactating Saanen goats | DTP | TMR (Tifton hay as forage) | 20%, 40%, and 60% | Increased DMI and milk yield up to 40% of DTP<br>Decreased DMI and milk yield at 60%<br>Increased total protein in blood, but no effects on other blood metabolites |
| [58] | Lactating Murciano-Granadina goats (50 DIM) | Cull tomatoes | Alfalfa hay:concentrate (50: 50) | 1.7% | No effects on milk yield and nutrient digestibility<br>Reduced lactose and increased PUFA concentrations in milk<br>Reduced methane production |
| [66] | Dry Murciano-Granadina goats | Cull tomatoes | Alfalfa hay:concentrate (50: 50) | 3.6% | Decreased diet digestibility and microbial nitrogen flow<br>Reduced methane production |

| | | | | | |
|---|---|---|---|---|---|
| [63] | Lactating Murciano-Granadina goats (60 DIM) | Ensiled cull tomatoes | TMR (oat hay and alfalfa hay as forage) | 20.2% and 19.4% | No effects on milk yield, but<br>Increased concentrations of milk fat, PUFA, UFA and CLA<br>Reduced methane production |
| [20] | Fleischaff by Romanov - Rasa Aragonesa growing lambs | DTP | Based on barley grain and straw | 10%, 20%, and 30% | Decreased DMI<br>Increased nutrient digestibility and N retention<br>No effects on growth performance |
| [91] | Growing Ossimi lambs | DTP | TMR (berseem hay as forage) | 5%, 10%, and 15% | Increased nutrient digestibility, final BW and ADG |
| [92] | Growing Hararghe lambs | DTP | Low-quality grass hay | 360 g/d | Increased DMI, diet digestibility, final BW and ADG<br>Decreased FCR |
| [93] | Growing Markhoz goat kids | DTP | TMR (alfalfa hay as forage) | 10%, 20%, and 30% | No negative effect on growth performance<br>Increased carcass protein and fat content<br>Greater production of greasy fiber; greater fiber diameter |
| [94] | Fattening Comisana lambs | DTP | Commercial pelleted diet | *Ad libitum* | No effects on total DMI, ADG and carcass weight<br>Increased meat concentrations of linoleic acid<br>Tended to increase meat PUFA concentrations<br>Increased concentrations of $\gamma$- tocopherol and retinol in meat |
| [18] | Growing Barbarine lambs | Wet TP | Wheat straw *ad libitum* and concentrate | 20.6% | No effects on final BW<br>Increased FCR |

ADG, average daily gain; BW, body weight; CLA, conjugated linoleic acid; DMI, dry matter intake; DTP, dried tomato pulp; MUFA, monounsaturated fatty acids; FCR, feed conversion rate; PUFA, polyunsaturated fatty acids; SFA, saturated fatty acids; TMR, total mixed ration; TP, tomato pulp; UFA, unsaturated fatty acids.

(PUFA) concentrations (especially, C18:2 and C18:3 acids). In addition, blood concentrations of albumin and total protein were increased when the diet containing DTP and dried beet pulp was supplemented with fibrolytic enzymes.

Finally, Ebeid et al. [83] fed multiparous Egyptian buffaloes (70 DIM) a diet with 40:60 concentrate:forage ratio (corn silage, clover, and rice straw as forage), and studied the effects of replacing the clover with TP silage on animals productive performance. The digestibility of all nutrients, excepting cellulose, was significantly increased by feeding TP silage. Fat-corrected (7%) milk production and milk fat concentration were also increased, and milk FA profile was shifted to a healthier profile, as concentrations of saturated FA (SFA) were reduced and those of PUFA were augmented. However, a reduction in milk amino acids concentrations was reported for the animals fed the diet containing TP silage, probably as consequence of the low CP digestibility of TP [15,20].

Studies evaluating tomato by-products for beef cattle feeding are scarce. Thus, Yuangklang et al. [33] analyzed the effects of replacing cassava chips and a small fraction of soybean meal with different levels of DTP (3.2,% 8.0%, and 11.2%) in TMR diets on diet digestibility, ruminal fermentation, and N balance in rumen-fistulated steers. No changes were observed on either feed intake or apparent nutrient digestibility, with the exception of CP digestibility that was decreased in agreement with that reported in other studies [56,80]. Steers fed DTP had increased concentrations of ruminal ammonia and blood urea N, which was attributed to the inclusion of additional urea in the DTP-diets to equilibrate the N content in all experimental diets. Additionally, N fecal excretion increased and urinary N decreased in the steers fed the DTP-diets, but there were no changes in the amount of retained N compared with the control group. However, total VFA concentrations were decreased by feeding the diets containing DTP. When the same experimental diets were fed to 2-year old Brahman x Native steers [79] for 120 days, the diets containing DTP significantly reduced DMI and final body weight (BW) of the animals, but differences in average daily gain (ADG; 1037 to 881 g/d for control and DTP at 11.2%, respectively) did not reach the significance level. The increased blood urea N concentrations observed in this study confirmed the results of the previous one conducted with the same experimental diets [33]. Contrary to these results, in a previous study by the same authors [84] it was observed that feeding DTP as the only roughage source to steers increased final BW compared with either hay or fresh grass. Similarly, D'Arleux et al. [85]

reported no negative effects on productive performance of Montbéliardes heifers when alfalfa hay in the diet was replaced by TP silage (29% of total diet). Chumpawadee et al. [86] also observed that DTP could be included at 50% of the diet (DM basis) in a TMR for Brahman-Thai steers without decreasing DMI compared with other by-products, such as dried brewer grains and soybean hulls, included at the same level in the diet.

Although results from some studies are controversial, it seems that tomato by-products can be included in moderate proportions in the diet of cattle and buffaloes with no negative effects on the performance and can improve the FA profile of milk. However, more research is needed, especially regarding the low digestibility of these by-products and the adequate level of inclusion.

### 2.3.2 Sheep and goats

The low DM content of tomato by-products has been identified as the main limitation to their utilization in small ruminant diets, and different approaches have tried to overcome this limitation such as silage [63] or their inclusion in multinutrient blocks [18,19,58,64]. However, the most common practice involves drying. Several studies have investigated the potential effects of DTP on milk yield and quality in small ruminants. Abbeddou et al. [87] reported that including 30% of DTP in the diet of lactating Awassi ewes had no influence on DMI and milk yield, but increased milk concentrations of lactose and PUFA. Likewise, Romano et al. [88] observed an increase in milk concentrations of PUFA and conjugated linoleic acid (CLA) when Comisana sheep were fed a TMR containing 20.4% of DTP, with no changes in milk yield. Similar results have been observed in dairy goats. Razzaghi et al. [89] reported increased concentrations of MUFA and CLA in the milk of Saanen goats fed a TMR including 24% of DTP compared with control-fed goats, without changes in DMI, milk yield, and ruminal concentrations of VFA. However, feeding high levels of DTP can negatively affect milk production. Thus, Mizael et al. [90] tested the effects of feeding Saanen goats with diets in which Tifton hay was replaced with increasing amounts of DTP (20%, 40%, and 60%), and reported decreases in DMI and milk yield by feeding the diet with 60% of DTP. In contrast, the diets containing up to 40% DTP increased both DMI and milk yield.

Other studies have focused on the nutritive value of cull tomatoes. Thus, Romero-Huelva et al. [58] analyzed the effects of partly replacing the concentrate in the diet of lactating Murciano-Granadina goats with multinutrient blocks containing cull tomatoes (125 g/kg DM block; 1.7% of cull tomatoes in total diet) and observed no effects on milk yield or composition (excepting a decrease in lactose content), but PUFA concentrations were augmented. Interestingly, methane production was decreased in the goats fed cull tomatoes, indicating that this could be a feeding strategy to reduce methane emissions by ruminants. The potential of cull tomatoes to reduce methane production in the rumen was further confirmed in a similar study [66] in which the multinutrient blocks including cull tomatoes were fed to nonlactating rumen-fistulated Murciano-Granadina goats, although a concomitant decrease in nutrient digestibility was reported. More recently, Romero-Huelva et al. [64] included dried cull tomatoes (7.1% of total diet) and other by-products in a 40:60 alfalfa hay:concentrate diet for lactating Murciano-Granadina goats and observed improvements in milk quality with no changes in milk yield. Although the reported effects might be attributed to the mixture of by-products, the observed reduction in methane production seems to confirm the antimethanogenic properties of cull tomatoes, despite the involved mechanism remains unknown. Arco-Perez et al. [63] observed that replacing 20.2% of oat hay with the same amount of ensiled cull tomatoes in the diet of lactating Murciano-Granadina goats resulted in increased milk concentrations of fat, total solids, and CLA, but milk yield remained unchanged. In this study, methane production was reduced by 14.6%, indicating that ensiling does not affect the antimethanogenic properties of tomatoes. Moreover, concentrations of SFA in milk were reduced and those of UFA were augmented by feeding the ensiled cull tomatoes, confirming the potential of tomato by-products to modify ruminant milk FA toward a more unsaturated profile [56,58,89].

The nutritive value of tomatoes by-products for growing and fattening small ruminants has also been assessed in a range of studies, and some of them focused on their influence on meat quality. Fondevila et al. [20] fed growing lambs with diets based on barley grain and straw and including increasing amounts of DTP (10%, 20%, and 30%) as replacement of barley straw and sunflower oil, and observed a lineal decrease of DMI with increasing amounts of DTP, but DM, OM and CP digestibility and N retention increased linearly without changes in either growth performance or feed conversion rate (FCR). Omer and Abdel-Magid [91] studied the use of lower levels of DTP (5%, 10%, and 15%) replacing berseem hay in the diet of growing lambs, and observed linear rises in OM, fat and CF digestibility, final BW, and ADG, and a 33% reduction of feeding costs. Other authors have assessed the possibility of using greater proportions of DTP in growing lambs feeding. Thus, Gebeyew et al. [92] observed that daily supplementation of 360 g of DTP to growing Hararghe lambs fed only a low-quality grass hay resulted in increased DMI, diet digestibility, final BW and ADG, and concluded that DTP could be a good supplement to poor quality hay. In this trial lambs were fed *ad libitum* and the intake of DTP accounted for 49.5% of total DMI.

Others have also analyzed the influence of tomato by-products on small ruminant's meat quality. Abdullahzadeh [93] replaced increasing amounts (10%, 20%, and 30%) of conventional ingredients with DTP in the diet of growing Markhoz goat kids and observed linear increases in protein and fat content of carcass, with final BW, carcass weight and dressing percentage being unaffected. In addition, the goats fed DTP produced greater amounts of greasy fiber with greater fiber diameter compared with those fed the control diet. More recently, Valenti et al. [94] offered DTP *ad libitum* to Comisana fattening lambs fed a commercial pelleted diet, and observed that consumption of DTP (daily average intake 85 g/lamb) did not change total DMI, final BW and carcass weight, but increased meat concentrations of linoleic acid, γ- tocopherol, and retinol and tended to raise PUFA proportions.

Ben Salem and Znaidi [18] investigated the potential of multinutrient blocks including 48.0% of wet TP to replace half of the concentrate in the diet of growing Barbarine lambs fed *ad libitum* wheat straw and a fixed amount of concentrate. Multinutrient blocks were well accepted by the lambs and final BW did not differ between groups, but FCR was greater for the TP-fed lambs and the authors suggested that formulation of multinutrient blocks could be improved by including either urea or an additional protein source.

In conclusion, tomato by-products can be successfully included in balanced diets at moderate proportions for both sheep and goats to improve milk and meat quality, mainly through modifications of FA profile, with no negative effects on animal performance. However, the results would depend on the feed ingredients being replaced in the diet.

### 2.3.3 Pigs

The use of tomato by-products in pigs feeding is limited by their high fiber content, but there is a body of research available on their potential use in the practice (Table 2.4). In some studies, the diets including tomato by-products were supplemented with fat to maintain the digestible energy concentration of the diets. Thus, Correia et al. [95] analyzed the effects of including either TP or wheat bran as fiber sources (5% and 10% of total

**Table 2.4** Effects of feeding tomato by-products to pigs on animal performance and meat quality.

| Reference | Animals | Tomato by-product | Diet characteristics | Inclusion level | Comments |
|---|---|---|---|---|---|
| [95] | 4-week old large white x Landrace pigs | TP | Based on wheat and soybean meal | 5% | No effect on final BW, DMI, and FCR<br>No influence on color and FA profile of *L. lumborum*<br>No effects on meat lipid oxidation |
| [96] | Landrace x Yorkshire x Duroc finishing pigs | TP | Based on corn, wheat, palm kernel meal and DDGS | 3% and 5% | Meat n-6/n-3 FA ratio tended to increase and meat tenderness increased linearly<br>Proline concentration in loin decreased linearly<br>Decreased shear force at 5% of TP |
| [97] | Landrace x Yorkshire x Duroc finishing pigs | TP | Based on corn and soybean meal | 1.7% and 3.4% | No effect on final BW, ADG, plasma metabolites related to lipid metabolism and meat FA profile<br>Improved oxidative stability of fresh belly meat |
| [98] | Growing pigs | TP | Based on corn and soybean meal | 5% and 10% | No effects on feed intake and final BW<br>Decreased ADG and increased FCR<br>Decreased nutrient digestibility |
| [99] | 7-month old Nero Siciliano smale pigs | TP | Based on corn, barley and soybean meal | 15% | No effects on DMI, ADG, final BW and carcass yield<br>Reduced intramuscular fat and increased PUFA in meat<br>No effect on fresh meat oxidative stability |
| [100] | 28-day old Duroc x York growing pigs | Ensiled cull tomatoes | Fermentable liquid diet based on corn grain and soybean meal | 30% | Increased ADG and DMI<br>No effect on meat pH and carcass characteristics, weight and yield |

ADG, average daily gain; BW, body weight; DMI, dry matter intake; FA, fatty acid; FCR, feed conversion rate; TP, tomato pomace; PUFA, polyunsaturated fatty acids.

diet, respectively), and two fat sources (lard or soybean oil; 5%) in the diet of 4-week old pigs (12.7 kg initial BW). In a 5-week growing trial, feed intake, final BW, FCR and meat color were not affected by the diet. As expected, pigs fed soybean oil had meat with greater UFA proportions than those fed lard, but meat FA profile was similar for TP and wheat bran groups. Meat lipid oxidation measured as thiobarbituric acid reactive substances (TBARS) concentration was not altered by the diet, despite that greater α-tocopherol concentrations were detected in the tissues of pigs fed the TP diets. Similarly, Chung et al. [96] analyzed the effects of TP (3 and 5%) in the diet of finishing pigs (75.7 kg initial BW) and after 7 weeks of trial observed no effects on meat color and chemical composition, excepting a decrease in proline concentration. However, TP tended to linearly increase meat n-6/n-3 FA ratio, increased meat tenderness, and decreased meat shear force when it was included at 5% of diet. Recently, An et al. [97] fed four diets, including no tomato by-product, TP (3.4%; equivalent to 10 ppm of lycopene), lycopene (20 ppm), or a mixture of both ingredients (1.7% of TP and 10 ppm of lycopene) to 18-week old finishing pigs (87.6 kg initial BW), and did not detect any difference among groups on growth performance, plasma metabolites related to lipid metabolism or meat FA profile. On the other hand, malondialdehyde concentrations in fresh belly meat was lower for the three diets with tomato by-products than for the control diet, indicating a beneficial effect of both TP and lycopene on meat oxidative stability. Altogether these results indicate that low levels of TP in the diet can improve meat characteristics and enhance its oxidative stability without hindering pig performance.

Yang et al. [98] tested two levels of TP (5% and 10%) in diets either supplemented or not with soybean oil to increase the energy content. When the experimental diets were fed to growing crossbred barrows and gilts (52.6 kg initial BW) for 28 days, no differences among groups were detected in feed intake or final BW, but the inclusion of TP decreased linearly the ADG and increased FCR probably due to the greater dietary fiber content in TP-diets. Additionally, nutrient digestibility decreased significantly with increasing amounts of TP, but this effect disappeared by increasing the energy content of the diet by adding soybean oil, which seem to counteract the negative effect of TP fiber on growth performance. Yang et al. [98] concluded that the optimal TP supplementation rate for growing pigs was 5%. However, greater dietary TP inclusion levels have been tested in other studies. Biondi et al. [99] tested the potential of TP as corn replacer by partly replacing corn by 15% of TP in the diet of growing pigs (42.7 kg initial BW). Although similar production parameters were observed in the two experimental groups after 86 days of trial, the TP-diet reduced intramuscular fat by 23%, increased vitamin A concentration in meat, and modified the meat FA profile towards a less saturated profile without influencing meat oxidative stability parameters. This study indicates that TP might be included in pig diets at greater levels than those reported previously.

Aguilera-Soto et al. [100] obtained positive results when a cull tomatoes silage (30% of diet) was tested as replacer of wet brewers grains in a fermentable liquid diet for growing finishing pigs (8.4 kg initial BW). Pigs fed the tomato silage had greater DMI and ADG than those fed brewers grains, needed 13 days less to reach the slaughter BW (95 kg), and showed similar meat pH and carcass characteristics. Similarly, Caluya et al. [101] reported that the inclusion of 35% of fresh TP in the diet of finishing pigs resulted in greater DMI, final BW and ADG compared with pigs fed a commercial diet, without affecting FCR.

All these results show the potential of tomato by-products as feed for grower and finisher pigs, but optimal levels of inclusion for each production stage have still to be identified. Supplementing the diets with an energy source can counteract the potential negative effects of tomato fiber and improve productive parameters. The incorporation of fresh tomato by-products into fermentable liquid diets is an alternative that should be further explored.

### 2.3.4 Poultry (broilers and laying hens)

The low energy content and high fiber level have been identified as the main limiting factors of tomato by-products for poultry feeding [102]. However, many studies have assessed the nutritive potential of these by-products for broilers and some of the results available are summarized in Table 2.5. Several authors have reported beneficial effects on growth performance of broilers when DTP was incorporated up to 20% of total diet. Al-Betawi [103] observed significant increases in ADG and final BW of broilers when 10% of DTP was included in the diet of 300-day old broilers, without changes in feed intake and FCR. Similarly, Yitbarek [104] reported increased ADG when 10% of DTP was included in the diet of 8-week broilers, with FCR remaining unchanged.

Interestingly, Cavalcante-Lira et al. [105] observed a reduction in feed intake of 1-day broilers during the first week of life when increasing amounts of DTP (5%, 10%, 15%, and 20%) were included in the diets, but afterwards neither feed intake nor growth performance was affected. These

**Table 2.5** Effects of feeding tomato by-products to broilers on growth performance and meat quality.

| Reference | Animals | Tomato by-product | Diet characteristics | Inclusion level | Comments |
|---|---|---|---|---|---|
| [103] | 300-day old Hubbard broilers | TP (dried or alkali-treated) | Based on corn and soybean meal | 10% | Significant increase in ADG and final BW<br>No effects on feed intake and FCR |
| [104] | 8-week old Rhode Island Red grower chicks | DTP | Based on corn, soybean meal and wheat by-products | 5%, 10%, 15%, and 20% | Increased feed intake and ADG (only at 10% DTP),<br>No effects on FCR<br>Decreased feed cost/kg gain |
| [105] | 1-day old Cobb broilers | DTP | Based on corn and soybean meal | 5%, 10%, 15%, and 20% | Decreased feed intake in the first week, no effects afterwards<br>No effects on slaughter, carcass, breast, abdominal fat, and organs (liver, heart, gizzard) weights |
| [106] | 1-day old Vencob broilers | DTP | NR | 5%, 10%, and 15% | No effects on feed intake and ADG<br>Decreased cholesterol levels in breast and thigh muscle<br>Decreased feed cost/kg gain |
| [107] | 21-day old Arian broilers | DTP | Based on corn and soybean meal | 3% and 5% | No effects at 3% of DTP<br>DTP at 5% improved BW and FCR, and reversed the changes caused by heat stress in antioxidant status, serum enzyme activities and immune response |

*(continued)*

**Table 2.5** (Cont'd)

| Reference | Animals | Tomato by-product | Diet characteristics | Inclusion level | Comments |
|---|---|---|---|---|---|
| [110] | Broilers | DTP | NR | 40%, 60%, 80%, and 100% | Increased meat color, texture, odor, taste, and overall acceptability up to 80% of tomato power,<br>Feeding only tomato power decreased meat taste |
| [111] | 21-day old Japanese quail | DTP | Based on corn and soybean meal | 5% and 10% | DTP at 5% reduced meat oxidation<br>DTP at 10% increased PUFA concentrations and UFA/SFA ratio in meat |
| [113] | Iranian native roosters | DTP | Based on corn and soybean meal | 15% and 30% | Decreased seminal volume<br>Increased concentration and proportion of live spermatozoa<br>Reduced spermatozoa lipid oxidation |
| [44] | 8-day old New Hampshire and Columbian Plymouth Rock cross broilers | Tomato seeds | Based on corn and soybean meal | 5%, 10%, 15%, and 20% | No effects on ADG and FCR up to 15%<br>Tomato seeds at 20% reduced significantly ADG |

ADG, average daily gain; BW, body weight; DTP, dried tomato pomace; FCR, feed conversion rate; NR, not reported; PUFA, polyunsaturated fatty acids; SFA, saturated fatty acids; TP, tomato pomace; UFA, unsaturated fatty acids.

authors concluded that the use of TP in early age chickens (1 to 21 days) should be avoided, as it can decrease feed intake and ADG, as well as worsen FCR. These negative effects were attributed to the high fiber content of DTP and the low fiber digestion capacity of young chicks, which increases after 21 days of life. Similarly, Pathakamuri [106] observed a reduction in feed efficiency when DTP was included up to 15% of diet in 1-day old Vencob broilers, but this adverse effect was overcame by supplementing the diet with an enzyme mixture including xylanase, α-amylase and α-galactosidase activities. These results support the hypothesis that negative effects of DTP in early age broilers are related to the high fiber content of this tomato by-product.

Feeding DTP to broilers can also influence animal metabolism. Hosseini-Vashan et al. [107] reported that including 5% of DTP in the diet of Arian chickens did not affect growth performance in a 42-day trial, but lessened the negative effects of thermal stress on oxidative status, serum enzyme activities, immune response and bone composition. The effectiveness of tomato by-products to reduce heat-stress in poultry has also been shown in other studies conducted with broilers [108] and quails [109]. The inclusion of TP in chicken diets can also improve meat quality. Thus, Akinboye et al. [110] detected that including up to 80% of tomato power in the diet of broilers improved color, texture, odor, taste and overall acceptability of meat, but feeding greater amounts affected negatively meat taste. Pathakamuri [106] observed that the incorporation of 15% of DTP in the diet of Vencob broilers reduced cholesterol concentrations in both breast and thigh muscles, and Botsoglou et al. [111] reported that feeding 10% of DTP to Japanese quails resulted in greater PUFA concentrations and UFA/SFA ratios in the meat compared to control quails, and lower levels (5%) reduced meat oxidation. Tomato by-products can likewise be used during poultry meat processing. For example, Skiepto et al. [112] observed that adding a tomato peel extract standardized for 5% of lycopene content to turkey breast muscles improved the color parameters compared with untreated meat, which was attributed to the antioxidant effects of lycopene.

Finally, it is worth to mention that tomato by-products can also have beneficial effects on poultry reproduction. Saemi et al. [113] observed that the inclusion of 15% or 30% of DTP in the diet of male chickens reduced the volume of ejaculate, but it increased significantly spermatozoa concentration and decreased both the percentage of abnormal spermatozoa and lipid oxidation of spermatozoa membranes, as indicated by the lower TBARS concentration compared with that in control chickens. The beneficial effects of TP on sperm quality were attributed to the antioxidant properties of some compounds, especially to lycopene. In fact, lycopene has been reported to exert favorable effects on sperm quality in rats [114] and to protect the membrane of frozen ram spermatozoa during freezing and thawing [115]. Similarly, Mangiagalli et al. [116] noticed that the supplementation of drinking water with lycopene increased significantly the volume of ejaculate, as well as spermatozoa concentration and viability, in broiler breeders.

The potential effects of other tomato by-products (i.e., peels and seeds) on broiler production has been less investigated. Persia et al. [44] reported that protein efficiency ratio quality (calculated as BW gain/protein intake) of tomato seeds was lower compared with soybean meal, but tomato seeds could be included up to 15%, replacing corn and soybean meal, in the diet of young chickens (8 to 21 days) without affecting negatively ADG or FCR when diets were formulated to contain equal amounts of digestible amino acids and true metabolizable energy. In contrast, tomato seeds at 20% reduced ADG, possibly due to the limited ingestion and digestion capacity in young chickens. Skin pigmentation was unaffected by tomato seeds feeding.

The potential value of TP as feed for laying hens has been analyzed in multiple studies, and most of them have reported no negative effects on production performance when used at dietary levels lower than 10% to 15% (Table 2.6). Abou Akkada et al. [117] fed three breeds of laying hens (Alexandria, Dokki, and Fayomi) with diets including 2%, 4%, or 6% of DTP for 6 weeks without detecting any effect of diet on egg production and weight compared to hens of the same breeds fed a diet without DTP. Dotas et al. [118] analyzed the effects of feeding DTP to Warren laying hens during two periods (either 4% or 8% of DTP for the first 135 days, and 6% or 12% DTP for the subsequent 111 days), and reported no changes in feed intake, FCR, egg production, egg weight, or shell thickness in any period. Jafari et al. [34] fed 5%, 10%, or 15% of DTP to laying hens receiving a conventional diet based on corn and soybean meal, and observed no changes in feed intake, egg weight and shell thickness. The dietary inclusion of DTP at either 5% or 10% increased egg production (by 3.4% and 2.1%, respectively) and mass (by 4.9% and 3.3%), and reduced FCR by 2.4% and 1.1%, respectively. On the contrary, feeding the diet with 15% of DTP did not affect FCR and decreased egg production and mass, which was attributed to the high fiber content of this diet, as high fiber diets can reduce nutrient availability in poultry [34].

**Table 2.6** Influence of tomato by-products feeding on laying hens performance and egg quality.

| Reference | Animals | Tomato by-product | Diet characteristics | Inclusion level | Comments |
|---|---|---|---|---|---|
| [117] | Alexandria, Dokki and Fayoumi hens | DTP | NR | 2%, 4%, and 6% | No effects on egg production and weight<br>Improved egg yolk color |
| [118] | Warren hens | DTP | Based on corn and soybean meal | 4% to 12% | No effects on feed intake, FCR, egg production and weight or shell thickness<br>Improved egg yolk color |
| [34] | 27-week old Hy-Line (W-36) hens | DTP | Based on corn and soybean meal | 5%, 10%, and 15% | No effects on feed intake, egg weight, shell thickness and yolk color<br>DTP at 5% and 10% increased egg production and mass and improved FRC, but at 15% decreased egg production and mass |
| [119] | 30-week old Dekalb White hens | DTP | Based on corn and soybean meal | 5, 10, 15 and 20% | Linear increase in feed intake; egg mass and yolk weight<br>Increased FCR at 10, 15 and 20%<br>No effects on egg, albumen and shell weight<br>Improved egg yolk color |
| [120] | 65-week old Lohman LsL-Lite hens | DTP | Based on corn and soybean meal | 15%, 17%, and 19% | No effects on feed intake, FCR, egg production and weight or shell thickness<br>Improved egg yolk color |
| [121] | 52-week old Harco SL hens | DTP | Based on corn and soybean meal | 8% and 15% | No effects on feed intake, FCR and egg production<br>Egg weight was increased at 15% of tomato meal<br>Improved egg yolk color and reduced number of blood and meat spots |

*(continued)*

**Table 2.6** (Cont'd)

| Reference | Animals | Tomato by-product | Diet characteristics | Inclusion level | Comments |
|---|---|---|---|---|---|
| [122] | Lohman brown hens | DTP | Based on corn and soybean meal | 10%, 15%, and 20% | Increased feed intake and final BW; increased FCR at 15% DTP<br>No effects on egg production and weight, and shell thickness and weight |
| [123] | 54-week old Single Comb White Leghorn, Hy-Line (W36) hens | DTP | Based on corn or wheat and soybean meal | 5% and 10% | No effects on feed intake, body weight changes, egg production, weight and mass, and shell thickness and weight<br>Improved egg yolk color<br>Cereal x DTP interactions on shell thickness and yolk color |
| [35] | 40-week Hy-Line (W-36) hens | DTP | Based on wheat and soybean meal | 5% | No effects on FCR, shell thickness, and egg production, weight and mass<br>Improved egg yolk color |
| [126] | Sussex hens | Tomato seeds | Based on corn and soybean meal | 5% | Decreased FCR and increased egg production<br>No effects on egg weight<br>Increased fertility and hatchability |
| | | Tomato peels | Based on corn and soybean meal | 7.2% | Decreased FCR and increased egg production<br>No effects on egg weight, fertility and hatchability |

ADG, average daily gain; BW, body weight; DTP, dried tomato pomace; FCR, feed conversion rate; NR, not reported.

Loureiro et al. [119] reported a linear rise in feed intake when Dekalb White hens were fed diets containing increasing DTP levels (5%, 10%, 15%, and 20%) for 9 weeks, but FCR was also augmented for all diets with more than 5% of DTP. The weight of egg, albumen and shell was unaffected, but egg mass and yolk weight increased significantly with all diets containing DTP. Interestingly, these authors noticed that effects of DTP on some parameters varied with the period of the study (30 to 33, 33 to 36, and 36 to 39 weeks), and in general were more marked in the last period. After evaluating all production parameters, Loureiro et al. [119] recommended an inclusion level of 5% of DTP for laying hens. These results contrast with those of Salajegheh et al. [120], who fed greater levels of DTP (15%, 17%, and 19%) to Lohman LsL-Lite hens and observed no negative effects on feed intake, FCR, egg production parameters, and serum metabolites (total protein, cholesterol, albumin, glucose, and triglyceride levels) compared with a control group fed a diet without DTP. Yannakopoulos et al. [121] also reported no changes in feed intake, FCR, and egg production when DTP was included at 8% or 15% in the diet of Harco DL hens, and egg weight was significantly increased in hens fed the diet with 15% of DTP. Calyslar and Uygur [122] also observed an increase in feed intake and final BW of 40-week old Lohman brown hens when they were fed 10%, 15%, or 20% of DTP, but FCR augmented at 15% of DTP. No effects on egg production, egg weight, shell thickness, or shell weight were detected in this study.

The contrasting results can be due to variability in the chemical composition of the DTP used in the different studies, but also to the composition of the control diet and the variable feed ingredients being replaced by this by-product. Thus, Mansoori et al. [123] analyzed the effects of 5% or 10% of DTP in the diet of hens fed diets with either corn or wheat as the main energy source, and observed significant cereal x DTP interactions on shell thickness and yolk color. For both diets, there were no effects of DTP on feed intake, body weight changes, egg production, weight and mass, and shell thickness and weight. More recently, Shahsavari [35] analyzed the effects of including 5% of DTP in the diet of Hy-Line (W-36) hens fed a diet based on wheat and soybean meal and observed no changes in FCR, shell thickness and egg production, weight and mass compared with hens fed a reference diet based on corn and soybean meal.

In general, shell thickness has not been affected by the inclusion of DTP in the diet of laying hens, and this was attributed to the similar levels of Ca, P, and vitamin $D_3$ in the experimental diets [34]. The reduction of the number of blood and meat spots in the eggs observed in some studies when tomato by-products were included in the diet of laying hens was attributed to the high level of carotenoids in these by-products [121].

The color of the yolk is an important criterion for consumers, and therefore many studies have evaluated the yolk-pigmenting properties of different agroindustrial by-products, including those from tomato. The inclusion of TP in the diet of laying hens at levels greater than 8% increased the yolk color in different studies [34,118,120,121], but this effect was also noticed at inclusion levels lower than 6% [35,117,123] in hens from different breeds. In contrast, no changes in yolk color were observed by Jafari et al [34] when feeding 5%, 10%, or 15% of DTP to 27-week laying hens receiving a diet based on corn and soybean meal, by Garcia and Gonzalez [124] when feeding tomato seeds meal to laying hens up to 33 weeks of age, and by Loureiro et al. [181] when DTP was introduced in the diet of laying hens at 5%, 10%, 15%, or 20%. Shahsavari [35] observed that yolk color in hens fed a diet without corn and supplemented with 5% of DTP was lower compared with that in hens receiving a diet, including corn. However, DTP increased significantly yolk color when compared with that in hens fed a diet without corn. These results indicate a positive effect of DPT, but also that yok pigmenting capacity of TP is lower than that of corn grains. As discussed by Shahsavari [35], the effect of DTP and other natural pigments sources that may influence yolk color would depend on the amount and nature of the pigments, especially of xanthophylls. Moreover, yolk color is also affected by other factors related to laying hens (genotype, age, health status, egg production rate, etc.) as well as by the production and management system [125].

The available information on the effects of tomato seeds and peels on laying hen's performance is limited. Tomczynski [126] analyzed the use of either tomato seeds (5% of diet) or tomato peels (7.2% of diet) as feed ingredients in Sussex hens for 8 months, and observed that both tomato by-products increased egg production without affecting egg weight compared with hens fed a control diet. In addition, tomato seeds increased fertility and hatchability. Knoblich et al. [12] investigated the transfer of carotenoids from tomato peels and seeds to the yolk by feeding diets containing one of each by-product (7.5% of diet) to White Leghorn hens for 14 days, and observed improved yolk color and augmented lycopene concentration in the yolk compared with the eggs from control hens (0.90, 0.86, and 0.0 μg/g yolk for tomato peels, seeds, and control diets, respectively). However, the concentrations of other carotenoids in the yolk (lutein, zeaxanthin, α-cryptoxanthin, and β-cryptoxanthin) remained unchanged, excepting the lutein content that was increased by feeding tomato peels (16.7 and 14.0 μg/g yolk for peels and control diets, respectively). The authors of this study concluded that both peels and seeds have low nutritive value for laying hens, but they can be included in the diet at low levels (<10%) to increase yolk pigmentation and lycopene concentrations.

As previously discussed, tomato by-products can be used as feed ingredients in poultry diets, but the inclusion level should be limited and feeding of these by-products in early chickens should be avoided. Using tomato by-products as an alternative to synthetic pigments seems to be a feasible nutritional strategy that can also reduce feeding costs.

### 2.3.5 Rabbits

The use of tomato by-products in rabbit feeding is supported by their high fat and CP content and by the ability of their lignified fiber to control digestive diseases. Peiretti et al. [127,128] assessed the effects of including 3 or 6% of TP in the diet of Hycole x Grimaud 38-day old rabbits on their growth performance, nutrient digestibility, carcass characteristics, and chemical, physical and sensorial characteristics of their meat. The TP was ensiled for 2 months and dried before being mixed with the rest of diet ingredients. The results showed no differences among diets in mortality, feed intake, ADG or FCR, but rabbits fed the TP-3% diet had greater final BW than those fed the control diet in a 50-day growth trial [127]. No differences were detected in nutrient digestibility between the control and TP-3% diet, but including TP at 6% decreased diet digestibility [127]. The diet with 3% TP resulted in improved BW at slaughter and carcass weight, but no effects were detected in carcass characteristics [128]. Feeding TP modified the meat FA profile, resulting in increased proportions of C18:2. Similar results were previously reported by Alicata et al. [129] when alfalfa hay was replaced with 20% of TP in growing rabbits, and were attributed to the high C18:2 content in TP. In the study of Peiretti et al. [128] feeding TP resulted in marked changes in FA profile of perirenal fat, with TP-fed rabbits showing lower proportions of SFA and greater of PUFA. Interestingly, a consumer test performed in this study by showed that the meat from rabbits fed the 6% of TP was the most preferred, followed by the meat from rabbits fed 3% of TP, and finally that from the control group. As observed in other animal species, feeding TP increased yellowness and Chroma values in meat, probably due to the greater content in lycopene and β-carotene in TP-diets.

Greater levels of TP in the diet of rabbits have been tested in other studies. Thus, Sayed and Abdel-Azeem [130] fed diets, including 10%, 20%, and 30% of DTP to growing New Zealand White rabbits (6-week old), and observed that rabbits fed 20% of DTP showed the best performance, but TP at 30% reduced feed intake and final BW and increased FCR. Similarly, Kavamoto et al. [131] observed that the inclusion of DTP at 21% of the diet of White Flemish Giant and Chinchilla rabbits resulted in the best growth performance compared with either a control diet or diets containing 42% and 63% of DTP, and both Ahmed et al. [132] and Caro et al. [133] reported a reduction in rabbits feed intake when 30% of DTP was included in the diet. In addition, several studies [130,132] have shown reductions in diet digestibility by feeding TP at 30% or greater levels. In contrast, Khadr and Abdel-Fattah [134] reported that DTP at 14%, 22%, or 30% in the diet of 5-week old growing rabbits did not produced any adverse effect on feed intake, ADG, FCR, carcass characteristics or liver and kidney function. However, it should be taken into account that in this study DTP replaced low-quality feeds (berseem hay and straw), whereas in other studies TP replaced medium to good quality feeds [130,133]. As already discussed, composition of both tomato by-products and feeds being replaced can influence the results obtained in different studies.

Regarding the use of other tomato by-products, Sayed and Abdel-Azeem [135] evaluated the performance and carcass quality of rabbits fed diets including dried tomato seeds at 10%, 20%, and 30% of total diet, and similarly to that observed for TP reported that tomato seeds at 30% reduced feed intake, ADG, and dressing percentage and increased FCR, but had no negative effect up to 20%. Battaglini et al. [136] noticed that tomato peels have low net energy value for rabbits (1.30 Mcal/kg DM) and are poorly digested, with digestibility coefficients of 36.9% and 51.4% for OM and CP, respectively. Consequently, inclusion levels lower than 10% of total diet are recommended for rabbits [136,137].

Altogether these results indicate that TP and tomato seeds can be included in rabbit diets up to 20% without adverse effects on growth performance and carcass characteristics. Using tomato by-products as feed ingredients can also improve the FA profile of rabbit meat and increase the economic efficiency of rabbit production. Thus, Sayed and Abdel-Azeem [130] observed that replacing conventional feed ingredients by 20% of DTP in the diet of growing rabbits increased economic efficiency (calculated as income/feed ratio) by 1.4 compared with a control diet.

### 2.3.6 Fish

Aquaculture production has steadily increased over the last decades, and in 2018 reached nearly $120 \times 10^6$ tons of fish in live weight [138]. The increase in production has led the sector to search for more efficient diets, and the potential use of different agroindustrial by-products, including those from tomato, has been investigated. Both Soltan [139] and Soltan and El-Laithy [140] tested DTP as feed for Nile Tilapia (*Oreochromis niloticus*). Soltan [139] observed that feeding up to 20% of DTP (replacing up to half of soybean meal and 20% of corn in the diet) did not affect feed intake, final BW, ADG or FCR compared with a control diet, but greater inclusion levels (24% to 32%) decreased growth performance and increased FCR. All tested levels of DTP decreased fat content in whole fish, but CP content was unchanged up to 28% of dietary DTP. Additionally, DTP up to 20% did not negatively influence nutrient digestibility, excepting for CP that was decreased by 4.0%. Feeding DTP at 20% reduced feeding cost by 11% and was economically profitable. In agreement with these results, Saad [141] observed that up to 44% of soybean meal and cotton seed meal in tilapia diets could be replaced by DTP, but greater replacement levels decreased growth performance. Soltan and El-Laithy [140] tested diets including up to 50% of a silage composed of fish, tomato, and potato by-products (40%, 30%, and 30%, respectively), and reported that feeding tilapia with diets including up to 30% of fermented silage (9% of DTP in total diet) had no significant effects on feed intake, ADG, FCR, or nutrient digestibility, and reduced by 22% the cost of feed per kg of weight gain. Azzaza et al. [142] also observed that DTP could be used in tilapia diets up to 20% (replacing soybean meal and corn) without affecting negatively either growth performance or carcass composition, and reduced feeding cost by 15.2%. However, fish growth was reduced at greater DTP inclusion levels.

Studies with other fish species, in which TP mainly replaced concentrate feeds, have reported similar results. Hoffman et al. [143] showed that partial replacing of fish meal with DTP resulted in similar ADG in African sharptooth catfish (*Clarias gariepinus*) and reduced by 34% the cost of feed per kg of weight gain. In common carp (*Cyprinus carpio* L.), Amirkolaie et al. [144] reported that using up to 20% of DTP in the diet (replacing soybean meal and wheat flour) had no negative effects on growth performance or flesh composition compared with a control diet, but the best growth performance was observed with the diet with 10% of DTP. As observed in other studies, CP digestibility was decreased by feeding DTP. In fact, Nengas et al. [145] measured the CP digestibility of different feeds in gilthead sea bream (*Sparus auratus* L.) and reported lower values for TP than for soybean meal and fish meal (20.1%, 90.9%, and 95.8%, respectively). In addition, the content in essential amino acids in TP is markedly smaller than that in soybean and fish meals [142].

In aquaculture, skin and muscle color is an indicator of quality influencing consumers' preferences, and tomato by-products might be used as a source of natural carotenoids and antioxidants. Montoya et al. [146] investigated the use of a tomato extract (1%, 3%, and 6%) and lycopene (0.1%, 0.3%, and 0.6%) in the diet of two aquarium fish, goldfish (*Carassius auratus*) and Southern platyfish (*Xiphophorus maculatus*), but failed to obtain improvements on either growth performance or pigmentation, which was attributed to the lack of enzymes to transform lycopene into carotenoids [146]. Likewise, Chatzifotis et al. [147] observed that lycopene supplementation of red porgy (*Pagrus pagrus*) diets had no influence on skin color. More recently, Nogueira et al. [148] showed the potential of a high ketocarotenid tomato line to increase by nearly twofold the color rainbow trout (*Oncorhynchus mykiss*) fillets without having any adverse effect on growth performance or flesh composition. These authors [148] also stated that producing this tomato variety is economically competitive and could be a feasible alternative to synthetic carotenoids.

## 2.4 Companion animals

Nowadays, pets are mostly considered companions or even family members, and increasing their quality of life is a priority for their owners. Nutrition is key aspect to improve health and longevity of pets, and the role of dietary fibers to support good gut health had gained interest in the pet food industry the last years [149]. Tomato by-products are rich in fiber and bioactive compounds, and therefore they could be used in the pet food industry as health enhancers. Furthermore, the high water content that is one of the main drawbacks of utilizing tomato by-products in animal feeding, could be advantageous in wet pet food diets, facilitating the control of their texture and rheological properties [150].

Allen et al. [151] observed that feeding English Pointers dogs a diet containing 8% of TP as replacer of extruded corn decreased DM, CP and energy digestibility compared with a control diet. Similar results were reported by Fahey et al. [21], who observed that the inclusion of 8.7% of TP in the diet of dogs as replacer of corn increased fat and fiber intake, but reduced nutrient digestibility and increased defecation rates compared with dogs fed a control diet without fibrous ingredients. Diets including TP usually have greater fiber content, which can reduce digesta retention time in the gut and nutrient digestibility [152]. In fact, it has been shown that increasing insoluble:soluble fiber ratio augments transit time and fecal output [153,154], which can decrease dietary energy supply [154]. Swanson et al. [155] hypothesized that TP may affect the production of VFA in the large bowel, and studied the *in vitro* fermentability of different vegetables and fruits pomace by dogs fecal microbiota. The intermediate OM degradability (35% in 24 h) and VFA production (1.71 mmol/g OM) values observed for TP were similar to those reported for medium to low-quality fiber sources. This study also showed clearly that both OM degradability and VFA production decrease as the insoluble:soluble fiber ratio in the diet increase.

Fahey et al. [21] concluded that TP could be used in dog diets having 12.5% of total dietary fiber at adequate levels to supply about half of total dietary fiber. More recently, Yuangklang et al. [156] observed lineal decreases in diet digestibility as the dietary amount of DTP increased, and recommended TP inclusion at 10% of total diet, although practical recommendations can vary depending on TP chemical composition.

Information regarding the inclusion of tomato by-products in cat diets is very scarce. Gray et al. [157] evaluated two vegan diets for cats including TP and observed that none meet the minimum nutrient requirements of cats. The diets were deficient in taurine that is an essential amino acid for cats [158], as taurine content in vegetables, including tomato, is null or negligible [159]. There is also evidence that taurine deficiency may cause health issues in dogs [160]. In addition, cats require other essential nutrients whose content is limited in vegetables such as arginine, lysine or vitamin D [157], and therefore including vegetable feeds in diets for cats should be done with caution to ensure that nutrients requirements are met.

Rats are widely used as models in experimentation, but they can be also kept as companion animals. Moreover, the pet food industry can profit from the information generated in trials using rats as models. Recently, Torales et al. [161] fed diets including either 12.8% or 25.5% of a tomato-spinach mixture as a source of dietary carotenoids to Sprague-Dawley rats (8-week old) with steatosis and observed a significant improvement of their antioxidant status, as well as reduced plasma concentrations of steatosis biomarkers. Similarly, Moreira et al. [162] reported an improvement in the antioxidant status of Wister rats fed a hyper-energetic diet when tomato powder was included in the diet as a source of lycopene.

Alvarado et al. [163] evaluated TP as source of dietary fiber in Sprague Dawley rats. Compared with a control diet, feed intake was greater for the experimental diets including 13.4%, 26.3%, or 38.7% of TP (replacing corn starch, sucrose and casein), and BW gain was unaffected. These results indicate that TP has no negative effect on palatability, as previously observed in dogs [151]. However, diet digestibility was reduced when TP was included at levels greater than 13.4%, confirming the results obtained in dogs. Sogi et al. [164] fed whole or defatted tomato seeds to albino rats and detected a reduction in feed and protein efficiency compared with casein, indicating a lower quality of tomato seeds CP.

In summary, TP has technological properties suitable for the pet food industry, is highly palatable, and might contribute to enhance animal health status, but its high content in insoluble fiber and the low CP digestibility limit its inclusion in pet diets. Finally, it is worth to mention that low levels (5%) of TP can be used in the diets of fox and mink to prevent diarrhea [165].

## 2.5 Conclusions

Chemical composition of tomato by-products (including tomato wastes), is highly variable, but in general they can be a good source of protein, fiber, sugars and carotenoids. Fresh tomato by-products can be used as raw material in animal diets as alternative to conventional feeds. However, their high moisture content limits their utilization and therefore they are mostly utilized after being dried, ensiled or included in multinutrient blocks. As widely reported in the literature, nutritional and sensory properties of milk, meat, eggs and fish can be positively influenced by including tomato by-products in animal diets due to improvements in the content in bioactive compounds and nutrients, fatty acid profile, lipid oxidation stability and yolk color, among other positive changes. Simultaneously, production performance and animal health status can be enhanced. The recommended inclusion levels for non-ruminant animals is usually below 10% to 15% of total diet, and is lower than those for ruminants due to the lower fiber utilization of these animal species compared with ruminants. Practical recommendations in ruminants depend greatly on the physiological stage of the animal and the type of tomato by-product being fed. Increasing the use of tomato by-products in livestock feeding would minimize the environmental effects produced by their management and disposal, reduce feeding costs, and contribute to circular bioeconomy [166,167], especially when used locally. Developing low-cost effective techniques to overcome the limitation of the moisture content in tomato by-products would greatly contribute to facilitate their use as animal feed ingredients. In addition, the potential antimethanogenic effect of cull tomatoes reported in some studies should be further explored.

## References


[1] FAO FAOSTAT Statistical Database. http://www.fao.org/faostat/en/#data/QC, 2020a (Accessed 12 January 2021).

[2] B. Burton-Freeman, K. Reimers, Tomato consumption and health: emerging benefits, Am. J. Lifestyle Med. 5 (2) (2010) 182–191.

[3] D. Bhowmik, K.P.S. Kumar, S. Paswan, S. Srivastava, Tomato - a natural medicine and its health benefits, J. Pharmacognosy Phytochem. 1 (1) (2012) 33–43.

[4] G.R. Beecher, Nutrient content of tomatoes and tomato products, Proc. Soc. Exp. Biol. Med. 218 (2) (1998) 98–100.

[5] L.H. Tonucci, J.M. Holden, G.R. Beecher, F. Khahick, C.S. Davis, G. Mulokozi, Carotenoid content of thermally processed tomato-based food products, J. Agric. Food Chem. 43 (3) (1995) 579–586.

[6] J. Shi, L.Le Maguer, Lycopene in tomatoes: chemical and physical properties affected by food processing, Crit. Rev. Biotechnol. 20 (4) (2000) 293–334.

[7] S.K. Vidyarthi, C.W. Simmons, Characterization and management strategies for process discharge streams in California industrial tomato processing, Sci. Total Environ. 723 (2020) 137976, doi:10.1016/j.scitotenv.2020.137976.

[8] E. Garcia, D.M. Barrett, Peelability and yield of processing tomatoes by steam, J. Food Process. Preserv. 30 (1) (2006) 3–14.

[9] M. Del Valle, M. Cámara, M.E. Torija, Chemical characterization of tomato pomace, J. Sci. Food Agric. 86 (8) (2006) 1232–1236.

[10] M.R. Ventura, M.C. Pieltain, J.I.R. Castanon, Short communication. Evaluation of tomato crop by-products as feed for goats, Anim. Feed Sci. Technol. 154 (3–4) (2009) 271–275.

[11] P.A. Silva, B.C. Borba, V.A. Pereira, M.G. Reis, M. Caliari, M.S. Brooks, T.A. Ferreira, Characterization of tomato processing by-product for use as a potential functional food ingredient: nutritional composition, antioxidant activity and bioactive compounds, Int. J. Food Sci. Nutr. 70 (2) (2019) 150–160.

[12] M. Knoblich, B. Anderson, D. Latshaw, Analyses of tomato peel and seed byproducts and their use as a source of carotenoids, J. Sci. Food Agric. 85 (7) (2005) 1166–1170.

[13] K. Szabo, A.F. Cătoi, D.C. Vodnar, Bioactive compounds extracted from tomato processing by-products as a source of valuable nutrients, Plant Food Hum. Nutr. 73 (1) (2018) 268–277.

[14] Z. Lu, J. Wang, R. Gao, F. Ye, G Zhao, Sustainable valorisation of tomato pomace: a comprehensive review, Trends Food Sci. Technol. 86 (1) (2019) 172–187.

[15] C.N. Marcos, T. de Evan, E. Molina-Alcaide, M.D. Carro, Nutritive value of tomato pomace for ruminants and its influence on *in vitro* methane production, Animals 9 (6) (2019) 343, doi:10.3390/ani9060343.

[16] V. Heuzé, G. Tran, P. Hassoun, D. Bastianelli, F. Lebas, Tomato pomace, tomato skins and tomato seeds, 2015. https://www.feedipedia.org/node/689 (Accessed 12 January 2021).

[17] M.P.S. Bakshi, H.Makkar M.Wadhwa, Waste to worth: vegetable wastes as animal feed, CAB Rev. 11 (012) (2016) 1–26.

[18] H. Ben Salem, I.E.A. Znaidi, Partial replacement of concentrate with tomato pulp and olive cake-based feed blocks as supplements for lambs fed wheat straw, Anim. Feed Sci. Technol. 147 (1–3) (2008) 206–222.

[19] M. Romero-Huelva, E. Molina-Alcaide, Nutrient utilization, ruminal fermentation, microbial nitrogen flow, microbial abundances, and methane emissions in goats fed diets including tomato and cucumber waste fruits, J. Anim. Sci. 91 (2) (2013) 914–923.
[20] M. Fondevila, J.A. Guada, J Gasa, C. Castrillo, Tomato pomace as a protein supplement for growing lambs, Small Rumin. Res. 13 (2) (1994) 117–126.
[21] G.C.Jr. Fahey, N.R. Merchen, J.E. Corbin, A.K. Hamilton, K.A. Serbe, D.A. Hirakawa, Dietary fiber for dogs: II. Iso-total dietary fiber (TDF) additions of divergent fiber sources to dog diets and their effects on nutrient intake, digestibility, metabolizable energy and digesta mean retention time, J. Anim. Sci. 68 (12) (1990) 4229–4235.
[22] M. Cámara, M. del Valle, M.E. Torija, C. Castilho, Fatty acid composition of tomato pomace, Acta Hortic. 542 (2001) 175–180, doi:10.17660/actahortic.2001.542.21.
[23] A. Zuorro, R. Lavecchia, Optimization of enzyme-assisted lycopene extraction from tomato processing waste, Adv. Mater. Res. 800 (7) (2013) 173–176.
[24] D.S. Sogi, A.S. Bawa, S.K. Garg, Sedimentation system for seed separation from tomato processing waste, J. Food Sci. Technol. 37 (5) (2000) 539–541.
[25] NRC (National Research Council), Nutrient Requirements of Dairy Cattle, seventh ed., National Academy of Sciences, Washington, 2001.
[26] P. Noziere, D. Sauvant, L. Delaby, INRA Feeding System for Ruminants, Wageningen Academic Publishers, The Netherlands, 2018.
[27] N. Maheri-Sis, M. Eghbali-Vaighan, A.R. Ali Mirza-Aghazadeh, A. Ahmadzadeh, A. Aghajanzadeh-Golshani, A.A. Mirzaei-Aghsaghali Shaddel-Telli, Effects of microwave irradiation on ruminal dry matter degradation of tomato pomace, Curr. Res. J. Biol. Sci. 3 (3) (2011) 268–272.
[28] D.C. Weakley, M.D. Stern, L.D. Satter, Factors affecting disappearance of feedstuffs from bags suspended in the rumen, J. Anim. Sci. 56 (2) (1983) 493–507.
[29] S. Abbeddou, S. Riwahi, L. Iñiguez, M. Zaklouta, H.D. Hess, M. Kreuzer, Ruminal degradability, digestibility, energy content, and influence on nitrogen turnover of various Mediterranean by-products in fat-tailed Awassi sheep, Anim. Feed Sci. Technol. 163 (2–4) (2011) 99–110.
[30] J. Gasa, C. Castrillo, J.A. Guada, Nutritive value for ruminants of the canning industry by-products: 1 Tomato pomace and pepper residues, Investigación Agraria, Producción y Sanidad Animales 3 (1–2) (1988) 57–73.
[31] W.P. Weiss, D.L. Frobose, M.E. Koch, Wet tomato pomace ensiled with corn plants for dairy cows, J. Dairy Sci. 80 (11) (1997) 2896–2900.
[32] A. Rahbarpour, V. Palangi, P. Eivazi, M. Jalili, Calculation of metabolizable protein and energy of tomato pomace by nylon bags and gas production data, J. Exp. Biol. 2 (6) (2012) 822–825.
[33] C. Yuangklang, K. Vasupen, S. Wongsuthavas, S. Bureenok, P. Panyakaew, A. Alhaidary, H.E. Mohamed, A.C. Beynen, Effect of replacement of soybean meal by dried tomato pomace on rumen fermentation and nitrogen metabolism in beef cattle, Am. J. Agric. Biol. Sci. 5 (3) (2010) 256–260.
[34] M. Jafari, R. Pirmohammadi, V. Bampidis, The use of dried tomato pulp in diets of laying hens, Int. J. Poult. Sci. 5 (7) (2006) 618–622.
[35] K. Shahsavari, Influence of different sources of natural pigments on the color and quality of eggs from hens fed a wheat based diet, Iran. J. Appl. Anim. Sci. 5 (1) (2015) 167–172.
[36] A.N. Grassino, J. Halambek, S. Djakovic, S.R. Brncic, M. Dent, Z. Grabaric, Utilization of tomato peel waste from canning factory as a potential source for pectin production and application as tin corrosion inhibitor, Food Hydrocoll. 52 (1) (2016) 265–274.
[37] X. Alibes, J.L. Tisserand, Tables of the nutritive value for ruminants of Mediterranean forages and by-products, Options Méditerr. 11 (4) (1990) 1–37.

[38] P.G. Herrera, M.C. Sanchez-Mata, M. Camara, Nutritional characterization of tomato fiber as a useful ingredient for food industry, Innov. Food Sci. Emerg. Technol. 11 (4) (2010) 707–711.
[39] I. Navarro-Gonzalez, V. Garcia-Valverde, J. Garcia-Alonso, M.J. Periago, Chemical profile, functional and antioxidant properties of tomato peel fiber, Food Res. Int. 44 (12) (2011) 1528–1535.
[40] C. Talens, J.C. Arboleya, M. Castro-Giraldez, P.J. Fito, Effect of microwave power coupled with hot air drying on process efficiency and physicochemical properties of a new dietary fibre ingredient obtained from orange peel, LWT Food Sci. Technol. 77 (7) (2017) 110–118.
[41] G.A. Piyakina, T.S. Yunusov, General characteristics of the proteins of tomato seed flour and tomato skin flour, Chem. Nat. Compd. 31 (7) (1996) 495–499.
[42] H.E. Brown, F.I. Meredith, G. Saldama, T.S. Stephens, Freeze peeling improves quality of tomatoes, J. Food Sci. 35 (2) (1970) 485–488.
[43] Q. Wang, Z. Xiong, G. Li, X. Zhao, H. Wu, Y. Ren, Tomato peel powder as fat replacement in low-fat sausages: Formulations with mechanically crushed powder exhibit higher stability than those with airflow ultra-micro crushed powder, Eur. J. Lipid Sci. Technol. 118 (1) (2016) 175–184.
[44] M.E. Persia, C.M. Parsons, M. Schang, J. Azcona, Nutritional evaluation of dried tomato seeds, Poultry Sci. 82 (1) (2003) 41–146.
[45] A.L. Winton, K.B. Winton, The Structure and Composition of Food, Cereals, Starch, Oil Seeds, Nuts, Oils, Forage Plants, John Wiley & Sons, Inc., New York /Chapman & Hall, London, 1950.
[46] A.M. Giuffrè, M. Capocasale, Physicochemical composition of tomato seed oil for an edible use: the effect of cultivar, Int. Food Res. J. 23 (2) (2016) 583–591.
[47] E. Yilmaz, B. Aydeniz, O. Guneser, E. Arsunar, Sensory and physico-chemical properties of cold press-produced tomato (*Lycopersicon esculentum* L.) seed oils, J. Am. Oil Chem. Soc. 92 (4) (2015) 833–842.
[48] C. Botineştean, A.T. Gruia, I. Jianu, Utilization of seeds from tomato processing wastes as raw material for oil production, J Mater. Cycles Waste Manag. 17 (1) (2014) 118–124.
[49] N.A. Kamazani, H. Tavakolipour, M. Hasani, M. Amiri, Evaluation and analysis of the ultrasound-assisted extracted tomato seed oil, J. Food Biosci. Technol. 4 (1) (2014) 57–66.
[50] E.O. Johns, E.E.F. Gersdorff, The proteins of the tomato seed, *Solanium esculentum*, J. Biol. Chem. 51 (2) (1922) 439–452.
[51] A. Anwar, H.A.El Alaily, M.F. Diab, Nutritive value of tomato seed meal as a plant protein supplement for growing chicks, Arch. für Geflügelkunde 42 (2) (1978) 56–58.
[52] M. Mechmeche, F. Kachouri, M. Chouabi, H. Ksontini, K. Setti, M. Hamdi, Optimization of extraction parameters of protein isolate from tomato seed using response surface methodology, Food Anal. Methods 10 (2017) 809–819.
[53] S.J. Latlief, D. Knorr, Tomato seed protein concentrates: effects of methods of recovery upon yield and compositional characteristics, J. Food Sci. 48 (11) (2010) 1583–1586.
[54] M. Valdez-Morales, L.G. Espinosa-Alonso, L.C. Espinoza-Torres, F. Delgado-Vargas, S. Medina-Godoy, Phenolic content and antioxidant and antimutagenic activities in tomato peel, seeds, and byproducts, J. Agric. Food Chem. 62 (11) (2014) 5281–5289.
[55] E. Elbadrawy, A. Sello, Evaluation of nutritional value and antioxidant activity of tomato peel extracts, Arab. J. Chem. 9 (2) (2016) S1010–S1018.
[56] C.A. Cassinerio, J.G. Fadel, J. Asmus, J.M. Heguy, S.J. Taylor, E.J. DePeters, Tomato seeds as a novel by-product feed for lactating dairy cows, J. Dairy Sci. 98 (7) (2015) 4811–4828.

[57] E.C. Soto, H. Khelil, M.D. Carro, D.R. Yañez-Ruiz, E. Molina-Alcaide, Use of tomato and cucumber waste fruits in goat diets: effects on rumen fermentation and microbial communities in batch and continuous cultures, J. Agric. Sci. 153 (2) (2015) 343–352.
[58] M. Romero-Huelva, E. Ramos-Morales, E. Molina-Alcaide, Nutrient utilization, ruminal fermentation, microbial abundances, and milk yield and composition in dairy goats fed diets including tomato and cucumber waste fruits, J. Dairy Sci. 95 (10) (2012) 6015–6026.
[59] EFSA Panel on Contaminants in the Food Chain (CONTAM), Risk assessment of glycoalkaloids in feed and food, in particular in potatoes and potato derived products, EFSA J 18 (8) (2020) e06222.
[60] C.B. Ammerman, R.H. Harms, R.A. Dennison, L.R. Arrington, P.E. Loggins, Dried tomato pulp, its preparation and nutritive value for livestock and poultry, Ala. Agric. Exp. Stn. Bull. 691 (1965).
[61] Dairy One, Interactive feed composition libraries. https://dairyone.com/services/forage-laboratory-services/feed-composition-library/interactive-feed-composition-libraries/, 2021 (Accessed 12 January 2021).
[62] F.G. Barroso, T.F. Martínez, A. Bernal, M.D. Megías, A. Martínez–Teruel, M.J. Madrid, F. Fernández, Alimentación de ovejas con ensilados basados en el fruto del tomate, Albeitar 120 (2008) 6–9. In Spanish.
[63] A. Arco-Pérez, E. Ramos-Morales, D.R. Yáñez-Ruiz, L. Abecia, A.I. Martín-García, Nutritive evaluation and milk quality of including of tomato or olive by-products silages with sunflower oil in the diet of dairy goats, Anim. Feed Sci. Technol. 232 (10) (2017) 57–70.
[64] M. Romero-Huelva, M.A. Ramírez-Fenosa, R. Planelles-González, P. García-Casado, E. Molina-Alcaide, Can by-products replace conventional ingredients in concentrate of dairy goat diet? J. Dairy Sci. 100 (6) (2017) 4500–4512.
[65] A. Arco-Pérez, E. Ramos-Morales, L. Abecia, D.R. Yáñez-Ruiz, A.I. Martín-García, Rumen in vitro fermentation of agriculture by-products with potential use in feeding goats, ITEA 1 (2013) 246–248.
[66] M. Romero-Huelva, I. Martin-García, R. Nogales, E. Molina-Alcaide, The effect of feed blocks containing tomato and cucumber by-products on in vitro ruminal fermentation, microbiota, and methane production, J. Anim. Feed Sci. 22 (3) (2013) 229–237.
[67] K.A. Johnson, D.E. Johnson, Methane emissions from cattle, J. Anim. Sci. 73 (8) (1995) 2483–2492.
[68] A. Lanzani, P. Bondioli, L. Folegatti, E. Fedeli, V. Bontempo, V. Chiofalo, Impiego di sanse di olive integrate nell'alimentazione della pecora da latte: Effetti sulle produzioni quali-quantitative di latte, Riv. Ital. Sostanze Gr. 70 (4) (1993) 375–383.
[69] W.P. Weiss, D.L. Frobose, M.E. Koch, Wet tomato pomace ensiled with corn plants for dairy cows, J. Dairy Sci. 11 (80) (1997) 2896–2900.
[70] R. Safari, R. Valizadeh, F.Efteljaro Shahroudi, A. Tahmasebi, J. Bayat, Effects of dried and ensiled tomato pomace on dry matter intake, milk yield and composition of dairy cows in Iran, Proc. Brit. Soc. Anim. Sci. (2007) 191.
[71] H. Tuoxunjiang, A. Yimamu, X.Q. Li, R. Maimaiti, Y.L. Wang, Effect of ensiled tomato pomace on performance and antioxidant status in the peripartum dairy cow, J. Anim. Feed Sci. 29 (2) (2020) 105–114.
[72] H. Tuoxunjiang, X.Q. Li, A. Yimamu, Mechanism research of fermented tomato pomace and its effect on oxidation resistance of transition dairy cows, Asian Agric. Res. 8 (9) (2017) 88–94.
[73] F. Abdollahzadeh, R. Pirmohammadi, F. Fatehi, I. Bernousi, Effect of feeding ensiled mixed tomato and apple pomace on performance of Holstein dairy cows, Slovak J. Ani. Sci. 1 (43) (2010a) 31–35.

[74] K. Toyokawa, Z. Saito, T. Inoue, S. Mikami, The effects of apple pomace silage on the milk production and the reduction of the feed cost for lactating cows, Bull. Fac. Agric. Hirosaki Univ. 41 (1984) 89–112.
[75] S.F. Ghoreishi, R. Pirmohammadi, A. Teimouri Yansari, Effects of ensiled apple pomace on milk yield, milk composition and DM intake of Holstein dairy cows, J. Anim. Vet. Adv. 9 (6) (2007) 1074–1078.
[76] M. Kennedy, D. Lis, Y. Lu, L.Y. Foo, R.H. Newman, I.M. Sims, P.J.S. Bain, B. Hamilton, G. Fenton, Apple pomace and products derived from apple pomace: uses, composition and analysis of plant waste materials, in: H.F. Linskens, J.F. Jackson (Eds.), Analysis of Plant Waste Materials, Springer-Verlag Berlin, Heidelberg, Germany, 1999, pp. 75–119.
[77] F. Abdollahzadeh, R. Pirmohammadi, P. Farhoomand, F. Fatehi, F.F. Pazhoh, The effect of ensiled mixed tomato and apple pomace on Holstein dairy cow, Ital. J. Anim. Sci. 41 (9) (2010b) 212–216.
[78] C. Yuangklang, K. Vasupen, S. Wittayakun, C. Sarnklong, S. Wongsuthavas, J. Mitchaothai, P. Srenanul, in: S. Aiumlamia, P. Rowlinson (Eds.), Proceedings of the AUNP SYMPOSIUM: new dietary strategies to improve health and food safety, Khon Kaen University, Thailand, 2005, 112–115.
[79] C. Yuangklang, K. Vauspen, S. Wongsuthavas, P. Panyakaew, A. Alhairdary, H.E. Mohamed, A.C. Beyen, Growth performance in beef cattle fed rations containing dried tomato pomace, J. Anim. Vet. Adv. 17 (9) (2010b) 2261–2264.
[80] M. Choubey, M. Wadhwa, M.P.S. Bakshi, Evaluation of urea molasses multi-nutrient blocks containing alternate feed resources in buffaloes, Buffalo Bull. 1 (34) (2015) 5–16.
[81] M. Wadhwa, M.P.S. Bakshi, Nutritional evaluation of urea molasses multi-nutrient blocks containing agro-industrial wastes in buffaloes, Indian J. Anim. Sci. 5 (84) (2014) 71–75.
[82] R.M.A. Gawad, M.A. Hanafy, A.E.M. Mahmoud, Y.H. Al-Slibi, Research article effect of tomato pomace, citrus and beet pulp on productive performance and milk quality of Egyptian buffaloes, Pak. J. Biol. Sci. 9 (23) (2020) 1210–1219.
[83] H.M. Ebeid, R.M.A. Gawad, A.E.M. Mahmoud, Influence of ration containing tomato pomace silage on performance of lactating buffaloes and milk quality, Asian J. Anim. Vet. Adv. 1 (10) (2015) 14–24.
[84] C. Yuangklang, K. Vasupen, P. Srenanul, S. Wongsuthavas, J. Mitchaothai, Effect of utilization of dried tomato pomace as roughage source on feed intake, rumen fermentation and blood metabolites in beef cattle, Proc. 44th Kasetsart University Annual Conference, Agricultural Science: Carrying Forward the Royal Bio-Energy Initiative, 2006 158–166.
[85] F.M. d'Arleux, G. Le Garff, J.P. Jillien, L.S. Lecompte, Utilisation de pulpe de tomate ensilée par des génisses laitières. In: Compte-rendu d'essai Institut de l'Elevage, Institut de l'Elevage, Paris, 1991.
[86] S. Chumpawadee, O. Pimpa, Effects of non forage fiber sources in total mixed ration on feed intake, nutrient digestibility, chewing behavior and ruminal fermentation in beef cattle, J. Anim. Vet. Adv. 8 (10) (2009) 2038–2044.
[87] S. Abbeddou, B. Rischkowsky, E.K. Richter, H.D. Hess, M. Kreuzer, Supplementing diets of Awassi ewes with olive cake and tomato pomace: on-farm recovery of effects on yield, composition and fatty acid profile of the milk, J. Dairy Sci. 94 (9) (2011) 4657–4668.
[88] R. Romano, F. Masucci, A. Giordano, S. Spagna Musso, D. Naviglio, A. Santini, Effect of tomato by-products in the diet of Comisana sheep on composition and conjugated linoleic acid content of milk fat, Int. Dairy J. 20 (12) (2010) 858–862.

[89] A. Razzaghi, A.A. Naserian, R. Valizadeh, S.H. Ebrahimi, B. Khorrami, M. Malekkhahi, R. Khiaosa-ard, Pomegranate seed pulp, pistachio hulls, and tomato pomaceas replacement of wheat bran increased milk conjugated linoleic acid concentrations without adverse effects on ruminal fermentation and performance of Saanen dairy goats, Anim. Feed Sci. Technol. 210 (12) (2015) 46–55.
[90] W.C.F. Mizael, R.G. Costa, G.R.B. Cruz, F.F. Ramos de Carvalho, N.L. Ribeiro, A. Lima, R. Domínguez, J.M. Lorenzo, Effect of the use of tomato pomace on feeding and performance of lactating goats, Animals 10 (9) (2020) 1574.
[91] H.A.A. Omer, S.S. Abdel-Magid, Incorporation of dried tomato pomace in growing sheep rations, Glob. Vet. 14 (1) (2015) 1–16.
[92] K. Gebeyew, A. Getachew, U. Mengistu, F. Teka, The effect of feeding dried tomato pomace and concentrate on nutritional and growth parameters of Hararghe highland sheep, Eastern Ethiopia, Adv. Dairy Res. 1 (3) (2015) 1–5.
[93] F. Abdullahzadeh, The effect of tomato pomace on carcass traits, blood metabolites and fleece characteristics of growing Markhoz goats, J. Am. Sci. 8 (8) (2012) 848–852.
[94] B. Valenti, G. Luciano, M. Pauselli, S. Mattiol, L. Biondi, A. Priolo, A. Natalello, L. Morbidini, M. Lanza, Dried tomato pomace supplementation to reduce lamb concentrate intake: effects on growth performance and meat quality, Meat Sci. 145 (11) (2018) 63–70.
[95] C.S. Correia, C.M. Alfaia, M.S. Madeira, P.A. Lopes, T.J.S. Matos, L.F. Cunha, J.A.M Prates, J.P.B. Freire, Dietary inclusion of tomato pomace improves meat oxidative stability of young pigs, J. Anim. Physiol. Anim. Nutr. 101 (6) (2017) 1215–1226.
[96] S.H. Chung, A.R. Son, S.A. Le, B.G. Kim, Effects of dietary tomato processing byproducts on pork nutrient composition and loin quality of pig, Asian J. Anim. Vet. Adv. 9 (12) (2014) 775–781.
[97] B.K. An, D.H. Kim, W.D. Joo, C.W. Kang, K.W. Lee, Effects of lycopene and tomato paste on oxidative stability and fatty acid composition of fresh belly meat in finishing pigs, Ital. J. Anim. Sci. 18 (1) (2019) 630–635.
[98] P. Yang, Y. Fan, M. Zhu, Y. Yang, Y. Ma, Energy content, nutrient digestibility coefficient, growth performance and serum parameters of pigs fed diets containing tomato pomace, J. Appl. Anim. Res. 46 (1) (2018) 1483–1489.
[99] L. Biondi, G. Luciano, D. Cutello, A. Natalello, S. Mattioli, A. Priolo, M. Lanza, L. Morbidini, A. Gallo, B. Valenti, Meat quality from pigs fed tomato processing waste, Meat Sci. 159 (1) (2020) 107940.
[100] J.I. Aguilera-Soto, F. Méndez-Llorente, M.A. López-Carlos, R.G. Ramírez, O. Carrillo-Muro, L.M. Escareño-Sánchez, C.A. Medina-Flores, Effect of fermentable liquid diet based on tomato Silage on the performance of growing finishing pigs, Interciencia 39 (6) (2014) 428–431.
[101] R.R. Caluya, R.R. Sair, B.B. Balneg, Fresh tomato pomace (FTP) as good feed for growing and fattening pigs, PCARRD Highlights '99: Summary Proceedings, Philippine Council for Agriculture, Forestry and Natural Resources Research and Development, Los Banos, Laguna, Philippines, 2000, 143.
[102] M.W. Squires, E.C. Naber, V.D. Toelle, The effects of heat, water, acid, and alkali treatment of tomato cannery wastes on growth, metabolizable energy value, and nitrogen utilization of broiler chicks, Poult. Sci. 71 (3) (2012) 522–529.
[103] N.A. Al-Betawi, Preliminary study on tomato pomace as unusual feedstuff in broiler diets, Pak. J. Nutr. 4 (1) (2005) 57–63.
[104] M.B. Yitbarek, The effect of feeding different levels of dried tomato pomace on the performance of Rhode Island Red (RIR) grower chicks, Int. J. Livest. Prod. 4 (3) (2013) 35–41.
[105] R. Cavalcante Lira, C. Bôa-Viagem Rabello, M.C.M. Marques Ludke, P.V. Ferreira, G.R. Quintão Lana, S. Roselí, V. Lana, Productive performance of broiler chickens fed tomato waste, R. Bras. Zootec. 39 (5) (2010) 1074–1081.

[106] K. Pathakamuri, Tomato pomace: alternative feed resource for poultry, J. Vet. Sci. Technol. 5 (3) (2014) 48.
[107] S.J. Hosseini-Vashan, A. Golian, A. Yaghobfar, Growth, immune, antioxidant, and bone responses of heat stress-exposed broilers fed diets supplemented with tomato pomace, Int. J. Biometeorol. 60 (8) (2016) 1183–1192.
[108] N.A. Selim, S.F. Youssef, A.F. Abdel-Salam, S.A. Nada, Evaluation of some natural antioxidant sources in broiler diets: 1-effect on growth, physiological, microbiological and immunological performance of broiler chicks, Int. J. Poult. Sci. 12 (10) (2013) 561–571.
[109] K. Sahin, C. Orhan, F. Akdemir, M. Tuzcu, S. Ali, N. Sahin, Tomato powder supplementation activates Nrf-2 via ERK/Akt signaling pathway and attenuates heat stress-related responses in quails, Anim. Feed Sci. Technol. 165 (3-4) (2011) 230–237.
[110] O.E. Akinboye, C.C. Nwangburuka, G.O. Tayo, A.O. Adeyemi, K.O. Oyekale, M.D. Olumide, G.O. Chioma, O.O. Akinboye, Sensory evaluation of meat of broiler poultry birds fed with tomato-supplemented feed, Am. Sci. Res. J. Eng. Technol. Sci. 47 (1) (2018) 145–150.
[111] N. Botsoglou, G. Papageorgiou, I. Nikolakakis, P. Florou-Paneri, V. Dotas I. Giannenas, E. Sinapis, Effect of dietary dried tomato pulp on oxidative stability of Japanese quail meat, J. Agric. Food Chem. 52 (10) (2004) 2982–2988.
[112] N. Skiepko, I. Chwastowska-Siwiecka, J. Kondratowicz, D. Mikulski, The effect of lycopene addition on the chemical composition, sensory attributes and physico-chemical properties of steamed and grilled Turkey breast, Braz. J. Poultry Sci. 18 (2) (2016) 319–330.
[113] F. Saemi, M.J. Zamiri, A. Akhlaghi, M. Niakousari, M. Dadpasand, M.M. Ommati, Dietary inclusion of dried tomato pomace improves the seminal characteristics in Iranian native roosters, Poultry Sci. 91 (9) (2012) 2310–2315.
[114] A. Ateşşahin, I. Karahan, G. Türk, S. Gür, S. Yılmaz, A.O. Çeribaşi, Protective role of lycopene on cisplatin-induced changes in sperm characteristics, testicular damage and oxidative stress in rats, Reprod. Toxicol. 21 (1) (2006) 42–47.
[115] O. Uysal, M.N Bucak, Effects of oxidized glutathione, bovine serum albumin, cysteine and lycopene on the quality of frozen-thawed ram semen, Acta Vet. Brno 67 (3) (2007) 383–390.
[116] M.G. Mangiagalli, P.A. Martino, T. Smajlovic, L. Guidobono, S.P. Cavalchini, Marelli effect of lycopene on semen quality, fertility and native immunity of broiler breeder, Br. Poult. Sci. 51 (1) (2010) 152–157.
[117] A.R. Abou Akkada, A. Khalil, MA. Kosba, M. Khalifah, Effect of feeding residues of tomato canning on the performance of laying hens, Alex. J. Agric. Res. 23 (1) (1975) 9–14.
[118] D. Dotas, S. Zamanidis, J. Balios, Effect of dried tomato pulp on the performance and egg traits of laying hens, Br. Poult. Sci. 40 (5) (1999) 695–697.
[119] R.R.S. Loureiro, C.B. Rabello, J.V. Ludke, W.M.Jr. Dutra, A.A. de Souza Guimarães, J.H. Vilar da Silva, Farelo de tomate (*Lycopersicum esculentum* Mill.) na alimentação de poedeiras comerciais, Acta Sci. 29 (4) (2007) 387–394.
[120] M.H. Salajegheh, S. Ghazi, R. Mahdavi, O. Mozafari, Effects of different levels of dried tomato pomace on performance, egg quality and serum metabolites of laying hens, Afr. J. Biotechnol. 87 (18) (2012) 5373–15379.
[121] A.L. Yannakopoulos, A.S. Tserveni-Gousi, E.V. Christaki, Effect of locally produced tomato meal on the performance and the egg quality of laying hens, Anim. Feed Sci. Technol. 36 (1–2) (1992) 53–57.
[122] S. Calyslar, G. Uygur, Effects of dry tomato pomace on egg yolk pigmentation and some egg yield characteristics of laying hens, J. Anim. Vet. Adv. 9 (1) (2010) 96–98.

[123] B. Mansoori, M. Modirsanei, Kiaei M., Influence of dried tomato pomace as an alternative to wheat bran in maize or wheat based diets, on the performance of laying hens and traits of produced eggs, Iran. J. Vet. Res. 9 (4) (2008) 341–346.

[124] M.E. Garcia, A. Gonzalez, Preliminary study on the use of tomato and pepper seed meals and excreta meal as pigments for egg yolk, Revista de Avicultura 28 (3) (1984) 155–163.

[125] H. Karunajeewa, R.J. Hughes, M.W. Mcdonald, F.S. Shenstone, A Review of factors influencing pigmentation of egg yolks, World Poultry Sci. J. 40 (1) (1984) 52–65.

[126] R. Tomczynski, Tomato seeds and skins for feeding of laying hens, Zeszyty Naukowe Akademii Rolniczo - Technicznej W Olsztynic 189 (2) (1978) 153–164.

[127] P.G. Peiretti, F. Gai, L. Rotolo, L. Gasco, Effects of diets with increasing levels of dried tomato pomace on the performance and apparent digestibility of growing rabbits, Asian J. Anim. Vet. Adv. 7 (6) (2012) 521–527.

[128] P.G. Peiretti, F. Gai, L. Rotolo, A. Brugiapaglia, L. Gasco, Effects of tomato pomace supplementation on carcass characteristics and meat quality of fattening rabbits, Meat Sci. 95 (2) (2013) 345–351.

[129] M.L. Alicata, A. Bonanno, P. Giaccone, Use of tomato skins and seeds in the feeding of meat rabbits, Riv. di Coniglicoltura 25 (1) (1988) 33–36.

[130] A.B.N. Sayed, A.M. Abdel-Azeem, Evaluation of dried tomato pomace as feedstuff in the diets of growing rabbits, Int. J. Agro Vet. Med. Sci. 3 (2009) 13–18.

[131] E.T. Kavamoto, M.M. Romeiro, A.A. Spers, By-product of the tomato industry in rations for growing and finishing rabbits, Bol. Ind. Anim. 27/28 (1) (1970) 463–473.

[132] S.S. Ahmed, K.M. El-Gendy, H. Ibrahim, A.A. Rashwan, M.I. Tawfeek, Growth performance, digestibility, carcass traits and some physiological aspects of growing rabbits fed tomato pomace as a substitution for alfalfa meal, Egypt. J. Rabbit Sci. 4 (1) (1994) 1–18.

[133] T.W. Caro, B.H. Manteroia, A.D. Cerda, Studies on the use of agroindustrial by-products in animal feeding. 5. Productive performance of growing rabbits fed with different levels of tomato pomace, Adv. Prod. Anim. 18 (1–2) (1993) 91–97.

[134] N.A. Khadr, F. Abdel-Fattah, Tomato waste as an unusual feedstuff for rabbit; response of growing rabbits to diets containing tomato waste, Vet. J. 36 (1) (2008) 29–48.

[135] A.B.N. Sayed, A.M. Abdel-Azeem, Evaluation of dried tomato seeds as feedstuff in the diets of growing rabbits, Int. J. Agro Vet. Med. Sci. 6 (2012) 263–268.

[136] M. Battaglini, F. Costantini, Byproducts from the tomato industry in diets for growing rabbits, Coniglicoltura 15 (10) (1978) 19–22.

[137] F. Lebas, Reflections on rabbit nutrition with a special emphasis on feed ingredients utilization, Proceedings of the 8th World Rabbit Congress, September 7-10, 2004, Pueblo, Mexico, World Rabbit Science Association (WRSA) Invited Paper, 2004, pp. 686–736.

[138] FAO. 2020 b. The State of World Fisheries and Aquaculture 2020. Sustainability in action. Rome. https://doi.org/10.4060/ca9229en (Accessed 12 January 2021).

[139] M.A. Soltan, Using of tomato and potato by-products as non-conventional ingredients in Nile tilapia, *Oreochromis niloticus* diets, Ann. Agric. Sci. Moshtohor 40 (4) (2002) 2081–2096.

[140] M.A. Soltan, S. El-Laithy, Evaluation of fermented silage made from fish, tomato and potato by-products as a feed ingredient for Nile tilapia, *Oreochromis niloticus*, Egypt. J. Aquat. Biol. Fisheries 12 (1) (2008) 25–41.

[141] F.A. Saad, Some studies on fish nutrition, Fac. Veterinary Medicine, Moshtohor, Zagazig University (Banha branch, Egypt, 1998 M. Sc. ThesisCited by Soltan (2002).

[142] M.S. Azaza, F. Mensi, I. Imorou Toko, M.N. Dhraief, A. Abdelmouleh, B. Brini, M.M. Kraiem, Effects of incorporation of dietary tomato feedstuff on nutrition of nile

tilapia (*Oreochromis niloticus*, L., 1758) reared in geothermal waters in southern Tunisia, Bulletin de l'Institut National des Sciences et Technologies de la Mer 33 (1) (2006) 47–58.

[143] L.C. Hoffman, J.F. Prinsloo, G. Rukan, Partial replacement of fish meal with either soybean meal, brewers yeast or tomato meal in the diets of African sharptooth catfish Clarias gariepinus, Water S.A. 23 (2) (1997) 181–186.

[144] A.K. Amirkolaie, F. Dadashi, H. Ouraji, K.J Khalili, The potential of tomato pomace as a feed ingredient in common carp (*Cyprinus carpio* L.) diet, J. Anim. Feed Sci. 24 (2) (2015) 153–159.

[145] I. Nengas, M.N. Alexis, S.J. Davies, G. Petichakis, Investigation to determine digestibility coefficients of various raw materials in diets for gilthead sea bream, Sparus auratus L, Aquac. Res. 26 (3) (1995) 185–194.

[146] C. Montoya, F. Vega, H. Nolasco, L. Espinosa, O. Carrillo, F. Olvera, Effects of dietary antioxidant of tomato extract and lycopene on *Carassius auratus* and *Xiphophorus maculatus*, Rev. MVZ Córdoba 19 (2) (2014) 4059–4071.

[147] S. Chatzifotis, M. Pavlidis, C.D. Jimeno, G. Vardanis, A. Sterioti, P. Divanach, The effect of different carotenoid sources on skin coloration of cultured red porgy (*Pagrus pagrus*), Aquac. Res. 36 (15) (2005) 1517–1525.

[148] M. Nogueira, E.M. Enfissi, Valenzuela M.E.M., G.N. Menard, R.L. Driller, P.J. Eastmond, W. Schuch, G. Sandmann, P.D Fraser, Engineering of tomato for the sustainable production of ketocarotenoids and its evaluation in aquaculture feed, Proceedings of the National Academy of Sciences, 114, 2017 10876–10881.

[149] M.R.C. de Godoy, K.R. Kerr, G.C. Fahey Jr., Alternative dietary fiber sources in companion animal nutrition, Nutrients 5 (8) (2013) 3099–3117.

[150] J. Fischer, Fruit Fibers, in: S.S. Cho, P. Samuel (Eds.), Fiber Ingredients: Food Applications and Health Benefits, CRC Press, Taylor & Francis Group, Boca Raton, FL, USA, 2009, pp. 427–438.

[151] S.E. Allen, G.C. Fahey Jr., J.E. Corbin, Evaluation of byproduct feedstuffs as dietary ingredients for dogs, J. Anim. Sci. 53 (6) (1981) 1538–1544.

[152] L. Bueno, F. Praddaude, J. Fioramonti, Y. Ruckebusch, Effect of dietary fiber on gastrointestinal motility and jejunal transit time in dogs, Gastroenterology 80 (1981) 701.

[153] M.H. Davidson, A. McDonald, Fiber: forms and function, Nutr. Res. 18 (1998) 617–624.

[154] C. Castrillo, F. Vicente, J.A. Guada, The effect of crude fibre on apparent digestibility and digestible energy content of extruded dog foods, J. Anim. Physiol. Anim. Nutr. 85 (8) (2001) 231–236.

[155] K.S. Swanson, C.M. Grieshop, G.M. Clapper, R.G.Jr Shields, T. Belay, N.R. Merchen, G.C. Jr Fahey, Fruit and vegetable fiber fermentation by gut microflora from canines, J. Anim. Sci. 79 (4) (2001) 919–926.

[156] C. Yuangklang, K. Vasupen, C. Wongnen, S. Wongsuthavas, A.C. Beynen, Digestibility of sundried tomato pomace in dogs, J. Appl. Anim. Sci. 8 (3) (2015) 35–42.

[157] C.M. Gray, R.K. Sellon, L.M. Freeman, Nutritional adequacy of two vegan diets for cats, J. Am. Vet. Med. Assoc. 225 (11) (2004) 1670–1675.

[158] P.D. Pion, M.D. Kittleson, Q.R. Rogers, J.G. Morris, Myocardial failure in cats associated with low plasma taurine: a reversible cardiomyopathy, Science 237 (4816) (1987) 764–768.

[159] D.H. Lee, *In Vitro* Analysis of Taurine as Anti-stress Agent in Tomato (*Solanum Lycopersicum*)-Preliminary Study, Experim. Med. Biol. 803 (2) 2015 75–85.

[160] R.C. Backus, G. Cohen, P.D. Pion, K.L. Good, Q.R. Rogers, A.J. Fascetti, Taurine deficiency in new found lands fed commercially available complete and balanced diets, J. Am. Vet. Med. Assoc. 223 (8) (2003) 1130–1136.

[161] L.I.E. Torales, I.M. González, J.R. García, J. Seva, J.G. Alonso, M.J.P. Castón, Consumption of spinach and tomato modifies lipid metabolism, reducing hepatic steatosis in rats, Antioxidants 9 (11) (2020) 1041.

[162] E.A.M. Moreira, R.L.M. Fagundes, D.W. Filho, D. Neves, F. Sell, F. Bellisle, E. Kupek, Effects of diet energy level and tomato powder consumption on antioxidant status in rats, Clin. Nutr. 24 (2005) 1038–1046.

[163] A. Alvarado, E.P. Delahaye, P. Hevia, Value of a tomato byproduct as a source of dietary fiber in rats, Plant Foods Hum. Nutr. 56 (2001) 335–348.

[164] D.S. Sogi, R. Bhatia, S.K. Garg, A.S. Bawa, Biological evaluation of tomato waste seed meals and protein concentrate, Food Chem. 89 (2005) 53–56.

[165] National Research Council, Underutilized Resources as Animal Feedstuffs, The National Academies Press, Washington, DC, 1983. https://www.nap.edu/catalog/41/underutilized-resources-as-animal-feedstuffs (Accessed 15 January 2021).

[166] A.A. Zorpas, Strategy development in the framework of waste management, Sci. Total Environ 716 (5) (2020). https://doi.org/10.1016/j.scitotenv.2020.137088 (Accessed 14 January 2021) 137088.

[167] P. Loizia, N. Neofytou, A.A. Zorpas, The concept of circular economy in food waste management for the optimization of energy production through UASB reactor, Environ. Sci. Pollut. Res. 26 (15) (2019) 14766–14773.

CHAPTER THREE

# Extraction and formulation of valuable components from tomato processing by-products

**Maya Ibrahim, Madona Labaki**
Lebanese University, Faculty of Sciences, Laboratory of Physical Chemistry of Materials LCPM/PR2N, Fanar, Lebanon

## 3.1 Introduction

Tomato (Lycopersicon esculentum) is considered one of the most vital vegetable crops, with a large worldwide consumption. In fact, the world production of tomatoes steadily increased during the last twenty-five years, going from 82 million tons in 1994 and reaching 180 million tons in 2019 [1]. It should be noted that a large amount of these yearly produced tomatoes is processed to supply various food products such as tomato juice, paste, purée, ketchup, sauce or salsa. Indeed, the estimated amount of processed tomato produced worldwide in 2019 was around 37 million tons [2].

However, commercial processing of tomato results in a large amount of waste, which can sum up from 10% to 40% of total processed tomatoes [3–5]. If no valorization strategies are adopted for the management of the tomato processing by-products, not only will they engender a disposal issue, but also exacerbate environmental pollution. Often, tomato processing waste is used as livestock feed or fertilizer or simply disposed of in landfills, whereas they can be utilized to provide high value-added compounds for food, cosmetics, or pharmaceutical applications [6,7].

Tomato pomace, shown in Fig. 3.1, is the solid waste derived from the industrial processing of tomatoes. It is mainly comprised of seed, skin (peel), and very little amount of pulp, present in proportions determined by the type of process from which the tomato pomace was created [8–10]. Tomato pomace can serve as a low-cost source of important bioactive and nutritional compounds namely: carotenoids, phenolic compounds, vitamins, dietary fibers, proteins, and essential oils. These bioactive constituents are associated with a large number of health benefits, such as anticarcinogenic, cardioprotective, antimicrobial, anti-inflammatory and antioxidants

*Tomato Processing by-Products*
DOI: https://doi.org/10.1016/B978-0-12-822866-1.00009-0

**Fig. 3.1** (A) Whole pomace, and (B) skin + pulp and (C) seed fractions of tomato pomace.

activities, to cite a few [7,11–13]. Moreover, seeing as there is an increasing demand by consumers nowadays for the use of natural additives in the food industry rather than chemically synthesized molecules, the recovery of functional ingredients from tomato processing by-products will prove to be a sustainable strategy for the recycling of agri-food wastes within the food chain. Thus, all of the above-mentioned reasons justify the recent growing interest in the development and optimization of economically viable extraction methods that will allow the retrieval of valuable compounds from tomato processing by-products.

Numerous extraction techniques have already been reported in literature for the recovery of bioactive compounds from tomato waste (namely carotenoids and phenolic compounds), like for instance conventional solvent extraction (CSE), enzyme-assisted extraction (EAE), ultrasound assisted extraction (UAE), microwave assisted extraction (MAE), high hydrostatic pressure extraction (HHPE), supercritical fluid extraction (SFE), pulsed electric field (PEF) and many others ... all aiming to maximize the product yield while limiting the environmental impact and economical cost of recovery.

Therefore, the following chapter will be dedicated towards first presenting the functional ingredients that are present in tomato processing by-products that can have high nutritive and biological values, followed by a report of the typically employed extraction methods for the recovery of these valuable compounds derived from tomato processing waste. Finally, an outline of essential properties and health-related applications found for the recovered compounds will be performed accompanied by various comparisons between the properties of the tomato processing by-products derived materials and market existing products in order to validate the economical appeal of the recovery approach. Nevertheless, it should be noted that while a brief description of all main findings regarding the extraction and formulation of valuable components from tomato processing by-products will be reported herein, the bulk of this chapter's content will be focused on carotenoids and phenolic compounds recovery and valorization.

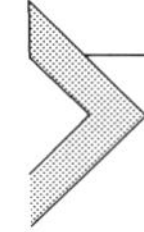

## 3.2 Valuable compounds present in tomato processing by-products

Tomato-processing by-products are rich sources of nutrients and bioactive compounds which can serve as potential nutraceutical resources. Indeed, carotenoids, phenolic compounds, vitamins, dietary fibers, proteins, and essential oils are typically found in tomatoes and their industrial by-products. However, their respective amounts within tomato-based materials will differ depending on numerous factors such as tomato variety, soil and climate conditions, degree of ripening, post-harvest storage conditions, the industrial processing parameters... It should be noted that while the skins of tomatoes are reported to be richer in carotenoids and phenolic compounds, tomato seeds were found to be good sources of high quality oil and were shown to have higher content of proteins and dietary fibers [14–17].

Presently, a brief description of the main valuable compounds present in tomato processing by-products will be carried out.

### 3.2.1 Carotenoids

Carotenoids are natural pigments responsible for the noted yellow, orange, and red hues of various fruits, vegetables, plants, birds, and marine species [4,14]. These colors are derived from the presence of conjugated double bonds, which also impart carotenoids with antioxidant characteristics. It has been reported that tomatoes and tomato-based foods constitute the main dietary sources of carotenoids. In fact, carotenoids are mainly found in tomato skin and seeing as the latter can represent more than 40% of tomato processing waste it becomes apparent that a recovery of these bioactive compounds from tomato pomace can be highly beneficial in terms of potential applications [18–20].

Based on the distinction in the structure of the polyisoprenoid chain, carotenoids can be divided into two main groups: hydrocarbon carotenoids, or carotenes, which are either cyclized (such as α-carotene and β-carotene) or linear (such as lycopene, phytofluene, and phytoene) and oxygenated carotenoids or xanthophylls (such as lutein, zeaxanthin, astaxanthin, and β-cryptoxanthin). Several of these carotenoids were found in tomato processing waste with lycopene being predominant (≈88% of total carotenoids), followed by β-carotene (2% to 3%) and lutein (≈1.5%) [6,21–23]. A chemical representation of the structure of these three carotenoids is reported in Fig. 3.2.

#### *3.2.1.1 Lycopene*

Lycopene (ψ,ψ-carotene) is composed of 40-carbon-atom chain containing 11 conjugated double bonds and two unconjugated double bonds [24]. This great degree of conjugation imparts lycopene with considerable antioxidant activity, making it one of the most effective antioxidant compounds, and justifies both its red hue and lipophilic character [25]. However, because of its linear structure and absence of a β-ionone ring, this carotenoid presents no pro-vitamin A activity [26]. The most thermodynamically stable form of lycopene is the all-trans-isomer, usually found in foods containing this pigment. Nonetheless, following thermal processing an isomerization occurs and all-trans isomers are converted to cis-isomers, which are more easily absorbed by body tissues, thereby improving the bioavailability of this bioactive compound [27]. Lycopene can be found in moderate concentrations in watermelon [28], pink and red grapefruit [29,30], guava [31], and rosehip [32], yet it is most abundant in tomatoes [33]. Furthermore, seeing as lycopene gradually accumulates in the outer skin layer of tomatoes throughout the ripening stage, the lycopene amounts

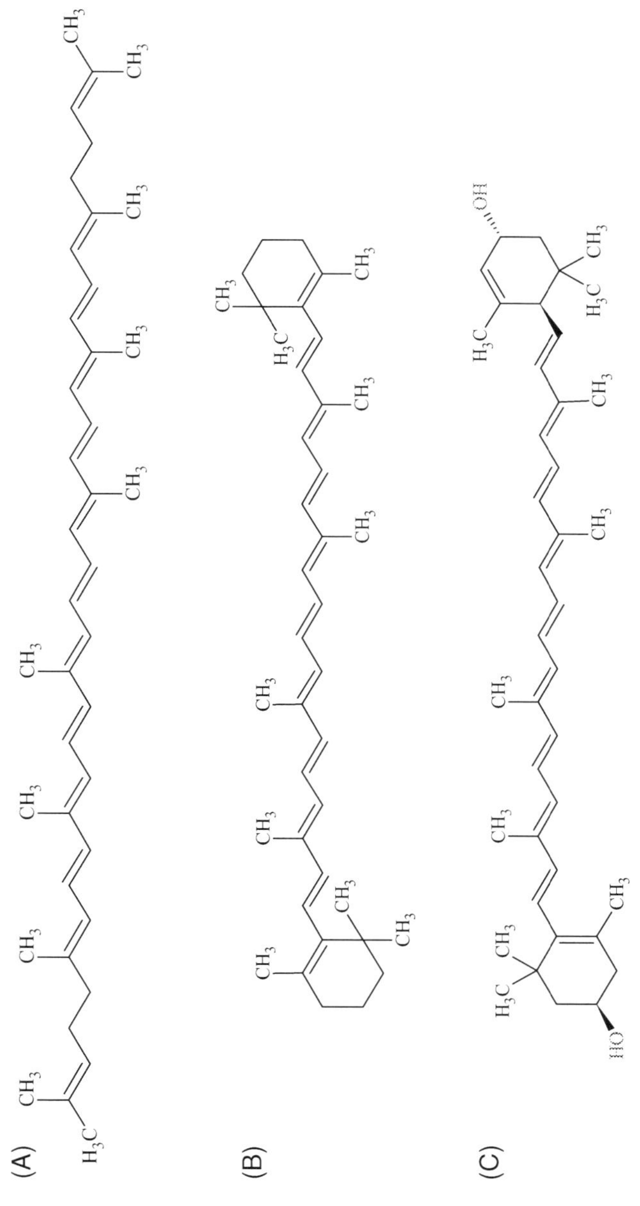

**Fig. 3.2** ***Chemical structure of*** (A) lycopene, (B) β-carotene and (C) lutein, present in tomato and tomato processing by-products.

at the end of the ripening process are about five times higher in the skin than in the pulp [20,34]. Thus, tomato pomace, comprised mostly of skin and seeds, is an especially rich source of this carotenoid.

#### 3.2.1.2 β-carotene

β-carotene is a well-known precursor of vitamin A. Indeed, an enzyme, β-carotene 15,15'-oxygenase, catalyzes the symmetrical cleavage of the β-carotene molecule to generate two molecules of retinol (vitamin A) [35,36]. Dietary sources of this carotenoid consist of red-orange fruits such as oranges [37], apricots [38] and tomatoes [39,40] as well as bright green vegetables among which figure carrots [41], kale [42], spinach [43] and turnip greens [44]. β-carotene is the second most preponderant carotenoid in tomato fruits and accounts for the observed yellow and orange hues [45]. As has been reported for lycopene, the β-carotene content was found to be greater in tomato processing by-products than in unprocessed tomatoes (14.9 vs 8.6 mg/100 g on a dry weight basis) [46].

#### 3.2.1.3 Lutein

Lutein can be found in foods like spinach [47], kale [48], egg yolk [49], which are consumed in little quantities. Very high levels of lutein are present as well in a number of plants, like for instance the marigold flowers, which represent nowadays the principal commercial source of nutritional lutein additives [50,51]. Extraction of lutein from marigold flowers is usually conducted by lixiviation with hexane. However, not only is the cultivation of marigold flowers economically expensive (low growth rate and high labor cost) [52] but the obtained extracts often need further purification and the recovery method is a multistep process which is also time-consuming [53]. Therefore, exploring novel low-cost sources of lutein is called for. Investigations revolving around the carotenoid levels in tomatoes by-products showed that it contains lutein in appreciable amounts which makes it a cheap potential source of lutein compound [54,55].

### 3.2.2 Phenolic compounds

Phenolic compounds represent one of the principal phytochemicals encountered in fruits and vegetables as well as in their industrial processing by-products. The major phenolic compounds identified in tomato processing waste include flavonoids (rutin, naringenin, naringenin chalcone, myricetin, and quercetin) and phenolic acids (ellagic, chlorogenic, salicylic, vanillic, gallic, p-coumaric, ferulic, catechin, and caffeic acids) [16,56–58].

While the amount of phenolic compounds was affected by the tomato variety, a constant observation reported by several studies was that the concentration of these phenolic compounds was higher in tomato skin than in seeds and pulp [56,59,60]. Furthermore, flavonoids were found to be the main components of total phenolic compounds present in tomato waste extracts with rutin being the most abundant followed by naringenin [58,60].

### 3.2.3 Vitamins

Tomato is an important source of vitamin C, a water-soluble compound with multiple biological properties. However, despite its presence in considerable concentrations in tomato fruits, a striking loss of vitamin C (ascorbic acid) is usually noted following industrial processing of tomatoes (potentially reaching 80% to 90%) because of the thermal lability and light-sensitivity of this vitamin [61–63].

Tomato also contains significant amounts of vitamin A, B, and E. The latter is a fat-soluble vitamin with important antioxidant characteristics, and it encompasses tocopherols (α-, β-,γ-, and δ), and tocotrienols (α-, β-,γ-, and δ). A considerable level of vitamin E is located in the seeds, comprised mainly of γ-tocopherol with a proportion of 97% to 98% along with a minor content of α-tocopherol (2% to 3%) [23,64].

### 3.2.4 Dietary fibers

The importance of fiber intake has been frequently reported over the years, especially in respect to its favorable effects in the digestive tract. Indeed, dietary fiber can impart a lot of functional characteristics when incorporated in human diet. It should be noted however that dietary fibers don't correspond to a singular and well-defined chemical component, but rather to a mixture of chemical substances. Thus are recognized insoluble dietary fibers (lignin, cellulose and hemicelluloses) and soluble dietary fibers (pectins, β-glucans, galactomannan gums, in addition to various indigestible oligosaccharides such as inulin) [58,65,66]. Tomato waste, namely tomato peel, is rich in dietary fibers which makes it a very attractive low-cost source for the elaboration of products with high fiber content [65]. In fact, the results of previous works showed the total dietary fiber content of tomato peel to be greater than that of other products such as red grape peels [67], carrot peels [68], and fiber-rich cocoa compounds [69].

Among the soluble dietary fiber, pectin which is a polysaccharide mainly composed of D-galacturonic acid, D-galactose and L-arabinose, is particularly interesting because of its potential industrial applications as gelling agent [70]. Pectins can correspond to either high-methoxy (HM) pectins characterized by a degree of methylation ≥50%, or to low-methoxy (LM) pectins if their degree of methylation ≤50%. Regarding the tomato-derived pectins, studies have revealed them to be of the HM pectin type [71].

### 3.2.5 Proteins

Tomato processing by-products contain considerable amounts of good quality protein, predominately present in tomato seeds [17,72]. Essential and nonessential amino acids were found in tomato by products. In fact, the study of Nour et al. [19] revealed that eight essential amino acids could be identified in dried tomato waste namely leucine, isoleucine, lysine, methionine, phenylalanine, threonine, arginine, and valine, amounting to 34.2% of total protein. Regarding nonessential amino acids, they include glutamic acid, alanine, glycine, tyrosine, cysteine, and aspartic acid [16]. In seeds as well as skin, glutamic and aspartic acid were the primary amino acids present [20]. In terms of essential amino acids, high lysine levels were found in tomato processing waste [20,73], making the latter suitable as food supplement to fortify products with low lysine content [74].

### 3.2.6 Essential oils

Tomato seeds are a rich source of fatty acids and can serve as raw materials for oil extraction [17,20]. In terms of fatty acid composition of tomato seed oil, reports have shown that linoleic acid is the predominant fatty acid with a percentage of 37% to 57%, succeeded by oleic acid with a concentration in the range of 18% to 30% [75]. These two compounds are known as unsaturated fatty acids, which along with many others (linolenic acid, palmitoleic acid ...), represent about 80% of the total fatty acid content [23]. In opposition, saturated fatty acids, comprised mainly by palmitic (7% to 24%), stearic (4% to 13%) and myristic acid (0.1% to 2.3%), amount to approximately 20% of total fatty acid content [16,23,75].

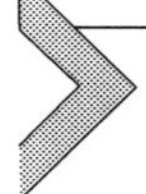

## 3.3 Extraction methods available for the recovery of valuable compounds from tomato processing by-products

A "5-stages universal recovery process" has been previously proposed for the recuperation of valuable compounds from agri-food wastes, such as tomato processing by-products for example, and it consists of: (1) macroscopic pretreatment, (2) separation of macro and micromolecules, (3) extraction, (4) purification/isolation, and (5) encapsulation or product formation [13]. As of late, many efforts have been focused on the optimization of the extraction step, considered as vital in the recovery operation.

Several extraction techniques have been so far reported for the retrieval of phytochemicals found in tomato processing waste, among which figure: CSE, EAE, UAE, MAE, HHPE, SFE, PEF, along with some other newly explored processes. The yields of valuable compounds recovered have been diversified, depending primarily on the tomato variety as well as the industrial processing methods employed along with the extraction technique selected and the parameters applied [5,7,17].

The present section will outline the extraction methods available for the recovery of valuable compounds from tomato processing waste, essentially carotenoids and phenolic compounds, in addition to presenting the effect of extraction conditions on retrieval efficiency as assessed through literature work.

### 3.3.1 Conventional solvent extraction

CSE is the most commonly employed technique for the recovery of valuable compounds from a wide range of fruits and vegetables. It consists in the dissociation of selected ingredients from their matrix, be it solid or liquid, according to the concept that solvent diffuses in the matrix and dissolves the soluble ingredients. Regarding tomato processing by-products, the majority of recovery studies have been focused on the extraction of carotenoids from the tomato waste [4,5,9,55,76–82]. Organic solvents are usually employed for efficient recovery of these bioactive pigments. However, as several research studies have shown, the carotenoid yield will alter depending on the following extraction parameters: nature of the solvent, solvent to solid ratio, number of extractions, extraction time, temperature and particle size.

#### *3.3.1.1 Effect of solvent nature*

Regarding the effect of solvent nature, it should be noted that many common organic solvents have been examined for the recovery of carotenoids from tomato by-products, namely hexane, ethyl acetate, acetone, ethanol, benzene ethyl ether, chloroform and petroleum ether [4,78,79], in addition to mixtures of polar and nonpolar solvents in varying ratios [4,55,76,82,83]. While oxygenated carotenoids, such as lutein, are more soluble in polar solvents, hydrocarbon carotenoids namely lycopene and β-carotene, are more soluble in nonpolar solvents. Indeed, as demonstrated by Vági et al. [18], the recovery yields of lycopene by Soxhlet extraction, a particular CSE technique, varied greatly depending on the choice of solvent. Thus, when hexane was used instead of ethanol, an almost 10-fold increase in lycopene yield was perceived. Similarly, Calvo et al. [80] also examined the effect of solvent nature on the lycopene extraction efficiency and found that when ethyl acetate was used as solvent the yield of all-trans lycopene was 20-fold higher than with ethanol. As for Strati et al. [79], their research indicated that the efficiency in lycopene extraction from dried tomato waste decreased according to the following order with respect to the nature of solvent used: acetone > ethyl acetate > hexane > ethanol.

In addition, it should also be noted that a mixture of polar and nonpolar solvents could prove highly useful for optimizing the recovery of total carotenoids present in tomato processing by-products. This was proven to be the case in the study conducted by Strati et al. [4], which revealed that a mixture of polar and nonpolar solvents that is ethyl acetate and hexane, resulted in good recovery of lipophilic lycopene and β-carotene, which represented 96% of total extracted carotenoids, and hydrophilic lutein, which amounted to 4% of total extracted carotenoids. Furthermore, when operating with a mixture of solvents, the proportion used for each solvent is very important. Hence, when it comes to lycopene extraction, Poojary et al. [76] discovered that between the three evaluated ratios of acetone/n-hexane mixtures (1:3, 2:2, and 3:1, v/v), the solvent mixture comprised of the greatest amount of n-hexane (acetone/n-hexane 1:3, v/v) resulted in the best lycopene yield.

#### *3.3.1.2 Effect of solvent to solid ratio*

A second element that affects the recovery efficiency in CSE is solvent to solid ratio. A balance should be found regarding the solvent to solid ratio, because while large volumes of solvent may increase the recovery yield, an excess of solvent quantity would lead to both high costs and solvent waste. In opposition, deficient solvent amount would result in poor mixing and risk of solvent saturation. Kaur et al. [9] who examined five different solvent/solid ratios (20:1, 30:1, 40:1, 50:1, and 60:1 v/w); reported maximum lycopene yield when the ratio value was equal to 30:1 v/w. Whereas Strati et al. [4] indicated that when it came to total carotenoid extraction from tomato waste the optimized solvent mixture (ethyl acetate + hexane)/solid ratio corresponded to 9.1:1 (v/w).

#### 3.3.1.3 Effect of number of extractions and extraction time

A number of investigations looked into the effect of number of extractions and extraction time on the recovery of carotenoids when using solvent extraction. Strati et al. [79] observed an increase in carotenoid yield with the number of extractions. Similarly, the study conducted by Kaur et al. [9] revealed a rise in lycopene yield with the increase of number of extractions from 1 to 4. As for the optimal extraction duration, a time of 8 min was found to give maximum lycopene amount.

#### 3.3.1.4 Effect of extraction temperature

Concerning the temperature impact, it should be reminded that while a rise in temperature can in fact improve extraction efficiency by enhancing the solvent ability to solubilize intended ingredients as well as diffusion rates, it can also cause a degradation of the targeted compound as a result of some unwanted transformation (isomerization, oxidation…). Therefore, the selection of extraction temperature value should take all these factors into consideration. In terms of lycopene recovery, Kaur et al. [9] indicated that between 5 temperatures tested (20, 30, 40, 50, and 60°C), a temperature of 50°C was best for achieving highest lycopene yield. As for Poojary et al. [76], following an evaluation of 3 different extraction temperatures (30, 40, and 50°C), they proposed a temperature 30°C for optimal lycopene recovery. The difference in the temperature values obtained by these authors can be due to the presence of other variables (such as solvent used, solvent to solid ratio, extraction time…) that will also influence the final product yield.

#### 3.3.1.5 Effect of particle size

A final factor that impacts recovery efficiency is particle size. A decrease in particle size improves the recovery yield owing to an increase in interfacial area contact between particles and solvent until an optimum is reached, after which the continuous decrease in particle size will be detrimental to carotenoid extraction because of induced packing of the extraction bed generating undesirable channeling effects. Kaur et al. [9] found that a decrease in particle size from 0.43 up till 0.15 mm was beneficial for lycopene extraction. Whereas Strati et al. [4] proposed, based on their experimental design, an optimized particle size of 0.56 mm for maximum carotenoids extraction from tomato by-products.

It's worth mentioning that while recovery of carotenoids from tomato processing waste represents the majority of studies revolving around CSE of valuable compounds from tomato fruit by-products, some investigations have been carried out on the extraction of phenolic compounds [56,60], pectins [84], and oil [75] from tomato waste using this conventional extraction process.

Last but not least, it is important to understand that while CSE exhibits good bioactive compound recovery yield it presents some limitations as well, such as the fact that it is time consuming, that it raises the risk of product degradation due extended time of heat exposure and that it necessitates the use of big volumes of solvents which can cause both environmental and health issues seeing as most organic solvents are considered to be toxic.

### 3.3.2 Enzyme-assisted extraction

EAE has been shown to be a very promising technique for the extraction of lycopene [8,15,85–88] and soluble dietary fiber [86] from tomato processing by-products. This is explained by the capacity of enzymes such as cellulases, pectinases and hemicellulases to degrade the cell walls and membranes, under mild process conditions, thus promoting solvent penetration and lycopene solubilization, which as a result enhances extraction efficiency and increases the yield of the targeted compound. EAE is often employed as pretreatment in combination with solvent extraction or other extraction techniques. In the study conducted by Lavecchia et al. [89], the authors found that enzymatic hydrolysis of tomato skin with enzymes characterized by pectinolytic, cellulolytic, and hemicellulolytic activities led to significant improvement in lycopene recovery compared to nonenzymatically treated samples. Similarly, the results reported by Zuorro et al. [8] indicated that an 8- to 18-fold increase in lycopene extraction yields is obtained following pretreatment of tomato skins with the mixed enzyme preparations with cellulolytic and pectinolytic properties. Furthermore, the investigation carried out by Gu et al. [86] revealed that enzymatic modification of tomato peels by 5% (w/v) Viscozyme L solution led to higher soluble dietary fiber and lycopene yield than original tomato peels. Thus, the recovery yield of soluble dietary fiber was increased by 72.3% while that of lycopene by 23.8% as a result of enzymatic treatment of tomato peels. Additionally, the quality and gelling ability of soluble dietary fibers recovered from enzymatically modified tomato skins was enhanced compared to those obtained from original tomato peels.

Even though EAE of tomato waste-derived bioactive compounds has a lot of potential in enhancing their recovery efficiency and final quality, it presents nevertheless some commercial and technical limitations. The first being that enzymes can be rather costly for the treatment of big amounts of tomato products and the second that enzyme behavior is altered when environmental parameters (i.e., the fraction of dissolved oxygen, the temperature and the nutrient disponibility) are modified which restricts its industrial use.

### 3.3.3 Ultrasound assisted extraction

UAE is a nonthermal extraction technique which can be especially beneficial for the recovery of heat labile compounds. The ability of ultrasound method in efficiently extracting targeted products is ascribed to acoustic cavitation, which involves the formation, growth, and collapse of microbubbles within a liquid due to the application of an acoustic field in the frequency range 20 to 100 kHz [90]. Microjetting and microstreaming effects credited to acoustic cavitation provoke disintegration of solid materials and disruption of cellular tissues, thereby promoting the release of bioactive compounds and the access of solvent into the cellular content [91]. Numerous works have examined the effect of ultrasound on the recovery of valuable components from tomato processing by-products, such as carotenoids [3,6,92,93], phenolic compounds [6,58,92,94], and pectin [10]. The advantage of UAE over CSE is that it provides high extraction efficiency while reducing both the amount of solvent used and the time of extraction, thus constituting an economically as well as environmentally attractive recovery process. In fact, Grassino et al. [10] showed that while pectin extraction from tomato waste using CSE resulted in high pectin yield, UAE managed to generate better quality pectin in a significantly less amount of time (15 min for ultrasound vs 36 h for CSE). Furthermore, an investigation conducted by Navarro-González et al. [58] revealed that phenolic compounds yield from tomato peel fiber was higher when ultrasound was used as an extraction method rather than enzymatic treatment. A following study on the potential of achieving a synergistic effect by combining ultrasound and EAE for the recovery of lycopene established the validity of this approach seeing as a 1.3-fold increase in lycopene yield could be attained when associating both techniques compared to single process of extraction.

### 3.3.4 Microwave assisted extraction

MAE is a rapid recovery technique which allows to overcome some of CSE limitations such as long extraction time and prolonged heat exposure that may cause the degradation of bioactive ingredients [95]. Thus, MEA employs microwave energy which heats the cells of the raw material inducing moisture evaporation and a rising pressure within these cells that results in the enhancement of bioactive ingredients migration in the extracting solvent, thereby generating better recovery yields of the targeted compound [96]. For this reason, several studies have focused on the optimization of the MEA process (microwave power, solvent/solid ratio, extraction time and temperature) in order to maximize valuable compounds yield such as lycopene, phenolic compounds, sugars and proteins [13,95,97]. One of these optimization investigations addressed the effects of coupling microwave with UAE on lycopene recovery efficiency. The authors found that combining MAE with UAE resulted in 97.4% lycopene yield under optimal conditions compared to 89.4% lycopene recovery in the case of UAE. Hence, these results showcase the benefit gained by combining these two extraction methods.

### 3.3.5 High hydrostatic pressure extraction

HHPE is a nonthermal process generally performed in the range of 100 to 1200 MPa in the aim of facilitating mass transport phenomena. Indeed, as a result of subjecting the raw material to elevated pressure, the cell permeability is enhanced and the dissolution rate of cell bioactive components is increased. Thus, when the pressure is raised, the solubility of the targeted compound is promoted leading towards greater recovery yields [98]. This was proven to be true in the study conducted by Grassino et al. [98], which indicated that HHPE improved tomato peel-derived pectin yield from 14% to 15% after 30 and 45 min of extraction in comparison with CSE performed for 180 min, while managing to maintain a similar pectin quality. Therefore, HHPE would not only increase compound recovery efficiency but also shorten the extraction time, which is of great industrial value. This notion was again reinforced when examining the appeal of using HHPE for the extraction of polyphenols from tomato peel waste [99]. This study showed that high polyphenol recovery can be achieved through HHPE despite applying very low extraction time (5 min).

### 3.3.6 Supercritical fluid extraction

SFE has lately acquired considerable interest, seeing as it represents an environmentally friendly technology that employs a fluid submitted to a pressure and a temperature that exceed its critical point [100]. The main benefits of SFE process is that it provides high selectivity, reduced extraction time and toxic solvent-free final products [14]. Several compounds can be employed as supercritical fluids (SFs) such as carbon dioxide, ethane, ethylene, methanol, and xenon to cite a few. Among these SFs, it is carbon dioxide that is generally favored due to its low critical temperature (31.1°C) and pressure (73.8 bar), factors which would lower the risk of degradation of thermolabile ingredients. Furthermore, $CO_2$ is available at low price and high purity. Thus, the majority of the studies on the extraction of bioactive ingredients from tomato processing by-products have been focused on the use of supercritical carbon dioxide (SC-$CO_2$) as means of product recovery namely, lycopene and β-carotene [14,18,85,101–108]. Several of these researchers invested their efforts in optimizing the technical parameters of SC-$CO_2$ process such as extraction temperature and time, pressure, particle size and $CO_2$ flow rate. Based on their findings, they concluded that it was mainly the extraction temperature that produced statistically significant impact on the process, with the recovery yield increasing along with the temperature value [18,106]. It should be mentioned nevertheless, that while SFE is faster and more efficient than CSE, it requires costly equipment and has a high-energy consumption.

### 3.3.7 Pulsed electric field

PEF is another nonthermal extraction technique that can provide better bioactive compounds recovery because of its capacity to induce electroporation of cytoplasmic membranes following the application of an electric field, which in turn causes an increased diffusion of intracellular substances, thus leading to improved valuable compounds extraction. Regarding tomato processing waste, PEF has been investigated for its potential to raise carotenoid [109–111] and phenolic compounds yields [111] and was shown to be a highly effective method.

### 3.3.8 Other newly explored techniques

It should be noted that in addition to the extraction techniques mentioned above, there have been some other explored recovery processes, namely microemulsion technique, a surfactant-based extraction approach that proved to be a promising lycopene extraction process in the study conducted by Amiri-Rigi et al. [112]; hydrocolloidal complexation that Nagarajan et al. [113] found to be highly beneficial for improving the extraction of carotenoids from tomato pomace; and finally ohmic heating (OH) technology, based on the generation of internal heat within materials by passage of electric currents, which was revealed as both an effective and selective recovery process for the extraction of bioactive compounds from tomato waste and it had the added advantage of increasing the extraction rate while producing comparable yields to CSE [114].

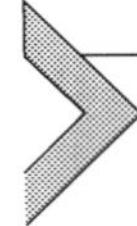

## 3.4 Potential applications for the valorization of tomato processing by-products

Besides mitigating the environmental pollution connected with the accumulation of tomato processing residues, the interest of extracting bioactive ingredients present in tomato waste lies within their potential use in various applications and field sectors among which figure, namely, food, cosmetic and pharmaceutical industries. However, in order to understand the reason behind the viability of these tomato waste-derived compounds in the formulation of food, cosmetic or pharmaceutical products, it is important to first outline the beneficial properties that are attributed to the valuable ingredients found in tomato waste, particularly carotenoids and phenolic compounds. This will be followed by a survey of their reported fields of application in relation to health and physical well-being. It should be noted that throughout this section the characteristics of the tomato-derived materials will be described and compared with those of market existing products.

### 3.4.1 Properties of the valuable compounds extracted from tomato processing by-products

#### *3.4.1.1 Antioxidant properties*

Tomato by-products contain a large variety of antioxidants such as carotenoids (lycopene, β-carotene, and lutein), phenolic compounds (phenolic acids and flavonoids) and vitamins (ascorbic acid and vitamin E) [21]. While lipophilic antioxidant activity of tomato processing waste is attributed to carotenoids, mainly lycopene, and vitamin E, hydrophilic antioxidant activity is ascribed to phenolic compounds and ascorbic acid [23,58]. Seeing as both carotenoids and phenolic compounds are mainly present in tomato skin, it would follow that the antioxidant activity of both lipophilic and hydrophilic extract of tomatoes should be highest in tomato skin rather than seed and pulp. This has indeed been the case in a study performed by Toor et al. [115] who examined the antioxidant activity in different fractions (skin, seeds, and pulp) of three tomato cultivars (Excell, Tradiro, and Flavourine) and found tomato skin to exhibit better antioxidant activity than the other two tomato fractions, as displayed in Fig. 3.3. However, it ought to be noted that the lipophilic antioxidant compounds, comprised for the most part of carotenoids, have been shown to constitute the major contributors to the total antioxidant activity of tomato by-products [22].

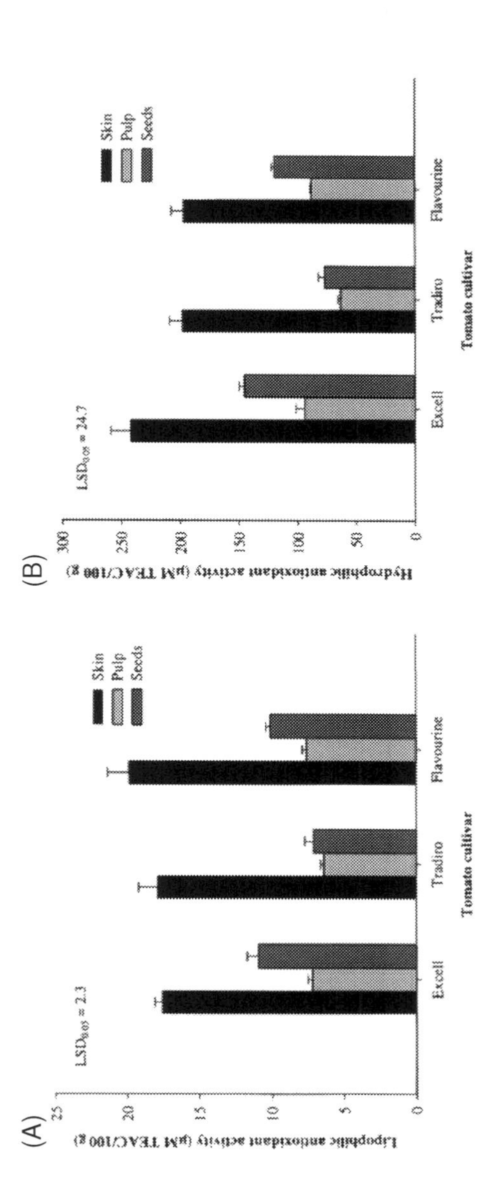


**Fig. 3.3** ***Antioxidant activity in the*** (A) lipophilic and (B) hydrophilic extracts of three varieties of tomatoes (Excell, Tradiro and Flavourine).

The antioxidant properties of carotenoids have been linked to their capacities of quenching singlet oxygen ($O_2^{\bullet}$) and trapping peroxyl radicals ($ROO^{\bullet}$) [116]. Among the carotenoids present in tomato processing by-products, studies have shown lycopene to be the most potent ROS (reactive oxygen species) scavenger and to be even more potent than other antioxidants, including vitamin E. Indeed, its singlet-oxygen-quenching ability has been revealed to be twice that of β-carotene and 10 times that of α-tocopherol (vitamin E) [11,117]. This has been attributed to the high degree of conjugation in lycopene structure [8,11]. It should be noted that the antioxidant ability of lycopene is further improved by the presence of the other bioactive ingredients present in tomato by-products, such as β-carotene, flavonoids, ascorbic acid, and tocopherols, owing to synergistic antioxidant activity between them [11,66,118].

In fact, in the work of Elbadrawy et al. [16] where the antioxidant activities of tomato peel extracts were assessed and compared to those of an artificial antioxidant, butylated hydroxy toluene (BHT), the authors founds that tomato peel extracts exhibited better antioxidant properties than the synthetic product. This was once again the case in a study conducted by Ćetković et al. [56] where the radical scavenging activity of tomato waste extracts proved to be greater than that of synthetic antioxidant, butylated hydroxyanisole (BHA). Thus, the use of natural antioxidants extracted from tomato processing by-products as substitutes of synthetic antioxidants can be highly advantageous.

#### *3.4.1.2 Chemoprevention and reduced risk of eye and cardiovascular diseases*

Consumption of tomato and tomato-based products has been associated over time with a lower risk of prostate, lung, colon, rectal and stomach cancers [117,119–121]. This cancer prevention property has been generally ascribed to the antioxidant components present in tomato-based materials [11], particularly carotenoids. Thus, an epidemiological study performed over 332 human subjects diagnosed with lung cancer and 865 control subjects revealed that the individuals with the biggest intake of a mixture of three carotenoids (β-carotene, α-carotene, and lutein) had the lowest risk of lung cancer [122]. It should be noted though, that in terms of the chemoprevention abilities of carotenoids, investigations are mostly centered on lycopene [76,117,120,123]. Thus, lycopene has been shown to slow the progression of prostate, breast, lung and endometrial cancers by retarding the cell cycle evolution from one growth phase to the next, thereby impeding the growth of tumor cells. Furthermore, several studies have indicated that people with greater consumption of lycopene are in lesser danger of developing prostate, cervical and breast cancer [76,117,124].

Besides its chemoprevention property, lycopene has also been shown to reduce the risk of age-related macular degeneration (ARMD), a frequently encountered form of blindness in elderly individuals [117]. That being said, when it comes to eye health, it is another tomato-derived carotenoid, lutein, that is most commonly known for its protective role of eyes, by keeping them shielded from oxidative stress and the high-energy photons of blue light [55,125–127].

Moreover, tomato-based products have been reported to reduce the risk of cardiovascular diseases (CVD), disorders which are at the origin of a high percentage of mortality worldwide [128,129]. Obesity, hypertension, high blood cholesterol, physical inactivity and smoking have been cited as some of the factors connected with the occurrence of CVD. However, a well-thought diet, rich in cardioprotective compounds, can help prevent CVD. Lycopene has once again been pointed out as beneficial to individuals with high cholesterol, atherosclerosis or coronary heart afflictions [76,117,130]. In fact, lycopene inhibits the oxidation of low-density lipoprotein, assisting in the decrease of blood cholesterol levels and diminishing the danger of arteries becoming thickened and blocked [76,117,131]. Furthermore, fiber-rich products such as tomatoes have also been revealed as highly favorable for CVD prevention through an induced reduction of cholesterol plasmatic concentration [65].

### 3.4.1.3 Antimicrobial properties

Phenolic compounds present in tomato-based products are well recognized for their antimicrobial activity [83,132–134]. Some researchers have ascribed the antimicrobial property of polyphenols to the hydroxyl groups that they contain, which would act as inhibitors for microorganisms. In fact, the hydroxyl groups would be able to interact with the cell membrane of bacteria, thereby disrupting the membrane composition which would lead towards the loss of cellular components [132]. Thus, numerous studies have shown that a rise in the concentration of phenolic compounds is directly correlated with an increase in antimicrobial activity [83,132,135]. An investigation by Szabo et al. [83] indicated that tomato waste extracts from different tomato varieties exhibited high antibacterial activity against Gram-positive bacteria, Staphylococcus aureus, and the authors were able to connect this antibacterial activity with the quantity of isochlorogenic acid present in each tomato variety. Furthermore, another research conducted by Silva-Beltrán et al. [132] regarding the antimicrobial property of tomato revealed that the Pitenza variety extracts exhibited the best antimicrobial activity against *E.coli* O157:H7, *Salmonella* Typhimurium, *Staphylococcus aureus*, and *Listeria ivanovii* owing to its high concentration in phenolic compounds. However, it should also be noted that a recent study examining the antimicrobial activities of tomato processing wastes derived from 10 different tomato varieties against six bacterial strains (three gram-positive and three gram-negative microorganisms) signaled that while all tomato peels extracts demonstrated good antimicrobial activity, it was the extract derived from Ţărăneşti roz variety which turned out to be the most effective due to a higher content of carotenoids [6]. Therefore, it would appear that both phenolic compounds and carotenoids present in tomato processing by-products might contribute towards the suitability of the latter as a source of low-priced antimicrobial agents.

#### 3.4.1.4 Natural colorant

As has already been mentioned previously, the carotenoids present in tomato-based products constitute natural pigments responsible for the coloring of tomato fruits. They can therefore be used as natural colorants in the dyeing of a wide range of food materials [4,5,7,14,21]. This natural pigmentation of carotenoids has been explained as being the consequence of the presence of conjugated double bonds [4,14]. It has been reported that the carotenoids that provide an attractive color to foods are based structurally on the α-, β-carotenes and lycopene [136]. In fact, lycopene which is the main carotenoid present in tomato products is considered as the chief coloring component of red tomato fruits seeing as it accounts for around 80% to 90% of the natural tomato-derived pigments [20,21].

#### 3.4.1.5 Gelling properties

Pectins, polysaccharides that are naturally present in tomato-based materials are well recognized as powerful gelling agents, thickeners and stabilizers. Indeed, the gelling property of pectin has been often employed in food formulation as well as in pharmaceutical industry [137–140]. Nonetheless, despite the many commercial applications of pectins, only few bio-resources are used to generate this valuable compound due to the fact that a lot of the pectic substances present in nature lack the functional characteristics, namely the capacity to gel acid sugar systems, which constitutes one of the key requirements of commercial pectins. This is why tomato-derived pectin is particularly attractive seeing as it not only displays good gelling properties but also presents similar, if not slightly better, physicochemical properties as those of the commercial high-methoxyl citrus-derived pectin (HMC) [71,141].

### 3.4.2 Applications in food formulation

The use of additives in food products is essential in order to preserve flavor and color, lengthen shelf life by avoiding oxidative degradation, and improve texture as well as consistency. However, considering that various studies have revealed a correlation between synthetic additives consumption and the occurrence of some health problems, like skin allergies, gastrointestinal tract troubles and even higher risk of cancer [142–146], more and more efforts are being turned nowadays toward the use of natural additives whose advantages are abundant: (1) they are easily embraced by consumers, (2) they are regarded as harmless, (3) no risk assessment tests are called for by legislation, and (4) they will provide food products with supplementary nutraceutical value compared to artificial additives.

Thus, the valorization of tomato processing by-products through their potential incorporation in food formulations appears as an attractive strategy. In the next section, several food products obtained by the inclusion of tomato processing waste will be presented and the impact of adding the tomato by-products will be exposed in detail. The results of all conducted investigations related to this subject have been reported in Table 3.1.

#### *3.4.2.1 Meat products*

Lipid oxidation represents the primary non-microbial origin of spoilage of meat products. Thus, oxidative processes worsen meat quality by lessening its nutritional value and generating various harmful entities that can bring about several disorders and damage the sensory property of meat [147]. In this sense, color is also a key factor in gaining buyers acceptance. This is why the addition of antioxidants and colorants is regarded as a necessity in the meat industry. Seeing as the bioactive compounds present in tomato by-products are characterized by remarkable antioxidant abilities and intense red coloring, their substitution of synthetic antioxidants and colcrants in the formulation of meat products appears to be highly beneficial and has been the subject of multiple papers.

A resourceful technique for producing dry fermented sausages with superior lycopene amount consisted in the insertion of dried tomato peels within the meat product. Indeed, four batches of traditional Spanish salchichón were manufactured, one without tomato peel and hence

**Table 3.1** Food materials reformulated with tomato by-products.

| Food supplemented with tomato by-product | Incorporated tomato by-product | Amount of incorporated by-product% (w/w) | Enhanced properties | References |
|---|---|---|---|---|
| Dry fermented sausages | Dry tomato peel | 0.6, 0.9, and 1.2 | High lycopene content<br>Good sensory properties<br>Good textural properties | [148] |
| Beef hamburgers | Dry tomato peel | 4.5 | High lycopene content<br>Good overall acceptability | [149] |
| Minced chicken meat | Tomato waste (skin + seeds) | 0.3 | Protection against lipid oxidation | [151] |
| Beef frankfurter | Tomato pomace | 1, 3, 5, and 7 | Better sensory properties | [152] |
| Beef ham | Tomato pomace | 1, 3, 5, and 7 | Better sensory properties | [152] |
| Lamb meat patties | Aqueous extracts from tomato by-products | 0.1 | Reduction of microbial counts | [153] |
| Flat bread | Tomato pomace | 1, 3, 5, and 7 | Higher moisture content<br>Softer texture<br>Delayed staling | [154] |
| Wheat bread | De-oiled seed meal | 10, 20, and 30 | Improvement of the overall protein content and quality<br>Delayed staling | [155] |
| Pasta and noodle products | Tomato peel | 15 | High levels of carotenoids<br>High level of dietary fiber | [156] |
| Soda crackers | Tomato pomace | 4, 8, and 12 | Higher content of protein, ash, fiber, minerals, total phenolics and antioxidant capacity | [157] |

| | | | | |
|---|---|---|---|---|
| Cookies | Tomato pomace | 5, 10, 15, 20, and 25 | Higher content of crude protein and ash | [158] |
| Traditional Tunisian butter | Tomato by-product extract | 0.4 and 0.8 | Antioxidant property | [159] |
| Olive oil | De-frosted and freeze-dried tomato by-product | De-frosted: 2.5 and 1.9 Freeze-dried: 0.6 and 0.45 | Enrichment in carotenoids | [161] |
| Refined olive oil, extra virgin olive oil and refined sunflower oil | Tomato peel | 5 and 10 | Upgrade of edible oils quality owing to enrichment in lycopene and β-carotene | [162] |
| Olive and sunflower oils | Tomato peels oleoresin | 2.5, 5, 10, and 20 | High oxidative stabilization of refined olive and sunflower oils due to an increased lycopene content | [163] |
| Low calorie jams | Tomato pomace | 13.4, 16.4, and 17.6 | Lower total carbohydrate content High dietary fiber content | [164] |
| Ice cream | Lyco-red extract from tomato peel | 1, 2, 3, 4, and 5 | Improved sensory properties | [165] |

considered as control sample, and three with different amounts of incorporated dry tomato peel: namely 6, 9 and 12 g/kg of lyophilized tomato peels within the meat mixture [148]. The tomato peel-supplemented sausages exhibited notable variances in terms of color and sensory characteristics in relation to the control batch but displayed satisfactory overall acceptability, thus revealing the potentiality of adding tomato peel to dry fermented sausages in the aim of generating a lycopene-rich meat product. The same was also true in the case of beef hamburgers enriched in lycopene through the utilization of dry tomato peel as an ingredient [149], in which the authors found that the incorporation of dry tomato peel up to 4.5% (w/w) led to hamburgers with good overall acceptability and high lycopene amount (4.9 mg of lycopene per 100 g of the meat product), close to the daily recommended intake of this carotenoid to protect against the impact of oxidative stress and inhibit some chronic afflictions [150].

Another study conducted over minced chicken meat supplemented with tomato waste comprised of skin and seeds indicated that the addition of 0.30% (w/w) tomato by-product resulted in a delay of 6 days in terms of formation of secondary oxidation products, thereby signaling the efficiency of tomato waste in protecting against lipid oxidation owing to its excellent antioxidant property [151].

Furthermore, an inquiry into the consequences of incorporating tomato pomace in the production of beef frankfurter and beef ham showed that following a sensory evaluation by a group of panelists, products containing tomato pomace obtained good sensory scores. This would suggest good consumer acceptability and maybe even preference for these tomato pomace-supplemented meat products which will also have the added-value of presenting higher fiber and protein contents [152].

Last but not least, an investigation of the antimicrobial potential of aqueous extracts obtained from tomato by-products when added to lamb meat patties was performed by Andrés et al. [153]. The authors found that microbial counts were reduced by the incorporation of the aqueous extracts within lamb meat patties, thereby demonstrating the efficiency of tomato waste in lengthening the shelf-life of this meat product.

Based on these different studies, it is clear that reformulation of meat products with tomato processing waste enhanced the nutritional value, retarded lipid oxidation as well as microbial growth and maintained or even improved the sensory characteristics and overall acceptability of the resulting material. Therefore, tomato by-products can be regarded as an attractive source of natural additives in the meat industry.

#### 3.4.2.2 Grain products

Bread constitutes a fundamental energy source worldwide. Fresh bread is savory and characterized by a crispy crust combined with a soft crumb. However, these attractive sensory attributes are short-termed and a couple of days after baking, a deterioration of these properties begins to take place leading to the phenomenon known as bread staling. Thus, discovering some means of improving bread shelf life is of tremendous consequence for bread manufacturers.

Studies revolving around the effects of using tomato processing by products in the preparation of bread have been performed by several authors [154,155]. Majzoobi et al. [154] found that bread supplemented with tomato by-products up to 5% (w/w) presented higher moisture content, softer texture and delayed staling, with no adverse repercussions on either color or taste compared to conventionally prepared bread. As to Sogi et al. [155], their research revealed that when 10% (w/w) deoiled seed meal derived from tomato processing waste is added to wheat flour, the resulting bread displayed good sensory properties and enhanced protein quality, rendering thereby the incorporation of tomato by-product in wheat flour suitable for bread making.

On another note, tomato by-products have also been examined as additives for pasta, a product whose consumption has increased greatly in recent years owing to the facility of its transportation, cooking and affordable price [156]. In fact, the ever increasing customer demands of better-quality food with higher nutritional value has pushed researchers towards investigating tomato by-products additives as a potential source of nutrient supplements. It was found that pasta enriched with tomato peels exhibited high amounts of carotenoids and dietary fibers. However, it was necessary to incorporate hydrocolloids as well to avoid any deterioration in the sensory quality of the final product.

Moreover, the effects of using tomato pomace, rich in bioactive compounds, in the formulation of soda crackers was looked into [157]. Results indicated that a partial substitution of wheat flour with dried tomato pomace (4%, 8%, and 12%) induced a rise in the content of protein, ash, dietary fiber, minerals, phenolic compounds and carotenoids which ultimately translated into higher antioxidant property. Furthermore, regarding sensory evaluation, crackers prepared with tomato pomace scored similarly in terms color, smell, flavor, crispiness, and overall acceptability to the control sample. Nevertheless, the replacement of more than 12% (w/w) wheat flour with tomato pomace was ill-advised according to sensory analysis.

Finally, the physicochemical properties of cookies prepared with different proportions of tomato pomace were probed [158]. Sensory evaluation indicated that cookies substituted up to 5% (w/w) with tomato pomace scored good overall desirability scores and were accepted by consumers.

#### 3.4.2.3 Dairy and oil products

Dairy products play an important role in the customary diet of people, seeing as they represent a major section of the basic food groups and that a large proportion of these products is prepared in farms by relying on traditional dairy technology. Traditional Tunisian butter, a highly valued dairy product in Tunisia, suffers from deteriorating quality following prolonged period of storage, due mostly to lipid oxidation reactions. Thus, the use of tomato processing by-products, characterized by high antioxidant activity, was regarded as a promising opportunity for protection against lipid peroxidation, allowing thus to increase butter storage stability. The study performed by Abid et al. [159] revealed that tomato by-products succeeded in extending the shelf-life of traditional Tunisian butter but only when the added amount didn't exceed 400 mg of tomato-derived material/kg of traditional Tunisian butter, otherwise the incorporation of tomato by-product would prove detrimental to the oxidative stabilization of butter.

Edible oils which are heavily consumed all around the world constitute important contributors to the renowned Mediterranean diet, but they can be subject to lipid oxidation in the course of various phases going from production, to storage, then purchase and finally consumption [160]. Hence, the recognized need for additives with antioxidant properties, of preferably natural origin, in order to avoid any synthetic additive-associated health problem. In this context, a study in which olives were co-milled with de-frosted or freeze-dried tomato by-products in the aim of obtaining an olive oil naturally enriched with antioxidants was conducted by Bendini et al. [161]. The authors found that the adopted method was effective in generating olive oil enriched in carotenoids, namely lycopene, acclaimed for their remarkable antioxidant abilities. This was again true in the case of a study conducted by Benakmoum et al. [162] who discovered that the incorporation of tomato peel in refined olive oil, extra virgin olive oil and refined sunflower oil resulted in an upgrade of edible oils quality owing to an enrichment in lycopene and β-carotene. Similarly, the work of Kehili et al. [163] revealed that the addition of tomato peels oleoresin to olive and sunflower oils led to high oxidative stabilization of these oils due to an increased lycopene content.

However, it is important to mention that tomato processing by products can serve on their own as natural sources of edible oil with physicochemical characteristics similar if not better than commercial seed oils [18,23,75]. In fact, Westphal et al. [23] showed that the amount of vitamin E in tomato seed oils was five times higher than it was in olive oil and twice as high as their amount in cold-pressed sunflower oil. Furthermore, by conducting a comparison of the fatty acid profiles of two tomato varieties (Waltinger and Red Currant) with those of commercial plant oils, as illustrated in Fig. 3.4, the authors found that the tomato seed oil was very alike soybean oil, with the latter presenting nevertheless higher proportions of α-linolenic acid. Accordingly, the fatty acid composition of tomato seed oil basically corresponds to that of low linolenic soybean oil.

#### 3.4.2.4 Sweet products

Jam is considered as one of the most sought-after fruit preserves. In the interest of providing low-calorie food products while maintaining similar textural and sensory properties as those of the traditional products, low calorie jams were developed from tomato pomace [164]. Results showed that tomato pomace jams held 15 to 20 times higher content of dietary fiber than what is usually found in commercial apricot jams, in addition to exhibiting inferior energy values and carbohydrate amounts.

Evaluation of functional ice cream manufactured with varied proportions of carotenoid extract derived from tomato peels was carried out by Rizk et al. [165]. Lyco-red, comprised of nine carotenoid compounds, extracted from tomato peels proved to be highly beneficial in terms of flavor, melting, color, body and texture of obtained ice creams when added in ratios of 2% and 3%. Hence, based on the improved sensory properties of the final product, ice creams supplemented with tomato by-product extract should have good consumer acceptability.

### 3.4.3 Applications in cosmetic and pharmaceutical industry

The demand for cosmetic products with bioactive ingredients or "cosmeceuticals" has been ever-growing these last few years. Indeed, consumers nowadays are becoming increasingly attentive to the compounds present in the formulation of their beauty products because of their awareness of the existence of nutritional products that can promote skin health while also reducing the risk of diseases occurrence. As an answer for this emerging need, nutricosmetics, new types of products resulting from the convergence between the cosmetic and food industries, have been recently explored.

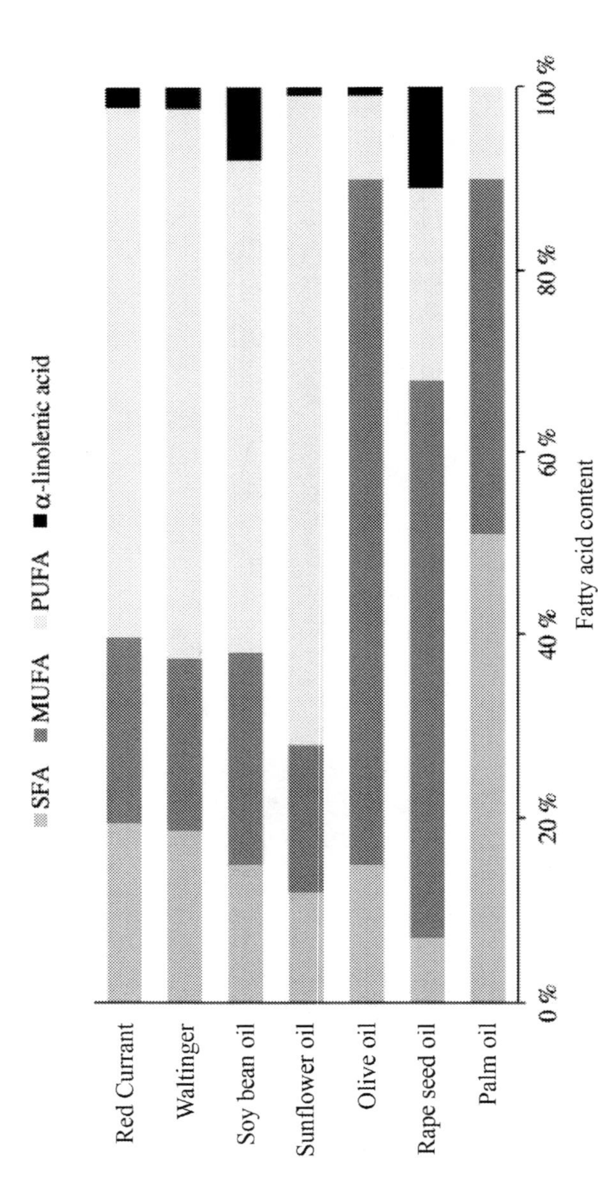


**Fig. 3.4** *Fatty acid profiles of different plant oils in comparison with the oil extracted from the two seed varieties, Red Currant and Waltinger (values standardized to 100%). SFA: saturated fatty acids, MUFA: monounsaturated fatty acids, PUFA: polyunsaturated fatty acids.*

These products, termed as well "beauty pills", consist in oral supplementation of nutrients with the aim of inducing an appearance improvement. Among the objectives fixed by nutricosmetics figures mainly the access to antiaging effects with a decrease of wrinkles, which will be achieved through the protection against free radicals generated by solar radiation. This is why antioxidants are regarded as most vital ingredients in the formulation of nutricosmetics. Carotenoids and phenolic compounds, both prominent antioxidants, are significantly present in tomato by-products, which makes this processing waste highly useful as a source of active ingredients in nutricosmetics [15,117,166–169]. In fact, clinical studies have shown carotenoids to play a crucial role in photoprotection against UV radiation. Mathews-Roth et al. [170] were among the firsts to report on the photoprotection function of β-carotene, which was later further explored by several groups of authors [171–173] who all found that high intake of β-carotene succeeds in protecting the skin against UV-caused erythema. Moreover, besides β-carotene, numerous papers have also been focused on the skin protection properties of lycopene, the main carotenoid of tomato, characterized by pronounced singlet-oxygen quenching ability [168,174–177]. The results of all these researches converged towards a similar conclusion stating that lycopene is able to reduce inflammation and assist in preventing skin damage induced by UV sun exposure. The same was also true for phenolic compounds which can act as protective agents against various skin disorders [169,178,179]. Therefore, the extensive potentialities of the extracts derived from tomato disposable wastes in the cosmetic field makes their recovery economically profitable, while providing at the same time inexpensive and bio-sustainable ingredients for the manufacturing of nutricosmetics. The remarkable properties that are at the origin of tomato processing by-products suitability for cosmetic-related applications, justify their additional use in pharmaceutical industries, where the antioxidant, antimicrobial, anti-cancer, cardioprotective and anti-inflammatory effects will prove to be highly valuable [179,180]. Furthermore, it should also be noted that the gelling property of tomato waste-derived pectin makes it very desirable as binding agent in tablet formulations as well as carrier of a number of drugs [84].

To sum up, in this third section of Chapter III, we have focused on what can be deemed as "health-related applications" of the valuable compounds derived from tomato waste, which essentially involve carotenoids and phenolic compounds. Nevertheless, it should be noted that many more applications for tomato processing by-products have been proposed in literature such as their use for the production of low-cost biosorbent that can efficiently remove cationic and anionic dyes [17]; their application in food packaging and as tin corrosion inhibitors [84,92,181]; their utilization for energy production by serving as feedstock for biogas production [12,182]; and their application as fertilizer for crop growth [12,183]. All of which goes to show the great interest that resides in the valorization of these tomato processing by-products.

## 3.5 Summary and perspectives

In the last decade, the yearly amount of industrially processed tomatoes has exceeded 30 million tons, resulting in a considerable quantity of tomato by-products which ultimately ended up as waste. Procuring answers for alternative applications of tomato by-products in response to increasing population number and recent waste management requirements is a fascinating topic for a lot of scientists and industrialists.

Lately, the extraction of valuable compounds presents in tomato processing by-products, primarily carotenoids and phenolic compounds, along with some other ingredients (vitamins, dietary fibers, proteins, essential oils...) aroused significant attention. The different extraction techniques (CSE, EAE, UAE, MAE, HHPE, SFE, PEF...) employed nowadays have been reported in this chapter, with their advantages and limitations clearly outlined. Based on the results of various researches, it was concluded that the efficiency of the recovery of bioactive ingredients from tomato processing by-products depended on the extraction process employed and the adopted parameters (solvent type, solvent/solid ratio, temperature, duration of the extraction, particle size of the sample ...). Therefore, an optimization of all these factors is required to maximize the recovery yield.

The remarkable properties of the valuable components present in tomato by-products, namely their antioxidant, antimicrobial, cancer and cardiovascular disease preventive activities offer them a wide range of utilization perspectives. Thus, valorization of tomato processing waste can be firstly accomplished through the incorporation of the nutritive components derived from tomato by-products extracts in several foods such as meat, grain, dairy, oil and sweet products. Secondly, the biological properties of tomato waste-derived ingredients make them perfectly suitable for the formulation of various cosmeceuticals such as UV photoprotection and anti-aging creams.

Currently, a rising number of clinical and pharmacologic investigations is paving the way for many more promising applications for the bioactive ingredients isolated from tomato processing by-products. This will contribute in providing a sustainable strategy that will promote both bio-economy and environmental management.

## References

[1] FAOSTAT, 2019. http://www.fao.org/faostat/en/#data/QC. (Accessed 4 January 2021).

[2] WPTC, World Production Estimate of Tomato for Processing (WPTC), Avignon (France), 2017. https://www.wptc.to/pdf/releases/WPTC%20crop%20update%20as%20of%2022%20October%202020. (Accessed 4 January 2021).

[3] E. Luengo, S. Condón-Abanto, S. Condón, I. Álvarez, J. Raso, Improving the extraction of carotenoids from tomato waste by application of ultrasound under pressure, Sep. Purif. Technol. 136 (2014) 130–136.

[4] I.F. Strati, V. Oreopoulou, Process optimisation for recovery of carotenoids from tomato waste, Food Chem. 129 (3) (2011) 747–752.

[5] I.F. Strati, V. Oreopoulou, Recovery of carotenoids from tomato processing by-products–a review, Food Res. Int. 65 (Part C) (2014) 311–321.

[6] K. Szabo, Z. Diaconeasa, A.-.F. Cătoi, D.C. Vodnar, Screening of ten tomato varieties processing waste for bioactive components and their related antioxidant and antimicrobial activities, Antioxidants 8 (8) (2019) 292.

[7] H.M. El Mashad, L. Zhao, R. Zhang, T. PZ, Integrated Processing Technologies for Food and Agricultural By-Products, Elsevier, London, 2019, pp. 107–131.

[8] A. Zuorro, M. Fidaleo, R. Lavecchia, Enzyme-assisted extraction of lycopene from tomato processing waste, Enzyme Microb. Technol. 49 (6–7) (2011) 567–573.

[9] D. Kaur, A.A. Wani, D.P.S. Oberoi, DS. Sogi, Effect of extraction conditions on lycopene extractions from tomato processing waste skin using response surface methodology, Food Chem. 108 (2) (2008) 711–718.

[10] A.N. Grassino, M. Brnčić, D. Vikić-Topić, S. Roca, M. Dent, SR. Brnčić, Ultrasound assisted extraction and characterization of pectin from tomato waste, Food Chem. 198 (2016) 93–100.

[11] S. Stajčić, G. Ćetković, J. Čanadanović-Brunet, S. Djilas, A. Mandić, D. Četojević-Simin, Tomato waste: carotenoids content, antioxidant and cell growth activities, Food Chem. 172 (2015) 225–232.

[12] C. Fritsch, A. Staebler, A. Happel, M.A. Cubero Márquez, I. Aguiló-Aguayo, M. Abadias, et al., Processing, valorization and application of bio-waste derived compounds from potato, tomato, olive and cereals: a review, Sustainability 9 (8) (2017) 1492.

[13] J. Pinela, M.A. Prieto, M.F. Barreiro, A.M. Carvalho, M.B.P. Oliveira, T.P. Curran, et al., Valorisation of tomato wastes for development of nutrient-rich antioxidant ingredients: A sustainable approach towards the needs of the today's society, Innov Food Sci. Emerg. Technol. 41 (2017) 160–171.

[14] T. Baysal, S. Ersus, DAJ. Starmans, Supercritical $CO_2$ extraction of β-carotene and lycopene from tomato paste waste, J Agric. Food Chem. 48 (11) (2000) 5507–5511.

[15] A. Zuorro, R. Lavecchia, F. Medici, L. Piga, Use of cell wall degrading enzymes for the production of high-quality functional products from tomato processing waste, Chem. Eng. 38 (2014) 355–360.

[16] E. Elbadrawy, A. Sello, Evaluation of nutritional value and antioxidant activity of tomato peel extracts, Arab. J. Chem. 9 (Supplement 2) (2016) S1010–S1018.

[17] S. Azabou, I. Louati, F.B. Taheur, M. Nasri, T. Mechichi, Towards sustainable management of tomato pomace through the recovery of valuable compounds and sequential production of low-cost biosorbent, Environ. Sci. Pollut. Res. 27 (31) (2020) 39402–39412.

[18] E. Vági, B. Simándi, K.P. Vásárhelyiné, H. Daood, Á. Kéry, F. Doleschall, et al., Supercritical carbon dioxide extraction of carotenoids, tocopherols and sitosterols from industrial tomato by-products, J. Supercrit Fluids 40 (2) (2007) 218–226.

[19] V. Nour, T.D. Panaite, M. Ropota, R. Turcu, I. Trandafir, AR. Corbu, Nutritional and bioactive compounds in dried tomato processing waste, CyTA-J Food 16 (1) (2018) 222–229.

[20] H. Al-Wandawi, M. Abdul-Rahman, K. Al-Shaikhly, Tomato processing wastes as essential raw materials source, J. Agric. Food Chem. 33 (5) (1985) 804–807.

[21] R. Domínguez, P. Gullón, M. Pateiro, P.E. Munekata, W. Zhang, JM. Lorenzo, Tomato as potential source of natural additives for meat industry. A review, Antioxidants. 9 (1) (2020) 73.

[22] S. Gharbi, G. Renda, L. La Barbera, M. Amri, C.M. Messina, A. Santulli, Tunisian tomato by-products, as a potential source of natural bioactive compounds, Nat Prod Res 31 (6) (2017) 626–631.

[23] A. Westphal, J. Bauerfeind, C. Rohrer, V. Böhm, Analytical characterisation of the seeds of two tomato varieties as a basis for recycling of waste materials in the food industry, Eur. Food Res. Technol. 239 (4) (2014) 613–620.

[24] K. Ray, T.N. Misra, Photophysical properties of lycopene organized in Langmuir-Blodgett films: formation of aggregates, J. Photochem Photobiol. Chem. 107 (1–3) (1997) 201–205.

[25] P. Di Mascio, S. Kaiser, H. Sies, Lycopene as the most efficient biological carotenoid singlet oxygen quencher, Arch. Biochem. Biophys. 274 (2) (1989) 532–538.

[26] S. Przybylska, Lycopene–a bioactive carotenoid offering multiple health benefits: a review, Int. J. Food Sci. Technol. 55 (1) (2020) 11–32.

[27] J. Shi, M.L. Maguer, Lycopene in tomatoes: chemical and physical properties affected by food processing, Crit. Rev. Food Sci. Nutr. 40 (1) (2000) 1–42.

[28] P. Perkins-Veazie, JK. Collins, Flesh quality and lycopene stability of fresh-cut watermelon, Postharvest Biol. Technol. 31 (2) (2004) 159–166.

[29] M.B. Matlack, Pigments of pink grapefruits, Citrus grandis (L., Osbeck), J. Biol. Chem. 110 (1935) 249–253.

[30] B.J. Lime, F.P. Griffiths, O.'. RT, D.C. Heinzelman, ER. McCall, Grapefruit pigment determination, spectrophotometric methods for determining pigmentation-beta-carotene and lycopene-in ruby red grapefruit, J. Agric. Food Chem. 5 (12) (1957) 941–944.

[31] A.G. Vasconcelos, A. das GN Amorim, R.C. dos Santos, J.M.T. Souza, L.K.M. de Souza, T. de SL Araújo, et al., Lycopene rich extract from red guava (Psidium guajava L.) displays anti-inflammatory and antioxidant profile by reducing suggestive hallmarks of acute inflammatory response in mice, Food Res. Int. 99 (Part 2) (2017) 959–968.

[32] V. Böhm, K. Fröhlich, R. Bitsch, Rosehip—a "new" source of lycopene? Mol Aspects Med. 24 (6) (2003) 385–389.

[33] J. Pól, T. Hyötyläinen, O. Ranta-Aho, M-L. Riekkola, Determination of lycopene in food by on-line SFE coupled to HPLC using a single monolithic column for trapping and separation, J. Chromatogr. A. 1052 (1–2) (2004) 25–31.

[34] S.K. Sharma, M. Le Maguer, Lycopene in tomatoes and tomato pulp fractions, Ital. J. Food Sci. 8 (2) (1996) 107–113.

[35] X. Gong, R. Marisiddaiah, LP. Rubin, Inhibition of pulmonary β-carotene 15, 15'-oxygenase expression by glucocorticoid involves PPARα, Plos one 12 (7) (2017) e0181466.

[36] D. Weber, T. Grune, The contribution of β-carotene to vitamin A supply of humans, Mol. Nutr. Food Res. 56 (2) (2012) 251–258.
[37] A. Ghazi, Extraction of β-carotene from orange peels, Food/Nahrung 43 (4) (1999) 274–277.
[38] B. İncedayi, C.E. Tamer, G.Ö. Sinir, S. Suna, ÖU. Çopur, Impact of different drying parameters on color, β-carotene, antioxidant activity and minerals of apricot (Prunus armeniacaL, Food Sci. Technol. 36 (1) (2016) 171–178.
[39] A.A. Abushita, E.A. Hebshi, H.G. Daood, PA. Biacs, Determination of antioxidant vitamins in tomatoes, Food Chem. 60 (2) (1997) 207–212.
[40] C.A. Tavares, D.B. Rodriguez-Amaya, Carotenoid composition of Brazilian tomatoes and tomato products, LWT-Food Sci. Technol. 27 (3) (1994) 219–224.
[41] S.A. Desobry, F.M. Netto, T.P. Labuza, Preservation of β-carotene from carrots, Crit. Rev. Food Sci. Nutr. 38 (5) (1998) 381–396.
[42] T. Muzhingi, K.-J. Yeum, A.H. Siwela, O. Bermudez, G. Tang, Identification of enzymatic cleavage products of β-carotene-rich extracts of kale and biofortified maize, Int. J. Vitam. Nutr. Res. 87 (5-6) (2018) 279–286.
[43] A. Altemimi, D.A. Lightfoot, M. Kinsel, D.G. Watson, Employing response surface methodology for the optimization of ultrasound assisted extraction of lutein and β-carotene from spinach, Molecules 20 (4) (2015) 6611–6625.
[44] M.W. Farnham, G.E. Lester, H.R. Collard, mustard and turnip greens: Effects of genotypes and leaf position on concentrations of ascorbic acid, folate, β-carotene, lutein and phylloquinone, J. Food Compos. Anal. 27 (1) (2012) 1–7.
[45] M. Baranska, W. Schütze, H. Schulz, Determination of lycopene and β-carotene content in tomato fruits and related products: comparison of FT-Raman, ATR-IR, and NIR spectroscopy, Anal. Chem. 78 (24) (2006) 8456–8461.
[46] N. Kalogeropoulos, A. Chiou, V. Pyriochou, A. Peristeraki, V.T. Karathanos, Bioactive phytochemicals in industrial tomatoes and their processing byproducts, LWT-Food Sci. Technol. 49 (2) (2012) 213–216.
[47] J.J. Castenmiller, C.E. West, J.P. Linssen, K.H. van het Hof, A.G. Voragen, The food matrix of spinach is a limiting factor in determining the bioavailability of β-carotene and to a lesser extent of lutein in humans, J. Nutr. 129 (2) (1999) 349–355.
[48] A. Becerra-Moreno, P.A. Alanís-Garza, J.L. Mora-Nieves, J.P. Mora-Mora, J.-V. DA, Kale: An excellent source of vitamin C, pro-vitamin A, lutein and glucosinolates, CyTA-J Food 12 (3) (2014) 298–303.
[49] E.R. Kelly, J. Plat, G.R. Haenen, A. Kijlstra, T.T. Berendschot, The effect of modified eggs and an egg-yolk based beverage on serum lutein and zeaxanthin concentrations and macular pigment optical density: results from a randomized trial, Plos. one 9 (3) (2014) e92659.
[50] R.L. Ausich, D.J. Sanders, Process for the formation, isolation and purification of comestible xanthophyll crystals from plants, Google Patents; (1997).
[51] R. Tsao, R. Yang, J.C. Young, H. Zhu, T. Manolis, Separation of geometric isomers of native lutein diesters in marigold (Tagetes erecta L.) by high-performance liquid chromatography–mass spectrometry, J. Chromatogr. A. 1045 (1–2) (2004) 65–70.
[52] J.-H. Lin, D.-J. Lee, J-S. Chang, Lutein production from biomass: marigold flowers versus microalgae, Bioresour Technol 184 (2015) 421–428.
[53] R. Kumar, W. Yu, C. Jiang, C. Shi, Y. Zhao, Improvement of the isolation and purification of lutein from marigold flower (Tagetes erecta L.) and its antioxidant activity, J. Food Process Eng. 33 (6) (2010) 1065–1078.
[54] M. Knoblich, B. Anderson, D. Latshaw, Analyses of tomato peel and seed byproducts and their use as a source of carotenoids, J. Sci. Food Agric. 85 (7) (2005) 1166–1170.
[55] D. Montesano, O. Gennari, S. Seccia, S. Albrizio, A simple and selective analytical procedure for the extraction and quantification of lutein from tomato by-products by HPLC–DAD, Food Anal. Methods 5 (4) (2012) 710–715.

[56] G. Ćetković, S. Savatović, J. Čanadanović-Brunet, S. Djilas, J. Vulić, A. Mandić, et al., Valorisation of phenolic composition, antioxidant and cell growth activities of tomato waste, Food Chem. 133 (3) (2012) 938–945.
[57] X.P. Perea-Domínguez, L.Z. Hernández-Gastelum, H.R. Olivas-Olguin, L.G. Espinosa-Alonso, M. Valdez-Morales, S. Medina-Godoy, Phenolic composition of tomato varieties and an industrial tomato by-product: free, conjugated and bound phenolics and antioxidant activity, J. Food Sci. Technol. 55 (9) (2018) 3453–3461.
[58] I. Navarro-González, V. García-Valverde, J. García-Alonso, MJ. Periago, Chemical profile, functional and antioxidant properties of tomato peel fiber, Food Res. Int. 44 (5) (2011) 1528–1535.
[59] M. Valdez-Morales, L.G. Espinosa-Alonso, L.C. Espinoza-Torres, F. Delgado-Vargas, S. Medina-Godoy, Phenolic content and antioxidant and antimutagenic activities in tomato peel, seeds, and byproducts, J. Agric. Food Chem. 62 (23) (2014) 5281–5289.
[60] S. Savatović, G. Ćetković, J. Čanadanović-Brunet, S. Djilas, Tomato waste: a potential source of hydrophilic antioxidants, Int. J. Food Sci. Nutr. 63 (2) (2012) 129–137.
[61] K. Jayathunge, A.C. Stratakos, G. Delgado-Pando, A. Koidis, Thermal and non-thermal processing technologies on intrinsic and extrinsic quality factors of tomato products: a review, J. Food Process Preserv. 43 (3) (2019) e13901.
[62] S. Georgé, F. Tourniaire, H. Gautier, P. Goupy, E. Rock, C. Caris-Veyrat, Changes in the contents of carotenoids, phenolic compounds and vitamin C during technical processing and lyophilisation of red and yellow tomatoes, Food Chem. 124 (4) (2011) 1603–1611.
[63] D. Pérez-Conesa, J. García-Alonso, V. García-Valverde, M.-.D. Iniesta, K. Jacob, L.M. Sánchez-Siles, et al., Changes in bioactive compounds and antioxidant activity during homogenization and thermal processing of tomato puree, Innov. Food Sci. Emerg. Technol. 10 (2) (2009) 179–188.
[64] A. Raiola, G.C. Tenore, A. Barone, L. Frusciante, MM. Rigano, Vitamin E content and composition in tomato fruits: beneficial roles and bio-fortification, Int. J. Mol. Sci. 16 (12) (2015) 29250–29264.
[65] P.G. Herrera, M.C. Sánchez-Mata, M. Cámara, Nutritional characterization of tomato fiber as a useful ingredient for food industry, Innov. Food Sci. Emerg. Technol. 11 (4) (2010) 707–711.
[66] K. Szabo, A.-.F. Cătoi, D.C. Vodnar, Bioactive compounds extracted from tomato processing by-products as a source of valuable nutrients, Plant Foods Hum. Nutr. 73 (4) (2018) 268–277.
[67] F. Saura-Calixto, Antioxidant dietary fiber product: a new concept and a potential food ingredient, J. Agric. Food Chem. 46 (10) (1998) 4303–4306.
[68] P. Chantaro, S. Devahastin, N. Chiewchan, Production of antioxidant high dietary fiber powder from carrot peels, LWT-Food Sci. Technol. 41 (10) (2008) 1987–1994.
[69] E. Lecumberri, R. Mateos, M. Izquierdo-Pulido, P. Rupérez, L. Goya, L. Bravo, Dietary fibre composition, antioxidant capacity and physico-chemical properties of a fibre-rich product from cocoa (Theobroma cacao L.), Food Chem. 104 (3) (2007) 948–954.
[70] K.H. Caffall, D. Mohnen, The structure, function, and biosynthesis of plant cell wall pectic polysaccharides, Carbohydr. Res. 344 (14) (2009) 1879–1900.
[71] M.M. Alancay, M.O. Lobo, C.M. Quinzio, LB. Iturriaga, Extraction and physico-chemical characterization of pectin from tomato processing waste, J. Food Meas. Charact. 11 (4) (2017) 2119–2130.
[72] G.N. Liadakis, C. Tzia, V. Oreopoulou, CD. Thomopoulos, Protein isolation from tomato seed meal, extraction optimization, J. Food Sci. 60 (3) (1995) 477–482.
[73] Y. PA Silva, B.C. Borba, V.A. Pereira, M.G. Reis, M. Caliari, M.S.-.L. Brooks, et al., Characterization of tomato processing by-product for use as a potential functional food ingredient: nutritional composition, antioxidant activity and bioactive compounds, Int. J. Food Sci. Nutr. 70 (2) (2019) 150–160.

[74] D. Brodowski, J.R. Geisman, Protein content and amino acid composition of protein of seeds from tomatoes at various stages of ripeness, J. Food Sci. 45 (2) (1980) 228–229.
[75] C. Botineştean, A.T. Gruia, I. Jianu, Utilization of seeds from tomato processing wastes as raw material for oil production, J. Mater. Cycles Waste Manag. 17 (1) (2015) 118–124.
[76] M.M. Poojary, P. Passamonti, Extraction of lycopene from tomato processing waste: kinetics and modelling, Food Chem. 173 (2015) 943–950.
[77] D.M. Phinney, J.C. Frelka, J.L. Cooperstone, S.J. Schwartz, D.R. Heldman, Effect of solvent addition sequence on lycopene extraction efficiency from membrane neutralized caustic peeled tomato waste, Food Chem. 215 (2017) 354–361.
[78] I.F. Strati, V. Oreopoulou, Recovery and isomerization of carotenoids from tomato processing by-products, Waste Biomass Valori. 7 (4) (2016) 843–850.
[79] I.F. Strati, V. Oreopoulou, Effect of extraction parameters on the carotenoid recovery from tomato waste, Int. J. Food Sci. Technol. 46 (1) (2011) 23–29.
[80] M.M. Calvo, D. Dado, G. Santa-María, Influence of extraction with ethanol or ethyl acetate on the yield of lycopene, β-carotene, phytoene and phytofluene from tomato peel powder, Eur. Food Res. Technol. 224 (5) (2007) 567–571.
[81] R.M. Ruiz, V. Mangut, C. Gonzalez, T.R. de la, A. Latorre, Carotenoid extraction from tomato by-products, VII International Symposium on the Processing Tomato, 2000 83–90.
[82] M.J. Periago, F. Rincon, M.D. Agüera, G. Ros, Mixture approach for optimizing lycopene extraction from tomato and tomato products, J. Agric. Food Chem. 52 (19) (2004) 5796–5802.
[83] K. Szabo, F.V. Dulf, Z. Diaconeasa, D.C. Vodnar, Antimicrobial and antioxidant properties of tomato processing byproducts and their correlation with the biochemical composition, LWT 116 (2019) 108558.
[84] A.N. Grassino, J. Halambek, S. Djaković, S.R. Brnčić, M. Dent, Z. Grabarić, Utilization of tomato peel waste from canning factory as a potential source for pectin production and application as tin corrosion inhibitor, Food Hydrocoll. 52 (2016) 265–274.
[85] M.S. Lenucci, M. De Caroli, P.P. Marrese, A. Iurlaro, L. Rescio, V. Böhm, et al., Enzyme-aided extraction of lycopene from high-pigment tomato cultivars by supercritical carbon dioxide, Food Chem. 170 (2015) 193–202.
[86] M. Gu, H. Fang, Y. Gao, T. Su, Y. Niu, LL. Yu, Characterization of enzymatic modified soluble dietary fiber from tomato peels with high release of lycopene, Food Hydrocoll. 99 (2020) 105321.
[87] G. Catalkaya, D. Kahveci, Optimization of enzyme assisted extraction of lycopene from industrial tomato waste, Sep. Purif. Technol. 219 (2019) 55–63.
[88] S. Azabou, Y. Abid, H. Sebii, I. Felfoul, A. Gargouri, H. Attia, Potential of the solid-state fermentation of tomato by products by Fusarium solani pisi for enzymatic extraction of lycopene, LWT-Food Sci. Technol. 68 (2016) 280–287.
[89] R. Lavecchia, A. Zuorro, Improved lycopene extraction from tomato peels using cell-wall degrading enzymes, Eur. Food Res. Technol. 228 (1) (2008) 153.
[90] H. Feng, W. Yang, Ultrasonic processing. In: Nonthermal Processing Technologies for Food, Wiley-Blackwell and IFT Press, UK, 2011, pp. 135–154.
[91] L.K. Ultrasound. Its chemical, physical, and biological effects. Kenneth S. Suslick, Ed. VCH, New York, 1988 xiv, 336 pp., illus. $65. Science. 1989;243(4897):1499.
[92] K. Szabo, B.-.E. Teleky, L. Mitrea, L.-.F. Călinoiu, G.-.A. Martau, E. Simon, et al., Active packaging–poly (vinyl alcohol) films enriched with tomato by-products extract, Coatings 10 (2) (2020) 141.
[93] M.R. Ladole, R.R. Nair, Y.D. Bhutada, V.D. Amritkar, A.B. Pandit, Synergistic effect of ultrasonication and co-immobilized enzymes on tomato peels for lycopene extraction, Ultrason Sonochem 48 (2018) 453–462.

[94] N. Sengkhamparn, N. Phonkerd, Phenolic compound extraction from industrial tomato waste by ultrasound-assisted extraction, IOP Conference Series: Materials Science and Engineering, IOP Publishing, Shenzhen, China, 2019 012040.
[95] K.K. Ho, M.G. Ferruzzi, A.M. Liceaga, M.F. San Martín-González, Microwave-assisted extraction of lycopene in tomato peels: Effect of extraction conditions on all-trans and cis-isomer yields, LWT-Food Sci. Technol. 62 (1) (2015) 160–168.
[96] W. Routray, V. Orsat, Microwave-assisted extraction of flavonoids: a review, Food Bioprocess Technol. 5 (2) (2011) 409–424.
[97] Z. Lianfu, L. Zelong, Optimization and comparison of ultrasound/microwave assisted extraction (UMAE) and ultrasonic assisted extraction (UAE) of lycopene from tomatoes, Ultrason Sonochem. 15 (5) (2008) 731–737.
[98] A.N. Grassino, J. Ostojić, V. Miletić, S. Djaković, T. Bosiljkov, Z. Zorić, et al., Application of high hydrostatic pressure and ultrasound-assisted extractions as a novel approach for pectin and polyphenols recovery from tomato peel waste, Innov. Food Sci. Emerg. Technol. 64 (2020) 102424.
[99] A.N. Grassino, S. Pedisić, V. Dragović-Uzelac, S. Karlović, D. Ježek, T. Bosiljkov, Insight into high-hydrostatic pressure extraction of polyphenols from tomato peel waste, Plant Foods Hum. Nutr. 75 (3) (2020) 427–433.
[100] L. Ciurlia, M. Bleve, L. Rescio, Supercritical carbon dioxide co-extraction of tomatoes (Lycopersicum esculentum L.) and hazelnuts (Corylus avellana L.): A new procedure in obtaining a source of natural lycopene, J. Supercrit Fluids 49 (3) (2009) 338–344.
[101] E. Sabio, M. Lozano, V. Montero de Espinosa, R.L. Mendes, A.P. Pereira, A.F. Palavra, et al., Lycopene and β-carotene extraction from tomato processing waste using supercritical $CO_2$, Ind. Eng. Chem. Res. 42 (25) (2003) 6641–6646.
[102] B.P. Nobre, A.F. Palavra, F.L. Pessoa, R.L. Mendes, Supercritical $CO_2$ extraction of trans-lycopene from Portuguese tomato industrial waste, Food Chem. 116 (3) (2009) 680–685.
[103] F. Venturi, C. Sanmartin, I. Taglieri, G. Andrich, A. Zinnai, A simplified method to estimate Sc-$CO_2$ extraction of bioactive compounds from different matrices: chili pepper vs. tomato by-products, Appl. Sci. 7 (4) (2017) 361.
[104] P. Tm, V. Sicari, L. Mr, M. Leporini, T. Falco, M. Poiana, Optimizing the supercritical fluid extraction process of bioactive compounds from processed tomato skin by-products, Food Sci. Technol. 40 (3) (2019) 692–697.
[105] G. Perretti, A. Troilo, E. Bravi, O. Marconi, F. Galgano, P. Fantozzi, Production of a lycopene-enriched fraction from tomato pomace using supercritical carbon dioxide, J. Supercrit Fluids 82 (2013) 177–182.
[106] M. Kehili, M. Kammlott, S. Choura, A. Zammel, C. Zetzl, I. Smirnova, et al., Supercritical $CO_2$ extraction and antioxidant activity of lycopene and β-carotene-enriched oleoresin from tomato (Lycopersicum esculentum L.) peels by-product of a Tunisian industry, Food Bioprod Process 102 (2017) 340–349.
[107] A.F. Silva, M.M. de Melo, C.M. Silva, Supercritical solvent selection ($CO_2$ versus ethane) and optimization of operating conditions of the extraction of lycopene from tomato residues: Innovative analysis of extraction curves by a response surface methodology and cost of manufacturing hybrid approach, J. Supercrit Fluids 95 (2014) 618–627.
[108] D. Urbonaviciene, P. Viskelis, The cis-lycopene isomers composition in supercritical $CO_2$ extracted tomato by-products, LWT-Food Sci. Technol. 85 (Part B) (2017) 517–523.
[109] E. Luengo, I. Álvarez, J. Raso, Improving carotenoid extraction from tomato waste by pulsed electric fields, Front Nutr. 1 (2014) 12.
[110] G. Pataro, D. Carullo, M.A.B. Siddique, M. Falcone, F. Donsì, G. Ferrari, Improved extractability of carotenoids from tomato peels as side benefits of PEF treatment of tomato fruit for more energy-efficient steam-assisted peeling, J. Food Eng. 233 (2018) 65–73.

[111] V. Andreou, G. Dimopoulos, E. Dermesonlouoglou, P. Taoukis, Application of pulsed electric fields to improve product yield and waste valorization in industrial tomato processing, J. Food Eng. 270 (2020) 109778.
[112] A. Amiri-Rigi, S. Abbasi, M.G. Scanlon, Enhanced lycopene extraction from tomato industrial waste using microemulsion technique: optimization of enzymatic and ultrasound pre-treatments, Innov. Food Sci. Emerg. Technol. 35 (2016) 160–167.
[113] J. Nagarajan, H. Pui Kay, N.P. Krishnamurthy, N.R. Ramakrishnan, T. Aldawoud, C.M. Galanakis, et al., Extraction of carotenoids from tomato pomace via water-induced hydrocolloidal complexation, Biomolecules 10 (7) (2020) 1019.
[114] M. Coelho, R. Pereira, A.S. Rodrigues, J.A. Teixeira, M.E. Pintado, Extraction of tomato by-products' bioactive compounds using ohmic technology, Food Bioprod Process 117 (2019) 329–339.
[115] R.K. Toor, G.P. Savage, Antioxidant activity in different fractions of tomatoes, Food Res. Int. 38 (5) (2005) 487–494.
[116] W. Stahl, H. Sies, Antioxidant activity of carotenoids, Mol. Aspects Med. 24 (6) (2003) 345–351.
[117] R.-.M. Stoica, C. Tomulescu, A. Căşărică, M-G. Soare, Tomato by-products as a source of natural antioxidants for pharmaceutical and food industries–a mini-review, Sci. Bull. Ser. F. Biotechnol. 22 (2018) 200–204.
[118] J. Shi, Y. Kakuda, D. Yeung, Antioxidative properties of lycopene and other carotenoids from tomatoes: synergistic effects, Biofactors 21 (1–4) (2004) 203–210.
[119] E.-.S. Hwang, P.E. Bowen, Effects of tomato paste extracts on cell proliferation, cell-cycle arrest and apoptosis in LNCaP human prostate cancer cells, Biofactors 23 (2) (2005) 75–84.
[120] P. Palozza, R.E. Simone, A. Catalano, M.C. Mele, Tomato lycopene and lung cancer prevention: from experimental to human studies, Cancers 3 (2) (2011) 2333–2357.
[121] T. Yang, X. Yang, X. Wang, Y. Wang, Z. Song, The role of tomato products and lycopene in the prevention of gastric cancer: a meta-analysis of epidemiologic studies, Med. Hypotheses 80 (4) (2013) 383–388.
[122] L. Le Marchand, J.H. Hankin, L.N. Kolonel, G.R. Beecher, L.R. Wilkens, L.P. Zhao, Intake of specific carotenoids and lung cancer risk, Cancer Epidemiol Prev. Biomark 2 (3) (1993) 183–187.
[123] Y. Sharoni, K. Linnewiel-Hermoni, G. Zango, M. Khanin, H. Salman, A. Veprik, et al., The role of lycopene and its derivatives in the regulation of transcription systems: implications for cancer prevention, Am. J. Clin. Nutr. 96 (5) (2012) 1173S–1178S.
[124] L.E. Kelemen, J.R. Cerhan, U. Lim, S. Davis, W. Cozen, M. Schenk, et al., Vegetables, fruit, and antioxidant-related nutrients and risk of non-Hodgkin lymphoma: a National Cancer Institute–surveillance, epidemiology, and end results population-based case-control study, Am. J. Clin. Nutr. 83 (6) (2006) 1401–1410.
[125] S. Carpentier, M. Knaus, M. Suh, Associations between lutein, zeaxanthin, and age-related macular degeneration: an overview, Crit. Rev. Food Sci. Nutr. 49 (4) (2009) 313–326.
[126] B.R. Hammond Jr, Possible role for dietary lutein and zeaxanthin in visual development, Nutr. Rev. 66 (12) (2008) 695–702.
[127] E. Loane, J.M. Nolan, O. O'Donovan, P. Bhosale, P.S. Bernstein, S. Beatty, Transport and retinal capture of lutein and zeaxanthin with reference to age-related macular degeneration, Surv. Ophthalmol. 53 (1) (2008) 68–81.
[128] SB. Kritchevsky, β-carotene, carotenoids and the prevention of coronary heart disease, J. Nutr. 129 (1) (1999) 5–8.
[129] Y. Ito, M. Kurata, K. Suzuki, N. Hamajima, H. Hishida, K. Aoki, Cardiovascular disease mortality and serum carotenoid levels: a Japanese population-based follow-up study, J. Epidemiol. 16 (4) (2006) 154–160.
[130] L. Arab, S. Steck, Lycopene and cardiovascular disease, Am. J. Clin. Nutr. 71 (6) (2000) 1691S–1695S.

[131] A.V. Rao, S. Agarwal, Role of antioxidant lycopene in cancer and heart disease, J Am Coll Nutr 19 (5) (2000) 563–569.
[132] N.P. Silva-Beltrán, S. Ruiz-Cruz, L.A. Cira-Chávez, M.I. Estrada-Alvarado, J. Ornelas-Paz J de, M.A. López-Mata, et al., Total phenolic, flavonoid, tomatine, and tomatidine contents and antioxidant and antimicrobial activities of extracts of tomato plant, Int J Anal Chem 2015 (2015) 10.
[133] Y. Bashan, Y. Okon, Y. Henis, Peroxidase, polyphenoloxidase, and phenols in relation to resistance against Pseudomonas syringae pv. tomato in tomato plants, Can J Bot 65 (2) (1987) 366–372.
[134] C. Cueva, M.V. Moreno-Arribas, P.J. Martín-Álvarez, G. Bills, M.F. Vicente, A. Basilio, et al., Antimicrobial activity of phenolic acids against commensal, probiotic and pathogenic bacteria, Res Microbiol 161 (5) (2010) 372–382.
[135] G.V. Smirnova, G.I. Vysochina, N.G. Muzyka, Z.Y. Samoylova, T.A. Kukushkina, ON. Oktyabrsky, Evaluation of antioxidant properties of medical plants using microbial test systems, World J Microbiol Biotechnol 26 (12) (2010) 2269–2276.
[136] NAM. Eskin, Plant Pigments, Flavors and Textures, Academic press, New York, 1979.
[137] AP. Imeson, Thickening and Gelling Agents for Food, Springer Science & Business Media, Hong Kong, 2012.
[138] W.G.T. Willats, J.P. Knox, J.D. Mikkelsen, Pectin: new insights into an old polymer are starting to gel, Trends Food Sci Technol 17 (3) (2006) 97–104.
[139] B.R. Thakur, R.K. Singh, A.K. Handa, M.A. Rao, Chemistry and uses of pectin—a review, Crit Rev Food Sci Nutr 37 (1) (1997) 47–73.
[140] S.M. Brejnholt, Pectin. In: Food Stabilisers, Thickeners and Gelling Agents, John Wiley & Sons, Ltd, 2009, pp. 237–265.
[141] B.E. Morales-Contreras, J.C. Contreras-Esquivel, L. Wicker, L.A. Ochoa-Martínez, J. Morales-Castro, Husk tomato (Physalis ixocarpa Brot.) waste as a promising source of pectin: Extraction and physicochemical characterization, J Food Sci 82 (7) (2017) 1594–1601.
[142] S.-.H. Jeong, B.-.Y. Kim, H.-.G. Kang, H.-.O. Ku, J-H. Cho, Effects of butylated hydroxyanisole on the development and functions of reproductive system in rats, Toxicology 208 (1) (2005) 49–62.
[143] A.B. Engin, N. Bukan, O. Kurukahvecioglu, L. Memis, A. Engin, Effect of butylated hydroxytoluene (E321) pretreatment versus L-arginine on liver injury after sub-lethal dose of endotoxin administration, Environ Toxicol Pharmacol 32 (3) (2011) 457–464.
[144] A.A.M. Botterweck, H. Verhagen, R.A. Goldbohm, J. Kleinjans, P.A. Van den Brandt, Intake of butylated hydroxyanisole and butylated hydroxytoluene and stomach cancer risk: results from analyses in the Netherlands cohort study, Food Chem Toxicol 38 (7) (2000) 599–605.
[145] S. Randhawa, S.L. Bahna, Hypersensitivity reactions to food additives, Curr Opin Allergy Clin Immunol 9 (3) (2009) 278–283.
[146] J.S. Kornienko, I.S. Smirnova, N.A. Pugovkina, J.S. Ivanova, M.A. Shilina, T.M. Grinchuk, et al., High doses of synthetic antioxidants induce premature senescence in cultivated mesenchymal stem cells, Sci Rep 9 (1) (2019) 1–13.
[147] R. Domínguez, M. Pateiro, M. Gagaoua, F.J. Barba, W. Zhang, J.M. Lorenzo, A comprehensive review on lipid oxidation in meat and meat products, Antioxidants 8 (10) (2019) 429.
[148] M.M. Calvo, M.L. Garcia, M.D. Selgas, Dry fermented sausages enriched with lycopene from tomato peel, Meat Sci 80 (2) (2008) 167–172.
[149] M.L. García, M.M. Calvo, M.D. Selgas, Beef hamburgers enriched in lycopene using dry tomato peel as an ingredient, Meat Sci 83 (1) (2009) 45–49.
[150] A.V. Rao, H. Shen, Effect of low dose lycopene intake on lycopene bioavailability and oxidative stress, Nutr Res 22 (10) (2002) 1125–1131.

[151] A.B. Alves, N. Bragagnolo, M.G. Da Silva, L.H. Skibsted, V. Orlien, Antioxidant protection of high-pressure processed minced chicken meat by industrial tomato products, Food Bioprod Process 90 (3) (2012) 499–505.
[152] S. Savadkoohi, H. Hoogenkamp, K. Shamsi, A. Farahnaky, Color, sensory and textural attributes of beef frankfurter, beef ham and meat-free sausage containing tomato pomace, Meat Sci. 97 (4) (2014) 410–418.
[153] A.I. Andrés, M.J. Petrón, J.D. Adámez, M. López, M.L. Timón, Food by-products as potential antioxidant and antimicrobial additives in chill stored raw lamb patties, Meat Sci. 129 (2017) 62–70.
[154] M. Majzoobi, F.S. Ghavi, A. Farahnaky, J. Jamalian, G. Mesbahi, Effect of tomato pomace powder on the physicochemical properties of flat bread (Barbari bread), J. Food Process Preserv. 35 (2) (2011) 247–256.
[155] D.S. Sogi, J.S. Sidhu, M.S. Arora, S.K. Garg, A.S. Bawa, Effect of tomato seed meal supplementation on the dough and bread characteristics of wheat (PBW 343) flour, Int. J. Food Prop. 5 (3) (2002) 563–571.
[156] L. Padalino, A. Conte, L. Lecce, D. Likyova, V. Sicari, T.M. Pellicanò, et al., Functional pasta with tomato by-product as a source of antioxidant compounds and dietary fibre, Czech J. Food Sci. 35 (1) (2017) 48–56.
[157] F. Isik, C. Topkaya, Effects of tomato pomace supplementation on chemical and nutritional properties of crackers, Ital. J. Food Sci. 28 (3) (2016) 525.
[158] M. Ahmad Bhat, H. Ahsan, Physico-chemical characteristics of cookies prepared with tomato pomace powder, J. Food Process Technol. 07 (2016) 1–4.
[159] Y. Abid, S. Azabou, M. Jridi, I. Khemakhem, M. Bouaziz, H. Attia, Storage stability of traditional Tunisian butter enriched with antioxidant extract from tomato processing by-products, Food Chem. 233 (2017) 476–482.
[160] M. Bouaziz, I. Feki, M. Ayadi, H. Jemai, S. Sayadi, Stability of refined olive oil and olive-pomace oil added by phenolic compounds from olive leaves, Eur. J. Lipid Sci. Technol. 112 (8) (2010) 894–905.
[161] A. Bendini, G.D. Lecce, E. Valli, S. Barbieri, F. Tesini, T.G. Toschi, Olive oil enriched in lycopene from tomato by-product through a co-milling process, Int. J. Food Sci. Nutr. 66 (4) (2015) 371–377 19.
[162] A. Benakmoum, S. Abbeddou, A. Ammouche, P. Kefalas, D. Gerasopoulos, Valorisation of low quality edible oil with tomato peel waste, Food Chem. 3 (110) (2008) 684–690.
[163] M. Kehili, S. Choura, A. Zammel, N. Allouche, S. Sayadi, Oxidative stability of refined olive and sunflower oils supplemented with lycopene-rich oleoresin from tomato peels industrial by-product, during accelerated shelf-life storage, Food Chem. 246 (2018) 295–304.
[164] M. Belović, A. Torbica, I. Pajić-Lijaković, J. Mastilović, Development of low calorie jams with increased content of natural dietary fibre made from tomato pomace, Food Chem. 237 (2017) 1226–1233.
[165] E.M. Rizk, A.T. El-Kady, A.R. El-Bialy, Characterization of carotenoids (lyco-red) extracted from tomato peels and its uses as natural colorants and antioxidants of ice cream, Ann. Agric. Sci. 59 (1) (2014) 53–61.
[166] T.P. Anunciato, P.A. da Rocha Filho, Carotenoids and polyphenols in nutricosmetics, nutraceuticals, and cosmeceuticals, J. Cosmet. Dermatol. 11 (1) (2012) 51–54.
[167] A. Barbulova, G. Colucci, F. Apone, New trends in cosmetics: By-products of plant origin and their potential use as cosmetic active ingredients, Cosmetics 2 (2) (2015) 82–92.
[168] W. Stahl, U. Heinrich, O. Aust, H. Tronnier, H. Sies, Lycopene-rich products and dietary photoprotection, Photochem. Photobiol. Sci. 5 (2) (2006) 238–242.
[169] W. Peschel, F. Sánchez-Rabaneda, W. Diekmann, A. Plescher, I. Gartzía, D. Jiménez, et al., An industrial approach in the search of natural antioxidants from vegetable and fruit wastes, Food Chem. 97 (1) (2006) 137–150.

[170] M.M. Mathews-Roth, M.A. Pathak, J. Parrish, T.B. Fitzpatrick, E.H. Kass, K. Toda, et al., A clinical trial of the effects of oral beta-carotene on the responses of human skin to solar radiation, J. Invest. Dermatol. 59 (4) (1972) 349–353.
[171] H.M. Gollnick, W. Hopfenmüller, C. Hemmes, S.C. Chun, C. Schmid, Systemic beta carotene plus topical UV-sunscreen are an optimal protection against harmful effects of natural UV-sunlight: results of the Berlin-Eilath study, EJD Eur. J. Dermatol. 6 (3) (1996) 200–205.
[172] J. Lee, S. Jiang, N. Levine, R.R. Watson, Carotenoid Supplementation reduces erythema in human skin after simulated solar radiation exposure (44476), Proc. Soc. Exp. Biol. Med. 223 (2) (2000) 170–174.
[173] W. Stahl, U. Heinrich, H. Jungmann, H. Sies, H. Tronnier, Carotenoids and carotenoids plus vitamin E protect against ultraviolet light–induced erythema in humans, Am. J. Clin. Nutr. 71 (3) (2000) 795–798.
[174] W. Stahl, U. Heinrich, S. Wiseman, O. Eichler, H. Sies, H. Tronnier, Dietary tomato paste protects against ultraviolet light–induced erythema in humans, J. Nutr. 131 (5) (2001) 1449–1451.
[175] A. Ascenso, H. Ribeiro, H.C. Marques, H. Oliveira, C. Santos, S. Simões, Chemoprevention of photocarcinogenesis by lycopene, Exp. Dermatol. 23 (12) (2014) 874–878.
[176] M.V. Butnariu, C.V. Giuchici, The use of some nanoemulsions based on aqueous propolis and lycopene extract in the skin's protective mechanisms against UVA radiation, J. Nanobiotechnology 9 (1) (2011) 1–9.
[177] M. Rizwan, I. Rodriguez-Blanco, A. Harbottle, M.A. Birch-Machin, R.E.B. Watson, LE. Rhodes, Tomato paste rich in lycopene protects against cutaneous photodamage in humans in vivo: a randomized controlled trial, Br. J. Dermatol. 164 (1) (2011) 154–162.
[178] M. Działo, J. Mierziak, U. Korzun, M. Preisner, J. Szopa, A. Kulma, The potential of plant phenolics in prevention and therapy of skin disorders, Int. J. Mol. Sci. 17 (2) (2016) 160.
[179] D. Tungmunnithum, A. Thongboonyou, A. Pholboon, A. Yangsabai, Flavonoids and other phenolic compounds from medicinal plants for pharmaceutical and medical aspects: an overview, Medicines 5 (3) (2018) 93.
[180] M. Taveira, L.R. Silva, L.A. Vale-Silva, E. Pinto, P. Valentao, F. Ferreres, et al., Lycopersicon esculentum seeds: an industrial byproduct as an antimicrobial agent, J. Agric. Food Chem. 58 (17) (2010) 9529–9536.
[181] A.N. Grassino, S. Djaković, T. Bosiljkov, J. Halambek, Z. Zorić, V. Dragović-Uzelac, et al., Valorisation of tomato peel waste as a sustainable source for pectin, polyphenols and fatty acids recovery using sequential extraction, Waste Biomass Valori 11 (9) (2020) 4593–4611.
[182] J. Bacenetti, D. Duca, M. Negri, A. Fusi, M. Fiala, Mitigation strategies in the agro-food sector: the anaerobic digestion of tomato puree by-products. An Italian case study, Sci. Total Environ. 526 (2015) 88–97.
[183] T. Suzuki, I. Usui, K. Tomita-Yokotani, S. Kono, H. Tsubura, Y. Miki, et al., Effects of acid extracts of tomato (Lycopersicon esculentum Mill.) and carrot (Daucus carota L.) wastes from the food industry on the growth of some crops and weeds, Weed Biol. Manag. 1 (4) (2001) 226–230.

CHAPTER FOUR

# Ingredients for food products

**George Liadakis, Tryfon Kekes, Georgia Frakolaki, Virginia Giannou, Constantina Tzia**
Laboratory of Food Chemistry and Technology, School of Chemical Engineering, National Technical University of Athens, Zografou, Greece

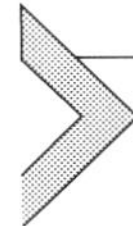

## 4.1 Introduction (nutritive constituents, bioactive compounds of tomato processing by-products)

Tomato processing industry is a dynamic food sector with financial interest for tomato growing countries. Various tomato products including canned tomatoes, tomato juice, puree and paste, ketchup, tomato soup, dried tomatoes or dehydrated pulp are intended for domestic use or food services industries. Industrial tomato processing involves the generation of a solid by-product, named tomato pomace, which amounts to 5%–10% (or 10%–30% according to other sources) of the raw fruit weight being processed. Thus, based on the annual amount of globally processed tomatoes (over 40 million tons), it can be estimated that remarkable amounts of tomato pomace are produced (approximately 4 [or 5.4–9.0] million tons). In Europe, as well, where tomatoes produced per year represent 12% of the total global production, a corresponding amount of tomato pomace is generated as the by-product of the tomato processing industries. An estimate of 410,000 tons of tomatoes are processed into tomato products annually in Greece generating approximately 19.5–39 × $10^6$ kg of tomato pomace.

Tomato pomace, consisting mainly of peels and seeds, contains valuable nutrients and bioactive components. The seeds represent a large portion of the pomace, approximately 50%–55% and are characterized by their nutritive potential. They contain high proportions of crude oil: 11%–20% and protein: 15%–22% in dry basis (d.b.), respectively. The oil exhibits a high unsaturated acid content, and the protein has a high lysine content and good functionality. In addition, tomato seeds lack antinutritional factors or any other toxic substances. The peels contain remarkable levels of carotenoids, in particular lycopene (288–734 mg/kg).

Food by-products, in general, create environmental pollution while they represent a loss of valuable nutrients. Nowadays, there is an increasing trend for their utilization to minimize the waste disposal problem as

*Tomato Processing by-Products*
DOI: https://doi.org/10.1016/B978-0-12-822866-1.00007-7

well as to increase the nutrient resources. With the existing shortage of high-quality, low-cost foods, there is a demand for the recovery of nutrients from valuable food by-products and for their conversion into high added value products.

Although tomato pomace contains components of high nutritive value (protein, lipids) and bioactive compounds of recognized functionality (carotenoids, lycopene) with health advantages, it is disposed in landfills or used as animal feed or as fertilizer. However, tomato pomace can be utilized by applying conventional or innovative technological methods in order to isolate the aforementioned valuable components for further use.

Tomato seeds could provide a valuable source for supplementing the proteins of cereal products; also, the functionality of tomato seed proteins can find many uses in food systems. The tomato seed oil can also be used by the oil industry to produce edible oil. The carotenoids and in particular lycopene are components with many applications in functional foods.

The purpose of this chapter is to analyze and present the vertical process for the utilization of tomato by-products emphasizing on the required stages for the isolation of useful components. It mainly focuses on tomato proteins and lipids, their role in the human nutrition, and the methods for their recovery. Additionally, it provides brief information on bioactive carotenoids and especially lycopene, their protective role against certain diseases and the methods for their recovery. Concluding, a sustainable utilization plan of tomato pomace is proposed since the resulting protein or oil products can be used as edible ingredients, while carotenoids and lycopene can be used as antioxidants or functional supplements in various food products.

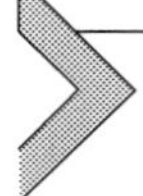

## 4.2 Valuable components (quantitative data, chemistry, role in nutrition, commercial product)

Tomato and tomato products are of interest because of their nutritional and health benefits and the antioxidant activity of lycopene and other carotenoids [33]. Tomato pomace comprises a valuable source of protein, lysine, oil rich in linoleic acid, carotenoids and minerals [55,84]. It is also rich in several other bioactive compounds and valuable antioxidants, such as tocopherols, polyphenols, terpenes, and sterols [41].

Tomato pomace resulting by the processing of tomato in the industry represents 5%–30% of the raw fruit. The wet residue obtained during tomato juice production is pressed to separate the holding juice and is then dried. Tomato pomace constitutes of peels and seeds, with the seeds representing a large portion of the dried pomace, approximately 50%–55%. The chemical compositions of peels and seeds are noticeably different from each other. The nutritive potential of tomato seeds has been reported by many authors. Tomato seeds contain high proportions of crude oil 11%–20%, and protein 15%–22% (d.b.). The peels contain high amounts of lycopene (288–734 mg/kg) nearly five times higher than the tomato pulp [41,86]. A significant part of tomato pomace consists of fibers most commonly found in the peels.

The approximate composition of tomato pomace, peels and seeds is presented in Table 4.1.

The amount of the tomato pomace generated during tomato processing, combined with the potentially beneficial characteristics of its bioactive components, justifies the interest of researchers and manufacturers in their isolation or extraction. A procedure for the sustainable tomato pomace utilization involves the recovery of protein and oil from the tomato seeds, and of carotenoids and lycopene from the peels and/or the pomace, with a fiber fraction remaining in both cases.

## 4.2.1 Nutritive components (protein, lipids)

### *4.2.1.1 Proteins*

Tomato pomace is rich in protein and thus may be utilized as a potential protein source. Tomato contains 1.1 g protein per 100 g of the edible part of the fruit, while tomato pomace contains 15.1%–22.7% protein. The tomato seeds, in particular, have a high protein content ranging from 16.6%–39.3% (d.b.), and the defatted dried seeds even higher protein content ranging from 24.5%–40.9%, respectively [32,55,65]. Apart from the high protein content the tomato seeds exhibit high protein quality exceeding the WHO/FAO/UNU [92] recommendations regarding their amino acids content [73].

**Table 4.1** Composition of tomato pomace, peels, and seeds in nutritive and bioactive components.

| Components (% w/w) | Tomato pomace | Peels | Seeds |
|---|---|---|---|
| Moisture | 6.6–8.5 | 7.0–10.1 | 3.1–11.7 |
| Protein | 15.1–22.7 | 5.7–20.0 | 16.6–39.3 |
| Oil | 8.4–16.2 | 1.7–3.8 | 6.4–36.9 |
| Fibers | 11.3–64.7 | 29.9–65.6 | 14.8–53.8 |
| Carbohydrates | 2.9–5.1 | 1.1–8.2 | 2.3–26.0 |
| Ash | 3.2–3.4 | 2.7–25.6 | 2.0–5.6 |
| Lycopene (mg/kg) | 413.7 | 734.0 | 130.0 |
| β-Carotene (mg/kg) | 149.8 | 29.3 | 14.4 |

The protein composition, in general, determines the functional properties required for its incorporation in a particular food matrix. The tomato seed proteins identified after fractionation and characterization consist of albumins, globulins, gliadins, and glutenins. More specifically, globulins constitute about 60%–70% of the protein content, whereas the minor fractions are mostly soluble glutenins and gliadins [66,80].

Regarding the biological evaluation of tomato seed proteins, they exhibit a high protein efficiency ratio (PER) value of 2.18–2.66 and net protein utilization (NET) value of 55 indicating the presence of high quality proteins [22,73].

The amino acid composition of proteins defines their nutritional quality. The typical amino acids composition of tomato seed proteins is presented in Table 4.2 [10,48,60,73,74,87,96]. Tomato proteins contain all the essential amino acids which constitute the 39.5% of the total protein content [73]. The proportions of hydrophilic amino acids (lysine, histidine, aspartic acid, glutamic acid and arginine) and hydrophobic amino acids (alanine, valine, leucine, isoleucine, and phenylalanine) are found 41.6% and 26.4%, respectively. The umami flavored amino acids (glutamic and aspartic acid) are the predominant ones, followed by arginine, lysine, valine, and leucine. The methionine and cysteine are found in low concentrations characterized

**Table 4.2** Amino acid composition of tomato seed proteins.

| Amino acid | Content (% w/w) |
|---|---|
| Glutamic acid | 13.5–24.7 |
| Aspartic acid | 5.1–12.0 |
| Lysine | 3.2–6.9 |
| Threonine | 1.9–7.2 |
| Valine | 2.3–7.9 |
| Leucine | 4.0–7.6 |
| Isoleucine | 2.4–4.9 |
| Phenylalanine | 2.6–6.6 |
| Methionine | 1.0–2.1 |
| Arginine | 5.6–11.3 |
| Glycine | 4.8–9.0 |
| Proline | 3.4–10.2 |
| Cysteine | 1.0–1.7 |
| Serine | 2.1–7.0 |
| Alanine | 2.4–8.4 |
| Tyrosine | 1.9–4.9 |
| Histidine | 2.5 |
| Tryptophan | 1.2 |

as the limited amino acids of tomato seed proteins. The protein quality of tomato processing by-products is lower than the respective one of animal proteins, but quite similar to that of most plant proteins. However, the high concentrations in glutamic and aspartic acid and particularly in lysine, compared to other plant proteins, make the use of tomato seed proteins favorable for the enrichment of foods (i.e., of flour with lysine) [19,66,96].

#### *4.2.1.2 Lipids*

Tomato pomace and particularly the seeds, contain significant amounts of oil, thus they are considered a utilizable source of oil. Tomato fruits contain 0.2 g oil per 100 g of the edible part of the fruit, while tomato pomace contains 8.4%–16.2% oil, with the seeds having a higher oil content ranging from 6.4%–36.9% (d.b.) [55,75].

The oil recovered from tomato seeds becomes edible after refining. The unsaturation level of fatty acids is an important quality parameter of the edible oils both in terms of a healthy nutrition and product stability. The fatty acid composition of the tomato seed oil is presented in Table 4.3; its composition is similar to that of cottonseed oil. The unsaturated fatty acids, which represent about 80% of the total lipids content, have been proved to lower blood triglyceride levels and blood pressure and exhibit protective properties against heart diseases. Tomato seed oil belongs to the linolenic-oleic acid group of oils having a high content in polyunsaturated fatty acids, especially in linoleic acid (up to 57%). It also contains monounsaturated fatty acids, such as oleic acid (up to 30%) and saturated fatty acids (palmitic, stearic, and arachidic acids) [8,9,16,59,66,75].

The tomato seed oil also contains phenolic acids, polyphenols and flavonoids which provide antioxidant and antibacterial properties to the oil [59]. It is rich in bioactive compounds such as tocopherol (282 mg/kg), lycopene

**Table 4.3** Fatty acid composition of tomato seed oil.

| Fatty acid | Concentration (%) |
|---|---|
| Linoleic acid (C18:2) | 37–57 |
| Palmitic acid (C16:0) | 7–24 |
| Oleic acid (C18:1) | 18–30 |
| Stearic acid (C18:0) | 4–13 |
| Linolenic acid (C18:3) | 1–6 |
| Myristic acid (C14:0) | 0.1–2.3 |
| Palmitoleic acid (C16:1) | 0.3–7 |
| Margaric acid (C17:0) | 0.1–0.3 |
| Arachidic acid (C20:0) | 0.2–3 |

(95 mg/kg), policosanol (70 mg/kg), phytosterol (11 mg/kg), β-carotene (4.5 mg/kg), and phenolic compounds (20 μg gallic acid/100 g) [56]. The tomato seed oil exhibits oxidative and thermal stability higher than that of other commercial oils with similar unsaturation level; this may be attributed to the antioxidants present in the tomato seed oil such as tocopherols and polyphenols [66,75].

The physicochemical characteristics of tomato seed oil that are useful for its utilization are shown in Table 4.4 [9,11].

The oil in the tomato peels is different from the tomato seed oil; it has high linoleic acid content, yellow-brown color, refractive index (60°C): 1.4684, iodine value: 77 g I/100 g oil, and saponification value: 133 mg KOH/g oil, while it contains high unsaponifiable matters and wax substances (17 mg/g) (liquid: 0.04 mg/cm$^3$ and solid: 36 mg/g) [50].

Tomato seed oil is characterized by a pleasant appearance and taste [66]. Its high nutritional value provides tomato seed oil a great potential as edible oil. Tomato seed oil exhibits antioxidant properties and health-promoting effects. It has been reported to decrease cholesterol levels in hamsters and guinea pigs fed with high-fat diets [35]. The high unsaturated fatty acids content of tomato seed oil provides anticarcinogenic properties, reduction of catabolic effects of immune stimulation, protection against heart diseases and decrease of blood triglyceride levels, blood pressure and atherosclerosis risk [76].

### 4.2.2 Bioactive compounds (carotenoids, lycopene)

Tomato waste is considered as an important source of natural carotenoids. Moreover, tomatoes are one of the main sources of carotenoids, especially of lycopene; it is the most abundant carotenoid in tomatoes with its content varying depending mostly on the variety and climate conditions. The carotenoids content of the tomato pomace shows further variation depending on the industrial processing methods. However, the peels have

**Table 4.4** Properties of tomato seed oil.

| Refractive index | 1.4710 |
|---|---|
| Density (kg/m$^3$) | 0.877–0.901 |
| Iodine value (g I/100 g oil) | 125.8–126.0 |
| Saponification value (mg KOH/g oil) | 177.0–192.5 |
| Peroxide value (meq/kg oil) | 0.63 |
| Acid value (mg KOH/g oil) | 0.22–1.34 |
| Unsaponifiable matter (%) | 1.747 |

approximately 5-fold more carotenoids than the seeds, with their content amounting to 793.2 μg/g against the seeds amounting to 157.9 μg/g, respectively [44].

Carotenoids are natural colorings that provide the natural yellow, orange, and red colors of various fruits and vegetables. They represent a large group of yellow-orange pigments that consist of eight isoprenoid units joined to form a conjugated double bond system in their molecule. The conjugated polyene structure is responsible for the characteristic color of each carotenoid. The two main subclasses of carotenoids are carotenes and xanthophylls. Some of them such as β-carotene and α-carotene exhibit provitamin A activity. As a result, the presence of conjugated double bonds, also provides carotenoids with antioxidant properties. Lycopene is a tetraterpenic hydrocarbon with 13 carbon-carbon double bonds, including 11 conjugated bonds, and is responsible for the deep red color of ripe tomatoes [33,77,86]. Lycopene, originally in the trans form, is subjected to degradation under intense conditions (light, oxygen, heating) and isomerizes to the cis structure resulting changes in color [78].

The main carotenoids found in tomato pomace include: α-carotene, β-carotene, lycopene, and lutein with lycopene prevailing and amounting to more than 85% of the total carotenoids [33,86]. The carotenoids content of tomato pomace [41], as well as of tomato peels and seeds [44] is presented in Table 4.5.

Carotenoids are well credited with important health-promoting functions based on scientific studies. In specific, the consumption of carotenoids through foods have been proven to possess high therapeutic effects on a wide range of diseases. Natural carotenoids that can be absorbed, transported, and deposited by the human body include β-Carotene, α-Carotene, lycopene, β-Cryptoxanthin, lutein, and zeaxanthin. In addition to provitamin A and antioxidant activity, carotenoids enhance the immune system and reduce the risk of degenerative diseases such as cataract and macular

**Table 4.5** Composition of carotenoids in peel, seeds, and tomato pomace.

| Carotenoids (mg/kg) | Tomato pomace | Peels | Seeds |
|---|---|---|---|
| Lycopene | 413.7 | 734.0 | 130.0 |
| β-Carotene | 149.8 | 29.3 | 14.4 |
| Lutein | | 14.5 | 6.5 |
| Zeaxanthin | | 3.7 | 1.0 |
| α-Carotene | | – | 0.4 |
| cis β-Carotene | | 11.7 | 5.6 |

degeneration. Moreover, lycopene presents exceptional antioxidant, anti-inflammatory, and immunomodulatory activity giving it extraordinary potential for protecting humans against a wide range of chronic disorders, including cardiovascular diseases (CVD) [68], hepatic fibrogenesis, solar light induced erythema, human papillomavirus persistence, different types of cancer (prostate, breast, digestive tract, lung), and oxidative stress. Several studies have linked the consumption of tomato products to a reduced risk of cardiovascular diseases and certain types of cancer [69].

## 4.3 Processing for isolation of valuable components

The tomato processing by-product, as presented above, exhibits high nutritive value mainly due to the oil and protein content of seeds as well as its bioactive potential due to carotenoids and lycopene content. The composition of the tomato pomace may differ in these utilizable components depending on the tomato variety, climate conditions and industrial processing. The common practices for tomato processing followed by the industry include the pressing of the separated by-product, its sun-drying for 3–6 days and its utilization in animal feeds.

Following, an integrated procedure is proposed in order for the tomato pomace to be sustainably utilized. According to this procedure, oil and protein isolates are produced from tomato seeds and carotenoids and lycopene from tomato peels and/or tomato pomace. The solid residues from the above processes can be utilized as a fiber source.

The tomato pomace can be treated either in a wet or in a dry form. In the first case, the excess of water is removed from the pomace by pressing in order to reduce storage requirements. In the second case, the drying method and conditions (temperature, time, vacuum, lyophilization) should be carefully selected in order to avoid the deterioration of the contained nutritive or bioactive components (protein denaturation, oil oxidation, carotenoids and lycopene degradation). Wet pomace should be then stored under refrigeration or freezing, while dried pomace in a dry, cool and protected environment. The preliminary treatment includes grinding in a wet or dry form and/or separation of the peels and seeds. The peels can be easily separated in wet pomace by flotation in water, while in dried pomace the peels can be removed by means of sieves and fans (or air cyclones).

The pretreatment process of the wet and dry tomato pomace is presented in Fig. 4.1 and the flow diagram for the valorization of tomato processing by-products is described in Fig. 4.2.

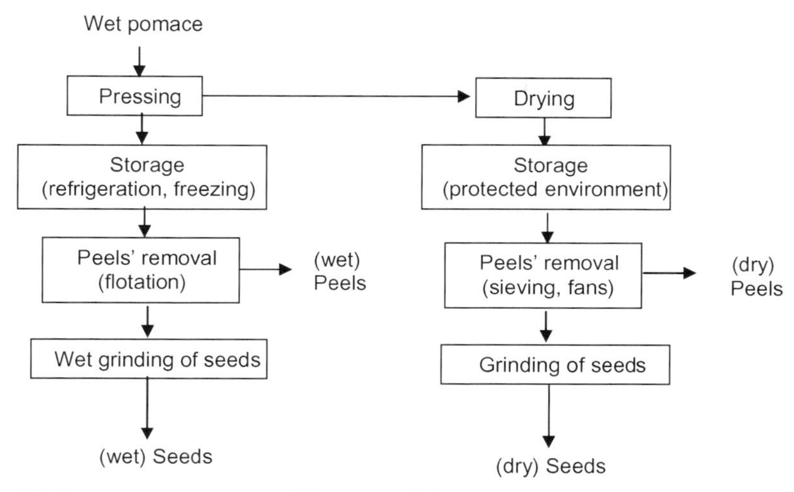


**Fig. 4.1** ***Pretreatment process of the wet and dry tomato pomace.***

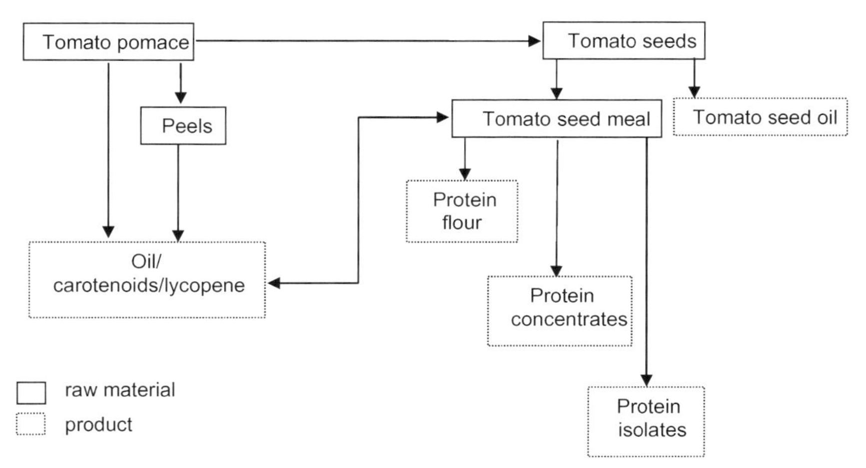


**Fig. 4.2** ***Flow diagram for the valorization of tomato processing by-products (tomato pomace) - Oil, protein products and carotenoids/lycopene.***

## 4.3.1 Tomato seed oil

Tomato seeds are mainly used as a raw material for oil recovery (tomato seed oil). The extraction of oil can be achieved by pressing (mechanical extraction) of the tomato seeds or by solvent extraction (conventional extraction) of the ground tomato seeds as commonly applied to oilseeds. The oil yield obtained by pressing is lower compared to solvent extraction. The oil yield varies reaching 20%–25% of the dry seeds. Innovative methods, such as microwave- or ultrasound-assisted solvent extraction have also been studied in tomato seed oil extraction.

#### 4.3.1.1 Conventional oil extraction methods

As mentioned above, tomato seed oil can be recovered through mechanical pressing using hydraulic or screw presses. The oil yield is affected by the operating pressure and time; higher values are achieved by continuous screw than hydraulic press [35,90]. The method is simple and involves low operating cost and high-quality flavored oil. Moreover, cold press techniques (<50°C) ensure better preservation of the natural compounds such as carotenoids, particularly lycopene and polyphenols providing yields of 85%–95% of the total oil content in tomato seeds. Thus, the resulting oil presents higher bioactive compounds concentration compared to other extraction methods, high oxidative stability and improved shelf life [35,94].

The oil can also be extracted from ground tomato seeds using suitable solvents usually n-hexane. The oil is obtained after the evaporation of the solvent, while the recovered solvent is reused for extraction and the defatted meal is further utilized for protein products. The yield and the fatty acid composition of the recovered oil may be affected by the solvent used (n-hexane, diethyl ether, petroleum ether) and the extraction conditions [8,14]. Solvent extraction provides higher extraction yields than mechanical pressing however it involves long time, higher cost and solvent losses. Moreover, due to waste solvents' disposal that may be toxic for the environment, and solvent residues within the extracted oil, novel approaches are investigated in order to partially or completely substitute the use of hazardous solvents [76].

#### 4.3.1.2 Innovative lipid recovery processes

In order to reduce the extraction time and the use of solvents, as well as increase the oil yield and improve its quality, alternative innovative processes have arisen. Supercritical fluid extraction (SFE) is a noteworthy alternative option for tomato seed oil. Supercritical fluids exhibit advantages as they have liquid-like density but superior mass transfer characteristics than liquid solvents as a result of their low viscosity and high diffusivity. Also, due to their low surface tension, they easily penetrate the pores of the solid matrix to release oil while they can be easy separated from the extracted oil by pressure reduction. The use of $CO_2$ as a unique extraction solvent makes this technique nontoxic, nonflammable, noncorrosive, highly selective, safer, and more environmentally friendly compared to the use of most conventional organic solvents. SFE-$CO_2$ is considered mostly effective, in particular, for the extraction of the valuable compounds of tomato seed, including phytosterols, tocopherols, β-carotene, and lycopene. Tomato seed oil has been successfully extracted using supercritical $CO_2$ obtaining higher purity than those obtained by organic solvent extraction [25,70,76]. The extraction yield is affected by the extraction time and pressure, thus Shen and Xu [76] optimized the extraction of tomato seed oil at 50°C and 30 MPa for 2 h achieving a yield of 96.3%. Microwave and ultrasound extraction of tomato seed can provide higher oil yields than the conventional extraction method [1].

### 4.3.2 Protein products

After the extraction of oil from the tomato seeds, a residue remains called "defatted tomato seed meal" or "cake." It can be used as animal feed or purified and standardized as a marketable product named "defatted tomato seed flour". The high protein content (38%–55% d.b.) and the high protein value of the tomato seed meal make it suitable for protein isolation [53,55]. Thus, tomato seed meal can be further utilized for the delivery of protein products using technological procedures similar to those applied to oilseeds.

The mean composition of the defatted tomato seed meal (flour) (in dry basis) is shown in Table 4.6 [53].

The utilization of tomato pomace and tomato seed meal has been examined by several researchers. Kramer and Kwee [45] have proposed a procedure for the complete utilization of tomato pomace leading to various protein products with different functionality and nutritional value. Protein extraction from defatted tomato seed flour [13,23,27,89], protein precipitation [48], as well as the process yield and functional properties of the protein products [21] have also been studied. Tchorbanov et al. [87] used tomato seeds (full fat) to obtain protein-oil products.

**Table 4.6** Mean composition of tomato seed meal.

| Components (% w/w) | Tomato pomace |
|---|---|
| Protein | 38–55 |
| Oil | 1.2–4.2 |
| Fibers | 15.1–32.5 |
| Carbohydrates | 5.4–29.1 |
| Ash | 3.9–9.6 |

#### *4.3.2.1 Technological procedures for protein products delivery*

Protein recovery from an oilseed meal, in general, can be achieved either by protein purification, through removal of nonprotein components, or by extraction and precipitation of the proteins (isolation procedure). The derived products are protein concentrates or protein isolates, respectively.

The flow chart of the complete industrial process for the utilization of tomato seeds as protein sources is presented in Fig. 4.3 [62].

Therefore, depending on their protein content, protein products are categorized as protein flours, concentrates, and isolates with about 50%, 70%, and higher than 90%–95% protein content, respectively.

Protein flours result directly upon the extraction of oil from the tomato seeds. In case the protein meals and flours are intended for human consumption, hygiene pretreatment of seeds is required (i.e., purification from foreign materials and impurities) in order to assure their quality. Protein flours are standardized in terms of their protein content and are available in the market for use in foods.

Protein concentrates are derived from defatted flakes of tomato seeds either after the extraction of carbohydrates, salts, and other low-molecular-weight

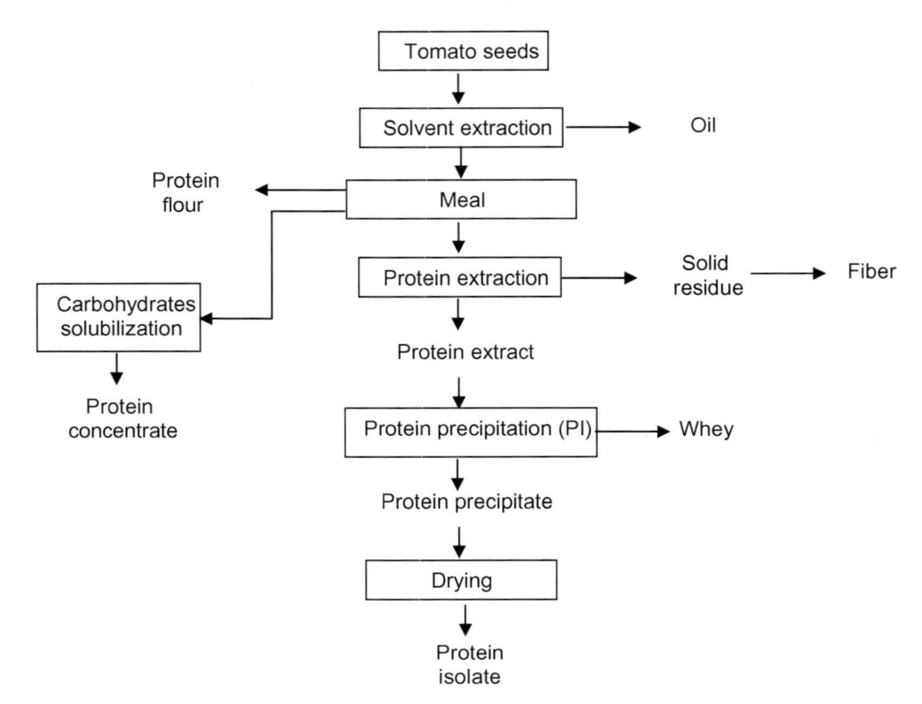


**Fig. 4.3** ***Complete utilization of tomato seeds as protein source - Protein products.***

substances so that the insoluble proteins are obtained, or after separation of the protein components by air cyclones. The resulting products present different functional properties depending on the method applied, and exhibit a protein content of 70%–72% and improved sensory properties. They have 2–2.5 times higher value on protein basis compared to protein flour.

The production of protein isolates is based on the extraction, purification, and recovery of the proteins from defatted flakes of tomato seeds. In contrary to protein concentrates, the isolation approach is obtained by dissolution of the proteins with the most impurities remaining in the solid fraction; defatted flakes are preferred against full-fat flakes to facilitate the extraction process. The protein content of the occurring protein isolates exceeds 90%, providing them with characteristic properties that make them suitable for food applications. In order to improve the functional properties of protein isolates, redissolution in water and neutralization with alkali or modification with proteolytic enzymes are also proposed [62].

#### 4.3.2.1.1 Protein concentrates

The production of protein concentrates, as previously mentioned, includes the extraction of carbohydrates, salts, and other low-molecular-weight substances from defatted flakes of tomato seeds. The most common methods applied for the production of protein concentrates include the use of (1) water-alcohol solutions, (2) dilute acid solutions, and (3) hot water after protein denaturation.

In the case of the extraction with water-alcohol solutions, methanol, ethanol, or isopropanol aqueous solutions (60%–70%, w/w) are usually proposed, due to their ability to extract the soluble carbohydrates without a negative impact on the dissolution of proteins. The removal or deactivation of other undesirable components may simultaneously be achieved. In order to ensure the maximal protein concentration in the insoluble matter, the extraction process should be repeated several times before the subsequent flash desolventizing for the removal of alcohol and the delivery of the protein concentrate.

The use of a diluted acid solution is based on the property of proteins to exhibit minimal solubility at their isoelectric point (pI), where carbohydrates still remain under solution. The pI of tomato seed proteins has been determined to be at pH 3.9; others also reported pI to be between 3.8 and 4.6 [55]. The defatted tomato seed meal is mixed with the diluted acid solution at the pI and the insoluble matter, also called "isoelectric concentrate", is separated by centrifugation from the whey containing the impurities and subsequently dried.

Alternatively, the toasting of tomato seeds that causes denaturation of the proteins, is proposed, followed by extraction with hot water (66–93°C) at adjusted pH. The method depends both on reducing proteins' solubility, due to their denaturation, and maximizing carbohydrates' dissolution.

Membrane technology has been also applied in the production of protein concentrates. The extract resulting from the aforementioned methods can be subjected to ultrafiltration through membranes of 100,000 molecular weight cut-off (MWCO). Thus, protein separation from low, molecular-weight compounds such as salts, carbohydrates, peptides, colorants, odorous materials, etc., is achieved and the protein concentrate can be obtained through spray-drying of the retentate [55,62].

#### 4.3.2.1.2 Protein isolates

The procedure specified for the protein isolation from tomato seeds includes three basic steps: protein extraction, protein precipitation, and drying of the protein precipitate. The first steps are the most important for efficient protein isolation, thus in each step the controlling factors that affect proteins' solubility should be taken into account. The economic feasibility of the whole process depends on the optimization of each of the main steps so as to maximize protein yield while maintaining the optimal quality of the protein isolates (in terms of protein content and functional properties) [13,27,45,46,48,53–55,63,72].

#### 4.3.2.1.3 Extraction

The extraction process is based on protein dissolution using certain solvents. Aqueous or non-aqueous organic solvents have been examined, depending on the solubility of the proteins; for example, albumins are soluble in water (pH 6-8), globulins in dilute salt solutions and glutelins in dilute acid or alkali solutions. Several factors should be taken into account in order to achieve the maximal extractability (dissolution) of proteins without affecting their structure and functionality, including extracting media, temperature, pH, agitation, liquid-to-solids ratio, particle size of ground solid material, flakes formation and extraction time [55,62]. Process conditions should be carefully designed to enhance protein extraction and maximize protein yield without denaturizing proteins or affecting their functional properties. As reported by Liadakis et al. [55] an extraction yield of about 66% of the total protein from tomato seeds can be achieved by applying the optimal extraction conditions (50°C, pH 11.5, 30/1v/w (liquid/solid), 20 min). Apart from the water-alkali system, salt solutions (NaCl or $Na_2CO_3$) were also examined during protein extraction from tomato seeds resulting in lower extraction yields but in isolates with increased protein content (93% by NaCl 5%). The separation of the aqueous protein extract from the solids is achieved by centrifugation, and polish filtration [54,62,87].

#### 4.3.2.1.4 Precipitation - drying

In order to recover the solubilized proteins contained in the extract, their precipitation is carried out. It involves the formation and aggregation of submicron particles and is affected by factors that decrease protein solubility. Organic solvents, divalent cations, heat, acid/bases (for pH adjustment), salts, nonionic polymers, and polyelectrolytes can be used for protein precipitation. Acid precipitation at the isoelectric point of proteins is one of the most commonly applied methods, due to the receipt of a concentrated product of high purity, as well as the recovery of the highest protein content. For this purpose, a food-grade acid may be used for acidification, such as acetic, hydrochloric, phosphoric, or sulfuric acid [54,62]. By precipitation of the extracted proteins at PI (3.8), the total protein yield was found 43.6% [53,55]. After precipitation, the occurring protein curd is collected by centrifugation or filtration, washed, and dried. The resulting protein isolate exhibits protein content of about 72% and good functional properties [53,55]. Spray-drying is preferred due to the reduced protein deterioration and its lower cost compared to other methods such as freeze-drying. Membrane technology can, also, be used for protein isolates, as in protein concentrates. In this case, membranes of 10,000–100,000 MWCO are used for separation, followed by drying of the protein isolate [62]. This way, an increased total protein yield (50%–54%) and a higher protein content of the isolate (72%–76%) can be achieved [53,55].

#### 4.3.2.1.5 Alternative processes - products

Alternative processes include the utilization of: i) the tomato seeds before their deoiling and/or ii) the wet pomace by applying a process similar to that used for protein isolation. Thus, by extracting at 50°C, pH = 11.5, 20 min, 30/1 v/w (liquid/solid) (the optimum conditions found for defatted tomato seeds) and precipitating with (hydrochloride) acid the following results were found: (1) protein extraction yield of 49%–54% and total protein yield of 35%–38%, and protein products with 44%–46% protein and 35%–36% oil level, (2) protein extraction yield of 62%–68% and total protein yield of 33%–35%, and protein products with 43%–48% protein and 25%–27% oil level (wet pomace required more water for its handling and the final product had red color) [53,55].

#### 4.3.2.1.6 Biological value of protein products

The tomato seed meal, protein concentrates and protein isolates have been evaluated for their biological value based on their essential amino acids content. Protein efficiency ratio (PER) (as weight gain in rats, along with standard and non-protein diets) was found 1.82 for whole meal, 1.93 for deoiled meal and 1.99 for protein concentrate as compared to 2.5 for casein. The net protein retention (NPR) was found 2.91 for casein, followed by 2.65 for whole meal, 2.52 for deoiled meal and 2.51 for protein concentrate [80,81]. Tomato seed protein isolate (92%) prepared from tomato seed meal was found to contain all essential amino acids (including lysine), meeting the minimum requirements of reference protein for preschool children aged 1–2 years (WHO/FAO/UNU) and had PER of 2.66. The tomato seed protein isolate showed negligible levels of phytate (3.48 μg/g) and trypsin inhibitory activity (2.655 TIU/mg) [73]. Therefore, tomato seeds are a rich source of high-quality plant proteins together with intrinsic polyphenols and antioxidant activities and could be regarded as a potential supplement in various food products.

### 4.3.3 Carotenoids - lycopene

The carotenoids and in particular lycopene, can be recovered from the whole tomato pomace or only from the tomato peels. Lycopene and β-carotene are the main carotenoids found in the tomato pomace. These natural carotenoids can be used as dyeing substances or antioxidants in various food products as well as bioactive components in innovative functional preparations [82]. Various extraction methods are available for the recovery of carotenoids from tomato processing by-products. These are discussed to a limited extent in this chapter and are analyzed in more detail in the previous chapter. Conventional organic solvent extraction and techniques have been used while innovative extraction methods have also been investigated to enhance the extraction of carotenoids. Additionally, supercritical fluid extraction (SFE), an environmentally safe technology is proposed for the recovery of carotenoids from industrial tomato processing by-products.

#### 4.3.3.1 Conventional/solvent extraction

Solvent extraction is the most widely used method for the recovery of bioactive compounds such as carotenoids from a broad range of plant materials. As far as the extraction process (solid-liquid) is concerned, the solvent type is of primary importance, as well as the solvent to solid ratio, particle size, temperature and extraction time. Focused optimization studies are useful to determine the optimal operation conditions which could provide the maximal yield (recovery) in total carotenoids and especially in the prevailing ones such as lycopene and β-carotene. Additionally, the purity and quality of the recovered carotenoids fraction, as evaluated by the coextracted impurities content and the target bioactive compounds content in the extract, should also be evaluated, as well as the safety and food grade nature of the final product (nonexistence toxic extracted compounds or solvent toxic residues) [28,62]. However, handling of the raw material and the extraction and purification process parameters should be carefully selected to preserve carotenoids' quality. Thus, raw tomato pomace or tomato peels may be used in wet form and be subjected to wet milling preserving their initial high carotenoid content. However, if they are to be dried, attention must be paid to the drying method (i.e., vacuum drying, freeze drying) and conditions and milling. The extraction and purification process should also avoid oxidation and degradation of carotenoids.

The solvent systems that have been commonly used for the extraction of carotenoids from various plant materials include hexane, ethanol, acetone, methanol, tetrahydrofuran, benzene, and petroleum ether or some of their mixtures (i.e., hexane with acetone, ethanol, ethyl acetate, or methanol). The suitable solvent system is selected based on the carotenoids present in the solid material. For example, hexane is extensively used for the nonpolar carotenes like β-Carotene or acetone, while ethanol, and ethyl acetate for the semipolar carotenoids like lutein [71]. Hexane and ethanol are also cheap and recognized as safe for use in food extraction processes. Strati and Oreopoulou [84] examined different solvent systems for the extraction of tomato pomace, evaluated them based on the total carotenoids yield, and resulted in the following order: hexane-ethyl acetate 1:1 (36.5 mg/kg) > acetone (33.4 mg/kg) > ethyl acetate (31.5 mg/kg) > hexane-acetone 1:1 (30.5 mg/kg) > hexane-ethanol 1:1 (28.1 mg/kg) > hexane (25.2 mg/kg) > ethanol (6.1 mg/kg). Especially with regard to the extraction of lycopene with various solvents (dichloromethane, hexane, ethanol, acetone, ethyl acetate, petroleum ether, and mixtures of polar or nonpolar solvents, such as acetone-chloroform (1:2) and hexane-acetone-ethanol (2:1:1), the hexane/acetone or hexane/ethanol systems were found more efficient [37]. Moreover, ethyl acetate was found more effective (120 mg/100 g) than ethanol (5 mg/100 g) in lycopene extraction [12].

Organic solvents used for the extraction of carotenoids and particularly lycopene may be toxic in traces which can makes them unsuitable for human consumption. In addition, conventional/solvent extraction processes may involve overexposure to heat, light and long times that could reduce carotenoids recovery and provide unstable products (color changes due to oxidations and lycopene isomerization, and reduction of the provitamin activity), while contributing to the increase of the greenhouse gas emissions. Toward this direction, continuous efforts are being made to develop sustainable, environmentally friendly and food safe quality ensuring approaches in order to overcome these shortcomings and address the increased ecological concern and the economic considerations. Thus, new extraction methods for carotenoids and/or lycopene aim to shorten the extraction time, reduce solvent consumption, reduce processing cost, and increase their functionality.

#### *4.3.3.2 Innovative processes for carotenoids recovery*

The alternative approaches for the extraction of carotenoids include the use of green solvents (ionic liquids) and assisted extraction methods such as supercritical carbon dioxide extraction, microwave-assisted extraction (MAE), ultrasound-assisted extraction (UAE), high hydrostatic pressure extraction (HHPE), enzyme-assisted extraction (EAE), pulsed electric field-assisted extraction (PEFE), accelerated solvent extraction (ASE), known as pressurized liquid extraction (PLE). Common objective of the above techniques is to reduce the volumes of the toxic, organic solvents used and/or to increase extraction yields and promote sustainability for the valorization of tomato pomace [14,71]. The principal mechanism on which the assisted extraction methods are based is their ability to cause physical damage of the cell walls (swelling and tissue softening) of the starting material allowing the higher diffusion of the extractants inside the food matrix in order to improve solvent penetration and solute solubilization.

##### 4.3.3.2.1 Ionic liquids as solvents for extraction

Ionic liquids (ILs), termed as green solvents, are proposed for the extraction of bioactive compounds such as carotenoids as they provide clean and sustainable processes. Ionic liquids are nonflammable and exhibit desirable properties such as low vapor pressure, good thermal stability, and ability to solubilize a wide range of solutes leading to elevated yields. Aqueous solutions of ammonium based ionic liquids have a great potential in extracting carotenoids from plants without utilizing n-hexane as extraction solvent. They can also be used in the assisted extraction methods [71].

#### 4.3.3.2.2 Supercritical fluid extraction with carbon dioxide

Supercritical fluid extraction (SCFE) with carbon dioxide ($CO_2$) is reported to provide the best extraction results for carotenoids including lycopene [15,37,91]. The $CO_2$ is nonhazardous, nonflammable, nontoxic, and inexpensive and is considered a perfect solvent for extracting food components like lycopene which are susceptible to light, heat and oxygen. Its extractability can be improved by mixing it with minute quantities of polar solvents like water, ethanol, methylene chloride, or hexane. The method has been applied both in tomato pomace and tomato peels. The proposed working conditions for carotenoids' extraction are: temperature 50–80°C and pressure 200–400 bars [37,79]. SCFE is more effective in the recovery of lycopene (72.9%) from tomato peels compared to the conventional extraction with hexane (60.9%), ethyl acetate (32.0%), and ethanol (28.4%) [42]. The use of an oil (i.e. olive oil) as a co-solvent improves the solubilization of carotenoids [79]. By SCFE at 275–460 bar and 80°C high recovery of carotenoids including lycopene (82.5%–90.0%) can be achieved providing a product of high stability (lower isomerization and degradation compared to other methods) [36,37].

#### 4.3.3.2.3 Microwave-assisted extraction

Microwave-assisted extraction (MAE) is an efficient method for carotenoids' extraction involved with reduced time, temperature, solvent consumption and energy input compared to conventional extraction methods. In MAE, the heat generated by microwaves provokes moisture evaporation inside the cells, thus increasing pressure on the cell wall and enhancing the extraction process [38]. MAE is reported to significantly improve the carotenoids' (lycopene and β-carotene) yields. Extraction efficiency is affected by the type of solvent; water, ethanol, and methanol can strongly absorb microwave energy compared to nonpolar solvents like hexane. Extraction yield increases when ethanol concentration increases to 50%, while the use of ethyl acetate results in a higher lycopene recovery compared to hexane. Using microwaves (400 W, 60 s) and ethyl acetate as a solvent, ensues a significant improvement in the total lycopene recovery (17.5%) [39]. However, limitations of the method are related to the nonpolar components' recovery and the chemical structure modification of the key components that may affect their bioactivity.

#### 4.3.3.2.4 Ultrasonic-assisted extraction

Ultrasonic-assisted extraction (UAE) is an efficient extraction method of bioactive compounds from food materials due to the cavitation phenomena occurring by ultrasounds, damaging the cell walls and enhancing their dissolution rate. UAE requires shorter total extraction time and reduced solvent consumption, and it is performed at lower temperatures avoiding thermal destruction of the extracts and minimizing the loss of bioactive compounds [95]. UAE is a promising technology for the extraction of carotenoids from tomato by-products (peels or seeds) as it significantly increases their total extraction yield by 43% compared to the conventional extraction without causing any degradation in carotenoids [47,57]. Ultrasounds of low frequencies (16–100 kHz) are proposed in this case. Moreover, by UAE the lycopene extraction is favored compared to the conventional solvent method, increasing yield by 26%. Additionally, a 75.93% extraction recovery of trans-lycopene was reported by Eh and Teoh [24] using ultrasonication, followed by minimal degradation and isomerization of lycopene. Limitations related to UAE concern the reduction of the in vitro bioaccessibility of the extracted lycopene and the handling of the solid tomato pomace (bubble or gelling formation) [3,37].

#### 4.3.3.2.5 High hydrostatic pressure extraction

HHPE is based on the cell membrane destruction or deformation, resulting in an increase in the mass transfer rate during extraction [40]. The method is advantageous for the extraction of carotenoids, since operating at ambient temperature protects them from heat degradation, while providing a sterilized extract that can be used in food preparations. Another advantage of HPPE is the higher yield obtained in a shorter time compared to the solvent extraction process. Experimental extractions of tomato pomace at 500 MPa with 75% ethanol for 1 min obtained a higher lycopene yield compared to UAE for 30 min [93], while HPPE extraction at 700 MPa resulted in a lycopene yield comparable to the conventional one with ethyl acetate in a shorter time (10 min against 30 min) using lower solvent volume (4/1 mL/g against 10/1 mL/g) [85]. Moreover, HPPE can provide a highly purified lycopene using tap water with minimal organic solvent use [37].

#### 4.3.3.2.6 Enzyme-assisted extraction

Enzymatic treatment is usually applied as a pretreatment prior to solvent extraction to accelerate the release of bioactive compounds. In tomato pomace or peels cellulase and pectinase enzymes have been used under mild conditions and/or ethyl lactate and significantly improved carotenoids or lycopene extractability and yields (i.e., up to 10-fold for lycopene) [17,30,49,67].

#### 4.3.3.2.7 Pulsed electric field extraction

Pulsed electric field, a nonthermal technology, enhances carotenoids' transfer rate by softening plant tissues and reducing cells' membrane integrity, as well as influencing the texture and electroporation of the plant material. It involves the application of high-voltage pulses (20–80 kV/cm) in liquid or semisolid foods between two electrodes for a short period of time. Carotenoids' accessibility has been reported to be high in samples treated with pulsed electric fields [2].

#### 4.3.3.2.8 Accelerated solvent extraction

Accelerated solvent extraction (ASE) or pressurized liquid extraction involves extraction at a constant high pressure (10–15 MPa) using ionic liquids. The high pressure improves cell permeability and facilitates penetration of the extracting solvent enhancing thus the extractability of carotenoids under reduced solvent consumption. This is also facilitated by the high temperatures (50–200°C) applied which ensure that the carotenoid-binding proteins are denatured [71].

#### 4.3.3.2.9 Final product formation

After the extraction process, the final product, apart from the carotenoids, may contain sub-extracted components such as lipids, fatty acids, and esters while some carotenoids like xanthophylls are often present in their esterified form. Consequently, carotenoids are purified, if necessary, by saponification to eliminate undesirable lipids and isolated in pure crystalline form. To increase their bioaccessibility they are incorporated in emulsions which are used in food products to enhance their bioavailability.

### 4.3.4 Fiber

Soluble dietary fibers can also be obtained (12.1%–12.9%) from tomato peels as well as tomato pomace pectin of high purity [34,51].

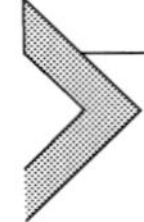

## 4.4 Uses of products in foods, feeds, pharmaceuticals - functional foods

The tomato processing by-products may have many applications, due to their high protein content including animal-feeds, functional ingredients for food products, and raw materials for nutraceuticals [56]. More specifically, dry tomato pressing residues can be used as a fodder for animals and especially poultry since they do not contain antinutritive substances. Tomato seed meal is richer in lysine by approximately 13%, compared to the soybean meal, which is commonly applied as a protein supplement in poultry feed. This difference in nutritional value due to the higher presence of lysine in the tomato seed meal can accordingly affect yolk weight and the rate of egg production in poultry [66].

Tomato seed proteins due to their high solubility can be used in soups to enhance flavor, texture, consistency and nutritional value. They can also be used in mayonnaise to increase consistency due to their high emulsification and water absorption capacity. The high percentage of lysine and threonine, also, make tomato seeds a remarkable alternative option for the increase of protein quality in cereal products [66]. Moreover, the tomato seed meal contains glutamic and aspartic acid at the levels of about 19 and 12 g/100g, respectively, thus being a good candidate for food products rich in umami flavors [16]. Also, the use of tomato seed flour (up to 10%) enriches the wheat flour in lysine and improves the breadmaking properties (dough stability, water holding and loaf volume). The dried tomato seeds exhibit antioxidant activity due to the tocopherols of the containing oil.

Tomato seed oil is recovered by extraction from dried seeds, and after refining is suitable to be used as an edible oil (salad oil) or as an ingredient in cosmetics [62]. The global tomato seed oil production can exceed 0.14 million tons per year. Its price is determined by the collection, transportation and processing costs. The integration of innovative extraction techniques may reduce processing cost and enhance the competitiveness and utilization of tomato seed oil. Tomato seed oil, being a significant source of essential fatty acid, can be included, as a dietary supplement, in diets deficient in essential fatty acids or in nutraceuticals and antiobesity formulations [32]. Due to its high content in linoleic acid, tomato seed oil can be employed in the food industry to enhance the nutritional value of various food products (i.e. shortenings), be incorporated into food products that are packed in oil (i.e. tuna or vegetables), or be included into the formulations of dried tomatoes and sauces [66]. Medical applications of this oil are also reported, due to its cholesterol-lowering properties or action against eczema, aging, UV damage of the skin, and psoriasis. Additionally, it can be an important ingredient for cosmetics (i.e. cream cleansers) as it has been classified as irritation and sensitizer safe. Other applications of tomato seed oil include lubricants, paints, varnishes, and soaps. Finally, tomato seeds oil is considered a good source for the production of biofuels due to its low sulfur and ash levels and its high density and iodine values [31,35].

Carotenoids exhibit significant health-promoting functions as previously presented and due to their fat-soluble properties, they are suitable to be incorporated into several food products. Besides their biological properties, they are utilized as natural antioxidants in food formulations; they are preferred by consumers over chemical antioxidants to elongate the shelf-life of food products [43,69]. Moreover, lycopene, due to its high conjugation degree, is considered as one of the most potent antioxidants. Lycopene and β-Carotene are also authorized natural pigments that are used to color various food products. Lycopene, an important nutrient, has been extensively studied as a result of its antioxidant properties and preventive activity against several pathologies (i.e., cardiovascular disease, cancer) [52]. Those characteristics make it an excellent component for a wide range of functional foods. Generally, carotenoids are used as natural colorants, antioxidants, nutraceutical supplements in foods as well as in feeds, medical and cosmetic applications.

Because of the versatile health-promoting properties of carotenoids, their global market share was estimated at about 1.8 billion US$ in 2019, plus presenting a continuously increasing trend, while many patents have lately been based on natural carotenoid sources [86]. Carotenoids have high market potential, with β-Carotene being predominant and followed by lutein, capsanthin, and zeaxanthin. Chemically synthesized carotenoids are also commercially available and are largely used for similar purposes [69]. These are specially formulated and distributed in the market as water-soluble and stable emulsions minimizing oxidation or isomerization. Nevertheless, since chemically synthesized carotenoids are linked to high toxicity, carcinogenicity, and teratogenicity properties, nowadays there is a demand from health-conscious consumers for carotenoids extracted from natural resources. This also justifies the interest of researchers for the recovery of carotenoids from tomato processing by-products and their use in food, cosmetic, and pharmaceutical products.

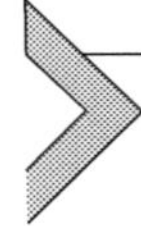

## 4.5 Utilization—added value products—nonpurified by-products

According to the above, tomato processing by-products (tomato pomace) may be used directly or further utilized for the delivery of food and high added value products (Fig. 4.2). Thus, tomato pomace, after pressing and sun-drying, can be used as animal feed due to its nutrient content (oil 11%–20% and protein 15%–22%) or further subjected to simple pretreatments in order to be purified, ground and standardized as "flour" for foods and feeds uses. An upgraded utilization of tomato pomace is proposed starting from the whole tomato pomace or even better from the seeds for the delivery of "tomato oil" and "defatted flour." Moreover, high added-value protein products (protein concentrates and protein isolates) may derive from the tomato seed meal (defatted flour). Tomato oil has a high unsaturated acid content and can be almost completely recovered by solvent extraction and used as an edible oil after refining. The tomato protein products have high lysine content compared favorably with those from soybeans and can be used as a valuable source of protein supplementation in several food products. Especially, regarding the protein isolates, although the protein recovery (yield) reaches 20%–35% of the meal's weight, it results in high added-value isolates (protein content >90%), thus justifying the sustainability of the protein isolation process. The bioactive components of tomato pomace (carotenoids including lycopene and β-Carotene) can be recovered from the peels (which are richer in carotenoids) and/or the whole pomace through suitable techniques (i.e., supercritical fluid extraction with $CO_2$ that provides high yields and high quality extracts) and further utilized in food and pharmaceutical products. Finally, the solid residue remaining from the production of protein isolates can be used as a fiber source in animal feeds or other applications.

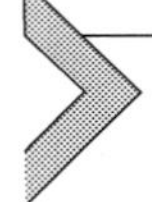

## 4.6 Future trends - innovative technologies (for isolation of valuable components)

As already mentioned, tomato processing is one of the most important sectors of the agricultural industry. However, at the same time, it involves the production of significant quantities of tomato pomace, containing nutritive components such as oil and protein, as well as multiple compounds, such as tocopherols, polyphenols, carotenoids, terpenes, and sterols, with bioactive or antioxidant properties, which is usually available as a fertilizer or dried and used as an animal feed.

In recent years, in the context of circular economy and in order to utilize the by-products of the food industry for the production of new products with high nutritional and commercial value, the isolation, receipt and exploitation of natural ingredients contained in tomato pomace has been proposed and implemented on a limited scale [20].

Nevertheless, the isolation of bioactive ingredients from tomato by-products is mainly based on conventional methods of solid-liquid extraction which involve the intensive use of organic solvents (e.g., methanol, ethanol, acetone), acid or alkali solutions, while exhibit long extraction times and relatively low extraction efficiency. In order to reduce the environmental impact caused by the use of organic or toxic solvents and to improve the quality characteristics of the ingredients received, the application of novel, upgraded and "green" pretreatment or isolation techniques is proposed. Toward this direction, the use of microwaves, ultrasounds, enzymes, supercritical fluids, pressurized liquids, high hydrostatic pressure, pulsed electric field, membrane technology, and cold plasma have been introduced to assist or replace conventional extraction. Moreover, the food industry inclines towards the use of ethanol and water as solvents for the separation of valuable compounds [7,20,88].

Microwave-assisted extraction (MAE) uses microwave energy to heat solvents in contact with the sample as well as to heat up the internal moisture of the plant cells, thus generating pressure inside the cell walls. This facilitates the penetration of the extraction solvent inside the cell walls which are finally ruptured allowing the easier delivery of the desirable substances. MAE exhibits efficient extraction yields and multiple advantages such as rapidity and reduced solvent and energy consumption [29,88]. In fact, processing times and solvent volumes can be reduced up to 10 times compared to the conventional extraction [83].

Ultrasound-assisting extraction (UAE) is based on the acoustic-induced cavitation in a liquid medium. This leads to the formation of continuously growing bubbles inside the liquid. When these bubbles finally collapse adiabatically, they result in a violent implosion that disrupts cell walls allowing the release of bioactive compounds. Furthermore, cavitation modifies the chemical processes in the system and enhances the reaction rates. UAE offers many advantages such as improved yields and extraction rates, limited solvent consumption, lower processing temperatures for heat sensitive components, low processing times, and the possibility of using alternative (GRAS) extraction solvents [26].

Enzyme-assisted extraction (EAE) involves the enzymatic pre-treatment of the plant material prior to conventional extraction. Hydrolytic enzymes (e.g., pectinase and cellulose) are ideal catalysts that facilitate the extraction, modification and/or synthesis of complex bioactive compounds of natural origin. EAE is based on the inherent ability of enzymes to catalyze specific reactions and the ability to act under mild treatment conditions in aqueous solutions. Enzymes can break down the structure of cell walls and membranes, thereby exposing the intracellular materials and allowing the more efficient release, diffusion and extraction of the bioactive compounds [61].

Supercritical fluid extraction (SFE) is a process based on the use of solvents above or near their critical temperature and pressure in order to facilitate the extraction of organic compounds from plant material. Under these conditions, the supercritical fluid is in a liquid-gas intermediate state, with a density similar to that of a liquid and a viscosity similar to that of a gas. The choice of a supercritical fluid, usually carbon dioxide, allows for a more selective extraction, and provides faster reaction kinetics compared to the use of conventional solvents. The solvent can also be easily removed from the mixture by pressure reduction and/or temperature adjustment. SFE is an environmentally friendly extraction method that exhibits high selectivity and short extraction times, as carbon dioxide is an inexpensive, nonflammable, and nontoxic solvent. In addition, supercritical fluids have relatively low surface tensions which prevent the deterioration of vulnerable components [18,29].

Pressurized liquid extraction (PLE), also referred to as accelerated solvent extraction (ASE), uses organic liquid solvents at temperatures of 50–200°C and pressures of 500–3000 psi. The high extraction temperature increases the ability of solvents to penetrate cell walls and dissolve bioactive compounds, and facilitates the rupture of the cohesive forces between those components and the plant material. PLE combines the benefits of high throughput, automation, and low solvent consumption, although expensive equipment is required [18].

The application of high hydrostatic pressure for the extraction of bioactive ingredients from natural sources is considered an emerging technique. The extraction is performed at high pressures (usually between 100–800 MPa) and low temperatures (usually up to 60°C), using small volumes of solvents, while providing yields similar to other techniques. The application of ultra-high pressure is directly related to the solubility of the bioactive components. In particular, the pressure causes changes in the membrane of tomato pomace cells, thus increasing their permeability, facilitating the penetration of the extraction solvent inside the cells and accelerating the mass transfer rate of the bioactive ingredients from the plant material to the solvent. The main advantages of HHPE are the short extraction times, the higher yields, the low energy consumption and the application of low temperatures that help maintain the activity and structure of the bioactive components [86].

Pulsed electric fields (PEF) assisted extraction is also considered a promising technology for the extraction of valuable compounds from plant sources. PEF is based on the pre-treatment of the plant tissues using an electric field of moderate intensity (<10 kV/cm) and relatively low energy (<10 kJ/kg) which induces the electroporation of cell membranes and thus enhances their permeability and enables the recovery of valuable intracellular compounds. PEF-assisted extraction has an advantage over conventional techniques as it involves a non-thermal process that preserves the quality characteristics of the extracted products. It can also be applicable in a continuous flow [58,64].

In addition, membrane technology can offer an alternative technique for the isolation of valuable components from plant material. Separation is achieved either through size exclusion, based on the membrane pore diameter, or on differences in the solubility and diffusion rates of the individual analytes in the membrane material [18]. Membrane technology involves low temperature processing and displays lower operating and energy cost and higher yields compared to conventional extraction. Ultrafiltration and nanofiltration have been studied for the separation of lycopene from tomato products [4].

Moreover, the use of cold plasma can induce surface modifications in the cellular membranes of plant tissues and therefore decrease the resistance to diffusion of internal molecules, thus facilitating the recovery of phenolic compounds from tomato pomace [7]. High voltage atmospheric cold plasmas generated with different working gases (air, argon, helium, and nitrogen) have been applied for this purpose [6].

Finally, an interesting approach has been proposed for the full valorization of tomato processing by-products. More specifically, the solid residues, which remain after the extraction of bioactive compounds from tomato peels and seeds, can be sufficiently utilized for the production of a low-cost biosorbent useful for the removal of dyes in various applications [5].

## References

[1] N. Ahmadi Kamazani, H. Tavakolipour, M. Hasani, M. Amiri, Evaluation and analysis of the ultrasound-assisted extracted tomato seed oil, J. Food Biosci. Technol. 4 (2) (2014) 57–66.

[2] V. Andreou, G. Dimopoulos, E. Dermesonlouoglou, P. Taoukis, Application of pulsed electric fields to improve product yield and waste valorization in industrial tomato processing, J. Food Eng 270 (April) (2020) 109778. https://doi.org/10.1016/j.jfoodeng.2019.109778.

[3] M. Anese, F. Bot, A. Panozzo, G. Mirolo, G. Lippe, Effect of ultrasound treatment, oil addition and storage time on lycopene stability and in vitro bioaccessibility of tomato pulp, Food Chem. 172 (April) (2015) 685–691.
[4] F.A. Arana Rodriguez, Membrane Separation of Bioactive Lycopene from Tomato Juice. Master of Science Thesis, Texas A&M University, 2004.
[5] S. Azabou, I. Louati, F. Ben Taheur, M. Nasri, T. Mechichi, Towards sustainable management of tomato pomace through the recovery of valuable compounds and sequential production of low-cost biosorbent, Environ. Sci. Pollut. Res. 27 (31) (2020) 39402–39412.
[6] Y. Bao, Innovative cold plasma-assisted extraction for bioactive compounds from agricultural byproducts, Master of science thesis. Purdue University, 2020.
[7] Y. Bao, L. Reddivari, J.-Y. Huang, Development of cold plasma pretreatment for improving phenolics extractability from tomato pomace, Innov. Food Sci. Emerg. Technol. 65 (1) (2020) 102445.
[8] C. Botineştean, A.T. Gruia, I. Jianu, Utilization of seeds from tomato processing wastes as raw material for oil production, J. Mater. Cycles Waste Manag. 17 (1) (2014) 118–124. https://doi.org/10.1007/s10163-014-0231-4.
[9] C. Botineştean, N.G. Hădărugă, D.I. Hădărugă, I. Jianu, Fatty acids composition by gas chromatography-mass spectrometry (GC-MS) and most important physical-chemicals parameters of tomato seed oil, J. Agroaliment. Process. Technol. 18 (1) (2012) 89–94.
[10] D. Brodowski, J.R. Geisman, Protein content and amino acid composition of protein of seeds from tomatoes at various stages of ripeness, J. Food Sci. 45 (2) (1980) 228–229. https://doi.org/10.1111/j.1365-2621.1980.tb02582.x.
[11] CAC, Codex Standard for Named Vegetable Oils, FAO, Rome, Italy, 1999 http://wenku.baidu.com/view/966a07ffc8d376eeaeaa3167.html (Accessed 2 July 2014).
[12] M.M. Calvo, D. Dado, G. Santa-Maria, Influence of extraction with ethanol or ethyl acetate on the yield of lycopene, beta-carotene, phytoene and phytofluene from tomato peel powder, Eur. Food Res. Technol 224 (5) (2007) 567–571.
[13] M. Canella, G. Castriotta, Protein composition and solubility of tomato seed meal, Lebensm. Wiss. Technol. 13 (1) (1980) 18–21.
[14] S. Chanioti, G. Liadakis, C. Tzia, Solid–liquid extraction, in: T. Varzakas, C. Tzia (Eds.), Food Engineering Handbook: Food Process Engineering, 2, CRC Press, Boca Raton, FL, USA, 2014, pp. 247–280.
[15] P. Choksi, V.Y. Joshi, A Review on lycopene-extraction, purification, stability and applications, Int. J. Food Prop 10 (2) (2007) 289–298.
[16] V. Coman, B.E. Teleky, L. Mitrea, G.A. Martău, K. Szabo, L.F. Călinoiu, D.C. Vodnar, Bioactive potential of fruit and vegetable wastes. In: Advances in Food and Nutrition Research 91, Academic Press, Cambridge, MA, USA, 2020 157–225. https://doi.org/10.1016/bs.afnr.2019.07.001.
[17] S. Cuccolini, A. Aldini, L. Visai, M. Daglia, D. Ferrari, Environmentally friendly lycopene purification from tomato peel waste: enzymatic assisted aqueous extraction, J. Agric. Food Chem 61 (8) (2013) 1646–1651.
[18] M. De la Guardia, S. Armenta, Greening sample treatments, Green Analytical Chemistry, Elsevier Inc, London UK, 2011, https://doi.org/10.1016/B978-0-444-53709-6.00005-7.
[19] G.S. Dhillon, S. Kaur, H.S. Oberoi, M.R. Spier, S.K. Brar, Agricultural-Based Protein By-Products: Characterization and Applications, Protein Byproducts: Transformation from Environmental Burden into Value-Added Products, Elsevier Inc, London, UK, 2016. https://doi.org/10.1016/B978-0-12-802391-4.00002-1.
[20] R. Domínguez, P. Gullón, M. Pateiro, P.E.S. Munekata, W. Zhang, G.M Lorenzo, Tomato as potential source of natural additives for meat industry. A review, Antioxidants. 9 (1) (2020) 73.

[21] G. Doxastakis, V. Kiosseoglou, D. Boskou, Emulsifying properties of tomato seed proteins obtained from processing waste, Sci. Aliments 8 (2) (1988) 259–267.

[22] N.J. Drouliscos, Nutritional evaluation of the protein of dried tomato pomace in the rat, Br. J. Nutr 36 (3) (1976) 449–456.

[23] L.K. Eggers, Some biochemical and electron microscopic studies of the protein present in seeds recovered from tomato cannery waste. In: Ph.D. dissertation, Ohio State Univ., Columbus, OH, USA, 1974.

[24] A.L.-S. Eh, S.-G. Teoh, Novel modified ultrasonication technique for the extraction of lycopene from tomatoes, Ultrason. Sonochem 19 (1) (2012) 151–159. https://doi.org/10.1016/j.ultsonch.2011.05.019.

[25] F.J. Eller, J.K. Moser, J.A. Kenar, S.L. Taylor, Extraction and analysis of tomato seed oil, J. Am. Oil Chem. Soc. 87 (7) (2010) 755–762. https://doi.org/10.1007/s11746-010-1563-4.

[26] M.D. Esclapez, J.V. García-Pérez, A. Mulet, J.A. Cárcel, Ultrasound-assisted extraction of natural products, Food Eng. Rev 3 (May) (2011) 108.

[27] G. Fazio, G. Arcoleo, L. Pirrone, Sul contenutoproteico del seme di pomodoro, Riv. Soc. Ital. Sci. Aliment. 12 (3) (1983) 195–200.

[28] C.M. Galanakis, Recovery of high added-value components from food wastes: conventional, emerging technologies and commercialized applications, Trends Food Sci. Technol. 26 (2) (2012) 68–87.

[29] C.M. Galanakis, Food Waste Recovery Processing Technologies and Industrial Techniques, Academic Press, London, UK, 2015, pp. 132–144.

[30] L. Gardossi, P. Poulsen, A. Ballesteros, K. Hult, V. Švedas, D. Vasić-Rački, et al., Guidelines for reporting of biocatalytic reactions, Trends Biotechnol. 28 (4) (2009) 171–180.

[31] P.N. Giannelos, S. Sxizas, E. Lois, F. Zannikos, G. Anastopoulos, Physical, chemical and fuel related properties of tomato seed oil for evaluating its direct use in diesel engines, Ind. Crops Prod. 22 (3) (2005) 193–199. https://doi.org/10.1016/j.indcrop.2004.11.001.

[32] A.M. Giuffrè, M. Capocasale, Physicochemical composition of tomato seed oil for an edible use: the effect of cultivar. Int, Food Res. J. 23 (2) (2016) 583–591.

[33] W.A. Gould, Tomato Production, Processing & Technology., 3rd edition, CTI Publications Inc., Baltimore, MD, USA, 1992.

[34] A.N. Grassino, J. Halambek, S. Djaković, S. RimacBrnčić, M. Dent, Z. Grabarić, Utilization of tomato peel waste from canning factory as a potential source for pectin production and application as tin corrosion inhibitor, Food Hydrocoll 52 (January) (2016) 265–274.

[35] Z.P. Gumus, Z. Ustun Argon, V.U. Celenk, S. Timur, Cold Pressed Tomato (Lycopersicon esculentum L.) Seed Oil, Cold Pressed Oils: Green Technology, Bioactive Compounds, Functionality, and Applications, Elsevier Inc, London, UK, 2020. https://doi.org/10.1016/b978-0-12-818188-1.00040-2.

[36] M. Haddadin, S. Haddadin, Lycopene extraction from tomato pomace with supercritical carbon dioxide: effect of pressures, temperatures and $CO_2$ flow rates and evaluation of antioxidant activity and stability of lycopene, Pak. J. Nutr 14 (12) (2015) 942–958.

[37] S. Haroon, Thesis, Master of Science (MSc), University of Waikato, Hamilton, New Zealand, 2014.

[38] B. Hiranvarachat, S. Devahastin, Enhancement of microwave-assisted extraction via intermittent radiation: extraction of carotenoids from carrot peels, J. Food Eng 126 (April) (2014) 17–26.

[39] K.K.H.Y. Ho, M.G. Ferruzzi, A.M. Liceaga, M.F.S Martin-Gonzalez, Microwave-assisted extraction of lycopene in tomato peels: Effect of extraction conditions on all-trans and cis-isomer yields, LWT - Food Sci. Technol 62 (1) (2015) 160–168.

[40] X. Jun, Ultrahigh pressure extraction of bioactive compounds from plants - a review, Crit. Rev. Food Sci. Nutr 57 (6) (2017) 1097–1106.
[41] N. Kalogeropoulos, A. Chiou, V. Pyriochou, A. Peristeraki, V.T. Karathanos, Bioactive phytochemicals in industrial tomatoes and their processing byproducts, LWT - Food Sci. Technol 49 (2) (2012) 213–216.
[42] M. Kehili, M. Kammlott, S. Choura, A. Zammel, C. Zetzl, I. Smirnova, E. Al., Supercritical $CO_2$ extraction and antioxidant activity of lycopene and beta-caroteneenriched oleoresin from tomato (Lycopersicum esculentum L.) peels by-product of a Tunisian industry, Food Bioprod. Process 102 (March) (2017) 340–349.
[43] M. Khalid, B. Saeed-ur-Rahman, I. M., M.N. Hafiz, D. Huang, Biosynthesis and biomedical perspectives of carotenoids with special reference to human health-related applications, Biocatal. Agric. Biotechnol 17 (January) (2019) 399–407.
[44] M. Knoblich, B. Anderson, D. Latshaw, Analyses of tomato peel and seed byproducts and their use as a source of carotenoids, J. Sci. Food Agric 85 (7) (2005) 1166–1170.
[45] A. Kramer, W.H. Kwee, Utilization of tomato processing wastes, J. Food Sci. 42 (1) (1977) 212–215..
[46] A. Kramer, W.H. Kwee, Functional and nutritional properties of tomato protein concentrates, J. Food Sci. 42 (1) (1977) 207–211.
[47] S. Kumcuoglu, T. Yilmaz, S. Tavman, Ultrasound assisted extraction of lycopene from tomato processing wastes, J. Food Sci. Technol. 51 (12) (2014) 4102–4107.
[48] S.J. Latlief, D. Knorr, Tomato seed protein concentrates: effects of methods of recovery upon yield and compositional characteristics, J. Food Sci. 48 (6) (1983) 1583–1586. https://doi.org/10.1111/j.1365-2621.1983.tb05036.x.
[49] R. Lavecchia, A. Zuorro, Improved lycopene extraction from tomato peels using cell-wall degrading enzymes, European Food Research and Technology 228 (1) (2008) 153–158.
[50] E.S. Lazos, P. Kalathenos, Technical note: composition of tomato processing wastes, Int. J. Food Sci. Technol 23 (6) (1988) 649–652. https://doi.org/10.1111/j.1365-2621.1988.tb01052.x.
[51] N. Li, Z. Feng, Y. Niu, L. Yu, Structural, rheological and functional properties of modified soluble dietary fiber from tomato peels, Food Hydrocoll 77 (April) (2018) 557–565.
[52] N. Li, X. Wu, W. Zhuang, L. Xia, Y. Chen, C. Wu, Z. Rao, L. Du, R. Zhao, M. Yi, Q. Wan, Y. Zhou, Tomato and Lycopene and Multiple Health Outcomes: Umbrella Review, Food Chem. 343 (May) (2021) 128396. https://doi.org/10.1016/j.foodchem.2020.128396.
[53] G.N. Liadakis, Utilization of tomato-processing by-products, PhD Thesis, National Technical University of Athens, Athens, Greece, 1999.
[54] G.N. Liadakis, C. Tzia, V. Oreopoulou, C.D. Thomopoulos, Isolation of tomato seed meal proteins with salt solutions, J. of Food Sci. 63 (3) (1998) 450–453.
[55] G.N. Liadakis, C. Tzia, V. Oreopoulou, C.D. Thomopoulos, Protein isolation from tomato seed meal, extraction optimization, J. Food Sci. 60 (3) (1995) 477–482. https://doi.org/10.1111/j.1365-2621.1995.tb09807.x.
[56] Z. Lu, J. Wang, R. Gao, F. Ye, G. Zhao, Sustainable valorisation of tomato pomace: a comprehensive review, Trends Food Sci. Technol. 86 (April) (2019) 172–187. https://doi.org/10.1016/j.tifs.2019.02.020.
[57] E. Luengo, S. Condon-Abanto, S. Condon, J. Raso, Improving the extraction of carotenoids from tomato waste by application of ultrasound under pressure, Sep. Purif. Technol. 136 (November) (2014) 130–136.
[58] E. Luengo, I. Álvarez, J. Raso, Improving carotenoid extraction from tomato waste by pulsed electric fields, Front. Nutr. 1 (12) (2014) 1–10.
[59] Y. Ma, J. Ma, T. Yang, W. Cheng, Y. Lu, Y. Cao, J. Wang, S. Feng, Components, antioxidant and antibacterial activity of tomato seed oil, Food Sci. Technol. Res. 20 (1) (2014) 1–6. https://doi.org/10.3136/fstr.20.1.

[60] M. Mechmeche, F. Kachouri, M. Chouabi, H. Ksontini, K. Setti, M. Hamdi, Optimization of extraction parameters of protein isolate from tomato seed using response surface methodology, Food Anal. Methods. 10 (August) (2017) 809–819. https://doi.org/10.1007/s12161-016-0644-x.
[61] J. Nagarajan, H.P. Kay, N.P. Krishnamurthy, N.R. Ramakrishnan, T.M.S. Aldawoud, C.M. Galanakis, O.C Wei, Extraction of carotenoids from tomato pomace via water-induced hydrocolloidal complexation, Biomolecules 10 (7) (2020) 1019.
[62] V. Oreopoulou, C. Tzia, Utilization of plant by-products for the recovery of proteins, dietary fibers, antioxidants, and colorants. Utilization of By-Products and Treatment of Waste in the Food Industry, Springer, Boston, MA, USA, 2006 209–232.
[63] S.H. Park, S.R. Bean, Investigation and optimization of the factors influencing sorghum protein extraction, J. Agric. Food Chem. 51 (24) (2003) 7050–7054. https://doi.org/10.1021/jf034533d.
[64] G. Pataro, D. Carullo, M.A. Bakar Siddique, M. Falcone, F. Donsì, G. Ferrari, Improved extractability of carotenoids from tomato peels as side benefits of PEF treatment of tomato fruit for more energy-efficient steam-assisted peeling, J. Food Eng 233 (September) (2018) 65–73.
[65] M.E. Persia, C.M. Parsons, M. Schang, J. Azcona, Nutritional evaluation of dried tomato seeds, Poult. Sci. 82 (1) (2003) 141–146. https://doi.org/10.1093/ps/82.1.141.
[66] S. Porretta, Tomato Chemistry, Industrial Processing and Product Development, Royal Society of Chemistry, London, UK, 2019.
[67] M. Puri, D. Sharma, C. Barrow, Enzyme-assisted extraction of bioactives from plants, Trends Biotechnol 30 (1) (2012) 37–44.
[68] A.V. Rao, Lycopene, tomatoes, and the prevention of coronary heart disease, Exp Biol Med 227 (10) (2002) 908–913.
[69] D. Rodriguez-Amaya, Food Carotenoids: Chemistry, biology and technology, John Wiley & Sons, Ltd., Chicago, IL, USA, 2016.
[70] B.C. Roy, M. Goto, T. Hirose, Temperature and pressure effects on supercritical $CO_2$ extraction of tomato seed oil, Int. J. Food Sci. Technol. 31 (2) (1996) 137–141
[71] R.K. Saini, Y.-S. Keum, Carotenoid extraction methods: a review of recent developments, Food Chem 240 (February) (2018) 90–103.
[72] B. Salcedo-Chávez, J.A. Osuna-Castro, F. Guevara-Lara, J. Domínguez-Domínguez, O. Paredes-López, Optimization of the isoelectric precipitation method to obtain protein isolates from amaranth (Amaranthus cruentus) seeds, J. Agric. Food Chem. 50 (22) (2002) 6515–6520. https://doi.org/10.1021/jf020522t.
[73] A. Sarkar, P. Kaul, Evaluation of tomato processing by-products: A comparative study in a pilot scale setup, J. Food Process Eng. 37 (3) (2014) 299–307. https://doi.org/10.1111/jfpe.12086.
[74] I. Seikova, E. Simeonov, E. Ivanova, Protein leaching from tomato seed-experimental kinetics and prediction of effective diffusivity, J. Food Eng. 61 (2) (2004) 165–171. https://doi.org/10.1016/S0260-8774(03)00083-9.
[75] D. Shao, C. Venkitasamy, X. Li, Z. Pan, J. Shi, B. Wang, H.E. Teh, T.H. McHugh, Thermal and storage characteristics of tomato seed oil, LWT - Food Sci. Technol. 63 (1) (2015) 191–197. https://doi.org/10.1016/j.lwt.2015.03.010.
[76] X. Shen, S. Xu, Supercritical $CO_2$ extraction of tomato seed oil, J. Food Technol. 3 (2) (2005) 226–231. https://doi.org/jftech.2005.226.231.
[77] J. Shi, M. Le Maguer, Lycopene in tomatoes: chemical and physical properties affected by food processing, Crit. Rev. Food Sci. Nutr 40 (1) (2000) 1–42.
[78] J. Shi, Y. Wu, M. Bryan, M. Le Maguer, Oxidation and isomerization of lycopene under thermal treatment and light irradiation in food processing, Prev. Nutr. Food Sci. 7 (2) (2002) 179–183.

[79] J. Shi, C.Yi, S.J. Xue,Y. Jiang,Y. Ma, D. Li, Effects of modifiers on the profile of lycopene extracted from tomato skins by supercritical $CO_2$, J. Food Eng. 93 (4) (2009) 431–436.

[80] D.S. Sogi, M.S. Arora, S.K. Garg, A.S. Bawa, Fractionation and electrophoresis of tomato waste seed proteins, Food Chem. 76 (4) (2002) 449–454. https://doi.org/10.1016/S0308-8146(01)00304-1.

[81] D.S. Sogi, R. Bhatia, S.K. Garg, A.S. Bawa, Biological evaluation of tomato waste seed meals and protein concentrate, Food Chem. 89 (1) (2005) 53–56.

[82] J. Song, X. Wang, D. Li, C. Liu, Degradation kinetics of carotenoids and visual colour in pumpkin (Cucurbita maxima L.) slices during microwave-vacuum drying, Int. J. Food Prop 20 (S1) (2017) S632–S643. https://doi.org/10.1080/10942912.2017.1306553.

[83] C. SparrEskilsson, E. Björklund, Analytical-scale microwave-assisted extraction, J. Chromatogr. A 902 (1) (2000) 227–250.

[84] I.F. Strati, V. Oreopoulou, Process optimisation for recovery of carotenoids from tomato waste, Food Chem. 129 (3) (2011) 747–752.

[85] I. Strati, E. Gogou, V. Oreopoulou, Enzyme and high pressure assisted extraction of carotenoids from tomato waste., Food Bioprod. Process 94 (April) (2015) 668–674.

[86] I.F Strati, V. Oreopoulou, Recovery of carotenoids from tomato processing by-products – a review, Food Res. Int 65 (November) (2014) 311–321.

[87] B. Tchorbanov, G. Ushanova, V. Litchev, Lipid-protein concentrations from tomato and paprika seeds, Nahrung 30 (3-4) (1986) 408–410. https://doi.org/10.1002/food.19860300359.

[88] C. Tsaltaki, M. Katsouli, T. Kekes, S. Chanioti, C. Tzia, Comparison study for the recovery of bioactive compounds from Tribulus terrestris, Panax ginseng, Gingko biloba, Lepidium meyenii, Turneradiffusa and Withaniasomnifera by using microwave-assisted, ultrasound-assisted and conventional extraction methods, Ind. Crops Prod 142 (December) (2019) 111875.

[89] M.T. Turakhozhaev, R.A. Alibekova, S.R. Rakhmetova, L.P. Zubkova, T.T. Shakirov, Isolation of protein from tomato seed meal, Chem. Nat. Compd. 15 (November) (1979) 734–736.

[90] C. Tzia, V. Giannou, V. Polychniatou, S. Chanioti, Fats and oils processing technology, in: T. Varzakas, C. Tzia (Eds.), Handbook of Food Processing: Food Safety, Quality, and Manufacturing Processes, 2, CRCPress, Boca Raton, FL, USA, 2015, pp. 381–424.

[91] E. Vági, B. Simándi, K.P. Vásárhelyine, H. Daood, Á. Kéry, F. Doleschall, B. Nagy, Supercritical carbon dioxide extraction of carotenoids, tocopherols and sitosterols from industrial tomato by-products, J. Supercrit. Fluids. 40 (2) (2007) 218–226.

[92] WHO/FAO/UNU. 2007. Protein and amino acid requirements of infants and children, Report of a Joint WHO/FAO/UNU Expert Consultation. WHO Technical Report Series. Geneva, Switzerland.

[93] J. Xi, Effect of high pressure processing on the extraction of lycopene in tomato paste waste, Chem. Eng. Technol. 29 (6) (2006) 736–739.

[94] E. Yilmaz, B. Aydeniz, O. Güneşer, E.S. Arsunar, Sensory and physico-chemical properties of cold press-produced tomato (Lycopersicon esculentum L.) seed oils. JAOCS, J. Am. Oil Chem. Soc. 92 (6) (2015) 833–842. https://doi.org/10.1007/s11746-015-2648-x.

[95] L. Zhang, L. Zelong, Optimization and comparison of ultrasound/microwave assisted extraction (UMAE) and ultrasonic assisted extraction (UAE) of lycopene from tomatoes, Ultrason. Sonochem 15 (5) (2008) 731–737.

[96] Y. Zhang, Z. Pan, C. Venkitasamy, H. Ma, Y. Li, Umami taste amino acids produced by hydrolyzing extracted protein from tomato seed meal, LWT - Food Sci. Technol. 62 (2) (2015) 1154–1161. https://doi.org/10.1016/j.lwt.2015.02.003.

CHAPTER FIVE

# Tomato wastes valorization for bio-based materials production

**Selsabil Elghazel Jeguirim**
Textile Research Unit of ISET of Ksar-Hellal, Ksar Hellal, Tunisia

## 5.1 Introduction

Tomato is the edible fruit of the plant *Lycopersicum esculentum*, which belongs to the nightshade family of Solanides (*Solanaceae*). Tomato is, according to the Food and Agriculture Organization, the second most important cultivated plants around the world, for commercial or domestic purposes [1]. In particular, tomato fruits production has increased in the last decades from $1.1 \times 10^8$ tonnes in 1999 to $1.8 \times 10^8$ tonnes [1]. The main tomato producers are China, India, Turkey, United States of America, and Egypt, accounting for 34.8%, 10.5%, 7.1%, 6.0%, and 3.7% of total world production in 2019, respectively [1].

Tomatoes cultivation, harvesting and industrial processing generate significant quantities of solid and liquid wastes that are usually discarded [2]. Cultivation and harvesting generate various plant residues including stem, leaves and roots as well as damaged or defected fruits that do not reach the required standards to be consumed. Industrial tomato processing, however, generates to the production of a significant volume of by-products known as tomato pomace. This latter is composed of a mixture of peels, seeds, and small amounts of pulp, and it can represent between 1.5% and 10% of the weight of the original fruits [3–5]. In addition, the tomato processing industry generates huge quantities of sludge (1.5% of the tomato) from the wastewater treatment [6]

The environmental management of the different tomato processing wastes is considered a problem for the tomato industry, since it has no commercial value and, in most cases, remains underutilized, is disposed as waste or used to a limited extent. Therefore, the exploitation of tomato processing by-products can provide an extra income for farmers and industrials. In addition, this valorization way can contribute in reducing the environmental problems caused by their accumulation, and transforming a waste into a usable resource.

*Tomato Processing by-Products*
DOI: https://doi.org/10.1016/B978-0-12-822866-1.00005-3

Recently, several investigations have focused of tomato waste conversion into renewable useful products such as biodiesel, biopolymers, and biochemical products using environmental friendly strategies [7–10].

The aim of this chapter is to identify the recent progress on the use of tomato by-products in bio-based materials production and their corresponding applications.

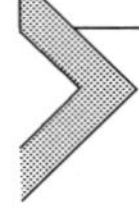

## 5.2 Tomato waste characterizations in bio-based materials synthesis

### 5.2.1 Chemical composition

Different analytical methods were used to characterize the chemical composition of different tomato by-products generated during cultivation and industrial processing.

1. Polymers content

   Table 5.1 summarizes the results of major polymers contents (hemicellulose, cellulose and lignin) in tomato stalks. These polymer contents are among the primary criteria for the selection of suitable agro wastes for bio-based materials applications.

   Table 5.1 shows that that tomato stalks have a fair cellulose content of 43.1% which in the same range of common agricultural wastes [14–15]. However, the cellulose content decreases when stalks are mixed with other tomato plant residues such as branches and leaves. Since cellulose concentration is a critical parameters for strength applications, Covino et al. have applied enzymatic treatment (ET) to increase remarkably the cellulose content from 28.10% to 53.05% [12]. Such value is interesting for the economic viability of particleboard application.

2. Elemental analysis

   The ultimate analysis of residues generated from tomato cultivation and processing industries are given in Table 5.2. These results values are comparable to values met in literature and show that tomato waste may be an interesting source of carbonaceous materials.

**Table 5.1** Polymers content of tomato waste (stalks and branches) (d.b.).

| Tomato waste | Hemicellulose | Cellulose | Lignin | References |
|---|---|---|---|---|
| Stalks | 7.9 | 43.1 | 12.3 | [11] |
| Stalks and branches | 21.4 | 28.1 | 18.5 | [12] |
| Enzymatic treated plant | 17.0 | 53.1 | 10.4 | |
| Post harvested plant | 10.5 | 33.1 | 7.8 | [13] |

**Table 5.2** Ultimate analysis of tomato waste (stalks and branches) (d.b.).

| | Mangut et al. (2006) [16] | Yargıç et al. (2015) [17] | Khiari et al. (2019a) [18] |
|---|---|---|---|
| Ash (%, db) | 4.6 | 4.5 | 8 |
| C (%, db) | 49.5 | 49.7 | 59.4 |
| H (%, db) | 6.7 | 7.4 | 7.6 |
| N (%, db) | 2.4 | 3.8 | 1.6 |
| O (%, db) | - | 39.1 | 23.4 |
| S (%, db) | 0.04 | - | 0.35 |

3. Mineral contents

   The minerals' content of different tomato processing industries including pomace and wastewater sludge is shown in Table 5.3. Among minerals, potassium presents the highest concentration 30.3 g/kg. The levels obtained for calcium, magnesium, sodium, iron, manganese and copper are low, while the sodium and nickel contents are quite high for pomace and wastewater sludge, respectively.

## 5.2.2 Morphological and textural properties

Scanning electronic microscopy (SEM) analyses are generally conducted to determine the physical morphology of biomass feedstocks. The images in Fig. 5.1 clearly show that tomato wastes presented a flat surface with no porosity on its surface. Thermal and chemical treatments are generally applied to collapse the surface and produced a macroporous structure.

Textural properties are assessed using nitrogen manometry. It is observed that specific surface are is 8.832 $m^2/g$, the pore volume is $8.13 \times 10^{-3}$ $cm^3/g$

**Table 5.3** Minerals content of tomato processing industry residues.

| Mineral (g/kg) | Wastewater sludge [19] | Tomato pomace [20] |
|---|---|---|
| Potassium | | 30.30 |
| Phosphorous | 0.94 | - |
| Magnesium | - | 2.11 |
| Sodium | - | 0.67 |
| Calcium | - | 1.32 |
| Zinc | 0.012 | 0.063 |
| Manganese | - | 0.014 |
| Copper | 0.006 | 0.005 |
| Iron | 0.97 | 0.06 |
| Nickel | 2.0 | - |

**Fig. 5.1** ***SEM Images of tomato wastes*** [21].

and the pore size is 1.841 nm [22]. These values increase for the chars and activated carbons prepared from tomato wastes

### 5.2.3 Water vapor sorption

Dynamic vapor sorption (DVS) was performed in order to evaluate the water uptake on tomato wastes.

Fig. 5.2 shows that the water uptake by PHT increases up to 29% upon increasing relative humidity to 90%. Such behavior could be related to the presence of hydrophilic carboxyl group as observed during the FTIR characterization of tomato residues in the literature [23].

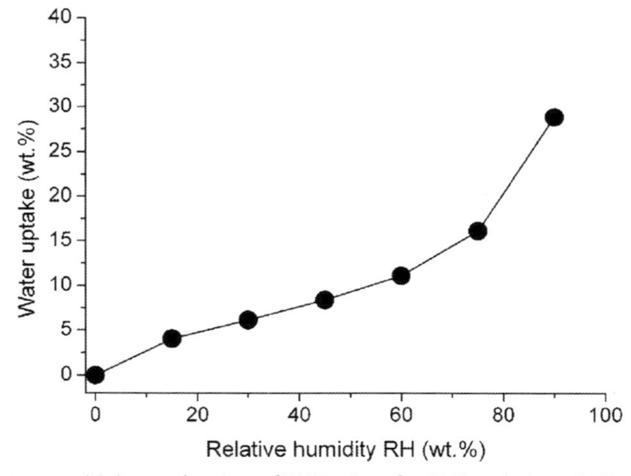


**Fig. 5.2** ***Water sensitivity evaluation of PHT using the DVS technique*** [**13**].

### 5.2.4 Thermal properties

Few investigations have examined the thermal properties of tomato processing wastes. Generally, thermal properties are assessed using thermogravimetric analysis and differential scanning calorimetry. Font et al. has performed thermal analysis under nitrogen atmosphere from room temperature to 800°C at 10°C/min.

Fig. 5.3 shows that the thermal behavior of tomato wastes has similar tendencies as the main lignocellulosic biomasses [25]. In particulier the DTG curve presents a shoulder at 550 K, a sharp peak at 600 K, and small peak at 720 K. Font et al. mentioned that the wide central band is a result of the overlapping of the three main polymers. In particular, the shoulder is attributed to the hemicellulose degradation, the intense peak to cellulose degradation and the small peak to lignin degradation.

## 5.3 Bio-based materials production

One major challenge of the food processing industry is the waste management of and their conversion into products with higher added values. Several environmental friendly strategies stimulate the use of food waste to obtain bio-based materials that can be used in different domains. Several attempts have been made for the development of bio-based materials prepared from tomato cultivation and processing industry by-products.

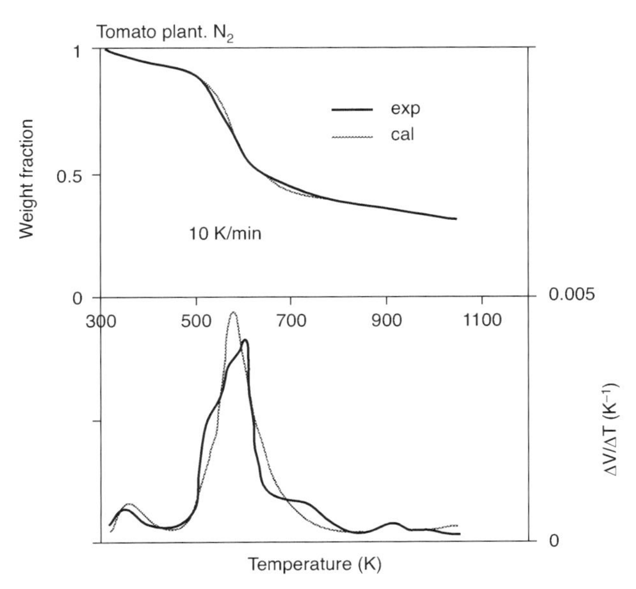


**Fig. 5.3** ***Weight fraction and its derivative during pyrolysis tests*** [24].

## 5.3.1 Biodegradable pots

In horticulture, transplanting the plant from the pot where it has been growing to another growing location (garden, field) is a worldwide cultural practice. Generally, gardeners and farmers use pots and cell trays of different materials, sizes and shapes. These containers are usually produced from petroleum derived materials, such as polystyrene, polyethylene and polypropylene [26]. After their usage, these pots are discarded causing environmental problems due to their wild neglecting in landfill or their uncontrolled combustion with the subsequent emission of toxic substances both into the atmosphere and into the soil. Therefore, an effective substitute for thermoplastic could be biodegradable pots since they are transplanted with the plantlets inside.

In this context, Schettini et al. have examined the development of innovative biodegradable pots prepared from biocomposites including sodium alginate or crosslinked calcium alginate, as polymeric matrix, and tomato and hemp fibers, as natural reinforcing dispersed phase [10]. In order to identify and characterize the blends to be used for the pots, two different laboratory-scale films, based on sodium alginate and calcium alginate, were prepared to perform physicochemical analysis helpful to predict the polymers performance as bonding agent inside the pots. Then, the pots and the sheets were obtained from the paste obtained after soaking 50.0 g of aggregate of fibers in 100 ml of a 2% (w/v) sodium/calcium alginate water solution. The pots were obtained by using a stainless-steel device appositely produced, made with pots shaped closed molds. They were characterized by a height of 40 mm, an end base diameter of 40 mm, a top base diameter of 55 mm, a thickness of 4 mm, and a weight of 9.0 g (Fig. 5.4).

During this investigation, three biocomposites were prepared with different compositions by varying the percentage of tomato and hemp fibers added to sodium alginate water solution. The first composite included 100% of tomato fibers (labeled ATH100), the second one consisted of 90% of tomato fibers and 10% of hemp fibers (ATH90) and the last one contained 70% of tomato fibers and 30% of hemp fiber (ATH70). Similarly, three other biocomposites labeled ATH100Cr, ATH90Cr, and ATH70Cr were prepared using crosslinked calcium alginate.

In order to predict the physicochemical behavior of the pots, mechanical tests, water vapor permeability tests, density and porosity tests, morphological analyses as well as water uptake and biodegradation tests were carried out on the laboratory films and sheets. Furthermore, the biodegradable pots for seedling transplanting were tested in real field condition inside a steel-constructed greenhouse.

**Fig. 5.4** *Experimental biodegradable pot.*

**Table 5.4** Flexural properties, maximum load, and displacement of the biocomposite sheets.

| Sheets | Maximum load (N) ±10% | Displacement (mm) ±5% | Young modulus (MPa) ±10 MPa | Tensile stress at break (MPa) ±0.2 MPa |
|---|---|---|---|---|
| ATH100 | 8.7 | 3.59 | 63.62 | 0.71 |
| ATH90 | 10.5 | 4.07 | 97.08 | 1.20 |
| ATH70 | 14.8 | 4.75 | 81.70 | 0.92 |
| ATH100Cr | 10.4 | 3.22 | 48.05 | 0.46 |
| ATH90Cr | 11.2 | 3.98 | 62.51 | 0.68 |
| ATH70Cr | 15.0 | 4.02 | 78.21 | 0.85 |

From the different mechanical characterization (Table 5.4), authors reported that the mechanical properties of the biocomposites made of tomatoes seeds and peels, combined to hemp fibers and bound together into the crosslinked alginate network, are very interesting for the biodegradable pots production. In addition, by increasing the hemp fibers content, a general enhancement of the mechanical parameters of both uncrosslinked and crosslinked samples was registered, since fibers from hemp strands are more rigid, stiff, and long in comparison to the more flexible and short fibers from tomato peels and seeds.

Concerning the biodegradation tests performed on ATH100Cr and ATH70Cr samples (Fig. 5.5), it is observed that during the first 60 days the ATH100Cr samples degraded more rapidly than the ATH70Cr samples due to the higher concentration of hemp fibres with comparing to the

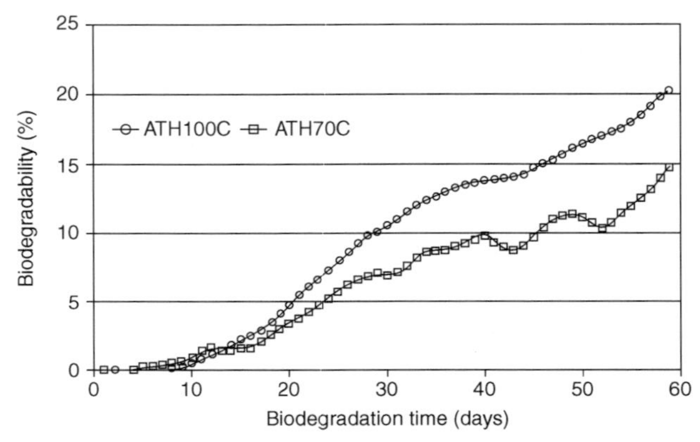


**Fig. 5.5** ***Biodegradation (%) in soil of the ATH100Cr and ATH70Cr sheets as a function of time.***

ATH100Cr. In particular, the ATH100Cr samples achieved about 20% of biodegradation whereas the ATH70Cr sample reached 15%. This finding could be explained by the fact that tomato peels and seeds, being short sized, are easily available to microbial population [27], thus providing a relative faster biodegradation. On the other hand, hemp fibres are longer and mostly based on microcrystalline cellulose [28] and for these reasons the hydrophilic macromolecular chains are not easily susceptible to microorganism attack.

In terms of horticultural performance, authors have observed that the biodegradable pots allowed to develop very dense and active root hair. Fig. 5.6 shows a significantly dense network of root hairs developed in ATH100 pots, a dense one in ATH70 pots and a less dense network in ATH90 pots. In PS pots, used as control, long roots dominated the root system, reducing overall root development.

Furthermore, during the transplanting operations, no transplant shock and root deformation were detected. Thus, at the end of crop cycle, the different pots influenced pepper height. The mean plant height was 0.76 m for the plants grown inside the ATH90 pots, 0.73 m inside the ATH100

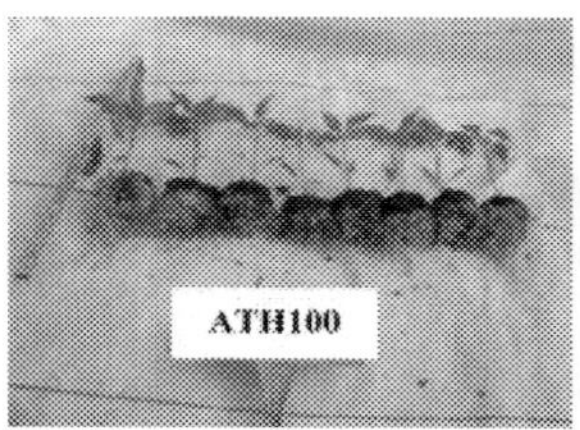

**Fig. 5.6** ***Roots development in the ATH100 and ATH90 biodegradable pots and in the polystyrene pot (PS) at seedlings transplanting.***

and ATH70 pots. The control plants were characterized by a mean plant height of 0.67 m. Moreover, the biodegradable containers degraded completely in 16 days allowing the passage of the roots through the container's walls and the growth of the plants. The roots spread in a radial fashion; no root rot or similar symptoms were observed.

From this investigation, Schettini et al. showed that tomato wastes could be converted in upgraded products used as biocomposites designed for the production of innovative biodegradable pots to be used for transplanting in gardening and agriculture.

These biodegradable pots, obtained after a low productive cost process, are environmentally friendly and easily available. Furthermore, these pots overstay and degrade into the soil with a consequent positive environmental impact. In addition, the use of these biodegradable pots implies the drastic reduction of man labour related to the pots recovering, cleaning and plants transplanting.

### 5.3.2 Particleboard

The global market for wood-based panels is increasing annually leading to a significant pressure on wood resources, especially in forest-lacking regions. Therefore, it seems essential to identify alternatives from agricultural residues. In this context, Taha et al. have evaluated the potential of tomato stalk as raw materials for single layers particleboards production. The selection of tomato stalks is motivated by the highest volume in Egypt and seasonal availability throughout the year.

During this investigation, particleboards were prepared by mixing chopped tomato stalks at varying resin concentrations (12, 14, and 16 wt% resin) and compression molding under different pressures (25, 29, and 35 $kg/cm^2$) at a constant temperature of 160°C. Fig. 5.7 gives the procedure for the particleboards production and more details could be found in [11].

Particleboards performance were evaluated through the analysis of physical and mechanical properties including density, tensile strength, modulus of rupture (MOR), and modulus of elasticity (MOE).

As expected, Fig. 5.8 shows that the increase in both of pressure and resin content increases density. Moreover, it is observed that the impact of pressure is more pronounced than the resin content. The particleboard density ranged from 0.66 to 0.95 $g/cm^3$ which is within the order of magnitude of common particleboards.

Similar trends were observed for the internal bond strength. In fact, as observed in Fig. 5.9 the internal bond strength increases with both increasing resin content and pressure. Such behavior is attributed to improved bonding based on enhanced wetting between particles as well as to better

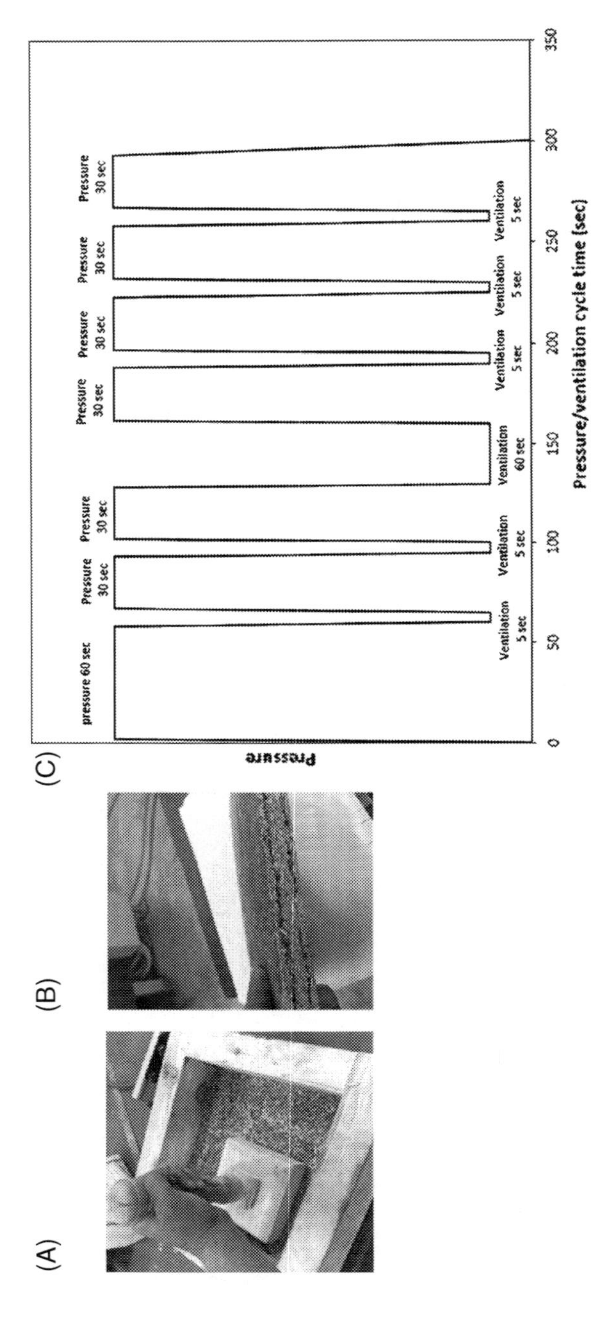


**Fig. 5.7** ***Fabrication of particleboards showing the*** (A) Pre-forming stage of the tomato-stalk-resin mix, (B) lofting phenomenon, (C) compression molding cycle.

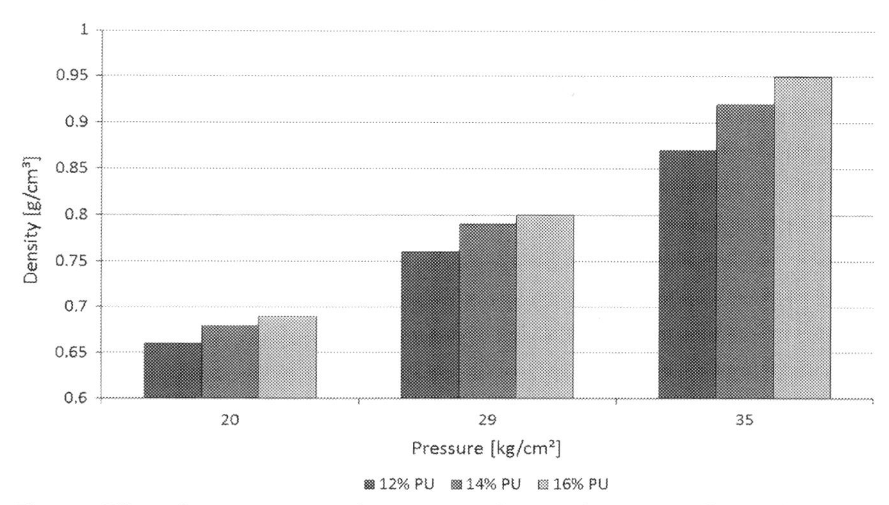


**Fig. 5.8** ***Effect of resin content and pressure on density of tomato stalk particleboards.***

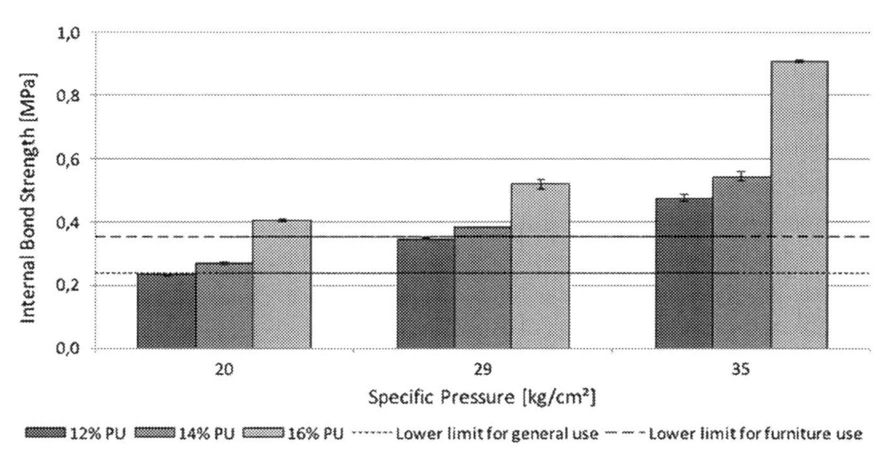


**Fig. 5.9** ***Effect of resin content and pressure on internal bond strength of tomato stalk particleboards.***

compaction of the particleboard mix, which also becomes evident in terms of increased board density. Measured internal bond strength meets the minimum requirements of 0.24 and 0.35 $N/mm^2$ for particleboard panels for both general use and furniture manufacturing in accordance to the European standard EN 312:2010. The sole exception is observed for the TSP prepared at a low pressure of 20 $kg/cm^2$ and low resin content (12%).

The mechanical performance of the different particleboards is assessed through both MOR and MOE. Fig. 5.10A shows that the MOR increases

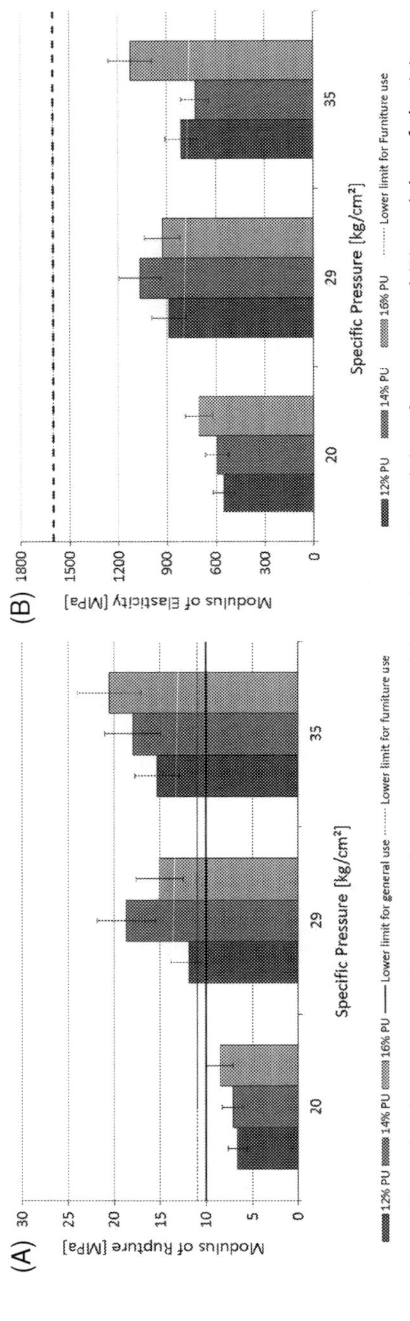


**Fig. 5.10** ***Effect of resin content and pressure of tomato stalk particleboards on*** (A) modulus of rupture, and (B) modulus of elasticity.

with increasing pressure. This behavior can be attributed to the increased compaction levels (density) at increased pressure levels, as also observed by Geimer [29]. Improved mechanical behavior is, however, also directly affected by proper wetting of the fiber by the matrix material leading to improved bonding. Hence, the MOR is also observed to increase with increasing resin content. The lower limits for the MOR, as given by the European standard EN 312:2010 for particleboard utilization for general use (10MPa) and furniture applications (11 MPa) are reached at pressures exceeding 29 $kg/cm^2$

Fig. 5.10B shows that the MOE for the prepared particleboards are below the minimum value of 1600 $N/mm^2$ prescribed by the guidelines of the European standard EN 312:2010 for furniture manufacturing applications. Johns et al. [30] related the generally reduced mechanical properties of woody particleboards to increased gelation time and incomplete curing of urea formaldehyde. They have suggested an optimum pH value around 4.0 for the fibers to allow adequate gelation behavior. Hence, the increased pH-value of tomato stalks (lying at 6.4) is assumed to negatively affects the curing behavior of the relatively acidic resin.

From this investigation, Taha et al. proves the potential of applying tomato stalks as a base material for particleboards. However, the particleboards development for broader applications, such as for the furniture industry for example, requires future investigations to improve the mechanical behavior of the panels.

### 5.3.3 Ceramic materials

The tomato processing industry produces huge quantities of sludge annually (1.5% of the tomato) representing an environmental concern. Therefore, the implementation of economically viable and environmentally friendly strategy for their recycle and reuse is required. Several investigations show that the use of sludge in ceramic matrix materials and cement-based materials (concrete and mortar) have beneficial effects for the industry such as sustainability, respect for the environment, and economic and energy savings. In this context, Sanchez et al. have proposed the substitution of clayey raw materials for tomato sewage sludge in the production of traditional ceramics. Therefore, different proportions (0%, 3%, 6%, 9% mud) of sewage sludge from the tomato processing industry have been added to the body of the clay to produce ceramic materials. From these mixtures, ceramic material was prepared according to the flowsheet shown in Fig. 5.11.

Then, the ceramic pieces were submitted to different tests in order to identify their mechanical behavior. These tests include linear cooking contraction, suction tests, flex resistance, absorption and surface color analyses.

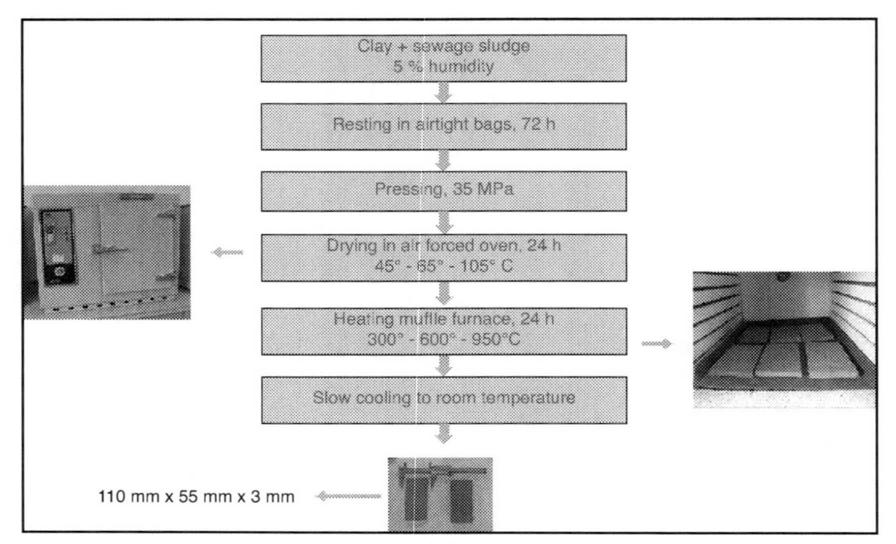


**Fig. 5.11** ***Flow sheet of ceramic pieces with sludge from tomato factory.***

Authors found that no clear correlation between the coefficient of linear shrinkage and the sludge/clay ratio is observed. The water suction increases with increasing the sludge the percentage in the ceramic material. The flexural strength of the material decreases as the sludge percentage increases. The decrease in resistance is 31% between the control sample and the one with the highest percentage of sludge. The color does not vary significantly once the material has been cooked.

Sanchez et al. found that considering that the trial has been developed at laboratory scale, the values can be considered as promising for future industrial manufacturing.

## 5.3.4 Paperboard

Lignocellulosic raw materials are being utilised in many industrial sectors as a natural source of interesting biopolymers. In this context, Covino et al. have examined the ET of tomato plant agri-waste in order to recover the polymeric matrices and to obtain good quality lignocellulosic fibres. These fibres were used in the paperboard preparation.

During this investigation, Covino et al. found that the use of enzymes during the pulping of the raw material from tomato plants has significantly affected the polymeric composition of solid fraction. In particular, chemical analyses revealed that the ET with an enzyme/solid (E:S) ratio of 1:80 (ET1) produced a remarkable increase in cellulose content from 28.10% to 53.05%, followed

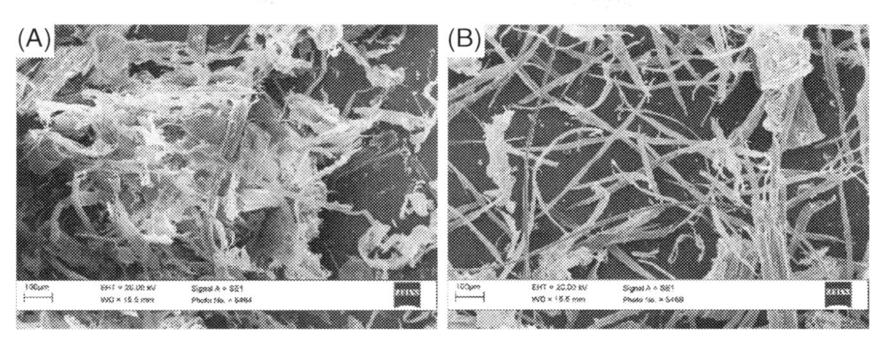

**Fig. 5.12** SEM images of UT (A) and ET1 (B) fibres after pulping process. Magnification: 250 ×.

by a decrease of hemicellulose and lignin contents from 21.40% to 16.95% and 18.45% to 10.40%, respectively compared to the control sample (UT).

Different characterizations were performed to assess the quality of the extracted fibers. SEM images presented in Fig. 5.12 show that UT fibers (control sample) are more compact with each other, and densely covered with hemicellulose-type cementing polysaccharides, while ET1(enzyme treated) fibers appear opened as thin and "cleaned" straws. Such behavior confirms that only the ET is able to promote the removal of aggregated hemicelluloses, thus facilitating the penetration of water molecules into the spaces within cellulose fibrils. Hence, in the presence of mechanical refining, the cellulosic fibers appear opened and dispersed (Fig. 5.12B) due to breaking of hydrogen bonds connecting cellulose chains that provoke the loss of 3D structure of the fibers.

Thermal analysis and physical properties including water holding capacity, viscosity and colloidal stability of suspended fibres and pulps were also examined. Authors found that the thermal profile, water absorption and pulp viscosity of fibres was strongly affected by the composition changes. As an example, shear viscosity of pulp suspensions was studied at 25°C and 80°C, at different percentage of fibers.

Fig. 5.13 shows that, as expected, apparent viscosity decreases with the increase of shear rate (Fig. 5.13A and B ) and with the increase of temperature (Fig. 5.13A vs Fig. 5.13B) for ET1 and UT samples. Furthermore, with the same concentration, ET1 suspensions always showed a higher viscosity than UT, at both analyzed temperatures. This result is in agreement the water holding capacity: ET1 fibers, having a higher WHC, are able to reduce the volume of free water, giving rise to a greater number of interactions with water molecules, which generate more stable and viscous

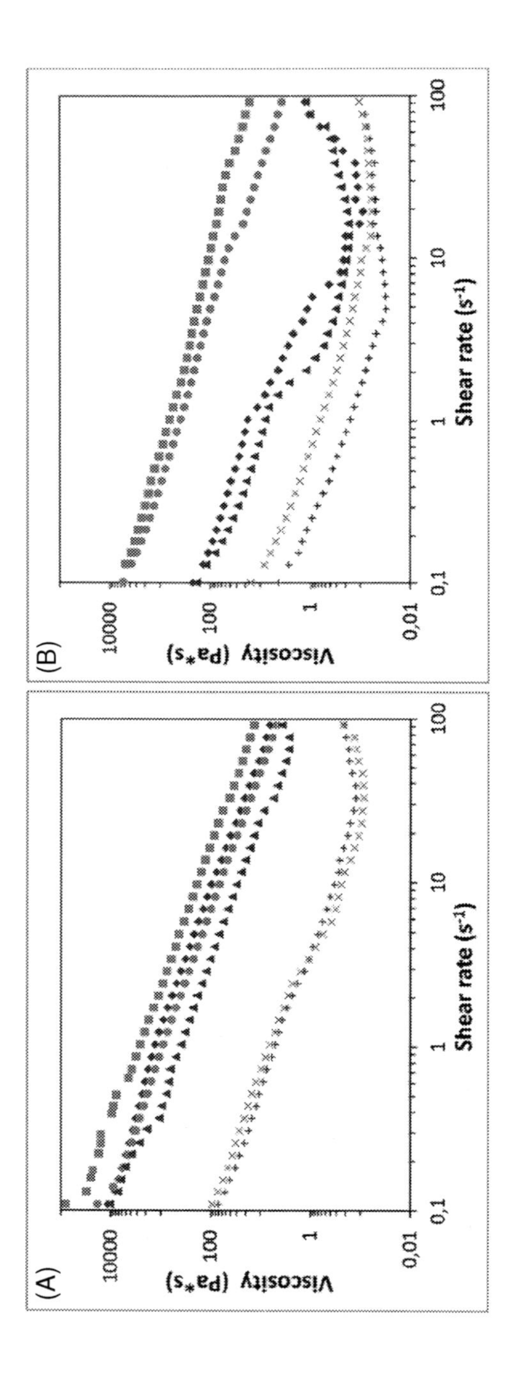


**Fig. 5.13** Viscosity at 25°C (A) and 80°C (B) of lignocellulosic suspensions prepared at different concentration of UT and ET fibres.

suspensions. Further characteristics obtained for the extracted fibres are given in details in [12].

The morphological and the mechanical properties of the prepared paperboard sheets were also determined. Fig. 5.14 shows the SEM images and highlights that UT paperboard has a compact surface filled with pectic and hemicellulosic polysaccharides, which completely covered cellulose fibres (Fig. 5.14A and C). However, in ET1 paperboard, cellulose fibres are clearly visible and "naked", leaving a glimpse of the interaction network that structured the paperboard sheet (Fig. 5.14B and D).

During mechanical properties characterization, ET paperboard showed increased stiffness when subjected to tensile testing respect to the control. In addition, a significant improving for maximum load, tensile strength at break and Young's modulus, for ET1 paperboard compared to UT (Table 5.5).

From the obtained results, authors suggest that a scale up of a green and sustainable process, leading to the recovery of fibres starting from a raw biomass, is possible. The obtained fibres could be used in the thermoforming of paperboards whose physical-mechanical characteristics are implementable to make them as similar as those of commercial cellulose ones.

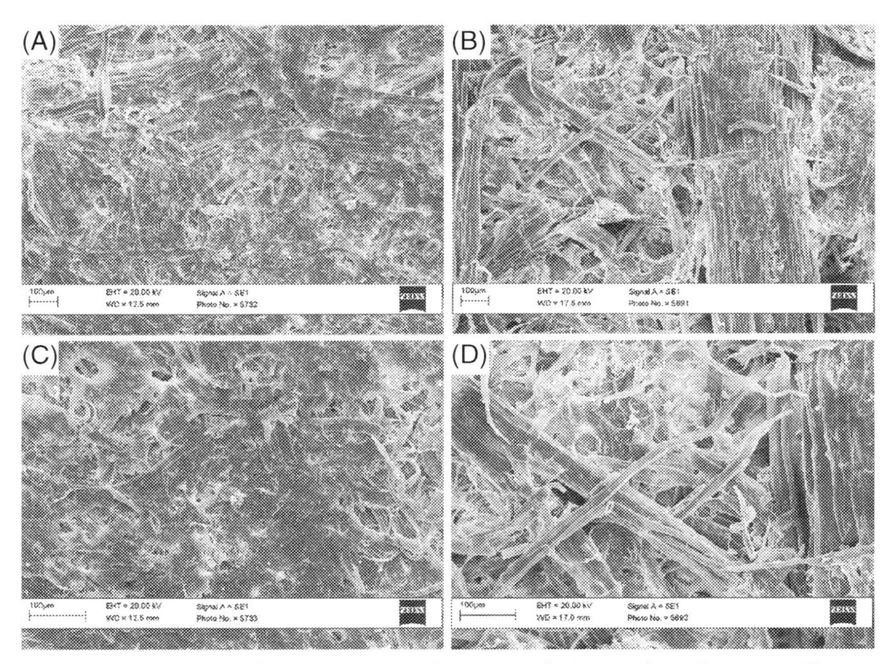


**Fig. 5.14** SEM images of UT (A and C) and ET1 (B and D) paperboards at magnificence of 250 × and 500 × respectively.

**Table 5.5** Tensile test of control (UT) and enzyme-treated (ET) paperboards.

| Paperboard | Maximum load (N) | Elongation at break (%) | Tensile strength at break (MPa) | Young's modulus (MPa) |
|---|---|---|---|---|
| UT | 5.36 ± 1.09 | 1.35 ± 0.46 | 0.17 ± 0.02 | 21.24 ± 3.64 |
| ET1 | 27.64 ± 4.14 | 1.25 ± 0.24 | 1.40 ± 0.42 | 264.49 ± 115.95 |

Values are expressed as mean ± SD (n = 10).

## 5.3.5 Plastic films

Packing of food materials with the biopolymer sheets is a better alternative to plastic packing to avoid environmental pollution. Nowadays polymer-based packing materials have been widely used because of their good mechanical strength and ensure a safe environment. Most of the biopolymers have the major drawback of being hydrophilic in nature and hence they have low barrier against water vapor when the food is packed hot.

Several studies have been carried through on PHT plants as source of BPs for manufacturing composite covering films to be used in agriculture. Mostly, a plastic film must be mechanically strong, flexible, elastic, easily processable and not expensive. Thus, no single polymer but composite films are made.

In this context, Nistico R. et al. [13] studied the manufacture and properties of composite plastic films obtained from a synthetic polymer Poly(vinyl alcohol-co-ethylene) with 2%–10% post-harvest tomato (PHT) plant.

The blend sheet dimensional data and thermal characterization were reported. Fig. 5.15 shows the aspect of the investigated films. With increasing the filler content up to 10%, the film becomes dark colored (Fig. 5.15),

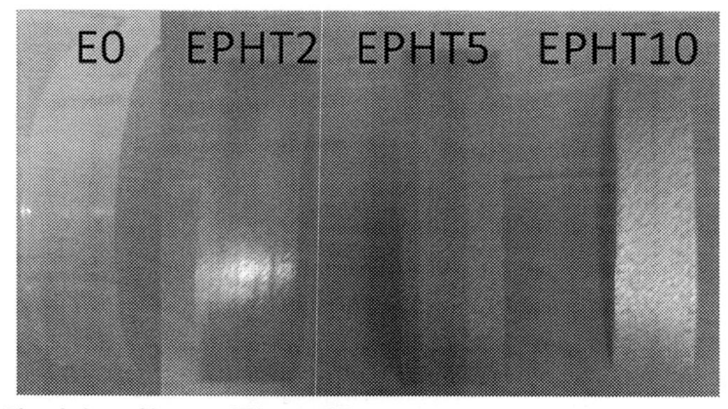


**Fig. 5.15** ***Blend sheet films at different filler contents.***

**Table 5.6** Blend sheet films dimensional data, and recorded pressure and temperature at the outlet die of the extruded material during single-screw extrusion.

| Sample | Length (m) | Width (cm) | Thickness (μm) | Weight (kg) | Pressure (bar) | Temperature (°C) |
|---|---|---|---|---|---|---|
| E0 | 9.1 | 2.9 | 130 ± 44 | 0.040 | 104 ± 1 | 198 ± 1 |
| EBP2 | 8.4 | 3.2 | 223 ± 25 | 0.066 | 116 ± 1 | 202 ± 1 |
| EBP5 | 8.5 | 3.5 | 250 ± 17 | 0.082 | 128 ± 6 | 203 ± 1 |
| EBP10 | 8.3 | 4.5 | 450 ± 35 | 0.078 | 116 ± 3 | 203 ± 1 |
| EPHT2 | 8.8 | 3.3 | 305 ± 21 | 0.072 | 114 ± 4 | 209 ± 2 |
| EPHT5 | 8.8 | 3.5 | 306 ± 27 | 0.070 | 114 ± 1 | 209 ± 1 |
| EPHT10 | 8.3 | 3.0 | 345 ± 21 | 0.040 | 84 ± 2 | 206 ± 2 |

Neat EVOH (E0), EVOH containing 2%, 5%, 10% BPs (EBP2, EBP5, and EBP10, respectively), EVOH containing 2%, 5%, 10% post-harvest tomato plant material (EPHT2, EPHT5, and EPHT10, respectively).

and the film thickness (Table 5.6) tends to increase. All films tended to become more and more rigid upon increasing the filler content.

Results showed also that the values of all indicators of mechanical performance tend to decrease significantly upon increasing the filler content. Nevertheless, there are no significant variation upon comparing values from PHT and BPs films at equal filler content. This behavior can be explained by the higher mechanical stiffness, and lower elasticity and flexibility of the blends, compared to the neat EVOH film.

Nistico et al. concluded that the present paper reports a case study showing, for the first time, that raw (i.e. not chemically pretreated) PHT plants may be suitable as fillers to manufacture composite films by coextrusion with synthetic polymers from fossil sources.

## 5.4 Conclusion

The focus on the production of renewable and eco-sustainable materials is constantly growing worldwide. Therefore, the conversion of agri-waste and agricultural residues into bio-based materials is promising strategy in the sustainable development context. Tomato cultivation and processing industry by-products have great potential to achieve such task. In particular, tomato wastes have proven their potential to be converted in biodegradable pots, particleboard, and ceramic materials. Furthermore, valuable components could be extracted and inserted during paperboard and plastic films production. Certainly, the development of new biobased materials using tomato residues expands the perspectives to realize cost-effective value-added biobased products for more sustainable end-of-life phase.

## References


[1] FAO FAOSTAT Statistical Database. http://www.fao.org/faostat/en/#data/QC, 2020a (Accessed 11 March 2021).

[2] S.K. Vidyarthi, C.W. Simmons, Characterization and management strategies for process discharge streams in California industrial tomato processing, Sci. Total Environ. 723 (2020) 137976. 10.1016/j.scitotenv.2020.137976.

[3] M. Del Valle, M. Cámara, M.E. Torija, Chemical characterization of tomato pomace, J. Sci. Food Agric. 86 (8) (2006) 1232–1236.

[4] M.R. Ventura, M.C. Pieltain, J.I.R. Castanon, Short communication. Evaluation of tomato crop by-products as feed for goats, Anim. Feed Sci. Technol. 154 (3-4) (2009) 271–275.

[5] P.A. Silva, B.C. Borba, V.A. Pereira, M.G. Reis, M. Caliari, M.S. Brooks, T.A. Ferreira, Characterization of tomato processing by-product for use as a potential functional food ingredient: nutritional composition, antioxidant activity and bioactive compounds, Int. J. Food Sci. Nutr. 70 (2) (2019) 150–160.

[6] R. Sánchez, R. de la Torre, R. Rodríguez, M. Gómez-Cardoso, J.L. Llerena-Ruiz, Elaboration of ceramic material with sludge from the wastewater treatment plant of the tomato processing industry, Acta Hortic. 1233 (2019) 185–192.

[7] N. Kraiem, M. Lajili, L. Limousy, R. Said, M. Jeguirim, Energy Recovery from Tunisian Agri-food wastes: evaluation of combustion performance and emissions characteristics of green pellets prepared from tomato residues and grape marc, Energy 107 (2016) 409–418.

[8] A.M. Giuffrè, M. Capocasale, C. Zappia, M. Poiana, Biodiesel from tomato seed oil: transesterification and characterisation of chemical-physical properties, Agron. Res. 15 (2017) 133–143.

[9] R. Mallampati, S. Valiyaveettil, Application of tomato peel as an efficient adsorbent for water purification-alternative biotechnology?, RSC Adv. 2 (2012) 9914–9920. https://doi.org/10.1039/c2ra21108d.

[10] E. Schettini, G. Santagatab, M. Malinconico, B. Immirzi, G.S. Mugnozza, G. Vox, Recycled wastes of tomato and hemp fibres for biodegradable pots: physico-chemical characterization and field performance, Resour. Conserv. Recycl. 70 (2013) 9–19.

[11] I. Taha, MS. Elkafafy, Hamed El Mously, Potential of utilizing tomato stalk as raw material for particleboards, Ain Shams Eng. J. 9 (2018) 1457–1464.

[12] C. Covino, A. Sorrentino, P. Di Pierro, G. Roscigno, A.P. Vece, P. Masi, Lignocellulosic fibres from enzyme-treated tomato plants: characterisation and application in paperboard manufacturing, Int. J. Biol. Macromol. 161 (2020) 787–796.

[13] R. Nistico, P. Evon, L. Labonne, G. Vaca-Medina, E. Montoneri, C. Vaca-Garcia, M. Negre, Post-harvest tomato plants and urban food wastes for manufacturing plastic films, J. Clean. Prod. 167 (2017) 68e74.

[14] M. Jeguirim, S. Dorge, A. Loth, G. Trouvé, Devolatilization kinetics of Miscanthus straw from thermogravimetric analysis, Inter, J. Green Energy 7 (2010) 164–173.

[15] A. Chouchene, M. Jeguirim, B. Khiari, G. Trouvé, F. Zagrouba, Study on the emission mechanism during devolatilization/char oxidation and direct oxidation of olive solid waste in a fixed bed reactor, J. Anal. Appl. Pyrolysis 87 (2010) 168–174.

[16] V. Mangut, E. Sabio, J. Gañán, J.F. González, A. Ramiro, C.M. González, S. Román, A. Al-Kassir, Thermogravimetric study of the pyrolysis of biomass residues from tomato processing industry, Fuel Process Technol. 87 (2) (2006) 109–115.

[17] A.Ş. Yargıç, R.Z. Yarbay Şahin, N. Özbay, E. Önal, Assessment of toxic copper(II) biosorption from aqueous solution by chemically-treated tomato waste, J. Clean. Prod. 88 (2015) 152–159.

[18] B. Khiari, M. Moussaoui, M. Jeguirim 2019a. Tomato-processing by-product combustion: thermal and kinetic analyses. Materials 12, 553.

[19] R. Sánchez, R. de la Torre, R. Rodríguez, M. Gómez-Cardoso, J.L. Llerena-Ruiz, Elaboration of ceramic material with sludge from the wastewater treatment plant of the tomato processing industry, Acta Hortic. 1233 (2019) 185-192.
[20] V. Nour, T.D. Panaite, M. Ropota, R. Turcu, I. Trandafir, AR. Corbu, Nutritional and bioactive compounds in dried tomato processing waste, CyTA – J. Food 16 (1) (2018) 222–229.
[21] A. Ayala-Cortés, C.A. Arancibia-Bulnes, H.I. Villafán-Vidales, D.R. Lobato-Peralta, D.C. Martínez-Casillas, A.K. Cuentas-Gallegos, Solar pyrolysis of agave and tomato pruning wastes: Insights of the effect of pyrolysis operation parameters on the physicochemical properties of biochar, AIP Conference Proceedings, 2019. https://doi.org/10.1063/1.5117681.
[22] E. Heraldy, W.W. Lestari, D. Permatasari, D.D. Arimurti, Biosorbent from tomato waste and apple juice residue for lead removal, J. Environ. Chem. Eng. (2018). https://doi.org/10.1016/j.jece.2017.12.026.
[23] E. Önal, N. Özbay, A.Ş. Yargıç, R.Z.Y. Şahin, Ö. Gök, 2014. Performance evaluation of the bio-char heavy metal removal produced from tomato factory waste, in: Progress in Exergy, Energy, and the Environment. https://doi.org/10.1007/978-3-319-04681-5-70.
[24] R. Font, J. Moltó, A. Gálvez, M.D. Rey, Kinetic study of the pyrolysis and combustion of tomato plant, J. Anal. Appl. Pyrolysis 85 (2009) 268–275.
[25] M. Jeguirim, Y. Elmay, L. Limousy, M. Lajili, R. Said, Devolatilization behavior and pyrolysis kinetics of potential Tunisian biomass fuels, Environ Prog. Sustain. Energy 33 (2014) 1452–1458.
[26] M.R. Evans, DL. Hensley, Plant growth in plastic, peat, and processed poultry feather fiber growing containers, HortScience 39 (5) (2004) 1012–1014.
[27] M. Knoblich, B. Anderson, D. Latshaw, Analyses of tomato peel and seed byproducts and their use as a source of carotenoids, J. Sci. Food Agric. 85 (2005) 1166–1170.
[28] K.M. Amar, M. Manjusri, TD. Lawrence, Natural Fibers, Biopolymers, and Biocomposites, Tailor & Francis: CRC Press, Boca Raton, Florida, United States, 2005.
[29] RL. Geimer, Data basic to the engineering design of reconstituted flakeboard, Proceedings of the W.S.U. symposium on particleboard, Pullman, Washington, 1979 104–125.
[30] W. Johns, K. Niazi, Effect of pH and buffering capacity of wood on the gelation time and urea-formaldehyde resin, Wood Fiber Sci. 12 (4) (1981) 255–263.

CHAPTER SIX

# Biochar production from the pyrolysis of tomato processing residues

Stylianou Marinos[a,b], Psichoula Terpsithea[b], Helmi Hamdi[c], Tsangas Michail[b], Antonis A. Zorpas[b], Agapiou Agapios[a]
[a]Department of Chemistry, University of Cyprus, Nicosia, Cyprus
[b]Open University of Cyprus, Faculty of Pure and Applied Sciences, Environmental Conservation and Management, Laboratory of Chemical Engineering and Engineering Sustainability, Latsia, Nicosia, Cyprus
[c]Center for Sustainable Development, College of Arts and Sciences, Qatar University, Doha, Qatar

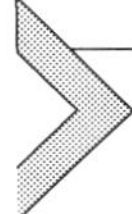

## 6.1 Production and management of food processing wastes

The Food and Agriculture Organization of the United Nations has been warning regarding the waste of food produced globally; it was estimated that one-third of food for human consumption is somehow lost [28]. According to the official EU data in 2012–3, around 88 million tons of food waste (either produce or prepared food) are generated annually, leading to associated costs estimated at 143 billion euros (with ~98 billion euros originated from household food waste). This raises ethical, economic, and environment concerns (natural resources depletion) [96]. The situation, however, seems to be worse as in a similar study in 2018 in EU households, the estimated percentage of wasted fruit and vegetables was 50%, leading to a waste of over 17 billion kg of fresh fruit and vegetables per year [20]. In parallel, food security remains a challenge of food demand that needs to be taken into consideration with various aspects (e.g., agricultural, safety, health, nutrition, processing, and supply chain), as well as with innovation technologies leading to personalised nutrition [19].

Modern societies are characterized by modern food habits, which result in large waste fractions that finally end up in landfills. It is estimated that worldwide one-third of the food produced for human consumption is lost; therefore, food waste practices and policies matter [89]. Since food production is primarily a resource-intensive effort, an array of serious environmental impacts (e.g., soil erosion, deforestation, water, soil and air pollution) are directly or indirectly correlated with food disposal. Reducing the food


DOI: https://doi.org/10.1016/B978-0-12-822866-1.00002-8

waste disposal in all phases, from farm to spoon (e.g., production, storage, transportation, management, waste) will result in social, environmental, and economic benefits. Having in mind, resource depletion on one hand and the large food waste fractions in landfills on the other, food management is a complex and multiphase issue, as consumers' behaviour appears to be driven by cultural, political, economic, and geographic barriers [89].

Management of food waste is considered as an emerging environmental issue of global concern. The various energetic constraints (e.g., fuels, chemicals, materials) can be covered through the appropriate processing of waste [62]. Traditional processes (e.g., dumping, burning, re-converting, volume reduction) need to be re-designed appropriately, so as to enable the production of higher value and marketable products, compared to the traditional and conventional approaches of incineration for energy recovery, feed or composting. The potentiality of food waste chain to become a valuable alternative feedstock for the chemical industry (e.g., production of speciality, consumer, and commodity chemicals) is being explored and widely discussed. In this regard, food and chemical industry, will unify and strengthen their power, towards the global challenges (e.g., over-population, global warming, scarcity of fossils and resources); toward this, the sustainable management of food waste through waste hierarchy is proposed [30,62].

Food processing industry is expected to see an upsurge in expansion as a result of population growth and increase in agricultural production worldwide [79]. Agro-food residues generated from the processing of fruits, vegetables and grains are mostly composed of lignin, cellulose and hemicellulose, with lignin conversion being more efficient in terms of char yield [63,92]. Considering food processing sectors globally and depending on the processing method, a significant amount of solid waste is generated. For instance, the amount of residues produced as percentage of processed crops in some of the main industries is as follows: apple 25%–35%, citrus 50%–60%, grapes up to 20%, banana 30%–40%, pineapple 40%–80%, and potato 15%–40% [106]. Exhausted olive pomace is also a major by-product of olive oil extraction in the Mediterranean region representing 55%–78% of the initial olive mass [101]. Besides, maize and wheat are two of the most widespread crops worldwide because of their importance for human consumption and livestock feed. In Europe only, agricultural residues composed of maize cobs and wheat chaff could provide an annual potential biomass of 9.6 Mt and 54.8 Mt, respectively [10]. Due to their continuous and inevitable generation during industrial processes, most of the above-mentioned lignocellulosic residues (among others) have increasingly been used as feedstock for biochar production destined for land application.

## 6.2 The case study of tomato

Tomato (*Solanum lycopersicum* L.) is a widespread and well-known berry, originated from South America (Andean regions). Historically speaking, it arrived in Europe through the Spanish colonies in the 1500s and soon afterwards has been cultivated in the Mediterranean region. Nowadays, it is widely used all over the word by food industry in processed (e.g., fried or raw canned tomatoes, as sauce, paste, juice) or prepared products as an extra ingredient. It is considered the most important crop in EU28 with 3569 varieties, 82% of which are hybrids [95], and the second most cultivated worldwide [85]. Nevertheless, in the Mediterranean area and especially in Italy (the second home canter of tomato), there is high concern for its future adaptation and sustainability in warmer and semi-arid environments due to the climate change; therefore, the adaptation of three Apulian tomato was examined in Puglia (Southern Italy) as a case study [85]. The main characteristics of the respective vegetable chain is the seasonality and the product overproduction. Factors such as the color, shape, maturity, and lesions contribute to tomato's commercial value. Lycopene is considered the driving force of tomato towards the market, as is being widely advertised for its health benefits. Nevertheless, in cases of unsuccessful production of good quality product, waste tomato end-up in landfill or as animal feed. It should be noted that only in Europe, the postharvest wastage of tomato is more than 3 million metric tons [67]. In a respective study, Norway, Belgium, Poland, and Turkey are examined as tomato-producing European case studies with different climate and volumes. Their efforts towards better quality production, the enabling technologies, the metabolic profiling, and the steps for valorization are therefore extensively discussed [5,67].

Tomato supply chain with around 37.8 million tons of tomatoes processed annually at the global level [115] plays significant role for the agri-sector, as well as for the economy of many countries such as Italy and Spain. Through the 13th World Processing Tomato Congress, it has been revealed that since 1994 the tomato production and processing has grown worldwide by 59%. Currently, California, Italy and China are the biggest tomato producers of the world followed by Spain, Turkey, and Portugal [48]. Essentially, the whole Mediterranean area is strongly characterized by tomato production and processing. At European level, Italy is the leading country with a contribution up to 48% of total production with 5.2 million metric tons of tomato produced [116]. Amongst Italian regions, 30%–40% (2 million tons/year) of the Italian tomato production comes from Apulia, especially from the province of Foggia, which represents the highest surface area dedicated to this cultivation (20,000 ha).

### 6.2.1 Tomato organoleptic characteristics

Tomato flavor is considered a characteristic quality criterion for consumers' acceptance, along with the rest of organoleptic characteristics such as appearance, texture, and color. Nevertheless, during recent years, there have been many complaints towards market tomatoes declined flavor. The respective aroma profile of tomatoes is due to the release of a mixture of volatile organic compounds (VOCs), consisting of various chemical groups; so far aldehydes, alcohols, ketones, esters, sulfur compounds, phenols, and terpenes have been reported [14,60,81,118]. The association of tomato taste and quality with its chemical synthesis was extensively explored in a series of Spanish local varieties [29].

Towards this effort, the non-invasive green analytical method of solid-phase microextraction-gas chromatography-mass spectrometry (SPME-GC-MS) was developed, optimized and applied for the direct determination of tomatoes aromatic volatiles. The parameters that were experimentally studied, included the internal standard usage (1-heptanol, out of 8 standards tested), the SPME coating material (85 μm carboxen/polydimethylsiloxane, among 4 commercial fibers), the optimum extraction temperature (60°C, 4 temperatures from 30–60°C were tested) and the addition or not of salt (no addition of NaCl was selected). Following the latter optimum parameters, 37 flavor components were overall identified; these were all grouped in seven chemical classes: aldehydes (11 compounds), hydrocarbons (7), alcohols (7), ketones (5), oxygen-containing heterocyclic compounds (3), esters (3), sulfur-and nitrogen-containing heterocyclic compounds (1) [60].

Nevertheless, in another study, using the same analytical methodology (SPME-GC-MS), the results were slightly different, showing the need for further research on the experimental parameters. In particular, in this study, the optimum experimental conditions for cherry tomatoes flavor analysis were the following: 50/30 μm DVB/CAR/PDMS fiber, 50°C extraction temperature, 50 min absorption time and 1.5 g salt addition in 8 g of tomato juice sample. Four different varieties were tested, and 50 VOCs were overall detected. Among the detected VOCs, hexanal and (E)-2-hexenal importance was noticed, next to that of other aldehydes (phenyl acetaldehyde, benzaldehyde, (E,E)-2,4-decadienal, alcohols (2-methyl-1-butanol, 1-pentanol, 1-hexanol, (Z)-3-hexen-1-ol, 1-octanol, and linalool), and sulfurs (2-isobutylthiazole and 2-propylthiazole). The role of ketones (1-penten-3-one, 6-methyl-5-hepten-2-one, geranylacetone, acetophenone, and 6-methylhepta-3,5-dien-2-one) and acids (acetic acid, octanoic acid, nonanoic acid, decanoic acid, and hexadecanoic acid) was also observed, followed by that of esters (ethyl acetate, butyric acid butyl ester, and benzyl benzoate) and phenolics (2-methoxyphenol). The further use of multivariate data analysis, such as principle component analysis (PCA) and partial least squares regression (PLSR), enhanced and highlighted hidden correlations on the sensory data. Some VOCs showed PLSR positive correlations (phenethyl alcohol, methylheptenone, phenyl acetaldehyde, phenol and 6-methylhepta-3,5-dien-2-on), while others were directly related with the fruity odor ((E)-2-hexenal, 2-pentylfuran, 2-methyl-1-butanol, and geranylacetone), the green/grassy aroma ((Z)-3-hexen-1-ol), hexanal, and 1-penten-3-one), the sweet (1-hexanol, pentanal), sour (acetic acid) and earthy/musty smell (2-propylthiazole, 3-methyl-1-butanol, 2-isobutylthiazole) [118].

Plants emit daily hundreds of VOCs as chemical signs interacting and communicating with the above and belowground environment mainly for recognition, attraction and defence purposes. The belowground microenvironment is known as rhizosphere and it entails the interface of plant roots with soil interactions [36]. However, due to various experimentation difficulties, it is still an open and a rather unexplored area of research. Therefore, healthy and infected (by the fungal pathogen *Fusarium oxysporum*) tomatoes roots were experimentally monitored based on their released odors. Differences on the emitted VOCs were noticed, next to VOCs role to attract beneficial bacteria for plants. Along with the previously reported tomato roots VOCs, as cymene, 3-carene, sabinene, myrcene and methyl salicylate, much more have been recently identified. In particular, from the healthy roots, n-alkanes, beclomethasone dipropionate, cymene, and decanal were reported, whereas from the unhealthy plant roots, benzonitrile, benzothiazol, dimethyl trisulfide, and formic acid. Note that the pathogen itself released decane, eicosane and napthalene into the soil. The origin of the majority of the above reported VOCs remains unclear; formic acid is released by many microorganisms affecting positively or negatively the involved microorganisms, while naphthalene and p-cymene showed antimicrobial function [36].

### 6.2.2 Tomato processing

The main stages of tomato production include cultivation and processing operations. In a related 10-year study (2005–15), the main environmental factors linked with tomato processing of diced and paste products included grid electricity, electricity for irrigation and cooling, and use of natural gas, water and diesel. The potential beneficial effects of fertilizer and pesticide selection, the employ of renewables (e.g., use of solar-powered irrigation pumps and its on-site generation at the supply chain), along with some changes at cultivation and processing stages, is expected to lead at a 9%–10% reduction of total global warming potential. The latter study, was performed for California processing, the biggest US producer (96%) and a main global tomato supplier (30%) [113].

## 6.3 Tomato-derived biochar

Fruit and vegetable waste processing is a subject under study by many researchers due to the need of waste minimization and of new resources under the circular economy concept. Tomato is a vegetable crop that is used as ingredient for a variety of products such as sauces, ketchup, juice, paste, etc. From the manufacturing processes of these products, large quantities of pomaces are obtained which contain seeds, skin, and pulp [98]. These fractions contain valuable products such as pectins, hemicellulose, and cellulose, which can be extracted and used by the nonfood sector, e.g., pharmaceutical and cosmetics industry. If this is not possible, then conversion to other useful material can be achieved through the production of compost or biochar from tomato pomaces. Tomato crop residues include stalks, leaves, root, and the pomace; their chemical composition consists of 22%–38% cellulose, 31%–48% hemicellulose, 6%–10% lignin, 0.5%–4% lipids, 10%–24% protein, and 2%–6% of ash, distributed in the above mentioned moieties [56]. Especially, the percentage fractions in tomato pomaces account of 19.77% for neutral detergent fiber fraction, 8.6% for cellulose, 5.33% hemicellulose, and 5.85% lignin [98]; however, this content depends on tomato variety and processing method. This is shown in Table 6.1, where tomatoes main chemical composition is presented. The respective by-products are promising feedstock ingredients for the production of biochars with good adsorption properties.

Nowadays, waste biomass is being valorized following three main processes: mechanical, biochemical and thermal treatment. The main wastes

**Table 6.1** Main chemical components measured in tomatoes [29].

| Parameter | Variety<br>*Borseta, Cherry, Cor, Penjar, Plana, Pruna, Redona, Valenciana* |
|---|---|
| Dry matter (%) | 5.81–10.11 |
| Soluble solids (%) | 4.97–7.68 |
| pH | 4.16–4.33 |
| Titratable acidity (%) | 0.35–0.71 |
| Taste index | 1.02–1.28 |
| Glucose (g/kg) | 7.15–10.22 |
| Fructose (g/kg) | 7.78–13.36 |
| Citric acid (g/kg) | 2.83–3.95 |
| Lycopene (mg/kg) | 25.25–58.73 |
| β-Carotene (mg/kg) | 6.33–18.70 |
| Ascorbic acid (mg/kg) | 151.6– 315.1 |
| Total phenolics (mg/kg) | 495.5–1171.4 |
| Antioxidant activity (mmol TE/kg) | 1.59–3.58 |

derived from agri-food industry include pomace, peels, and shells; all these entail high water content. Therefore, a low-temperature pyrolysis or high-temperature drying (known as torrefaction) are widely applied. Torrefaction thermal treatment is implemented without oxygen, in the temperature range of 200–300°C at atmospheric pressure, for around 90 min. This process results in moisture depletion and subsequent lack of the hydroxyl groups; overall, torrefied products gain better physical and chemical properties. In such food waste treatment pathway, black currant pomace, apple pomace, orange peels, walnut shells, and pumpkin seeds undergo a torrefaction procedure; overall, the increase of temperature impacts positively on their hydrophobic properties. Nevertheless, an optimization process is always necessary for food waste valorization applications as economic (costs for energy) and final result balance need to be reached [26]. In Table 6.2, details of the main thermal treatments are summarised.

Biochar is a product rich in carbon, which can be produced from materials with high organic content (biomass) using pyrolysis (200–1000°C) in oxygen-limited or anoxic conditions. Converting biomass into biochar involves raw biomass drying, grinding, pyrolysis, and separation. The technologies used for biomass thermochemical conversion include fast, medium-speed, and slow pyrolysis. The properties of produced biochar depend on feedstock materials and pyrolysis conditions including

**Table 6.2** Main thermochemical processes and their respective parameters.

| Thermochemical process | Temperature range (°C) | Yield (%) | Residence time | Heating rate |
|---|---|---|---|---|
| Slow pyrolysis | 100–1000 | 15–40 | Minutes to hours | Slow |
| Fast pyrolysis | 300–1000 | 10–25 | <2 s | Very fast |
| Torrefaction | 200–300 | 61–77 | Minutes to hours | Slow |
| Gasification | 700–1500 | ~10 | Seconds to minutes | Moderate—very fast |

temperature and reaction time. Apart from the process conditions, the basic characteristics of the feedstock such as lignin, cellulose and hemicellulose content affect the product properties. These organic compounds behave differently during the heat treatment and hence, the composition of the biomass directly influences biochar yield and properties [9]. Hemicellulose decomposes at temperatures of about 220–315°C, in relation to cellulose that is more stable, decomposing at temperatures between 280 and 400°C. The lower structural stabilities of hemicelluloses and cellulose lead to earlier thermal degradation of these components and contribute significantly to the yield of condensable gases, while lignin being more stable, contributes to the biochar yield. Furthermore, lignin decomposes over a broad temperature range due to the large number of functional groups with different thermal stabilities degradation; it starts from 200°C and may require temperatures as high as 900°C to be completed, depending on the residence time [87,112].

Biochar main advantages are the abundant existence of feedstock and its efficiency next to the low production cost (20 USD per ton); however, among its disadvantages are pore blocking, flammability, and hygroscopicity [122].

### 6.3.1 Laboratory production of biochar

In a study carried out by Laifi [58], tomato pomaces were collected from a fruit market (no data exist on variety, localization and agriculture processing) and oven dried at 100 °C for 2:30 h in order to remove water content, as shown in Fig. 6.1.

Furthermore, they were pyrolyzed in a laboratory Nabertherm GmbH kiln at two different temperatures in order to compare their properties, as indicated in Fig. 6.2. The conditions used were: 350°C and 550°C with a rate of 6°C/min and residence time of 1:30 h with a nitrogen flow of 50 L/h [94].

**Fig. 6.1** ***Tomatoes feedstock prior to pyrolysis*** [58].

In order to increase the porosity and specific surface area (SSA) of the produced biochars, an activation of both materials was performed as presented in Fig. 6.3. In general, activation can be achieved either by treating the material with gases such as $CO_2$ or stream (physical activation) or by using different agents such as KOH, $H_3PO_4$, $ZnCl_2$, etc. (chemical activation) [50]. Especially for the absorption of VOCs, the respective modification technologies include $CO_2$ or $H_2O$ (physical activation), whereas chemical activation entails acid, alkali treatment, nitrogen, metal/metal oxide, and organic polymer doping [122]. Nowadays, VOCs are targeted for their negative environmental and health impact since they are emitted from the abundant anthropogenic activities. Among the most addressed compounds for sorption are benzene, toluene, ethyl benzene and xylene, widely known as BTEX, due to their hazardous and toxic effects. Other target groups include the chlorinated VOCs of methyl chloride, chloroform, and carbon tetrachloride released mainly from landfills [57].

Taking into consideration the importance of activation step, it was decided to proceed towards its implementation. Thus, the activation process included mixing into a solution of 20 g of both biochars (350°C, 550°C) with 200 ml 3M KOH (1:10 ratio) for 1 h at 65°C. Then the mixture was filtrated (DP 400 125) and the solid material washed with deionized water until reaching a steady pH. Furthermore, it was dried at 100°C. The dried material was then pyrolyzed at 800°C with a temperature ratio of 6°C/min under $N_2$ flow [77,94].

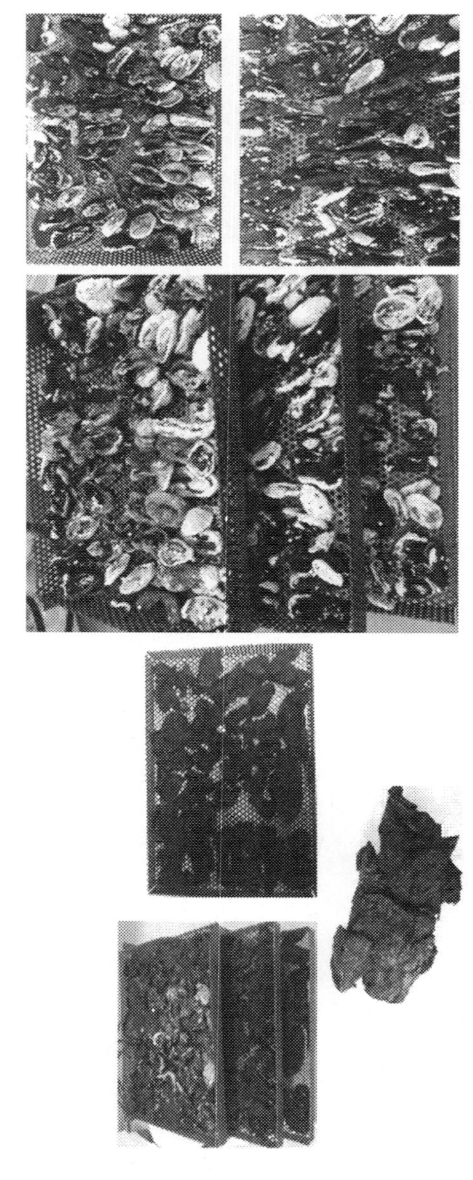

**Fig. 6.2** Tomatoes after pyrolysis at 350 °C (left) and 550 °C (right) [58].

**Fig. 6.3** Activation of tomatoes biochar at 350 °C (left) and 550 °C (right) with KOH at 800°C [58].

The produced and activated biochars can be analysed for their physicochemical properties for the determination of various parameters such as SSA (using Brunauer-Emmett-Teller method), surface morphology (by scanning electron microscope [SEM] and energy dispersive X-ray analysis [EDX]), functional groups (through Fourier transform infrared, FTIR), mass lose (by thermogravimetric analysis, TGA), elemental synthesis (C, H, N, S), minerals presence (X-ray diffraction [XRD] and X-ray fluorescence [XRF]) as well as for their respective physical properties (e.g., pH, density, conductivity, *etc.*). Biochar full characterization highlights the prevailing adsorption mechanisms.

The biochar yield is decreased as pyrolysis temperature increases. At 350°C, the yield for tomato-derived biochar was 7.99% and at 550°C was 4.91%, respectively. The pH of tomato biochar was alkaline; at 350°C it was estimated 10.61, whereas at 550°C it was 10.85, respectively. The density at 350°C was 1.51 $g/cm^3$ and at 550°C 1.45 $g/cm^3$. It should be noted that the initial moisture content and pH of tomato pomace were 94% and 4.4, respectively.

Tomato processing waste biochars could be used as organic amendments/enhancers to improve the fertility of agricultural soils as climate change and intensive agriculture may result in soil physicochemical degradation and fertility losses. Soil amendments with biochar could absorb large amounts of organic wastes to be converted. Consequently, agricultural valorization can be a potential alternative to manage these valuable organic wastes.

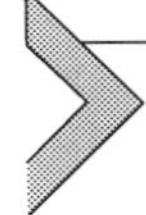

## 6.4 Agricultural valorization of tomato waste biochar as a sustainable recycling practice

### 6.4.1 Biochars as soil organic amendments or hydroponic substrates

The current trends of sustainable agricultural practices, including the adoption of circular economy and close-loop recycling concepts, entail the use of biochar as a soil amendment for open-field cropping systems or cultivation substrate for hydroponics. In this regard, two interconnected benefits are sought by the agricultural reuse of biochars namely, soil improvement and crop growth/productivity [79]. The most attractive qualities of biochar for agricultural reuse are the high carbon content, adsorption capacity and the porous structure, which will potentially improve structural stability, water retention and fertility of amended soils [38,51]. As such, biochars have been considered among the environmentally friendly fertilizers for sustainable and modern agriculture [18]. In this context, extensive data have been published addressing the reuse of biochars for agricultural purposes [12,54,61].

As a substrate for hydroponics in greenhouses, few studies have shown the beneficial effects of biochars alone or mixed with commercial substrates on crop yield and quality. For instance, Awad et al. [4] found that an equimixture of perlite and rice husk biochar (1:1) as a hydroponic substrate could be an alternative and effective technology for the better management of unwanted algal growth in nutrient solutions and high production of leafy vegetables. Yu et al. [120] demonstrated that replacing bark-based commercial substrates by mixed hardwood biochar (at 50%) or sugarcane bagasse biochar (at 70%) had the same effect on lettuce and basil yields, as did the commercial substrate alone. This implies also reductions in production costs and positive effects on carbon sequestration. Besides, Ding et al. [22] and Lone et al. [65] reported that biochars can significantly contribute to the improvement of the physical (granulometry, density, and porosity), chemical (pH, electrical conductivity, cation exchange capacity), hydrodynamic (hydraulic conductivity, water storage capacity), and biological properties of amended soils. When mixed with agricultural soils, biochar generally increases the nutrient release rates as compared to unamended soils [38,45]. For example, Chan et al. [17] studied nutrient availability to radish at pot-scale by mixing a raw poultry manure (RPM) biochar generated at 550°C with an agricultural soil at a dose of 50 t/ha. They showed that compared to unamended soil, nutrient uptake (g/pot) for N, P, Na, K, and Ca increased by approximately 141%, 263%, 223%, 94%, and 124%, respectively. In the same time, other biochars are likely to have significant adsorption/retention capacity for some nutrients. When studying nutrient retention under dynamic conditions in a typical Plinthudult soil amended with sludge biochars (1%) produced between 300°C and 700°C, Yuan et al. [121] demonstrated that biochar addition reduced the release of $NH_4^+$, P and K by 6.8%–35.9%, 8.5%–23.7%, and 7.9%–23.4%, respectively. Depending on biochar and soil nature, nutrients could present opposite dynamics and behaviour in amended soils [76, 83]. For instance, Pratiwi et al. [83] studied the impact of rice husk biochar added to a loamy soil at 4% (w/w) under dynamic conditions. They demonstrated that compared to control assays, biochar addition reduced the leaching of total ammonium and nitrates by about 11% and 23%, respectively. On the contrary, an increase of phosphates leaching by about 72% was observed. These discrepancies in physicochemical characteristics after biochar application have mainly been imputed to both the properties of soils/biochars, as well as the experimental conditions [17,83].

In any case, it is generally admitted that the improvement of soil physicochemical and biological properties, positively influenced seed germination, plant growth, and ultimately crop yield [39,108,119]. Table 6.3 summarizes the agronomic effects of selected biochars resulting from the pyrolysis of various feedstocks. Doan et al. [23] observed a positive effect of bamboo biochar applied alone or mixed with buffalo manure compost on maize growth and yield on the long term. They attributed this improvement to reduced water runoff and retention of bioavailable nitrogen forms in amended soils. For the same biochar and crop, Uzoma et al. [105] found that maize grain yield significantly increased by 150% and 98% as compared to control after biochar application at 15 and 20 t/ha, respectively. The higher maize yield at 15 t/ha was explained by the lower P availability in the soil amended with biochar at 20 t/ha because higher biochar application rates favor P fixation by calcium. Enhanced nitrogen uptake efficiency due to amendments with rice straw biochar in Chinese paddy fields was also reported by Huang et al. [44], which showed a progressive effect on the yields of rice. Houben et al. [43] noted a three-fold increase in biomass production of rapeseed crop with an addition of 10% biochar mass fraction in a heavy metal contaminated soils. Moreover, the yield of cherry tomatoes during a pot experiment was 64% higher than control pots, along with significant changes in nutrient uptake (P and K) and chemical properties of amended soils [42].

**Table 6.3** Agronomic impact of biochars produced from various feedstocks at different experimental conditions.

| Feedstock type | Pyrolysis conditions | Reuse conditions | Agronomic impact | Literature |
|---|---|---|---|---|
| RPM | 450°C or 550°C with activation using high temperature steam | 0, 10, 25, and 50 t/ha with or without N fertilizer (100 kg/ha) | Nutrient uptake by radish proportional to biochar dose and presence of N fertilizer | [17] |
| Sewage sludge | 300, 400, 500, 600, and 700°C for 3 h under a flow of $N_2$ of 1000 mL/min | Added to a typical Plinthudult soil at 1% | Biochars produced at 500 and 700°C reduced the soil leaching of $NH_4^+$, $NO_3^-$, $PO_4^{-3}$, and $K^+$ | [121] |
| Rice husk | Commercial grade char | Added to paddy loamy soil at 4% | Reduction of total $NO_3^-$ and $NH_4^+$ leaching by 11% and 23%, respectively. Increase of $PO_4^{-3}$ leaching by 72%, compared with the control soil | [83] |
| Bamboo fragments | 600°C for 8–10 h | Added once to a sandy soil at 7 t/ha during a 3-year trial | Positive long-term influence on maize yield and growth. Further improvement in presence of buffalo vermicompost. | [23] |
| Cow manure | 500°C for 5 h | Added to a sandy soil at 0, 10, 15, and 20 t/ha in a greenhouse | Biochar at 15 and 20 t/ha significantly increased maize grain yield by 150% and 98%. Water use efficiency of maize plants increased with biochar addition. | [105] |
| Giant miscanthus (*Miscanthus × giganteus*) straw | 30 min pyrolysis and an end temperature of 600°C. | Added to a metal-contaminated soil at 1%, 5%, and 10% | Decrease of bioavailable Cd, Zn, and Pb with biochar dose. More efficient than soil liming. Rapeseed plants grew normally in soil amended with 10% biochar. | [43] |

Depending on pedoclimatic conditions, the positive effects of biochar may become more evident on the long term, providing benefits over several cropping seasons. Accordingly, one biochar application of 20 t/ha did not result in a significant increase of maize grain yield during the first year compared to control, but increases of 28%, 30%, and 140% were later observed for the second, third, and fourth year, respectively [69]. This shows that such organic materials may need time to interact with soil components until reaching the capacity of improving soil properties and nutrient release [45]. In general, lignocellulosic biomasses and sewage sludge have relatively lower capacity in releasing nutrients [37,59,117,121], when compared to animal biomasses [38,111].

### 6.4.2 Properties and impact of lignocellulosic biochars

According to Uchimiya et al. [104], the SSA of biochars generated from crop residues range from values as low as 1.2 and 6.3 $m^2/g$ for olive pomace and wheat straw, respectively to very high SSA of about 220 $m^2/g$ for palm and coconut shells. However, nutrient content in biochar has generally no relationship with its physical properties, but depends often on the original composition of the feedstock [114]. For instance, total P content varies from less than 0.1 g/kg for palm and coconut shell biochars up to 4.2 g/kg for biochar of olive pomace. Moreover, total K is 10-fold higher in wheat straw biochar (60 g/kg) than palm shell biochar [104]. It is important, however, to note that total levels of nutrients in biochars do not necessarily imply proportional bioavailability in soil after application [16,45]. In any case, a lignocellulosic biochar of good quality should show a balance between physical properties, nutrient content and most importantly a regulated adsorption/desorption capacity. Consequently, improved soil properties, contaminant retention and nutrient slow release would be expected after biochar application [45].

Piscitelli et al. [82] applied a microbial charged biochar generated by slow pyrolysis of olive mill pomace to a sandy soil. After a 7-day incubation, SEM images of amended soils showed the presence of biofilm, suggesting an early stage of microbial colonization. Despite the short time of incubation, the microbial biomass C and N increased indicating a beneficial effect of this amendment. Manolikaki et al. [72] tested the effect of lignocellulosic biochars on P release and ryegrass yield. Rice husk and grape pomace biochars showed significant effects on plant yield only in slightly acidic soil in second and third harvests. Besides, P uptake by ryegrass was higher in presence of grape pomace biochar at the third harvest in slightly acidic soil while during first and second harvests, significant differences were observed in alkaline soil. The authors suggest that these biochars may act as good sources of available P in croplands and improve plant growth, although soil conditions may play a significant role. In addition to soil amendments, biochar may play the role of substrate to grow crops in hydroponics and constitute a sustainable input in certified organic production systems [109]. When substituted for peat moss in potting substrate mixtures in rates up to 15% (v/v), biochars produced from hardwood pellets and pelletized wheat straw were similar in performance to mixtures containing only peat moss [110]. In another experiment, Vaughn et al. [109] found that combining potato peel anaerobic digestate with acidified wood pellet biochar resulted in an increased growth of tomato plants as compared to the conventional peat:vermiculite control. Besides, a digestate:acidified wheat straw pellet biochar was equal to the control for marigold growth. All these data, point out the suitability of lignocellulosic biochar for agricultural applications despite the lower nutrient content (mainly nitrogen) as compared to those generated from animal wastes [38].

### 6.4.3 Case of tomato residues

Tomato crop systems produce tomatoes as the main marketable product to be sold fresh or processed. Two main residues are generated by the whole production system namely, plant debris and tomato pomace [68,99]. For instance, greenhouse-grown tomato may produce about 4.5 t/ha/week of tomato leaf waste [24], and 15 t/ha/year of fresh plant residues after harvest [78]. According to statistics from the World Processing Tomato Council, about 37 million tons of fresh tomatoes were processed worldwide in 2019 to produce tomato juice, ketchup, canned dices or paste, and many other products [116]. Tomato pomace represents approximately 5% by weight of the processed fresh tomatoes and consists mainly of peels, pulp residues, and seeds. Tomato pomace has no commercial value and is mostly disposed of as a solid waste or used to a limited extent for animal feeding or the extraction of bioactive compounds [68,91]. However, soil enhancement has been proposed as a valorization pathway for tomato waste by its incorporation under different forms: (1) fresh residues (stems, leaves and roots), especially in open field cultivation systems [31], (2) composted fresh or processed residues [78], (3) digestate generated during anaerobic processes [71], and (4) biochar [73].

Most of reported data described the reuse of fresh or composted tomato residues as soil amendments, the latter being the main incorporated form.

García-Raya et al. [31] applied tomato plant residues for biosolarization (biodisinfection) and fertilization purposes. They observed no significant differences with chemical fertilization in terms of tomato yield under greenhouse conditions (7.9 and 8.3 kg/m$^2$, respectively). In their study on weed control, Achmon et al. [1] effectively inactivated the germination of *Brassica nigra* and *Solanum nigrum* by soil biosolarization with tomato pomace amendments in as little as 5 days of treatment. This was a much shorter time than the multiple weeks usually required for conventional solarization to control weeds. Roussis et al. [88] used fresh tomato pomace mixed with biocyclic humus soil as an alternative to conventional chemical fertilization to promote organic farming. Despite higher tomato yields using N fertilizers (163.4 t/ha vs 150.7 t/ha), the organic mixture improved total soluble solids in fruits, which influence quality (sweetness, sourness and flavor) and marketability. As for most of lignocellulosic biomasses, the composting of tomato residues alone results in a low quality end product that has a high moisture content and lacks nitrogen [99]. Therefore, tomato residues are generally co-composted with various organic materials to equilibrate the C:N ratio and ameliorate the fertilization potential after land application [90,97]. For instance, mixing tomato plant residues with sheep manure improved the composting process, reduced N loss, decreased electrical conductivity, and increased the concentration of humic acid and macroelements in the compost [99]. Likewise, by using a feedstock composed of 50% tomato biomass residues, 48% woodchips and 2% mature compost as starter, Pane et al. [78] obtained a compost that improved soil biological activities and marketable tomato yield (88 t/ha), as compared to a commercial compost (55 t/ha).

Available literature describing the production and reuse of tomato waste biochar in agriculture is considerably scarce and has different purposes. As lignocellulosic materials, tomato wastes and subsequent biochars are generally characterized by a higher C fraction and lower N content [25], as compared to biochars deriving from animal wastes [84] or urban sludge [121]. For instance, $CO_2$ emission was measured from soil plots amended with different biochars of tomato harvest wastes added at 30 t/ha [73]. It was shown that biochar application resulted in lesser cumulative soil $CO_2$ emissions when produced at 600 and 700°C rather than 300, 400, and 500°C under 40-min holding time. Besides, the lowest cumulative $CO_2$ emission over a total period of 130 days (533 g/m$^2$) was recorded for biochars produced at 500°C for a holding time of 240 min. Llorach-Massana et al. [64] addressed the technical feasibility and carbon footprint of biochar co-production with tomato plant residue. Biomass conversion yields of over 40%, 50% carbon stabilization and low pyrolysis temperature conditions (350–400°C) would be required for biochar production to sequester carbon under urban pilot scale conditions. In another experiment, Dunlop et al. [25] used a closed-loop recycling system by evaluating the suitability of biochar produced from tomato crop green waste (440–550°C) as a substrate for soilless, hydroponic tomato production. Results indicate that many of the relevant properties of substrates containing or consisting entirely of tomato biochar represent potentially valuable characteristics of a typical hydroponic substrate. Accordingly, tomatoes grown in this biochar did not differ significantly in terms of growth, yield, or fruit quality with those grown in pine sawdust, a substrate which is currently used for commercial tomato production. The same authors assume that producing a biochar at 550°C would convert 10–60 t/ha of tomato green waste and meet 13% to 50% of the grower's substrate requirements, on a per hectare basis.

In the framework of cyclic economy, the green waste derived from tomato were properly processed as biochar towards the production of soilless hydroponic greenhouse tomato. Toward this effort, different biochar mixtures (e.g., pine, sawdust) were produced, characterized, and tested. Important soil parameters (e.g., electric conductivity, pH) were monitored and adjusted so as to enable the fruit cultivation [25]. In another agricultural challenge, biochar addition to tomato was studied and compared along with pepper (*Capsicum annuum L.*). There were both positive (enhanced plant height and leaf size) and negative results (no effect on flower and fruit yield). Emphasis was also given to understand the relative mechanisms based on the biochar extracts. Biochar either contributed indirectly to the plant growth through hormesis or indirectly through the related microorganisms (e.g., rhizobacteria or fungi) [35]. In another application, different irrigation regimes were also examined (e.g., full, deficit and partial rootzone drying irrigation) along with specific biochar levels (e.g., 0% and 5% by weight) and provided mixed results. The advantages and disadvantages of this approach towards water productivity and quality of tomato are discussed and presented [2].

Another important challenge is that of tomato production in arid climates. The beneficial effects of the use of biochar (e.g., wood-derived, cow manure, and coamendment) and manure as soil amendments on tomato biomass and yield (yield, heavy metal, soil nutrients) were examined compared to fertilizers [40]. Saline water irrigation is another modern challenge; towards this, tomato growth was implemented using wheat straw biochar as soil amendment. Despite the relevant salt stress, tomato was enable to survive. The ability of biochar to withstand such extreme agriculture challenges is discussed and the prevailing mechanisms are revealed [93].

## 6.5 Life cycle analysis

Using life cycle analysis (LCA) to illustrate the benefits of biochar production and use to mitigate carbon emissions, can be useful to the several stakeholders such agriculturist, policymakers, competent authorities, and the public when the direct and indirect cost of the production and wastes of fruits and vegetables, as well as other edible foods are considered [103]. There have been many studies reporting the LCA of biochar production. Most are difficult to compare based on the functional unit (FU) or based on the scope or system boundaries of the LCA. In regard to the choice of a FU [41], it was suggested that carbon abatement per unit of delivered energy is not an appropriate unit for comparing different biochar systems. Additionally, it was concluded that the $CO_2$ eq. per oven dry ton (odt) of biomass feedstock, is an appropriate FU for comparing different bioenergy systems. The FU, $CO_2$ eq. per odt of biochar product, is the best for comparing different biochar systems. The results revealed that a starting estimate for the climate mitigation potential of a biochar system was equal to one metric ton of $CO_2$ eq. per oven dry ton of biomass. Roberts et al. [86] choose one metric ton of dry biomass as the FU for their biochar-pyrolysis system, which compared corn stover, yard waste, and switchgrass feedstocks used in a bioenergy facility. The net climate change impact was calculated as the sum of the net greenhouse gases (GHG) reductions (biochar sequestered carbon and avoided emissions) and the net GHG emissions.

LCA is broadly implemented for agriculture and agri-food products environmental analyses and improvement. It has been applied to evaluate the use of energy and the related environmental impacts of pistachio cultivation in Aegina island in Greece, where the current production and two alternative scenarios were investigated and improvement opportunities were detected [6]. Furthermore, it has been used to compare different cultivation methods of the same agri-product. For example, Longo et al. [66] applied the respective methodology to examine the supply chain of organic and conventional apples, including raw materials and energy sources input, the farming step, the post-harvest processes and the apples distribution to the final users, evaluating which of the two products is better from an energy and environmental point of view. The method can also be used as a tool to address sustainable production and consumption patterns of local policies, and to get knowledge for environmental assessment of a wide agricultural production area [15]. Furthermore, it is a suitable method to evaluate how the management of an agro-product growing in a well-defined farming system, e.g., for apples production, influences environmental impacts [74]. It has also been used for managerial decisions justification within agricultural holdings, in order to develop them sustainably in a case study for Romanian viticulture [13].

LCA is a technique that reports the environmental aspects and potential environmental impacts (e.g., use of resources and the environmental consequences of releases) during the life cycle of any product or services including the cultivation (if exist), the production, the use, the end-of-life treatment, recycling, as well as final disposal (i.e., cradle-to-grave) [49]. It has been developed fast over the past three decades from simple energy analysis to complete life cycle impact assessment (LCIA), life cycle costing and social-LCA and recently to a more comprehensive life cycle sustainability analysis, which broads the traditional environmental evaluation [33]. The LCA approach is fully applicable for any development and improvement of any product or service, including strategic planning and assessment of environmental performance, marketing approach and public policy making, etc. [103].

LCA must followed specific steps which according to [49] are considered to be four [103]. The goal and scope definition phase, in which the boundaries of the system under study, the FU and the level of detail of the LCA are specified. The life cycle inventory (LCI) analysis, which contains the gathering of the necessary data with respect to the considered system, in order to encounter the goals of the defined study. The LCIA phase, with purpose to provide information for LCI results of a product's system by assessing the impacts in order to understand their environmental importance and the life cycle interpretation phase at which, the results of the inventory and impact assessment phase are debated, as well as the assumptions, recommendations in relation with the goal and scope, being designed [33,103].

Comparative LCA of three water intensive tree cultivation systems, i.e., almond, pistachio, and apple, identified the "hotspots" for the crops, exhibiting the most significant environmental impacts and consumption of energy. Additionally, sensitivity analysis was achieved in order to examine actions for water necessities reduction and energy conservation promotion [7]. Tsangas et al. [103] applied LCA models to assess the agricultural strategic development planning for Balkan region (approach covered the entire production line of each product, respectively), as a useful tool to identify and quantify potential environmental impacts from the production of wine, apple juice and pesto produced from peppers on behalf of Greece, North Macedonia and Bulgaria. Gases emitted from agricultural sector consist primarily of $CO_2$, $N_2O$, and $CH_4$ and are primarily emitted from food production [52]. The sector is the major contributor to climate change impact through energy use, livestock, fertilizer production, pesticide production, machineries, land use change and soil degradation. Based on the LCA results [103], several suggestions were made in order to reduce the footprint of those three cultivations. The results indicate that changes in the cultivation and the production must be considered in order to optimize the environmental footprint.

Conventional crop production strategies had the main target, which is dealing with the maximum production while at the same time paying little or no attention to the related environmental issues and impacts [46]. In fact, the food and agricultural sectors are considered as two of the main priority areas within the European Sustainable Production and Consumption policies [27]. This is due to the fact that their activities contribute to human health and prosperity [47], as well as on waste strategies development [123] as they are main contributors to environmental impacts like acidification, eutrophication, land degradation, global warming, and water depletion [8,34,75,100].

As previously mentioned, cultivation of any of agri-product has several environmental dimensions [103]. Assessing and improving those impacts with a life cycle approach would make it possible to enhance sustainability and management quality issues, so being in line with the concept of environmental supply chain management. Potentially, LCA is one of the most instructive management tools to gain insights into product-related environmental impacts and stimulate implementation of cleaner production and management systems. Several LCAs were found [46] that address environmental issues related to tomato supply chain considering two different types of cultivation, *i.e.*, greenhouse or open field cultivation. Most of tomato LCA studies are related to greenhouse tomato cultivation. According to Jolliet (1993), the energy consumption and environmental emissions were evaluated [53]. The use of LCA approach from Van Worden [107] in a Dutch glasshouse, indicates that the use of energy was responsible for about 75% of the total environmental impact of the crop. Torellas et al. [102] evaluated the environmental impacts of tomato crop in a multitunnel greenhouse in Almeria, to suggest different cleaner productions within greenhouse. Anton et al. [3] during an environmental impact assessment of tomatoes cultivation, introduced in the LCA model new impact categories linked to land use, water consumption, as well as pesticides and fertilisers. LCA was also applied by Payen et al. [80] to compare local and imported tomatoes, while a life cycle perspective on the sustainability of Ontario greenhouse tomato production was proposed by Dias et al. [21].

Moreover, Karakaya and Oziligen [55] considered energy utilization and $CO_2$ emissions during the manufacture of tomato products such as fresh, peeled, diced and juiced tomatoes and was found that transportation to distribution centers is considered as the most important source of $CO_2$ emissions, highlighting the logistic phase as the most impactful stage. On the other hand, Manfredi andVignali [70] on glass jar packaged of tomatoes products, displayed that packaging, cultivation and processing were the main responsible for any of the main environmental impacts (such water and energy consumption, $CO_2$ emissions, etc). Garofalo et al. [32] using LCA approach on canned, whole peeled tomato indicated that most significant impact were that of landfill disposal, packaging, processing, and cultivation.

Bosona and Gebresenbet [11] considered the case of tomato production and consumption within Sweden. It was assumed that both fresh and dried produces were cultivated in the same area, southern Sweden. The system boundary of this LCA study starts from agricultural production and ends at consumers' gate. Bosona et al. (2018) showed that the main impact categories were washing, packaging, and retail cooling. Also, the fresh tomato value chain (FTVC), represents the 5% of the cumulative energy demand (CED) equal with 2.17 GJ. Moreover, packaging was one of the main contributors; with cardboard inclusion, the impact of packaging increases more and improvement of packaging contributes to reduction of environmental impacts. LCA also shows that in dried tomato value chain (DTVC), the demand of energy at post-harvest stage increased from 2.17 GJ to 7.61 GJ due to drying process. The considered GHG emission values at post-harvest stage were 88.12 kg $CO_2$ eq and 48.64 kg $CO_2$ eq for FTVC and DTVC, respectively.

Being the most consumed fresh vegetable in France, tomato production requires much water. Off-season tomatoes are either produced locally in heated greenhouses or imported from Morocco and Spain. Morocco (North Africa) is the primary supplier of the French market, with 68% of the imported off-season tomatoes (French customs); production for export is located in the Souss-Massa region (West Southern Morocco). The environmental impacts of Moroccan tomato production system for export have never been assessed. According to Payen et al. [80], the impacts of water use for vegetable production are crucial in the choice of vegetable sourcing. The LCA from Payen et al. [80], indicates that transportation to France was the main contributor to terrestrial acidification, with 50% of the impact, followed by tomato cultivation (39%) and packaging (10%). Impact was dominated by NOx emission during the truck transportation, SOx which was associated with fertigation (fertiliser production and energy consumption), and $NH_3$ emissions occurring after N fertiliser field application. Also, it was found out that 60% of the tomato cultivation contributes to freshwater eutrophication, while the cultivation was the main contributor to all ecotoxicity impact categories (96% of terrestrial ecotoxicity, 59% of freshwater ecotoxicity, and 54% of marine ecotoxicity).

## 6.6 Conclusions

To adopt the circular economy concept for more sustainability, modern economic challenges and guidelines imply the maximum usage of every available resource; including the huge amounts of agri-food wastes currently generated in modern societies. In the current chapter, biochar production from the raw and processed wastes of a well-known and abundant commodity largely cultivated worldwide, that of tomato crop, is being examined as a case study of waste valorization. As a first step, food supply chain losses (food loss and waste) is being reviewed for highlighting the quantities of the respective raw feedstock. Then, the chemical composition of tomato fruit along with its lignocellulosic components (e.g., cellulose, hemicellulose and lignin) is explored, as it is considered of paramount importance for biochar production. Thirdly, the appropriate oxygen-free pyrolysis process required for an optimal transformation of biomass to biochar is addressed as well. Key-role parameters include that of final temperature, heating rate, residence time, *etc.*, largely affecting the quality and quantity of the final product. After the thermochemical conversion of tomato waste, biochar physicochemical characterization is also a crucial step using a variety of analytical techniques. Biochar quality could be further enhanced and improved, using various activation methods, as well as the addition of nanoparticles depending on the final application. For valorization purposes, a summary on biochar agricultural reuse as soil organic amendments or hydroponic substrate and its corresponding impacts are also presented and discussed. Finally, the utility and importance of LCA in the agricultural sector and more specifically of tomato supply chain is depicted contributing mainly to environmental and financial issues, resources, as well as energy saving. Overall, the generated knowledge can be further expanded and applied to other large food waste commodities, assisting in mitigating the negative impacts of food waste contributing to water, soil and air pollution.

## References

[1] Y. Achmon, J.D. Fernández-Bayo, K. Hernandez, et al., Weed seed inactivation in soil mesocosms via biosolarization with mature compost and tomato processing waste amendments, Pest. Manag. Sci. 73 (2017) 862–873, doi:10.1002/ps.4354.

[2] S.S. Akhtar, G. Li, M.N. Andersen, F. Liu, Biochar enhances yield and quality of tomato under reduced irrigation, Agric. Water Manag. 138 (2014) 37–44, doi:10.1016/j.agwat.2014.02.016.

[3] A. Anton, M. Torellas, M. Nunez, et al., Improvement of agricultural life cycle assessment studies through spatial differentiation and new impact categories: case study on greenhouse tomato production, Environ. Sci. Technol. 48 (16) (2014) 9454–9462.

[4] Y.M. Awad, S.E. Lee, M.B.M. Ahmed, et al., Biochar, a potential hydroponic growth substrate, enhances the nutritional status and growth of leafy vegetables, J. Clean Prod. 156 (2017) 581–588, doi:10.1016/j.jclepro.2017.04.070.

[5] S. Baldina, M.E. Picarella, A.D. Troise, et al., Metabolite profiling of Italian tomato landraces with different fruit types, Front Plant Sci. 7 (2016) 1–13, doi:10.3389/fpls.2016.00664.

[6] G. Bartzas, K. Komnitsas, Life cycle analysis of pistachio production in Greece, Sci Total Environ. 595 (2017) 13–24, doi:10.1016/j.scitotenv.2017.03.251.

[7] G. Bartzas, D. Vamvuka, K. Komnitsas, Comparative life cycle assessment of pistachio, almond and apple production, Inf. Process Agric. 4 (2017) 188–198, doi:10.1016/j.inpa.2017.04.001.

[8] M. Beccali, M. Cellura, M. Iudicello, M. Mistretta, Resource consumption and environmental impacts of the agrofood sector: Life cycle assessment of italian citrus-based products, Environ. Manage 43 (2009) 707–724, doi:10.1007/s00267-008-9251-y.

[9] M. Belhachemi, B. Khiari, M. Jeguirim, A. Sepúlveda-Escribano, Characterization of biomass-derived chars, in: M Jeguirim, L Limousy (Eds.), Char and Carbon from Biomass - Production, Characterization and Applications, Elsevier, Amsterdam, 2019, pp. 69–108. http://dx.doi.org/10.1016/B978-0-12-814893-8.00012-2.

[10] S. Bergonzoli, A. Suardi, N. Rezaie, et al., An innovative system for Maize Cob and wheat chaff harvesting: simultaneous grain and residues collection, Energies 13 (2020), doi:10.3390/en13051265.

[11] T. Bosona, G. Gebresenbet, Life cycle analysis of organic tomato production and supply in Sweden, J. Clean Prod. 196 (2018) 635–643.

[12] P. Brassard, S. Godbout, V. Lévesque, et al., Biochar for soil amendment, in: M Jeguirim, L Limousy (Eds.), Char and Carbon Materials Derived from Biomass, Elsevier, 2019, pp. 109–146.

[13] C. Burja, V. Burja, Decisions in sustainable viticulture using life cycle assessment, J. Environ. Prot. Ecol. 13 (2012) 1570–1577.

[14] R. Buttery, Quantitative and sensory aspects of flavor of tomato and other vegetables and fruits, in: T.E. Acree, R. Teranishi (Eds.), Flavor Science: Sensible Principles and Techniques, American Chemical Society, Washington, DC., 1993, pp. 259–286.

[15] M. Cellura, S. Longo, M. Mistretta, Life cycle assessment (LCA) of protected crops: an Italian case study, J. Clean Prod. 28 (2012) 56–62, doi:10.1016/j.jclepro.2011.10.021.

[16] K.Y. Chan, L. Van Zweiten, I. Meszaros, et al., Assessing the agronomic values of contrasting char materials on Australian hardsetting soil, Proceedings of the Conference of the International Agrichar Initiative, Terrigal, NSW, Australia, 2007.

[17] K.Y. Chan, L. Van Zwieten, I. Meszaros, et al., Using poultry litter biochars as soil amendments, Aust. J. Soil Res. 46 (2008) 37–444.
[18] Y. Chen, J. Xu, Z. Lv, et al., Impacts of biochar and oyster shells waste on the immobilization of arsenic in highly contaminated soils, J. Environ. Manag. 217 (2018) 646–653.
[19] M.B. Cole, M.A. Augustin, M.J. Robertson, J.M. Manners, The science of food security, NPJ Sci. Food 2 (2018) 1–8, doi:10.1038/s41538-018-0021-9.
[20] V. De Laurentiis, S. Corrado, S. Sala, Quantifying household waste of fresh fruit and vegetables in the EU, Waste Manag. 77 (2018) 238–251, doi:10.1016/j.wasman.2018.04.001.
[21] G.M. Dias, N.W. Ayer, R.V. Acker, et al., Life cycle perspectives on the sustainability of Ontario greenhouse tomato production: benchmarking and improvement opportunities, J. Clean Prod. 140 (2017) 831–839.
[22] Y. Ding, Y. Liu, S. Liu, et al., Potential benefits of biochar in agricultural soils: a review, Pedosphere 27 (2017) 645–661, doi:10.1016/S1002-0160(17)60375-8.
[23] T.T. Doan, T. Henry-Des-Tureaux, C. Rumpel, et al., Impact of compost, vermicompost and biochar on soil fertility, maize yield and soil erosion in Northern Vietnam: A three year mesocosm experiment, Sci. Total Environ. 514 (2015) 147–154, doi:10.1016/j.scitotenv.2015.02.005.
[24] M. Dorais, Y. Dubé, Managing greenhouse organic wastes: a holistic approach, Acta Hortic 893 (12) (2011) 183–197.
[25] S.J. Dunlop, M.C. Arbestain, P.A. Bishop, J.J. Wargent, Closing the loop: use of biochar produced from tomato crop green waste as a substrate for soilless, hydroponic tomato production, Hort Science 50 (2015) 1572–1581, doi:10.21273/hortsci.50.10.1572.
[26] A. Dyjakon, T. Noszczyk, M. Smędzik, The influence of torrefaction temperature on hydrophobic properties ofwaste biomass from food processing, Energies 12 (2019), doi:10.3390/en12244609.
[27] EC (2008) Communication from the Commission of the European Parliament, the Council, the European Economic and Social Committee and the Committee of the Regions on the Sustainable Consumption and Production and Sustainable Industrial Policy Action Plan. COM, Europe.
[28] FAO (2011) Global food losses and food waste – extent, causes and prevention. Rome.
[29] M.R. Figàs, J. Prohens, M.D. Raigón, et al., Characterization of composition traits related to organoleptic and functional quality for the differentiation, selection and enhancement of local varieties of tomato from different cultivar groups, Food Chem. 187 (2015) 517–524, doi:10.1016/j.foodchem.2015.04.083.
[30] G. Garcia-Garcia, E. Woolley, S. Rahimifard, et al., A methodology for sustainable management of food waste, Waste Biomass Valori 8 (2017) 2209–2227, doi:10.1007/s12649-016-9720-0.
[31] P. García-Raya, C. Ruiz-Olmos, J.I. Marín-Guirao, et al., Greenhouse soil biosolarization with tomato plant debris as a unique fertilizer for tomato crops, Int. J. Environ. Res. Public Health 16 (2019) 1–11, doi:10.3390/ijerph16020279.
[32] P. Garofalo, L. D'Andrea, M. Tomaiuolo, et al., Environmental sustainability of agri-food supply chains in Italy: the case of the whole-peeled tomato production under life cycle assessment methodology, J. Food Eng. 200 (2017) 1–12.
[33] G. Goudouva, P. Loizia, V. Inglezakis, A.A. Zorpas, Quarries environmental footprint in the framework of sustainable development: The case study of Milos island, Desalin Water Treat 133 (2018) 307–314.
[34] G.T. Goudouva, A.A. Zorpas, Water Footprint determination by quarry operation in island regions», Desalin Water Treat 86 (2017) 271–276.
[35] E.R. Graber, Y.M. Harel, M. Kolton, et al., Biochar impact on development and productivity of pepper and tomato grown in fertigated soilless media, Plant Soil 337 (2010) 481–496, doi:10.1007/s11104-010-0544-6.

[36] S. Gulati, M.B. Ballhausen, P. Kulkarni, et al., A non-invasive soil-based setup to study tomato root volatiles released by healthy and infected roots, Sci. Rep. 10 (2020) 1–11, doi:10.1038/s41598-020-69468-z.
[37] K. Haddad, S. Jellali, M. Jeguirim, et al., Investigations on phosphorus recovery from aqueous solutions by biochars derived from magnesium-pretreated cypress sawdust, J. Environ. Manage. 216 (2018) 305–314, doi:10.1016/j.jenvman.2017.06.020.
[38] S. Hadroug, S. Jellali, J.J. Leahy, et al., Pyrolysis process as a sustainable management option of poultry manure: characterization of the derived biochars and assessment of their nutrient release capacities, Water (Switzerland) 11 (2019) 1–18, doi:10.3390/w11112271.
[39] H. Hamdi, S. Benzarti, I. Aoyama, N. Jedidi, Rehabilitation of degraded soils containing aged PAHs based on phytoremediation with alfalfa (Medicago sativa L.), Int. Biodeterior Biodegrad 67 (2012) 40–47, doi:10.1016/j.ibiod.2011.10.009.
[40] Hameeda, S. Gul, G. Bano, et al., Biochar and manure influences tomato fruit yield, heavy metal accumulation and concentration of soil nutrients under wastewater irrigation in arid climatic conditions, Cogent. Food Agric. 5 (2019), doi:10.1080/23311932.2019.1576406.
[41] J. Hammond, S. Shackley, S. Sohi, P. Brownsort, Prospective life cycle carbon abatement for pyrolysis biochar systems in the UK, Energy Policy 39 (2011) 2646–2655, doi:10.1016/j.enpol.2011.02.033.
[42] M.K. Hossain, V. Strezov, K. Yin Chan, P.F. Nelson, Agronomic properties of wastewater sludge biochar and bioavailability of metals in production of cherry tomato (Lycopersicon esculentum), Chemosphere 78 (2010) 1167–1171, doi:10.1016/j.chemosphere.2010.01.009.
[43] D. Houben, L. Evrard, P. Sonnet, Beneficial effects of biochar application to contaminated soils on the bioavailability of Cd, Pb and Zn and the biomass production of rapeseed (Brassica napus L.), Biomass Bioenergy 57 (2013) 196–204, doi:10.1016/j.biombioe.2013.07.019.
[44] M. Huang, L. Yang, H. Qin, et al., Quantifying the effect of biochar amendment on soil quality and crop productivity in Chinese rice paddies, F. Crop. Res. 154 (2013) 172–177, doi:10.1016/j.fcr.2013.08.010.
[45] A. Ibn Ferjani, S. Jellali, H. Akrout, et al., Nutrient retention and release from raw exhausted grape marc biochars and an amended agricultural soil: Static and dynamic investigation, Environ. Technol. Innov. 19 (2020) 100885, doi:10.1016/j.eti.2020.100885.
[46] C. Ingrao, N. Faccilongo, F. Valenti, et al., Tomato puree in the Mediterranean region: an environmental life cycle assessment, based upon data surveyed at the supply chain level, J. Clean Prod. 233 (2019) 292–313, doi:10.1016/j.jclepro.2019.06.056.
[47] C. Ingrao, A. Matarazzo, C. Tricase, et al., Life cycle assessment for highlighting environmental hotspots in Sicilian peach production systems, J. Clean Prod. 92 (2015) 109–120, doi:10.1016/j.jclepro.2014.12.053.
[48] Ismea (2017) Report - I numeri della filiera del pomodoro da industria. Roma.
[49] ISO 14040, Environmental management–life cycle assessment–principles and framework, Int. Organ. Stand (2006).
[50] M. Jeguirim, M. Belhachemi, L. Limousy, S. Bennici, Adsorption/reduction of nitrogen dioxide on activated carbons: Textural properties versus surface chemistry – A review, Chem. Eng. J. 347 (2018) 493–504, doi:10.1016/j.cej.2018.04.063.
[51] P. Jha, A.K. Biswas, B.L. Lakaria, A. Subba Rao, Biochar in agriculture - prospects and related, Curr. Sci. 99 (2010) 1218–1225.
[52] D. Johansson, Life Cycle Assessment (LCA) of Apples - a Comparison between Apples Produced in Sweden, Italy and Argentina. Master's thesis, Swedish University of Agricultural Sciences, Uppsala, 2015.

[53] Jolliet, Ökobilanz thermischer, mechanischer und chemischer, Kartoffelkrautbeseitigung Landwirtschaft Schweiz 6 (1993) 675–682.
[54] M. Kamran, Z. Malik, A. Parveen, et al., Ameliorative effects of biochar on rapeseed (Brassica napus L.) growth and heavy metal immobilization in soil irrigated with untreated wastewater, J. Plant Growth Regul. 39 (2020) 266–281, doi:10.1007/s00344-019-09980-3.
[55] A. Karakaya, M. Oziligen, Energy utilization and carbon dioxide emission in the fresh, paste, whole-peeled, diced and juice tomato production processes, Energy 36 (2011) 5101–5110.
[56] Z. Kheiralla, N.S.E.l. Gendy, H. Ahmed, et al., Upgrading of tomato (Solanum lycopersicum) agroindustrial wastes, J. Microb. Biochem. Technol. 10 (2018) 46–48, doi:10.4172/1948-5948.1000394.
[57] A. Kumar, E. Singh, A. Khapre, et al., Bioresource technology sorption of volatile organic compounds on non-activated biochar, Bioresour Technol. 297 (2020) 122469, doi:10.1016/j.biortech.2019.122469.
[58] T. Laifi, Development of a methodology for the treatment of organic potato and tomato waste by pyrolysis for the production of Biochars. In: MSc Thesis, Faculty of Pure and Applied Sciences, Environmental Conservation and Management Programme, Open University of Cyprus, 2020.
[59] D. Laird, P. Fleming, B. Wang, et al., Biochar impact on nutrient leaching from a Midwestern agricultural soil, Geoderma 158 (2010) 436–442, doi:10.1016/j.geoderma.2010.05.012.
[60] J. Li, Y. Fu, X. Bao, et al., Optimization of solid phase microextraction combined with gas chromatography-mass spectrometry (GC-MS) to analyze aromatic compounds in fresh tomatoes, J. Food Biochem. 43 (2019) 1–12, doi:10.1111/jfbc.12858.
[61] S. Li, C.Y. Chan, M. Sharbatmaleki, et al., Engineered biochar production and its potential benefits in a closed-loop water-reuse agriculture system, Water (Switzerland) (2020) 12, doi:10.3390/w12102847.
[62] C.S.K. Lin, L.A. Pfaltzgraff, L. Herrero-Davila, et al., Food waste as a valuable resource for the production of chemicals, materials and fuels. Current situation and global perspective, Energy Environ. Sci. 6 (2013) 426–464, doi:10.1039/c2ee23440h.
[63] Y. Lin, P. Munroe, S. Joseph, et al., Water extractable organic carbon in untreated and chemical treated biochars, Chemosphere 87 (2012) 151–157, doi:10.1016/j.chemosphere.2011.12.007.
[64] P. Llorach-Massana, E. Lopez-Capel, J. Peña, et al., Technical feasibility and carbon footprint of biochar co-production with tomato plant residue, Waste Manag. 67 (2017) 121–130, doi:10.1016/j.wasman.2017.05.021.
[65] A.H. Lone, G.R. Najar, M.A. Ganie, et al., Biochar for Sustainable Soil Health: A Review of Prospects and Concerns, Pedosphere 25 (2015) 639–653, doi:10.1016/S1002-0160(15)30045-X.
[66] S. Longo, M. Mistretta, F. Guarino, M. Cellura, Life cycle assessment of organic and conventional apple supply chains in the North of Italy, J. Clean Prod. 140 (2017) 654–663, doi:10.1016/j.jclepro.2016.02.049.
[67] T. Løvdal, B. Van Droogenbroeck, E.C. Eroglu, et al., Valorization of tomato surplus and waste fractions: a case study using Norway, Belgium, Poland, and Turkey as examples, Foods 8 (2019), doi:10.3390/foods8070229.
[68] Z. Lu, J. Wang, R. Gao, et al., Sustainable valorisation of tomato pomace: a comprehensive review, Trends Food Sci. Technol. 86 (2019) 172–187, doi:10.1016/j.tifs.2019.02.020.
[69] J. Major, M. Rondon, D. Molina, et al., Maize yield and nutrition during 4 years after biochar application to a Colombian savanna oxisol, Plant Soil 333 (2010) 117–128, doi:10.1007/s11104-010-0327-0.

[70] M. Manfredi, G.Vignali, Life cycle assessment of a packaged tomato puree: a comparison of environmental impacts produced by different life cycle phases, J. Clean Prod. 73 (2014) 275–284.
[71] A. Manfredini, A. Chiariotti, E. Santangelo, et al., Assessing the biological value of soluble organic fractions from tomato pomace digestates, J. Soil Sci. Plant Nutr. (2020), doi:10.1007/s42729-020-00361-4.
[72] I.I. Manolikaki, A. Mangolis, E. Diamadopoulos, The impact of biochars prepared from agricultural residues on phosphorus release and availability in two fertile soils, J. Environ. Manage 181 (2016) 536–543, doi:10.1016/j.jenvman.2016.07.012.
[73] M. Memici, K. Ekinci, Pyrolysis of tomato harvest waste as a function of temperature and duration: characteristics, production energy, and carbon dioxide emission in field conditions, Soil Tillage Res. 202 (2020) 104652, doi:10.1016/j.still.2020.104652.
[74] P. Mouron, T. Nemecek, R.W. Scholz, O. Weber, Management influence on environmental impacts in an apple production system on Swiss fruit farms: Combining life cycle assessment with statistical risk assessment, Agric. Ecosyst. Environ. 114 (2006) 311–322, doi:10.1016/j.agee.2005.11.020.
[75] Notarnicola B., Tassielli G., Renzulli P.A., Lo Giudice A. Life Cycle Assessment in the Agri-Food Sector: An Overview of its Key Aspects, International Initiatives, Certification, Labelling Schemes and Methodological Issues. In: Notarnicola, B., Salomone, R., Petti, L., Renzulli, P.A., Roma, R., Cerutti, A.K. (Eds.) Springer International Publishing, Switzerland, (2015).
[76] J.M. Novak, W.J. Busscher, D.L. Laird, et al., Impact of biochar amendment on fertility of a southeastern coastal plain soil, Soil Sci. 174 (2009) 105–112, doi:10.1097/SS.0b013e3181981d9a.
[77] O. Oginni, K. Singh, G. Oporto, et al., Influence of one-step and two-step KOH activation on activated carbon characteristics, Bioresour. Technol. Reports 7 (2019) 100266, doi:10.1016/j.biteb.2019.100266.
[78] C. Pane, G. Celano, A. Piccolo, et al., Effects of on-farm composted tomato residues on soil biological activity and yields in a tomato cropping system, Chem. Biol. Technol. Agric. 2 (2015), doi:10.1186/s40538-014-0026-9.
[79] A. Parmar, P.K. Nema, T. Agarwal, Biochar production from agro-food industry residues: a sustainable approach for soil and environmental management, Curr. Sci. 107 (2014) 1673–1682.
[80] S. Payen, C. Basse-Mens, S. Perret, LCA od local and imported tomato: an energy and water trade-off, J Clean Prod 87 (2015) 139–148.
[81] M. Petro-Turza, Flavor of tomato and tomato products, Food Rev. Int. 2 (1986) 309–351, doi:10.1080/87559128609540802.
[82] L. Piscitelli, A. Shaaban, D. Mondelli, et al., Use of olive mill pomace biochar as a support for soil microbial communities in an Italian sandy soil, Soil Horizons 56 (2015), doi:10.2136/sh15-02-0006.
[83] E.P.A. Pratiwi, A.K. Hillary, T. Fukuda, Y. Shinogi, The effects of rice husk char on ammonium, nitrate and phosphate retention and leaching in loamy soil, Geoderma 277 (2016) 61–68, doi:10.1016/j.geoderma.2016.05.006.
[84] J. Qin, S. Qian, Q. Chen, et al., Cow manure-derived biochar: Its catalytic properties and influential factors, J. Hazard Mater. 371 (2019) 381–388, doi:10.1016/j.jhazmat.2019.03.024.
[85] M. Renna, M. D'Imperio, M. Gonnella, et al., Morphological and chemical profile of three tomato (Solanum lycopersicum L.) landraces of a semi-arid Mediterranean environment, Plants 8 (2019), doi:10.3390/plants8080273.
[86] K.G. Roberts, B.A. Gloy, S. Joseph, et al., Life cycle assessment of biochar systems: estimating the energetic, economic, and climate change potential, Environ. Sci. Technol. 44 (2010) 827–833, doi:10.1021/es902266r.

[87] F. Ronsse, D. Dickinson, R. Nachenius, W. Prins, Biomass pyrolysis and biochar characterization. In: 1st FOREBIOM Workshop: Potentials of Biochar to Mitigate Climate Change, Vienna, 2013, pp. 1–24.
[88] I. Roussis, I. Kakabouki, A. Folina, et al., Effects of tomato pomace composts on yield and quality of processing tomato (Lycopersicon esculentum Mill.), Bull UASVM Hortic 76 (2019) 250–257, doi:10.15835/buasvmcn-hort.
[89] K. Schanes, K. Dobernig, B. Gözet, Food waste matters - A systematic review of household food waste practices and their policy implications, J. Clean Prod. 182 (2018) 978–991, doi:10.1016/j.jclepro.2018.02.030.
[90] F. Sevik, I. Tosun, K. Ekinci, Composting of olive processing wastes and tomato stalks together with sewage sludge or dairy manure, Int. J. Environ. Sci. Technol. 13 (2016) 1207–1218.
[91] R. Sharma, H.S. Oberoi, G.S. Dhillon, Fruit and vegetable processing waste: renewable feed stocks for enzyme production, in: G.S. Dhillon, S. Kaur (Eds.), Agro-Industrial Wastes as Feedstock for Enzyme Production, Academic Press, 2016.
[92] R.K. Sharma, J.B. Wooten, V.L. Baliga, et al., Characterization of chars from pyrolysis of lignin, Fuel 83 (2004) 1469–1482, doi:10.1016/j.fuel.2003.11.015.
[93] D. She, X. Sun, A.H.D. Gamareldawla, et al., Benefits of soil biochar amendments to tomato growth under saline water irrigation, Sci. Rep. 8 (2018) 1–10, doi:10.1038/s41598-018-33040-7.
[94] T.A. Sial, Z. Lan, M.N. Khan, et al., Evaluation of orange peel waste and its biochar on greenhouse gas emissions and soil biochemical properties within a loess soil, Waste Manag. 87 (2019) 125–134, doi:10.1016/j.wasman.2019.01.042.
[95] A. Signore, M. Renna, P. Santamaria, Agrobiodiversity of vegetable crops: aspect, needs, and future perspectives, Annu. Plant Rev. 2 (2019) 41–64, doi:10.1002/9781119312994.apr0687.
[96] Å. Stenmarck, C. Jensen, T. Quested, G. Moates, Estimates of European Food Waste Levels, Fusions Eu Project, European Commission, Stockholm, Sweden, 2016.
[97] K. Sülük, İ. Tosun, K. Ekinci, Co-composting of two-phase olive-mill pomace and poultry manure with tomato harvest stalks, Environ. Technol. 38 (2017) 923–932, doi:10.1080/09593330.2016.1217279.
[98] M. Szymanska-Chargot, M. Chylinska, K. Gdula, et al., Isolation and characterization of cellulose from different fruit and vegetable pomaces, Polymers (Basel) (2017) 9, doi:10.3390/polym9100495.
[99] I. Tabrika, K. Azim, E.H. Mayad, M. Zaafrani, Composting of tomato plant residues: improvement of composting process and compost quality by integration of sheep manure, Org. Agric. 10 (2020) 229–242, doi:10.1007/s13165-019-00268-0.
[100] E. Tamburini, P. Pedrini, M.G. Marchetti, et al., Life cycle based evaluation of environmental and economic impacts of agricultural productions in the Mediterranean area, Periodical 7 (2015) 2915–2935, doi:10.3390/su7032915.
[101] K. Tawarah, Specifications of raw olive pomace as an energy source: a statistical approach, Chem. Sci. Int. J. 22 (2019) 1–20, doi:10.9734/csji/2018/40045.
[102] M. Torellas, A. Anton, J.C. Lopez, et al., LCA of a tomato crop in a multi-tunnel greenhouse in Almeria, Int. J. Life Cycle Assess 17 (2012) 863–887.
[103] M. Tsangas, I. Gavriel, M. Doula, et al., Life cycle analysis in the framework of agricultural strategic development planning in the Balkan region, Sustainability 12 (2020) 1813, doi:10.3390/su12051813.
[104] M. Uchimiya, L.H. Wartelle, K.T. Klasson, et al., Influence of pyrolysis temperature on biochar property and function as a heavy metal sorbent in soil, J. Agric. Food Chem. 59 (2011) 2501–2510, doi:10.1021/jf104206c.
[105] K.C. Uzoma, M. Inoue, H. Andry, et al., Effect of cow manure biochar on maize productivity under sandy soil condition, Soil Use Manag. 27 (2011) 205–212, doi:10.1111/j.1475-2743.2011.00340.x.

[106] J.S.Van Dyk, R. Gama, D. Morrison, et al., Food processing waste: problems, current management and prospects for utilisation of the lignocellulose component through enzyme synergistic degradation, Renew. Sustain Energy Rev. 26 (2013) 521–531, doi:10.1016/j.rser.2013.06.016.

[107] S. Van Worden, The application of life cycle analysis in glasshouse horticulture, International Conference LCA in Foods. Gothenburg, 2001 136–140.

[108] D. Vasu, S.K. Singh, S.K. Ray, et al., Soil quality index (SQI) as a tool to evaluate crop productivity in semi-arid Deccan plateau, India, Geoderma 282 (2016) 70–79, doi:10.1016/j.geoderma.2016.07.010.

[109] S.F.Vaughn, F.J. Eller, R.L. Evangelista, et al., Evaluation of biochar-anaerobic potato digestate mixtures as renewable components of horticultural potting media, Ind. Crops. Prod. 65 (2015) 467–471, doi:10.1016/j.indcrop.2014.10.040.

[110] S.F. Vaughn, J.A. Kenar, A.R. Thompson, S.C. Peterson, Comparison of biochars derived from wood pellets and pelletized wheat straw as replacements for peat in potting substrates, Ind. Crops. Prod. 51 (2013) 437–443, doi:10.1016/j.indcrop.2013.10.010.

[111] Y. Wang, Y. Lin, P.C. Chiu, et al., Phosphorus release behaviors of poultry litter biochar as a soil amendment, Sci. Total Environ. 512–513 (2015) 454–463, doi:10.1016/j.scitotenv.2015.01.093.

[112] K. Weber, P. Quicker, Properties of biochar, Fuel 217 (2018) 240–261, doi:10.1016/j.fuel.2017.12.054.

[113] K. Winans, S. Brodt, A. Kendall, Life cycle assessment of California processing tomato: an evaluation of the effects of evolving practices and technologies over a 10-year (2005–2015) timeframe, Int. J. Life Cycle Assess 25 (2020) 538–547, doi:10.1007/s11367-019-01688-6.

[114] J.H. Windeatt, A.B. Ross, P.T. Williams, et al., Characteristics of biochars from crop residues: potential for carbon sequestration and soil amendment, J. Environ. Manage. 146 (2014) 189–197, doi:10.1016/j.jenvman.2014.08.003.

[115] WPTC (2018) World production estimate of tomato for processing. http://www.wptc.to. October 2018.

[116] WPTC, Crop Update and World Production Estimate, World Processing Tomato Council. Tomato News i, 2020, pp. 0–3.

[117] H. Wu, K. Yip, Z. Kong, et al., Removal and recycling of inherent inorganic nutrient species in mallee biomass and derived biochars by water leaching, Ind. Eng. Chem. Res. 50 (2011) 12143–12151, doi:10.1021/ie200679n.

[118] Z. Xiao, Q. Wu, Y. Niu, et al., Optimization of headspace solid-phase micro-extraction and its application in analysis of volatile compounds in cherry tomato by gas chromatography, Food Anal. Methods 10 (2017) 596–609, doi:10.1007/s12161-016-0622-3.

[119] G. Xu, Y. Lv, J. Sun, et al., Recent advances in biochar applications in agricultural soils: benefits and environmental implications, Clean - Soil, Air, Water 40 (2012) 1093–1098, doi:10.1002/clen.201100738.

[120] P. Yu, L. Huang, Q. Li, et al., Effects of mixed hardwood and sugarcane biochar as bark-based substrate substitutes on container plants production and nutrient leaching, Agronomy 10 (2020), doi:10.3390/agronomy10020156.

[121] H. Yuan, T. Lu, Y. Wang, et al., Sewage sludge biochar: nutrient composition and its effect on the leaching of soil nutrients, Geoderma 267 (2016) 17–23, doi:10.1016/j.geoderma.2015.12.020.

[122] L. Zhu, D. Shen, K.H. Luo, A critical review on VOCs adsorption by different porous materials: species, mechanisms and modification methods, J. Hazard Mater. 389 (2020) 122102, doi:10.1016/j.jhazmat.2020.122102.

[123] A.A. Zorpas, Strategy development in the framework of waste management, Sci. Total Environ. 716 (2020) 137088, doi:10.1016/j.scitotenv.2020.137088.

CHAPTER SEVEN

# Vermicomposting of tomato wastes

María Desamparados Soriano Soto[a], Antonis A. Zorpas[b], Jose Navarro Pedreño[c], Ignacio Gómez Lucas[c]

[a]School of Agricultural Engineering and Environment, Polytechnic University of Valencia, Camino de Vera, Valencia, Spain

[b]Open University of Cyprus, Faculty of Pure and Applied Sciences, Environmental Conservation and Management, Laboratory of Chemical Engineering and Engineering Sustainability, Latsia, Nicosia, Cyprus

[c]Agrochemical and Environment Department, Miguel Hernández University of Elche, Avd. de la Universidad sn, Elche, Alicante, Spain

## 7.1 Introduction

Taking into account the requirement of the European Union Landfill Directive 1999/31 focusses on the disposal of waste, as well as the requirement to decrease all the biodegradable waste being dumped [33], recycling and especially composting (including vermicomposting), are considered as appropriate methodologies to treat the biodegradable wastes. Indeed, the composting process gives a by-product that can be useful as soil amendment and this is part of the agricultural practices that can be an effective climate adaptation strategy by increasing the organic carbon content of soils [66]. The production of the main fresh agricultural products and their processed products can generate large amounts of wastes. Consequently, recycling of biodegradable agro-industrial residues and the organic fraction of municipal solid waste (after source separation) could also reduce the problems relating to the increasing production of wastes and the difficulties of locating new landfills.

### 7.1.1 The forecast of tomato cultivation

In the case of fresh tomato (*Lycopersicon esculentum* Mill.) production in the framework of the European Union, Italy and Spain are the major producers, together accounting for a little under two thirds (62.9 %) of the EU total [47]. In 2017, according to FAOSTAT [51], 6015868 t in Italy, 5,163,466 t in Spain, 1,747,634 t in Portugal, and 910,000 t in the Netherlands were harvested in 2018 and the related cultivation area in Europe was about 270,053 ha. This provide a key performed indicator related with the cultivation land and the total production (for those countries = 13,836,968 t) equal with 51.23 t/ha. However, China, India, USA, Turkey, and Egypt are the main world producers (Fig. 7.1).

*Tomato Processing by-Products*
DOI: https://doi.org/10.1016/B978-0-12-822866-1.00010-7

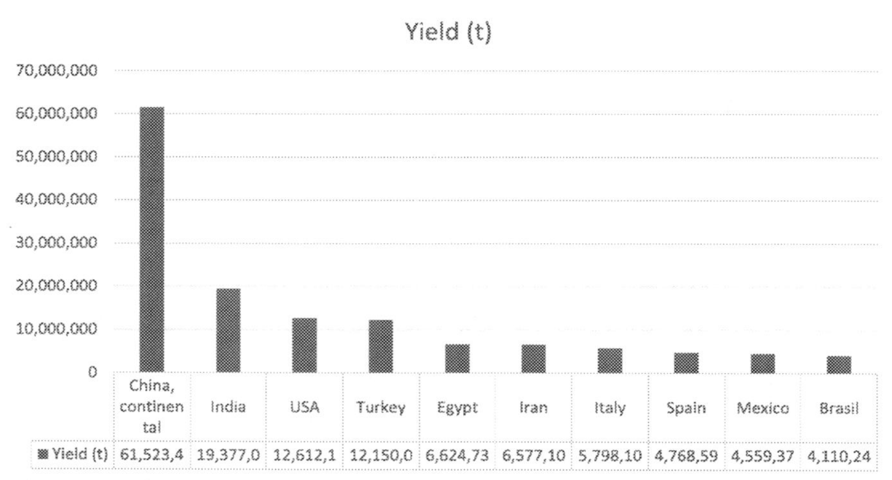


**Fig. 7.1** ***Fresh tomato harvested in 2018 in the ten major producers*** *(source FAOSTAT, http://www.fao.org/faostat/es/#home).*

Intensive irrigated agriculture is the main system used for tomato cultivation in Spain. The cultivated area of fresh tomato was of 56,629 ha irrigation and 308 ha rainfed in 2019 [91]. The increment in some countries of intensive agriculture in recent years increased the waste production, especially in crops under greenhouse conditions, which cause a large production of many types of vegetables throughout the year. In Europe, the yield per hectare of fresh tomato in the Netherlands is between five to six times higher than in the one reported for Spain due to the intensification of agriculture, generating an important economic benefit due to the higher volume of production compared to traditional crops facilitated by the high degree of technology (automated control of irrigation and fertilization, weather, pest control, optimization of the crop substrate, etc.). As a result, it is expected that the tomato cultivation to increase under greenhouse conditions, according to the data given by FAOSTAT [51] and MAPA [91]. There are studies that estimate the global protected agriculture area of 5,630,000 ha, the majority being tomato cultivation with China having almost 55% of the total world's plastic greenhouse acreage (including large plastic tunnels] and over 75% of the world's small plastic tunnels [32].

### 7.1.2 Fresh tomato wastes

According to IFAPA [64], the residual biomass generated by intensive horticultural crops is a serious problem for the sustainability of the sector if they are not removed from the farms and given an outlet that is economically and environmentally viable. In the case of horticultural crops on the Almería coast (Spain), where intensive agriculture is practiced, the production of plant residues is estimated at 1,751,242 t of fresh material per year, about 25 t/ha [128]. FAO [49] distinguished five system boundaries in the food supply chains that produce wastes from vegetables: agricultural production, post-harvest handling and storage, processing, distribution, and consumption. Moreover, FAO [49] showed that about 40%–45% percent of the initial production lost or wasted at different stages of the food supply change for fruits and vegetables in different regions. According with this and the recent data of FAO [50], at worldwide scale, about 70,000,000 tonnes of tomato wastes can be derived from the food supply chains.

In general, horticultural residues have a high C content, a C/N ratio from 15 to 30), a high moisture content, which favour their easy biodegradability. As indicated in Table 7.1, their hemicellulose, cellulose, and lignin contents range from 5% to 15%, 10% to 40%, and 1% to 10%, respectively [8,67,78,83].

The reuse of this waste opens the field to new local economic activities associated with the composting process, also guaranteeing a source of organic matter and nutrients that is easily accessible to farmers. On the other hand, and from the environmental point of view, it facilitates the reduction of pollution produced by the abandonment and accumulation of plant wastes that constitute a source of spread of diseases, pests, and bad smells [103].

The peculiarity of greenhouse crops is that they generate large amount of organic wastes in a small space in each harvest cycle, which needs to be treated or revalued in the form of organic fertilizer or energy since they

**Table 7.1** Horticultural wastes composition [9,84].

| | Tomato | Artichoke | Lupine | White cabbage | Peas | Beans | Potato | Pepper |
|---|---|---|---|---|---|---|---|---|
| | Stem/leaf | Stem | Stem | Leaves | Stem | Stem | Stem | Stem |
| TC (%) | 28.3/39.2 | - | 44.614 | 40.13 | 45.78 | 41.114 | 45.78 | - |
| OC (%) | - | - | - | - | - | - | 29.213 | - |
| OM (%) | - | - | - | - | - | - | 50.24 | - |
| TN (%) | 3.3/1.2 | 3.114 | 1.63 | 1.38 | 3.414 | 2.61 | 2.313 | - |
| C/N | 49/30.8 | - | - | - | - | - | 12 | - |
| Hemicellulose (%) | 4.4 | 14.82 | - | 5.42 | 24.62 | - | 10.41 | - |
| Lignin (%) | 13.3 | 15.21 | - | 0.78 | 0.78 | - | 6.58 | 19.01 |
| Cellulose (%) | 28.2 | 21.81 | - | 11.48 | 38.78 | - | 27.58 | - |

C/N: carbon-to-nitrogen; OC, organic carbon; OM, organic matter; TC, total carbon; TN, total nitrogen.

can generate a serious environmental problem. They are usually characterized by high moisture contents and low ratios between organic carbon and nitrogen (C/N ratio) mainly due to their lignocellulosic matrix that gives them great resistance to chemical and biological degradation and makes their treatment difficult [84], reflected by their composition (Table 7.1).

### 7.1.3 Tomato processing by-products

The agricultural industries and the transformation of agricultural products are a fundamental part of the productive fabric of developed countries. They are responsible for the production of waste that is located and therefore can be easily treated and give new by-products. Processing fresh tomato gives a residue which is mainly composed of the skins and seeds of the fruit, along with the remains of waste branches. According to Askar [12] seeds represent the 10% of the fruit and 60% of the total waste. Furthermore, are considered to be a valuable source of protein (35%) and fat (25%), as tomato seed oil is rich in unsaturated fatty acids, especially in linoleic acid.

Tomatoes red colour is result of the presence of lycopene, which is the main carotenoid and constitutes approximately 80% to 90% of the total carotenoid content [114,115]. Most of the lycopene is associated with the water-insoluble fraction and with the skin [113]. In this sense, the extraction of this substance for other purposes would be applied, for example to be used as dye.

However, recycling of organic residues (which is obligatory from the concept of Circular Economy Strategy and European Green Deals) [88,123,162] such as tomato wastes (which also produced from household level) [157], by composting and/or vermicomposting for land application would contribute to the sustainable development and reduces the problems derived from the management and disposal of wastes in origin and/or after processing the tomato.

## 7.2 Composting and vermicomposting processes

The goal set from waste framework directive (WFD) ([36]/98/EC), indicates that composting is considered as prevention activities to minimize the organic waste ends to landfill [161,162]. Moreover, the target set from United Nation Sustainable Development Goals (UNSDGs - goal No 12) that emphases on the sustainable reduction of food wastes through prevention [4,87,132,133,152,154,158–160,162], reuse and recycling [129], and the goal set from the Circular Economy Strategy [88,162] to decrease up to 10% the organic waste going to landfill [161], home composting and more general composting [79,146,150,151] and vermicomposting seems to be a green compulsory widely accepted strategy. Furthermore, due to limited spaces in insular communities as the example of Cyprus or Crete, composting (in general as an in-situ process) contributes to the reduction and recycling at home of the organic matter [52].

As a definition according to many researchers (27,38–41,57,58,63, 127,130,131,137,138,141–151,153,155,156,161]; A. [5,104]), composting is the aerobic process in which indigenous microorganisms, both mesophile and thermophile, convert organic matter creating stabilized and disinfected compost. Sanitization or disinfected of composts is generally related to the early thermophilic phase of composting, when temperatures reach 45–70°C. During the mesophilic phase, also known as maturation, the remaining more recalcitrant organic compounds are degraded at a slow pace in a process similar to humification in soils, which entails the disappearance of residual phytotoxicity. Composting usually involves aeration, to maintain aerobiosis, and watering.

On the other hand, vermicomposting [58,71] is similarly a bio-oxidative process which involves microorganisms and earthworms. The microorganisms, both in the earthworm guts and in the feedstock, are responsible for the biochemical degradation of the organic matter whilst the earthworms are responsible for the fragmentation of the substrate, which increases the surface area exposed to the microorganisms. Vermicomposting has also been defined as "*biooxidation and stabilization of organic material involving the joint actions of earthworms and (mesophilic) microorganisms*" according to Aira et al., [13]. In contrast to traditional compost, vermicompost never heats up much above ambient temperatures. Many feedstocks can be used directly in vermicomposting systems; however, animal manures can drive temperatures into the thermophilic range during decomposition, which can kill compost earthworms. According to Frederickson et al., [59] both vermicompost and traditional compost can be combined, initially doing a partial precomposting at high temperatures followed by a finishing stage of vermicomposting. The phases in vermicomposting involve a period where worms acclimatize to the substrate that they are placed in. There is a hydrolytic phase in which readily degradable organic matter is broken down, and then a curing phase where the more recalcitrant organic matter is broken down [14,19]. Vermicomposting and composting are processes by which, with or without earthworms, organic waste is transformed into vermicompost or compost that can be used as a substrate for plant growth due to the beneficial effects, which could help to unite management waste and agriculture [22].

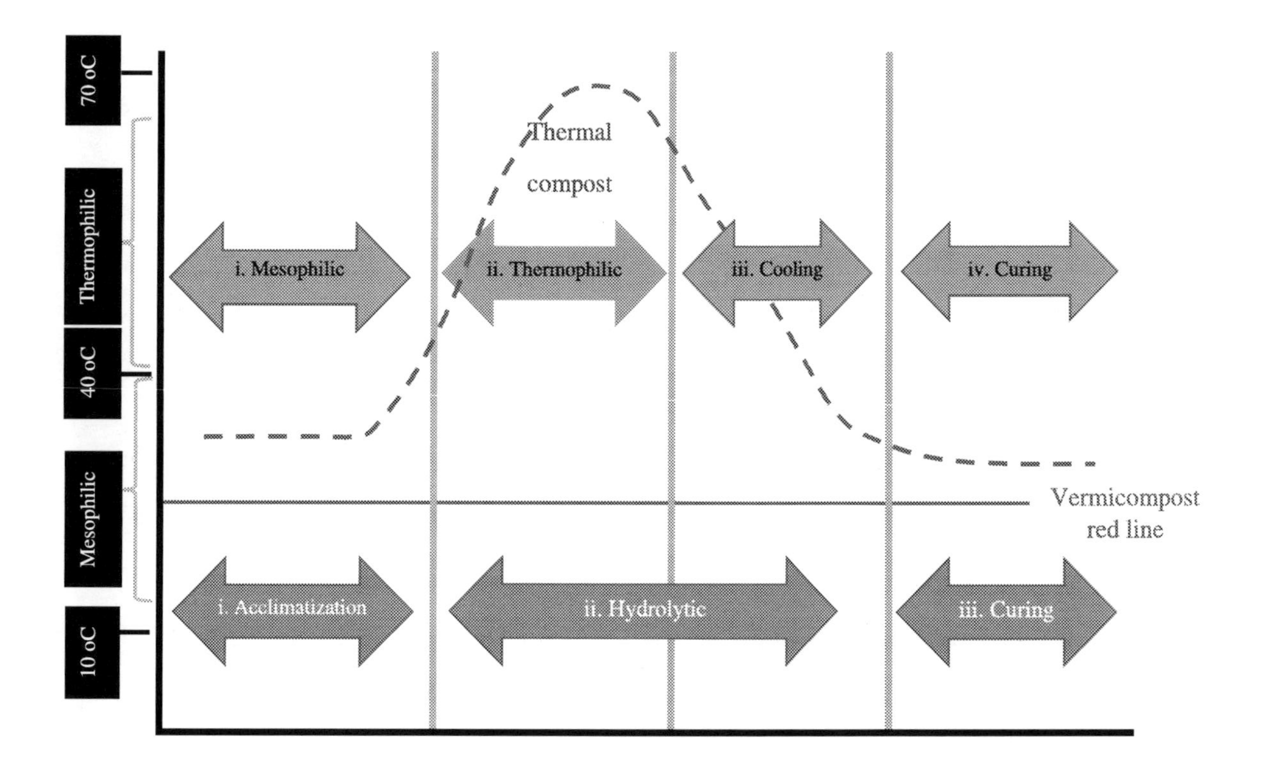


**Fig. 7.2** *The theoretical time vs. temperature curves for thermogenic (thermal) compost and vermicompost.*

Fig. 7.2 indicates the theoretical time vs temperature curves for thermogenic (thermal) compost and vermicompost. Arrows represent major phases in the composting process. Phases for thermal compost are adapted from Chefetz et al. [30] and those for vermicompost are adapted from Benítez et al. [19].

There are several practices in place to convert waste into value-added products such as in the form of compost (organic fertilizer), biogas [89], animals feed chemicals [5,15,161]. Consequently, composting (in general) is indicated as an eco-friendly biochemical and sustainable method for the management of organic wastes [70,104,144,150,161]. Composting organic wastes is currently the most widely used treatment technique behind landfill disposal [80,81].

When organic waste is not properly treated, it can lead to adverse effects on the environment and the health of animals and humans. In this case, the two techniques for treating organic wastes (e.g., plant residues and biosolids) that contribute to management, recycling and sustainability are composting and vermicompostings [2,3,10, 17,18,31,37,68,96,97,107,111,112,126].

The possibility of obtaining through these two techniques organic amendments, substrates or fertilizers for crops, makes them suitable for recycling and recovery. For this reason, in recent years the regulations on waste have been modified. In this context, the [36]/98/EC of the European Parliament and of the Council, as well as the new amendment of the WFD, Directive [35] 2018/851, refer, among other aspects, to the recent waste policy, which must contemplate, reduce the use of resources and favour the practical application of the waste hierarchy [162].

## 7.2.1 Composting transformations

During composting, it is well known that microorganisms (including bacterial and fungi) break down the complex organic matter into simpler products and finally into a mature compost [60,99,121,139,145]. The reaction that takes places during the composting process is given by the following equation [5]:

$$C_xH_wO_zN_k \cdots \alpha H2O + \beta O_2 \rightarrow C_dH_eO_fN_g \cdots \gamma H_2O + nH_2O + CO_2$$

(Where this can be translated into: organic matter + oxygen → compost with evaporated water + produced water + produced carbon dioxide)

While the entire process, as several researchers indicated, is completed in 3-8 months, some other composting techniques offer options to minimize the processing period [15], like the frequent turnings and shredding of feedstock, the use of effective microorganisms [121], the chemical nitrogen activators, the use of worms (vermicomposting), natural minerals and various additives and amendments such as zeolites, food waste, biochar, waste paper, etc. [26,34,85,142,149,161].

The effectiveness of the transformation by composting and vermicomposting to obtain a substrate or organic amendment that helps the elimination and recovery of these residues, has been the subject of numerous investigations [22,46,72,77,105]. The final objective is to convert a waste into a resource. López et al. [84] indicate the utility of the transformation of agricultural residues for their reuse in food production and that favours the management of residues, minimizing the increasing problems given by the growth of the world population. Recycling organic waste in food production amendments is an opportunity to partially solve this double challenge.

Regarding the composting process, it is a process of transformation of organic matter carried out by microorganisms in aerobic conditions. The related metabolic activity generates heat that must be preserved for the disinfection of the product obtained by carrying out the process in closed piles [84,99]. In addition, in order to facilitate this process, it is convenient that the material has adequate conditions regarding the size of the incorporated residue (1–5 cm), the wetting (40%–50%), the C/N ratio (25%–30%), and oxygen content of 8%–12%, as this produces the optimal conditions for the development of complex populations of microorganisms to act sequentially on organic matter facilitating its transformation [96].

This process takes place in several phases according to the microbial metabolic intensity: initial phase, thermophilic phase, and maturation phase. The initial phase, with a high proportion of biodegradable organic matter and intense microbial activity. Period where sugars and other easily degradable simple compounds are used, and mesophilic microorganisms (fungi and acidifying bacteria) predominate and the temperature rises to about 40°C. The thermophilic phase, where the highest temperatures are reached (60–80°C) permits the degradation of the complex polymers and that can last several weeks or months depending on the content of cellulose and hemicellulose in the starting material [99]. Thermophilic microorganisms predominate in it, and exceeding 60°C the thermophilic fungi responsible for the decomposition of cellulose and hemicellulose cease their activity [73,74]. However, since a temperature gradient occurs, fungi remain active in the peripheral areas. As for thermophilic bacteria, they also behave similar to fungi [45], initially breaking down proteins, carbohydrates, and lipids, and finally hemicellulose and lignin [44]. When the temperature drops, the phase with the highest degradation rate is reached, reaching the cooling and maturation phase, where the temperature drops to 35-40°C, and the decomposition rate decreases, with less biodegradable organic matter and they begin to the complex polymers decompose, the microbial activity decreases and the temperature of the material equals the ambient temperature (Fig. 7.1).

The composting process reduces the weight and volume of the waste, ensuring that the material is sanitized, obtaining a compost of a great agronomic interest, with a high content of stabilized organic matter that can later be incorporated into the soil, using it as a substrate or fertilizer, obtaining useful organic fertilizers that can be used as amendments for crops.

## 7.3 Vermicomposting

Organic matter plays a key role to achieve sustainability in agricultural production because it possesses many desirable properties such as high-water holding capacity, cation exchange capacity (CEC), ability to sequester contaminants (both organic and inorganic) and beneficial effects on the physical, chemical and biological characteristics of soil [102,108]. The role of organic residues in the soil goes beyond improving the substrate, since it is a source of nutrients and energy for soil organisms. Such is the case of the worms because plant and animal wastes are important food sources for earthworms. They eat soil living organisms such as nematodes, protozoans, rotifers, bacteria, but also feed on the decomposing remains of other animals.

Vermicomposting comprise the bio-oxidation and stabilization of organic matter under aerobic and mesophilic conditions through the joint action of microorganisms and earthworms [54]. Lorimor et al. [86] mentioned that the mass-reduction capacity is higher in vermicomposting than composting and the production of humus content is higher as well the phytotoxicity is significantly lower [86]. Comparing vermicompost and compost products in the senesce of which is consider as more marketable, vermicompost is more gorgeous as has more nutrients and microbial activity [98]. Bansal and Kapoor, [20], Suthar, [124] and Fernandez-Gomez et al. [54] state that that vermicomposting is effective for handling of crop wastes from conventional farming systems through numerous earthworms like *Eisenia fetida, Eisenia andrei, Perionyx excavatus, and Eudrilus eugeniae*. The *E. fetida* and *E. andrei* earthworm species, due to their high tolerance in relation to environmental variables such as pH, moisture, and temperature [98], are consider as alternative solution in vermicomposting of organic waste under temperate climates. The investigation of the activities of enzyme could play significant role when evaluating the stability of vermicomposted end products according to Benítez et al. [21]. A more recent study Sen and Chandra [125] indicate that a new fingerprinting method is crucial to be used to optimize composting and vermicomposting technology. They have proved that DGGE (denaturing gradient gel Electrophoresis) is capable of evaluating the relationship between variations in physicochemical and microbial community structures occurring during the vermicomposting process due to the presence of earthworms. Fernandez-Gomez et al. [54] using the above technique, evaluate the viability of using *E. andrei* to vermicompost the enormous quantities of heterogeneous-plant, tomato-plant, and damaged tomato-fruit vegetable waste produced by Mediterranean greenhouse crops. They have mentioned that, all earthworms died in the initial materials containing heterogeneous- plant or tomato-plant wastes after 24 h [54]. Additionally, C/N varies between 16% and 23% for the samples T, T/D 2:1, T/D 4:1, and D, and by over 35% for T/S 2:1 and T/S 4:1 (where P, tomato-plant waste; T, tomato-fruit waste; D, cow dung; S, wheat straw). Table 7.2 presents the deviations that occur in P, K, Ca, Mg, Na for several samples using vermicomposting as treatment method.

One of the most significant parameters in composted material is the germination index (GI). According to Zorpas et al., [140,143,145,149,161] if the $0 < GI < 26$ the substrate is characterized as very phytotoxic, $27 < GI < 66$ the substrate is characterized as phytotoxic, $67 < GI < 100$ the substrate is characterized as nonphytotoxic and if the $GI > 101$ then the substrate is characterized as phytonutrient. According to Fernandez-Gomez et al. [54] the GI values were close to zero in the initial materials containing tomato-fruit wastes. At the end of the process the GI were increased by over 50% except from samples (which according to the same researcher this could be explained by the higher EC values in the samples). However, Zucconi et al. [163], indicate deferent a GI value to declare the level of the substrates. More specific if GI is greater than 50% are considered no phytotoxic and stable for agricultural application.

Fernandez-Gomez et al. [54] was to test the efficiency of *E. fetida* to vermicompost tomato-fruit wastes from greenhouse crops using a pilot-scale continuous-feeding vermicomposting system. A reactor was applied and tomato-fruit wastes were added over a 150-day (organic loading rate was 13.6 kg $TOC/m^3/wk^1$) vermicomposting process

**Table 7.2** Physicochemical variations through vermicomposting process.

| Treatments[a] | TKN | | | P | | | K | | |
|---|---|---|---|---|---|---|---|---|---|
| | I | F | *t*-test | I | F | *t*-test | I | F | *t*-test |
| D | 15.0 a | 14.1 a | 0.30 | 5.1 bcd | 6.9 a | 0.06 | 28.4 a | 34.1 a | 0.01[b] |
| T | 22.8 b | 20.4 bc | 0.11 | 4.9 bc | 9.7 c | 0.00[b] | 23.6 b | 28.8 b | 0.02[b] |
| T/S (2:1) | 18.9 a | 23.0 c | 0.17 | 3.9 a | 8.1 ab | 0.01[b] | 21.8 b | 30.3 b | 0.00[b] |
| T/S (4:1) | 18.0 a | 21.8 bc | 0.02[b] | 4.4 ab | 7.9 ab | 0.00[b] | 21.7 b | 28.6 b | 0.00[b] |
| T/D (2:1) | 18.0 a | 20.3 bc | 0.06 | 5.8 d | 8.4 b | 0.00[b] | 24.7 b | 33.5 a | 0.01[b] |
| T/D (4:1) | 20.1 ab | 19.4 b | 0.61 | 5.6 cd | 8.2 b | 0.01[b] | 24.2 b | 34.8 a | 0.01[b] |
| | Ca | | | Mg | | | Na | | |
| D | I | F | *T*-test | I | F | *T*-test | I | F | *T*-test |
| T | 15.8 a | 20.6 b | 0.02 | 8.5 a | 10.8 a | 0.04[b] | 2304 a | 2829 a | 0.03[b] |
| T/S (2:1) | 7.7 c | 16.7 cd | 0.00[b] | 3.3 c | 6.1 c | 0.00[b] | 798 c | 1522 cd | 0.00[b] |
| T/S (4:1) | 6.5 c | 14.7 de | 0.01[b] | 2.4 c | 5.1 d | 0.01[b] | 556 d | 1371 d | 0.00[b] |
| T/D (2:1) | 7.0 c | 14.2 e | 0.00[b] | 2.8 c | 5.3 cd | 0.00[b] | 663 cd | 1198 d | 0.00[b] |
| T/D (4:1) | 11.6 b | 18.6 bc | 0.00[b] | 5.4 b | 8.5 b | 0.00[b] | 1363 b | 1864 c | 0.04[b] |
| | 11.6 b | 23.1 a | 0.01[b] | 5.2 b | 10.6 a | 0.00[b] | 1164 b | 2344 b | 0.00[b] |

TKN, total Kjeldhal nitrogen.
Means in the same column followed by same letters are not significantly different from each other ($P < 0.05$).
*T*-test, P values of paired-sample *t*-test.
[a]Refer to text, for explanation of treatment abbrev iations.
[b]Significant difference between the initial material and final vermicompost.

(which the controlled temperature was 25°C). The reactor was consisting of a four-sided metal pilot-scale (0.6 m × 0.9 m × 0.2 m). A 0.1 cm mesh was engaged at the bottom of the reactor. A 5 cm layer containing 15 kg dry weight of mature sheep manure (36% moisture, pH 8.6, EC 1.8 dS/m$^1$ EC, TOC 138 g/kg$^1$, TKN 9.6 g/kg$^1$, and with C/N to be 14), was placed to the mesh to offer a preliminary habitat for the earthworms. 15 days after the earthworms were added, the vermireactor was fed with 10 kg of the liquid-paste of tomato-fruit wastes (92% moisture, pH 3.9, EC 1.4 dS/m$^1$, TOC 459 g/kg, TKN 23 g/kg$^1$, and C/N ratio 20). The role of earthworms and the stabilization of the organic matter during the process were calculate through physicochemical parameters, the activity of enzymes, the increasing of the bacterial communities etc. Earthworm biomass improved after 90 days, and then dropped due to increasing pH, EC and $NH_3$ concentration. From day 90 to 120, although more tomato-fruit was added in the reactor, the total earthworm biomass reduced shortly and continued to reduce until the end of the process, which last for 150 d. This trend of earthworm biomass reflected the reduction in the number of earthworms, and not the individual worm weights, which remained basically constant throughout the process. The temporal patterns of dehydrogenase, b-glucosidase, protease and urease were related to earthworm growth and the stabilization of organic matter. Bacterial DGGE profiles differed between the periods of degradation of labile substrates and the maturation step. Fungal communities at the stage of maximum earthworm biomass differed most, suggesting a gut passage effect. The final product was chemically stable and enhanced in nutrients, indicating that the waste from tomato-fruit can be effectively vermicomposted into a valuable soil conditioner. TKN, TOC, and water-soluble carbon (WSC) were increased by 96%, 100%, and 35%, respectively taking into account the initial values of the parameters with in the 150 d of the process. According to Fernandez-Gomez et al. [54] this was expected as the fresh organic matter was constantly incorporated by the addition of tomato-fruit waste at a high loading rate.

### 7.3.1 Conditions for vermicomposting

Vermicomposting is a simple and unique method based on the nutritional activity of the worms. It is a low-cost biotechnological process. In the soil, earthworms may be accelerated the transformation of the wastes into stabilized organic matter (humification) preparing the soil and litter mixtures composed of fragmented and macerated leaves and fine soil particles for microbial attack [23]. Biodegradable organic wastes are inoculated with earthworms (i.e., the redworm *Eisenia fetida)* which break down and fragment the organic wastes [95]. This transformation implies a many microorganism associated with the worm diet and activity.

The vermicomposting process has been studied regarding its chemical, biochemical, and microbiological levels. The organic matter fragmentation through combination of the worm and the associated organisms' activities produces a by-product called vermicompost [24]. Adult worms could weigh from 0.2 to 1.5 gr. Depending on the specie, they could consume a daily ration that tends to their own weight [61]. This process transforms organic wastes into vermicompost and permits the wastes volume reduction until 50% [118]. The main difference during the transformation processes in vermicomposting compared with the composting is the temperature (Fig. 7.2) where no thermophilic phase is observed. The main species used in vermicomposting are: *Eisenia fetida*, *Eisenia Andrei*, *Dendrobaena veneta, Lumbricus rubellus* for temperate climate *Eisenia hortensis*, and for tropical climate *Perionyx excavatus*, *Eudrilus eugeniae,* among others [101].

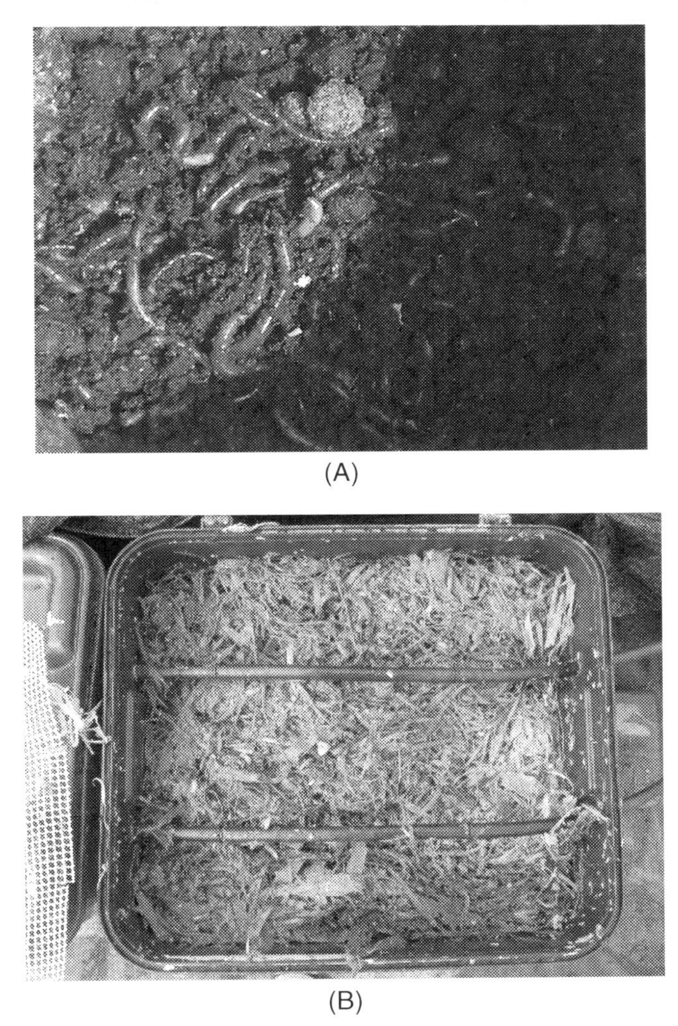

(A)

(B)

**Fig. 7.3** *Eisenia fetida* (A) and bed with organic wastes for vermicomposting (B).

Vermicomposting is relatively labor intensive and for large-scale production, requires large land areas to cool the waste in order to avoid high temperatures which are detrimental to earthworms. The temperature suitable for earthworms growing varies from 22–27°C [134] or 20–30°C [28]. As a result, in traditional open vermicomposting systems, the waste is placed in beds or windrows only up to a height of about 0.5–1.2 m (Fig. 7.3). At elevated temperatures, worms are only found in a fairly localized zone towards the outside of the compost pile [82]. Vermicomposting is only implemented for

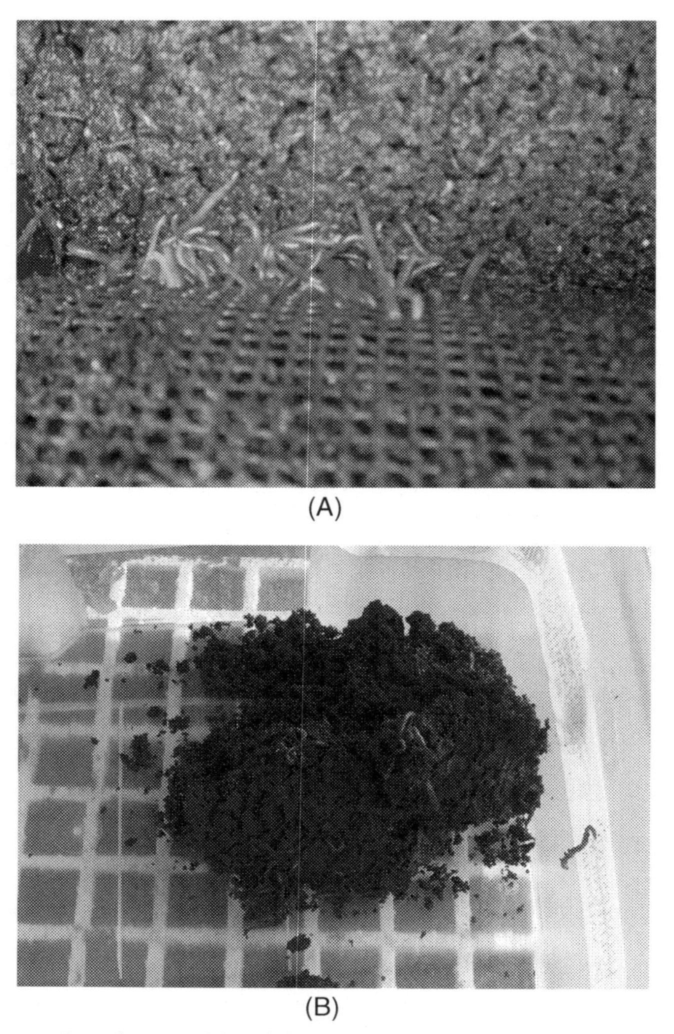


**Fig. 7.4** Separation of worms (A) and the composted material (B).

curing where temperatures are low, ranging between 28–35°C during the composting period [28]. Fig. 7.4

Another reason for combining high-rate degradation (3–15 days) without worms with vermicomposting for stabilization and curing is the need for a previous pathogen reduction [109]. At the end of the process, the worms are separated from the castings by screening. The end product is a fine, peat-like material [82]. The odour of wet soil's smell and a dark colour are indicators of the maturity of a vermicompost. Some authors indicated that in solid state, the ratio C/N will be under 20 [101].

Earthworms are key biological agents in the degradation of organic wastes [7,69,122], recognised as "ecosystem engineers" [23], as well as have been long recognized by farmers as advantageous to soil [43,117] and, as one of the major soil macro fauna, constitute an important group of secondary decomposers and improvers of many soil functions [23].

Vermicomposting technology using earthworms as versatile natural bioreactors for effective recycling of organic wastes to the soil is an environmentally acceptable means of converting wastes into nutritious composts for crop production [42]. In addition, it is considered as homogenous process, with desirable aesthetics, even with plant growth hormones and high levels of soil enzymes production, while enhancing microbial populations and tending to hold more nutrients over longer periods without adverse impacts on the environment [100]. It can also be used as a bioremediation measure to reclaim soils, especially acid soils, because of the near-neutral to alkaline pH of vermicompost and the suppression of labile aluminium [94].

After composting and vermicomposting, the obtained by-product is a less heterogeneous product. Table 7.3 illustrate the comparison between vermicompost and compost main physicochemical characteristics.

Although, both compost and vermicompost has similar physicochemical characteristics, those are very important. According to Zorpas [158] waste criteria are crucial for any composted material. Creating end of waste criteria (EWC) for any compost provide, provide financial and environmental

**Table 7.3** Values of the physical and chemical characteristics of vermicompost obtained from agricultural residues.

| Parameter | Vermicompost | Compost |
|---|---|---|
| Moisture % | 31–38 | 31–35 |
| Density (g/cm$^3$) | 0.8–1 | 0.9–1.1 |
| WHC %Water capacity retention | 70–89 | 60–80 |
| EC (dS.m-1) 1:10 | 1.0–3.0 | 1.0–4.0 |
| pH 1:10 | 7.2–7.6 | 7.0–8.0 |
| Total carbon (%) | 15.0–30 | 15.0–25 |
| C/N | 10–14 | 10–14 |
| P (% $P_2O_5$) | 1.2–1.8 | 1.0–1.5 |
| Ca (% CaO) | 1.0–4.0 | 1.0–4.0 |
| K (% $K_2O$) | 1.0–2.0 | 1.0–2.0 |
| Mg (% MgO) | 1.5–3.8 | 1.5–3.8 |
| Na (% $Na_2O$) | 0.2–0.5 | 0.2–0.8 |
| CEC (cmolc.kg$^{-1}$) | 60–90 | 60–80 |

**Table 7.4** Proposed EWC for compost.

| Parameter | Rate |
|---|---|
| Minimum organic matter content | 20% in weight |
| Minimum stability | To be proposed |
| No content of pathogens | No Salmonella sp. in 50 g sample |
| Limited content on viable weeds | To be proposed |
| Limited content of impurities | 0.5% of dry matter weight |
| Limited content of heavy metals | mg/kg (dry weight) |
| Zn | 400 |
| Cu | 100 |
| Ni | 50 |
| Cd | 1.5 |
| Pb | 120 |
| Hg | 1 |
| Cr | 100 |

benefits. Compost production with high and reliable quality and facilitates its use, by avoiding unnecessary regulatory burden improves harmonisation and legal certainty. Based on the JRC report [48,75,76], the EU Commission has assigned the JRC to work out a proposal for EWC digestate and compost. EWC are all the "requirements that have to be content by a material derived from waste to confirm that the quality of the material is such that its use is not detrimental for human health or the environment" [11,75, 76]. As a general principle, EWC would reproduce that a waste substance has reached a phase of processing, whereby it has essential value, so that it is unlikely to be rejected as a waste and it has been promoted to a point at which its use does not provide any environmental risk, which would then merit regulating the material as a waste. Zorpas [158] proposed some general EWC for compost (even though is produced from vermicomposting) which are presented in Table 7.4.

## 7.3.2 Vermicomposting of tomato wastes

Technologies such as vermicomposting are widely used to convert agricultural residues into organic fertilizers. Generally, these processes need to meet some initial requirements such as balanced carbon-nitrogen ratio, good physical structure of the mixtures that favour oxygenation, etc., including the environmental temperature and moisture, to be more efficiently Mazuela [92], evaluating different compost of horticultural residues, showed that the mean particle size was between 0.25 and 2.5 mm, while in vermicompost the percentage of particles greater than 0.2 mm was less than 50%. Moreover, bulk density is inversely proportional to particle size and low in vermicomposted samples of horticultural waste with optimal values of water retention capacity between 50% and 70% [1].

In the specific case of tomato waste (both fresh or processed wastes), it has been shown that they need to be complemented with other wastes to carry out vermicomposting [53]. Yang et al., [136] carried out a vermicomposting treatment (*Eisenia fetida*) for a period of 50 days, taking tomato stems and using cow manure as a nutrient substrate and investigated the maturity of the product obtained against a control. They concluded that vermicomposting could shorten the period required to reach compost maturity, producing mature compost while reducing gaseous emissions.

The best found efficient alternative was to mix them with other biodegradable wastes that contribute to the improvement of the physical, chemical, and biological properties of the substrates to be vermicomposted. Some authors choose as a complementary waste the pulp produced in the paper industry, in other cases the association with rabbit manure, cow dung and others, is usually favorable to obtain a good vermicompost quality [16]. For instance, Férnandez-Gómez et al. (2013) found that a mixture of tomato plant waste with paper-mill sludge at a ratio of 2:1 or 1:1 improved earthworm development during the vermicomposting of paper-mill sludge alone, and also allowed the vermicomposting of tomato-plant waste, which is otherwise harmful for *E. fetida.*

The vermicompost obtained from tomato residues and their mixtures usually present alkaline pH, being necessary a previous treatment for their use as an amendment. The same happens with the electrical conductivity (EC) of these materials indicating high content of salts that can be eliminated if they are washed by leachate with water. Our results (unpublished) are similar that those found by other authors using horticultural wastes [16].

The data obtained when analysing the biological activity of the metabolism of the main nutrients in the vermicomposting of these residues (enzymatic activities mainly related to carbon, nitrogen and phosphorus) are favourable for these mixtures, giving good values of the Germination Index test, indicating the absence of phytotoxicity.

Fernández-Gómez et al. [53] studied the evolution of microbial populations after vermicomposting using worms of the genus *Eisenia foetida*, and indicating a reduction in the number of microorganisms present in the final product due to the stability and maturation of the produced vermicompost. Authors such as Menéndez-Yuffá [93] indicate the importance of conducting biological studies of the substrates and compare composted and vermicomposted tomato residues studying the physical and chemical properties of the products obtained. They showed a more adequate environmental behaviour in composting, given the lower quantity of water consumed and lower losses of carbon and nitrogen. However, they indicate that a leaching step is necessary to reduce the higher EC in composting. Then, it would later be necessary to consume water similar to vermicomposting and greater losses of nutrients due to leachate. These authors indicated that both composting and vermicomposting are valid options, both being aerobic processes, the differences of which lie in the combined participation of microorganisms in mesophilic and thermophilic phases in the case of composting, compared to the combined participation of microorganisms and worms, which is an exclusively mesophilic process.

It has been described that in the sequential application of composting followed by vermicomposting of organic tomato crop residues, salinity is diminished, been quite lower in the final product.

Vermicomposting is an effective technique for treating tomato residues from both field and greenhouse cultivation, and the wastes produced along the food supply chain, giving by-products with a high degree of maturity that can be applied in agriculture.

### 7.3.3 Advantages of vermicomposting

Disposal of wastes from fruit and vegetables industries has been a problem, due to high transportation costs and limited availability of landfills. Aggarwal et al. [6] mentioned that improper disposal of mango kernel and peel appreciably increase environmental pollution. In this regard, a higher level of biological oxygen demand (BOD) and chemical oxygen demand (COD) in mango solid waste creates a real disposal problem. To a certain extent, use of vermicomposting technology could help with the disposal of mango industrial waste in a safe, economic, and useful manner. However, valorization through different techniques will certainly eliminate the disposal problem.

Several authors [62,82,90,116] mentioned that vermicomposting has been applied for several organic waste coming from agricultural sector, forestry sector, food waste and manure. Hand et al. [62] mentioned that vermicomposting could considered as a very cheap method for processing or treatment of organic waste.

Hence, some researchers mentioned that (comparing with composting process), vermicomposting optimized several physicochemical characteristics such as humic acids contents. Ferreire and Cruz [55] and Padmavathiamma et al. [102] indicated that humic acid in vermicompost was 28% higher than that of conventional compost. This finding was attributed to the more rapid materials decomposition by the earthworms which result in finer particles that can react and retain humic acid in considerably larger proportion. Moreover, Padmavathiamma et al. [102] mentioned that the ratio HA/FA (humic acid/fulvic acid) may reflect better due to the neutral to alkaline nature and high microbial activity of vermicompost and the advanced humifying degree of vermicompost may be is related with acceleration of the humification process of microflora. Additionally, the same researchers found out that the application of the compost in cropping situations with increased C/N ratio indicates possible temporary immobilization of N in the soil and suggests the need for small amounts of nitrogenous fertilizers to overcome this temporary immobilization. Hence, vermicompost produced by *Eudrillus* (in the vermicomposting) had higher C/N ratio than those produced by local worms and conventional method. Furthermore, vermicompost application to agricultural soils, improves the soil environment, promoting the proliferation of roots, which in turn draw more water and nutrients from larger areas. It was found in the same research that a maximum height of 75 cm recorded after application of *Eudrillus* compost enriched with *Azospirillum* and P-solubilising organisms after 95 days of planting. Similar research was carried out by Rao and Gushanlal [106].

### 7.3.4 Application as soil amendment

The application of compost and vermicompost to soils must be carried out under different criteria and cautions (Molina et al., 2014). Firstly, these organic amendments must meet specific characteristics and requirements for their use, such as the content of organic matter, pH, N, and C/N ratio among others as specified in Table 7.5 [65,110,119,120]. Moreover, several authors have agreed that plant height increased, after applying N-enriched compost [106] indicating the need to balance C/N ratio.

Furthermore, in these products the microbial composition of the vermicompost is another aspect of its quality since it is very important to know and evaluate the quantity and quality of the microbiota present in these amendments, as well as its effects on the soil-plant system [25,29,56,135]. In this sense, the vermicomposting process depends not only on earthworms, but in the microbiota presented along the process.

**Table 7.5** Requirements for group six organic amendments (BOE-A-2017-14332) [24].

| Type | Information on how to obtain it and the essential components | Minimum and maximum content (percentage by mass) Other requirements | Other information on the type designation or labelling | Nutrient content to be declared and guaranteed. Nutrient forms and solubility. Other criteria |
|---|---|---|---|---|
| Organic Compost Manure Amendment | Sanitized and stabilized product, obtained by aerobic biological decomposition (including thermophilic phase), exclusively from manure, under controlled conditions | Total organic matter: 35% Maximum humidity: 40% C/N <20. It may not contain impurities or inert of any kind such as: stones, gravel, metals, glasses, or plastics | pH Electric conductivity. C/N ratio Minimum and maximum humidity. Treatment or manufacturing process, according to the description indicated in column 3 | Total organic matter C organic Total N (if it exceeds 1%). Organic N (if it exceeds 1%) N ammoniacal (if it exceeds 1%). Total $P_2O_5$ (if it exceeds 1%). Total $K_2O$ (if it exceeds 1%). Humic acids. Granulometry |
| Vermicompost organic amendment | Stabilized product obtained from organic materials, by digestion with earthworms, under controlled conditions | Total organic matter: 30%. Maximum humidity: 40%. C/N <20. 90% of the particles will pass through the 25 mm mesh | pH Electric conductivity. C/N ratio. Minimum and maximum humidity. The usual names in commerce may be added | Total organic matter C organic. Total N (if it exceeds 1%). Organic N (if it exceeds 1%). Total $P_2O_5$ (if it exceeds 1%) Total $K_2O$ (if it exceeds 1%). Humic acids. Granulometry. Type or types |

## 7.4 Conclusion

The main conclusion is that tomato, as one of the major horticultural products of the world, is responsible of a great amount of wastes due to the food supply chain. Indeed, both at field and processing factories, the wastes are locally and can be easily used to produce an adequate by-product, a compost, that can help to combat climate chain by adding organic matter to the soils. Moreover, the vermicomposting technology can help to facilitate the transformation of these wastes to a useful product. However, it may be interesting to combine these residues with other organic wastes in order to enhance the worm activity and to obtain a good vermicompost quality. It is very important in the case of tomato wastes management because the residues can inhibit or difficult per se the worm activity and the vermicomposting process can be affected.

Sustainable strategies like this, can help to minimize the impact of agriculture and food processing companies in the carbon footprint, adapting the circular economy and the green deal in their environment.

## References

[1] M. Abad, P.F. Martinez, M.D. Martínez, J. Martínez, Evaluación agronómica de los sustratos de cultivo, Actas de Horticultura 23 (1992) 45–61.

[2] E.A. Abou Hussien, A.M. Elbaalawy, M.M. Hamad, Chemical properties of compost in relation to calcareous soil properties and its productivity of wheat, Egyptian J. Soil Sci. 59 (1) (2019) 85–99.

[3] E.A. Abou Hussien, B.Y. El-Koumey, F.S. El-Shafiey, Effect of composted plant residues on newly reclaimed soils properties and its productivity, Menoufia J. Agric. Res. 37 (2012) 231–245.

[4] K. Abeliotis, C. Chroni, A. Kyriacou, K. Boikou, A.A. Zorpas, K. Lasaridi, Household food waste in Greece - estimation through a self-reported food waste diary, 9th International Scientific Conference ORBIT 2014, New challenges, new responses in the 21st Century, Gödöllő, Hungary, 2014 18 held 26-28, June, 2014.

[5] A. Agapiou, A. Vasileiou, M. Stylianou, K. Mikedi, A.A. Zorpas, Waste aroma profile in the framework of food waste management through household composting, J. Clean. Prod. 257 (2020) 1203.

[6] P. Aggarwal, A. Kaur, A. Bhise, Value-added processing and utilization of mango by-products, in: M. Siddiq, J.K. Brecht, J.S. Sidhu (Eds.), Handbook of Mango Fruit: Production, Postharvest Science, Processing Technology and Nutrition, John Wiley and Sons, New York, 2017, pp. 279–293.

[7] E. Albanell, J. Plaixats, T. Cabero, Chemical changes during vermicomposting (*Eisenia foetida*) of sheep manure mixed with cotton industrial wastes, Biol. Fert. Soil. 6 (3) (1988) 266–269.

[8] R. Albiach, R. Canet, T. Montoya, A. Pérez, A. Quiñones, P. Rojo, Gestión Integral de Residuos Orgánicos Poniendo en Marcha la Economía Circular en la Sociedad, Red Española de Compostaje, Valencia, 2018.

[9] P.C. Anderson, F.M. Rhoads, S.M. Olson, K.D. Hill, Carbon and nitrogen budgets in spring and fall tomato crops, Hort Science 34 (4) (1999) 648–652.

[10] Z.I. Antoniolli, G.P.K. Steffen, R.B. Steffen, Utilizacao de casca de arroz e esterco bovino como substrato para a multiplicacao de Eisenia fetida Savigny (1826), Revista Ciencia e Agrotecnologia 33 (3) (2009) 824–830.

[11] N. Antoniou, A.A. Zorpas, Quality protocol development to define end-of-waste criteria for tire pyrolysis oil in the framework of Circular Economy Strategy, Waste Manag 95 (2019) 161–170.

[12] A. Askar, Importance and characteristics of tropical fruits, Fruit Process. 8 (1998) 273–276.

[13] M. Aira, F. Monroy, S. Mato, How earthworm density affects microbial biomass and activity in pig manure, Eur. J. Soil Biol. 38 (1) (2002) 7–10.

[14] E. Aranda, I. Barois, P. Arellano, S. Frisson, T. Salazar, J. Rodríguez, J.C. Patron, Vermicomposting in the tropics, in: P. Lavelle, L. Brussaard, P. Hendrix (Eds.), Earthworm Management in Tropical Agroecosystems, CABI Publishing, New York, 1999, pp. 253–287.

[15] M.K. Awasthi, A.K. Pandey, P.S. Bundela, J.W.C. Wong, R. Li, Z. Zhang, Co-composting of gelatin industry sludge combined with organic fraction of municipal solid waste and poultry waste employing zeolite mixed with enriched nitrifying bacterial consortium, Bioresour. Technol. 213 (2015) 181–189.

[16] R. Balachandar, L. Baskaran, A. Yuvaraj, R. Thangaraj, R. Subbaiya, B. Ravindran, S.W. Chang, N. Karmegan, Enriched pressmud wermicompost productions with green manure plants using *Eudrilus eugeniae*, Biores. Technol. 299 (2020) 122578.

[17] A.C. Bassaco, Z. Antoniolli, Brum J. I., S. Berilo, Chemistry characterization from animal origin residues and *Eisenia andrei* behaviour. Caracterizacao quimica de residuos de origen animal e comportamento de *Eisenia andrei*, Ciencia e Natura 37 (1) (2015) 45–51.

[18] B. Bian, X. Hu, S. Zhang, X.R. Hu, S.P. Zhang, C.X. Lv, Z. Yang, W.B. Yang, L.M. Zhang, Pilot-scale composting of typical multiple agricultural wastes: Parameter optimization and mechanisms, Bioresour. Technol. 287 (2019) 121482.

[19] E. Benítez, R. Nogales, G. Masciandaro, B. Ceccanti, Isolation by isoelectric focusing of humic-urease complexes from earthworm (Eisenia fetida)-processed sewage sludges, Biol. Fertil. Soils 31 (2000) 489–493.

[20] S. Bansal, K.K. Kapoor, Vermicomposting of crop residues and cattle dung with Eisenia foetida, Bioresour. Technol. 73 (2000) 95–98.

[21] E. Benítez, R. Nogales, C. Elvira, G. Masciandaro, B. Ceccanti, Enzyme activities as indicators of the stabilization of sewage sludges composting with Eisenia foetida, Bioresour. Technol. 67 (1999) 297–303.

[22] M. Blouin, J. Barrere, N. Meyer, S. Lartigue, S. Barot, J. Mathieu, Vermicompost significantly affects plant growth. A meta-analysis, Agron. Sustain. Dev. 39 (4) (2019) 34.

[23] M. Blouin, M.E. Hodson, E.A. Delgado, G. Baker, L Brussaard, K.R. Butt, L. Dendooven, G. Péres, E.J. Tondoh, J.J. Brun, A review of earthworm impact on soil function and ecosystems services, Eur. J. Soil Sci. 64 (2013) 161–182.

[24] BOE-A-2017-14332, 2017. Real Decreto 999/2017, de 24 de noviembre, por el que se modifica el Real Decreto 506/2013, de 28 de junio, sobre productos fertilizantes.

[25] L. Cai, X. Gong, X. Sun, S. Li, X. Yu, Comparison of chemical and microbiological changes during the aerobic composting and vermicomposting of green waste, PLoS One 13 (11) (2018) e0207494.

[26] M.T. Chan, A. Selvam, J.W.C. Wong, Reducing nitrogen loss and salinity during "struvite" food waste composting by zeolite amendment, Bioresour. Technol. 200 (2016) 838–844.

[27] C.W. Carry, J.F. Stahl, B.E. Hansen, P.L. Friess, Sludge management and disposal practices of the country sanitation districts of Los Angeles (USA), Water Sci. Technol. 22 (12) (1990) 23–32.

[28] L.J. Chanu, S. Hazarika, B.U. Choudhury, T. Ramesh, A. Balusamy, P. Moirangthem, A. Yumnam, P.K. Sinha, Guide to vermicomposting-production process and socio economic aspects. Extension bulletin 81, ICAR Research Complex for NEH Region, Meghalaya (2020).

[29] C.Q. Chen, P. Ray, K.F. Knowlton, A. Pruden, K. Xia, Effect of composting and soil type on dissipation of veterinary antibiotics in land-applied manures, Chemosphere 196 (2018) 270–279.

[30] B. Chefetz, Y. Chen, Y. Hadar, P. Hatcher, Chemical and biological characterization of organic matter during composting of municipal solid waste, J. Environ. Qual. 25 (1996) 776–785.

[31] M. Ciavatta, L. Govi, P. Pasotti, P. Sequi, Changes in organic matter during stabilization of compost from municipal solid wastes, Bioresour. Technol. 43 (2) (1993) 141–145.
[32] J.M. Costa, E. Heuvelink, N. Botden, Greenhouse horticulture in China: situation and prospects, Ponsen & Looijen BV, Wageningen (2004).
[33] Council of the European Union, Council Directive 1999/31/EC of 26 April 1999 on the landfill of waste, Official Journal L 182 (1999) 0001–0019 16/07/1999 P.
[34] W. Czekała, K. Malinska, R. Cáceres, D. Janczak, J. Dach, A. Lewicki, Co-composting of poultry manure mixtures amended with biochare-The effect of biochar on temperature and C-$CO_2$ emission, Bioresour. Technol. 200 (2016) 921–927.
[35] Directive (EU) 2018/851 of the European Parliament and of the Council of 30 May 2018 amending Directive 2008/98/EC on waste.
[36] Directive, 2008/98/EC of the European Parliament and of the Council of 19 November 2008 on waste and repealing certain Directives.
[37] J. Doublet, C. Francou, M. Poitrenaud, Influence of bulking agents on organic matter evolution during sewage sludge composting; consequences on compost organic matter stability and N availability, Bioresour. Technol. 102 (2) (2011) 1298–1307.
[38] K.M. Doula, K. Elaiopoulos, P. Kouloumbis, A.A. Zorpas, In situ application of clinoptilolite to improve compost quality produced from pistachio bio-wastes, Fresenius Environ.l Bull 27 (3) (2018) 1312–1318.
[39] M. De Bertoldi, G. Vallini, A. Pera, The biology of composting: a review, Waste Manag. Res. 1 (1983) 157–176.
[40] H. Engeli, W. Edelmann, Combined digestion and composting of organic industrial and municipal wastes in Switzerland, Water Sci. Technol. 27 (2) (1993) 169–182.
[41] E. Epstain, The Science of Composting, Technomic Pup. Co, Pennsylvania, USA, 1997.
[42] C.A. Edward, I. Burrows, K.E. Fletcher, B.A. Jones, The use of earthworms for composting farm wastes. In: Gasser, J.K.R. (Ed.), Composting of Agricultural and Other Wastes, London, 1985, pp. 229–242.
[43] C.A. Edward, J.R. Lofty, Biology of Earthworms, Chapman and Hall, London, 1977.
[44] F. El Ouaqoudi, L. El Fels, P. Winterton, P. Lemée, A. Amblès, M. Hafidi, Study of humic acids during composting of ligno-cellulose waste by infra-red spectroscopic and thermogravimetric/thermal differential analysis, Compost Sci. Util. 22 (3) (2014) 188–198.
[45] F.Z. El Ouaqoudi, L El Fels, L. Lemee, A. Ambles, M. Hafidi, Evaluation of lignocelullose compost stability and maturity using spectroscopic (FTIR) and thermal (TGA/TDA) analysis, Ecol. Eng. 75 (2015) 217–222.
[46] NA. El-Tayeh, FM. Salama, N. Loutfy, A. Abou, F. Mona, Effect of sandy soil amendment with filter mud cake on growth and some ecophysiological parameters of *Daucus carota* and *Beta vulgaris* Plants, Int. J. Environ. Sci. 18 (1) (2019) 97–104.
[47] Eurostat, 2020. https://ec.europa.eu/eurostat/statistics-explained/index.php/Agricultural_production_-_crops#Vegetables (Accessed 14 September 20).
[48] ECN, European Compost Network. End of waste criteria for compost and digestate, Org. Resour. Biol. Treat. 1 (2011) 1–10.
[49] FAO, 2011. Global food losses and food waste – Extent, causes and prevention. Rome.
[50] FAO, The State of Food and Agriculture 2019: Moving Forward on Food Loss and Waste Reduction, Rome, 2019.
[51] FAOSTAT, 2020. http://www.fao.org/faostat/es/#home (Accessed 18 September 2020).
[52] J. Faverial, J. Sierra, Home composting of household biodegradable wastes under the tropical conditions of Guadeloupe (French Antilles), J. Clean. Prod. 83 (2014) 238–244.
[53] M.J. Fernández-Gómez, M. Díaz-Raviña, E. Romero, R. Nogales, Recycling of environmentally problematic plant wastes generated from greenhouse tomato crops through vermicomposting. Inter, J. Environ. Sci. Technol. 10 (4) (2013) 697–708.
[54] J.M. Fernandez-Gomez, E. Romero, R. Nogales, Feasibility of vermicomposting for vegetable greenhouse waste recycling, Bioresour. Technol. 101 (2010) 9654–9660.

[55] M.E. Ferreire, M.C.P. Cruz, Effect of compost from municipal wastes digested by earthworms on the dry matter production of maize and soil properties, Cientifica 20 (1) (1992) 217–226.
[56] K. Fogler, G.K. Guron, L.L. Wind, I.M. Keenum, W.C. Hession, L.A. Krometis, L.K. Strawn, M.A. Ponder, A. Pruden, Microbiota and antibiotic resistome of lettuce leaves and radishes grown in soils receiving manure-based amendments derived from antibiotic-treated cows, Front. Sustain. Food Syst. 3 (2019) 220–221.
[57] M. Fang, J.W.C. Wong, K.K. Ma, M.H. Wong, Co-compostnig of sewage sludge and coal fly ash: nutrient transformations, Biores. Technol. 67 (1999) 19–24.
[58] F. Formes, D. Memdoza-Hernandez, R. Garcia-de-la-Fuente, M. Abad, M.R. Belda, Composting versus vermicomposting: a comparative study of organic matter evolution through straight and combined processes, Biores. Technol. 118 (2012) 296–305.
[59] J. Frederickson, K.R. Butt, R.M. Morris, C. Daniel, Combining vermiculture with traditional green waste composting systems, Soil Biol. Biochem. 29 (1997) 725–730.
[60] M.S. Finstein, F.C. Miller, P.F. Strom, Waste Treatment Composting as a Controlled System, Department of Environmental Science Cook College, Rutgers University, 1987, pp. 361–398.
[61] V.K. Garg, R. Gupta, A. Yadav, Potential of vermicomposting technology in solid state management, in: A Pandley, C.R. Soccol, E. Larroche (Eds.), Current Development in Solid-State Fermentation, Springer, New York, 2008, pp. 468–511.
[62] P. Hand, W.A. Hayes, J.E. Satchell, J.C. Frankland, C.A. Edwards, E.F. Neuhauser, Vermicomposting of cow slurry, in: C.A. Edwards, E.F. Neuhauser (Eds.), Earthworms in Waste and Environmental Management, SPB Academic Publishing, The Hague, 1988, pp. 49–63.
[63] M. Hamidpour, M. Afyuni, E. Khadivi, A.A. Zorpas, V. Inglezakis, Composted municipal waste effects on forms and plant availability of Zn and Cu in a calcareous soil, Int. Agrophys. 26 (2012) 365–374.
[64] IFAPA, 2018. Memoria Anual. Consejería de Agricultura, Ganadería, Pesca y Desarrollo Sostenible. Junta de Andalucía, Sevilla.
[65] F. Ingelmo, M.J. Molina, Soriano Soto, D. Mª, J. Llinares, Influence of organic matter transformations on the bioavailability of heavy metals in a sludge based compost, J. Environ. Manage. 95 (2012) 104–109.
[66] IPCC, 2018. Global warming of 1.5°C. Available at: https://www.ipcc.ch/site/assets/uploads/sites/2/2019/06/SR15_Full_Report_High_Res.pdf (Accessed 13 September 2020).
[67] M. Iranzo, J.V. Canizares, L. Roca-Perez, I. Sainz-Pardo, S. Mormeneo, R. Boluda, Characteristics of rice straw and sewage sludge as composting materials in Valencia (Spain), Bioresour. Technol. 95 (1) (2004) 107–112.
[68] M. Iranzo, M. Gamon, R. Boluda, S. Mormeneo, Analysis of pharmaceutical biodegradation of WWTP sludge using composting and identification of certain microorganisms involved in the process, Sci. Total Environ. 640 (2018) 840–848.
[69] H.A. Jambakar, Use of earthworms as potential source to decompose organic wastes, Proc. of the National seminar on Organic Farming M.P.K.V., Pune, 1992 53–54.
[70] J. Jara-Samaniego, M.D. Pérez-Murcia, M.A. Bustamante, A. Pérez-Espinosa, C. Paredes, M. López, D.B. López-Lluch, I. Gavilanes-Terán, R. Moral, Composting as sustainable strategy for municipal solid waste management in the Chimborazo region, Ecuador: suitability of the obtained composts for seedling production, J. Clean. Prod. 141 (2017) 1349–1358.
[71] A.L.H. Jack, J.E. Thies, Compost and vermicompost as amendments promoting soil health, pp, 454-466, in: N. Uphoff (Ed.), Biological Approaches to Sustainable Soil Systems, CRC PressEditors, Boca Ratón, 2006, pp. 453–466.

[72] I. Jemai, N.B. Aissa, T. Gallali, Effects of municipal reclaimed wastewater irrigation on organic and inorganic composition of soil and groundwater in Souhil Wadi Area (Nabeul, Tunisia), Hydrol. Current Res. 04 (04) (2013).
[73] R.G. Joergensen, F. Wichern, Quantitative assessment of the fungal contribution to microbial tissue in soil, Soil Biol. Biochem. 40 (2008) 2977–2991.
[74] I. Jorge-Mardomingo, M.E. Jiménez-Hernández, L. Moreno, A. de la Losa, M.T. de la Cruz, M.A. Casermeiro, Application of high doses of organic amendments in Mediterranean agricultural soil: an approach for assessing the risk of groundwater contamination, Catena 131 (2015) 74–83.
[75] JRC, Scientific and technical report; end of waste criteria, institute for prospective and technological studies, European Commission (2008) http://susproc.jrc.ec.europa.eu/documents/Endofwastecriteriafinal.pdf (Accessed 18 September 2020).
[76] JRC, Final report; end of waste criteria, institute for prospective and technological studies, European Commission (2009). http://susproc.jrc.ec.europa.eu/documents/Endofwastecriteriafinal.pdf (Accessed 18 September 2020).
[77] M.B. Khan, X. Cui, G. Jilani, U. Lazzat, A. Zehra, Y. Hamid, B. Hussain, L. Tang, X. Yang, Z. He, *Eisenia fetida* and biochar synergistically alleviate the heavy metals content during valorization of biosolids via enhancing vermicompost quality, Sci. Total Environ. 684 (2019) 597–609.
[78] U. Krogmann, I. Korner, Technologies and Stratgies for Composting, in: H.-J. Rehm, G. Reed, A. Ptihler, P. Stadler (Eds.), Technologies and Stratgies for Composting, Biotechnology Second, Completely Revised Edition (2000) 127–150.
[79] K. Lasaridi, C Chroni, K. Abeliotis, A. Kyriacou, A.A. Zorpas, Environmental assessment of composting in the context of sustainable waste management, in: Antonis Zorpas (Ed.), Sustainability Behind Sustainability, 11788, Nova Science Publishers, 400 Oser Avenue, Suite 1600, Hauppauge, NY, 2014, pp. 229–242.
[80] A.M. Litterick, P. Harrier, C. Wallace, A. Watson, M. Wood, The role of uncomposted materials, composts, manures, and compost extracts in reducing pest and disease incidence and severity in sustainable temperate agricultural and horticultural crop production. A review, Crit. Rev. Plant. Sci. 23 (2004) 453–479.
[81] X. Liu, Q. Chen, Z. Wang, L. Xie, Z. Xu, Allelopathic effects of essential oil from *Eucalyptus grandisx*, e. urophylla on pathogenic fungi and pest insects, Front. Fores. China 21 (1) (2008) 25–33.
[82] G. Logsdon, Worldwide progress in vermicomposting, BioCycle 35 (1994) 63–65.
[83] M. López, R. Boluda, Residuos Agrícolas, Compostaje, Ediciones Mundi-Prensa, Madrid, 2008, Moreno, J., Moral, R. (Eds.), pp. 489–518
[84] M.J. López, A. Masaguer, C. Paredes, L. Roca, M. Ros, M. Salas, R. Boluda, Residuos orgánicos y agricultura intensiva, in: J. Moreno, R. Moral, J.L. García-Morales, J.A. Pascual, M.P. Bernal (Eds.), De residuo a recurso. El camino hacia la sostenibilidad, Mundi-Prensa, Madrid, 2015, vol. III. Ed. pp. 41–67
[85] I. López-Cano, A. Roig, M.L. Cayuela, J.A. Alburquerque, M.A. Sánchez-Monedero, Biochar improves N cycling during composting of olive mill wastes and sheep manure, Waste Manag 49 (2016) 553–559.
[86] J. Lorimor, C. Fulhage, R. Zhang, T. Funk, R. Sheffield, C. Sheppard, G.L. Newton, Manure management strategies/technologies, in: J.M. Rice, D.F. Caldwell, F.J. Humenik (Eds.), White Paper on Animal Agriculture and the Environment for National Center for Manure and Animal Waste Management, ASABE, Michigan, 2001, p. 52.
[87] K. Lasaridi, K. Abeliotis, O. Hatzi, G. Batistatos, C. Chroni, N. Kalogeropoulos, N. Chatzieleftheriou, N. Gargoulas, A. Mavropoulos, A. Zorpas, M. Nikolaou, D. Anagnostopoulos, Waste prevention scenarios using a web-based tool for local authorities, Waste and Biomass Valorisation 6 (2015) 625–636.

[88] P. Loizia, I. Voukkali, A.A. Zorpas, J. Navarro-Pedreño, G. Chatziparaskeva, J.V Inglezakis, I. Vardopoulos, Measuring environmental performance in the framework of waste strategy development, Sci. Total Environ. 753 (2021) 141974.
[89] P. Loizia, N. Neofytou, A.A. Zorpas, The concept of circular economy in food waste management for the optimization of energy production through UASB reactor, Environ. Sci. Pollut. Res. 26 (2019) 14766–14773.
[90] M. Madan, S. Sharma, R. Bisaria, R. Bhamidimarri, Recycling of organic wastes through vermicomposting and mushroom cultivation, in: R. Bhamidimarri (Ed.), Alternative Waste Treatment Systems, Elsevier Applied Science, London, 1988, pp. 132–141.
[91] MAPA, 2020. Superficies y producciones anuales de cultivos, https://www.mapa.gob.es/es/estadistica/temas/estadisticas-agrarias/agricultura/superficies-producciones-anuales-cultivos/ (Accesed 18.09.20).
[92] P. Mazuela, MC. Salas, M. Urrestarazu, Vegetable Waste Compost as Substrate for Melon, Commun. Soil Sci. Plant Anal. 36 (11–12) (2005) 1557–1572.
[93] A. Menéndez-Yuffá, D. Barry-Etienne, B. Bertrand, F. Georget, H. Etienne, A comparative analysis of the development and quality of nursery plants derived from somatic embryogenesis and from seedlings for large-scale propagation of coffee (*Coffea arabica* L.), Plant Cell Tissue Organ Cult 102 (2010) 297–307.
[94] A. Mitchell, D. Alter, Suppression of labile aluminium in acidic soils by the use of vermicompost extract, Commun. Soil Sci. Plant Anal. 24 (11–12) (1993) 1171–1181.
[95] MJ. Molina, MD. Soriano, J. Llinares, Stabilisation of sewage sludge and vinasse biowastes by vermicomposting with, Bioresour. Technol. 137 (2013) 88–97.
[96] J. Moreno, R. Moral, Compostaje, Mundi-Prensa, Madrid, 2008.
[97] B. Moya, A. Parker, R. Sakrabani, Challenges to the use of fertilisers derived from human excreta: the case of vegetable exports from Kenya to Europe and influence of certification systems, Food Policy 85 (2019) 72–78.
[98] R. Nogales, J. Domínguez, S. Mato, Vermicompostaje, in: J. Moreno, R. Moral (Eds.), Compostaje, Ediciones Mundi Prensa, Madrid, 2008, pp. 187–208.
[99] J. Navarro-Pedreño, R. Moral, I. Gómez, J. Mataix, Residuos Orgánicos y Agricultura, Servicio de Publicaciones de la Universidad de Alicante, Alicante, 1995.
[100] P.M. Ndegwa, S.A. Thompson, Integrating composting and vermicomposting in the treatment of bioconversion of biosolids, Bioresour. Technol. 76 (2001) 107–112.
[101] R. Nogales Vargas-Machuca, E. Romero Taboada, M.J. Fernández Gómez, Vermicompostaje: procesos, productos y aplicaciones, in: J. Moreno, R. Moral, J.L. García Morales, J.A. Pascual, M.P. Bernal (Eds.), De residuo a recurso: Recursos orgánicos: aspectos agronómicos y medioambientales, Ediciones Paraninfo, Madrid, 2014, vol. 5 p. 172.
[102] K.P. Padmavathiamma, Y.L. Li, R.U. Kumari, An experimental study of vermi-biowaste composting for agricultural soil improvement, Bioresour. Technol. 99 (2008) 1672–1681.
[103] M.B. Parra, Alimentación de las plantas y compostaje, Fertilidad de la tierra: revista de agricultura ecológica 15 (2004) 42–45.
[104] A. Pérez-Gimeno, J. Navarro-Pedreño, M.B. Almendro-Candel, I. Gómez, A.A. Zorpas, Characteristics of organic and inorganic wastes for their use in land restoration, Waste Manag. Res. 37 (5) (2019) 502–507.
[105] T. Rangaraj, E. Somasundaram, M.M. Amanullah, Effect of agro-industrial wastes on soil properties and yield of irrigated finger millet (*Eleusine coracana* L. Gaertn) in coastal soil, Res. J. Agric. Biol. Sci. 3 (3) (2007) 153–156.
[106] E.H. Rao, L. Gushanlal, Response of chilli (*Capsicum annum* L.), variety pant C-1 to varying levels of nitrogen and spacing, Veg. Sci. 13 (1) (1986) 17–21.
[107] F.M. Rashad, W.D. Saleh, M.A. Moselhy, Bioconversion of rice straw and certain agro-industrial wastes to amendments for organic farming systems: 1. Composting, quality, stability and maturity indices, Bioresour. Thechnol. 101 (2010) 5952–5960.
[108] J.R. Rico Hernández, I. Gómez, J. Navarro-Pedreño, M.M. Jordán, J. Bech, V.M. Nieto Asencio, N. Portell Iñiguez, Environmental consequences from the use

of sewage sludge in soil restoration related to microbiological pollution, J. Soils Sediments 18 (2018) 2172–2178.
[109] D. Riggle, H. Holmes, Expanding horizons for commercial vermiculture, BioCycle 35 (1994) 58–62.
[110] L. Roca Pérez, C. Gil, J.J. Ramos, MD. Soriano, R. Boluda, Efecto de la aplicación de compost de lodo de depuradora y residuo del arroz sobre el pH, nitrógeno, materia orgánica y las sustancias húmicas de un Luvisol cálcico dedicado al cultivo de cítricos, Spanish J. Rural Develop. 1 (2010) 101–109.
[111] L. Roca-Pérez, C. Martínez, P. Marcilla, R. Boluda, Composting rice straw with sewage sludge and compost effects on the soil–plant system, Chemosphere 75 (2009) 781–787.
[112] R.N. Schubert, T.B.G.A. Morselli, S.M. Tonietto, J.M.O. Henriquez, R.D. Trecha, R.P. Eid, D.P. Rodriguez, S.R. Piesanti, M.R.S. Maciel, A.P.F. Lima, Edaphic macrofauna in degradation of animal and vegetable residues, Braz. J. Biol. 79 (4) (2019) 589–593.
[113] S.K. Sharma, M.L. Maguer, Lycopene in tomatoes and tomato pulp fractions, Ital. J. Food Sci. 2 (1996) 107–113.
[114] J. Shi, M. Le Maguer, B. Mike., Lycopene from Tomatoes, in: J. Shi, G. Mazza, M. Le Maguer (Eds.), Functional Foods: Biochemical and Processing Aspects2nd ed., New York CRC Press, New York, 2002, pp. 134–167.
[115] J. Shi, C. Yi, X. Ye, S. Xue, Y. Jiang, Y. Ma, D. Liu, Effects of supercritical $CO_2$ fluid parameters on chemical composition and yield of carotenoids extracted from pumpkin, Food Sci. Technol. 43 (1) (2010) 39–44.
[116] A. Singh, S. Sharma, Composting of a crop residue through treatment with microorganisms and subsequent vermicomposting, Bioresour. Technol. 85 (2002) 107–111.
[117] J. Singh, K.S. Pillai, The world beneath us, Sci. Rep. (1973) 318–321.
[118] R.K. Sinha, S. Agarwal, K. Chauhan, V. Chandran, B.K. Soni, Vermiculture technology: reviving the dreams of Sir Charles Darwin for xcientific use of earthworms in sustainable development programs, Technol. Invest. 1 (2010) 155–172.
[119] MD. Soriano, L. Garcia-España, R. Ruiz, R. Boluda, Desarrollo de especies aromáticas sobre residuos orgánicos y vegetales estabilizados, Ae Agricultura y ganadería ecológica 15 (2014) 18–19.
[120] MD. Soriano, L. Garcia-España, R. Boluda, Utilización de compost y vermicompost obtenidos a partir de residuos de especies leguminosas como cubiertas vegetales para jardinería, Ae Agricultura y ganadería ecológica 1 (2016) 30–31.
[121] D. Sun, Y. Lan, E.G. Xu, J. Meng, W. Chen, Biochar as a novel niche for culturing microbial communities in composting, Waste Manag 54 (2016) 93–100.
[122] J.K. Syres, A.N. Sharpley, D.R. Keeney, Cycling of nitrogen by surface casting earthworms in a pasture ecosystem, Soil Biol. Biochem. 11 (1979) 181–185.
[123] D. Symeonides, P. Loizia, A.A. Zorpas, Tires Waste Management System in Cyprus in the Framework of Circular Economy Strategy, J. Environ. Sci. Pollut. Res. 26 (35) (2019) 35445–35460.
[124] S. Suthar, Bioconversion of post harvest crop residues and cattle shed manure into value-added products using earthworm Eudrilus eugeniae Kinberg, Ecol. Eng. 32 (2008) 206–214.
[125] B. Sen, T.S. Chandra, Do earthworms affect dynamics of functional response and genetic structure of microbial community in a lab-scale composting system? Bioresour. Technol. 100 (2009) 804–811.
[126] D. Tedesco, E.A. Doriana, C. Conti, E. Biazzi, J. Bacenetti, Bioconversion of fruit and vegetable waste into earthworms as a new protein source: The environmental impact of earthworm meal production, Sci. Total Environ. 683 (2019) 690–698.
[127] H. Taherisoudejani, H. Kazemian, J.V. Inglezakis, A.A. Zorpas, A review on the application of zeolites in organic waste composting, Biocatal. Agric. Biotechnol. 22 (2019) 101396.

[128] A. Tolón, X. Lastra, La agricultura intensiva del poniente almeriense. Diagnóstico e instrumentos de gestión ambiental, M+A. Revista Electrónic@ de Medio Ambiente 8 (2010) 18–40.
[129] UNDP, Sustainable Development Goals, United Nations Development Program, 2019. https://www.undp.org/content/undp/en/home/sustainable-development-goals.html (Accesed 18 September 2020).
[130] G.Vlyssides, A.A. Zorpas, P.K. Karlis, A.G. Zorpas, Description of a pilot plant for the co-composting of the solid residue and wastewaters from the olive oil industry, Hung. J. Ind. Chem. 2 (2000) 59–64.
[131] G. Vlyssides, M. Loizidou, A.A. Zorpas, Characteristics of solid residues from olive oil processing as a bulking material for co-composting with industrial wastewater, J. Environ. Sci. Health A 34 (3) (1999) 737–748.
[132] I. Voukkali, P. Loizia, A.A. Zorpas, K. Lasaridi, C. Chroni, C. Abeliotis, A. Georgiou, K. Fanou, D. Pyrilli, P. Goumenou, L. Fitiri, P. Trisokka, N. Bikaki, P. Voukkali, 2014a. Evaluation of The Implementation of Waste Prevention Plant in Local Authorities, Crete 2014, 4th International Conference, Industrial and Hazardous Waste Management, Sep 2–5, Chania, Crete, Oral Presentation.
[133] I. Voukkali, P. Loizia, A.A. Zorpas, S.M. Miliotou, C. Kaliroe, 2014b. Proposed prevention actions in relation to specific waste stream and target groups from Cyprus Strategic Plan, Crete 2014, 4th International Conference, Industrial and Hazardous Waste Management, Sep 2–5, Chania, Crete, Oral Presentation.
[134] W. Wesemann, Wurmkompostierung, in: F. Amlinger (Ed.), Handbuch der Kompostierung - Ein Leitfaden für Praxis,Verwaltung, Forschung, Bundesministerium fur Land- und Forstwirtschaft, Bundesministerium fur Wissenschaft und Forschung, Vienna, 1992, pp. 106–110.
[135] G. Xue, M. Jiang, H. Chen, M. Sun, Y.F. Liu, X. Li, P. Gao, Critical review of ARGs reduction behavior in various sludge and sewage treatment processes in wastewater treatment plants, Critic. Rev. Environ. Sci. Technol 49 (18) (2019) 623–1674, doi:10.1080/10643389.2019.1579629.
[136] FYang, GX. Li, B. Zang, ZY. Zhang, The maturity and CH4, N2O, NH3 emissions from vermicomposting with agricultural waste, Compost Sci. Util. 25 (n 4) (2017) 262–271.
[137] A.A. Zorpas, V. Stamatis, G. A. Zorpas, A. G. Vlyssides, M. Loizidou, Compost characteristics from sewage sludge and organic fraction of municipal solid waste, Fresenius Environ. Bull. 8 (3–4) (1999a) 154–162.
[138] A.A. Zorpas, A. Apostolos, G. Vlyssides, M. Loizidou, Dewater anaerobically stabilized primary sewage sludge composting. metal leach ability and uptake by natural clinoptilolite, Commun. Soil Sci. Plant Anal. 30 (11–12) (1999b) 1603–1614.
[139] A.A. Zorpas, T. Constantinides, G. Vlyssides, I. Haralambous, M. Loizidou, Heavy metal uptake by natural zeolite and metals partitioning in sewage sludge compost, Bioresour. Technol. 72 (2000) 113–119.
[140] A.A. Zorpas, D. Arapoglou, G. Chou, G. Vlyssides, Physicochemical parameters, which affect the phytotoxicity in the final compost product, 1st International Conference in Ecological Production of the Planet Earth, University of Trace, Xanthi, 2001, pp. 1005–1012 2–5 June, Vol 2 Oral Presentation.
[141] A.A. Zorpas, V. Inglezakis, M. Loizidou, H. Grigoropoulou, Particle size effects on the uptake of heavy metals from sewage sludge compost using natural eolite Clinoptilolite, J. Colloid Interface Sci. 205 (2002) 1–4.
[142] A.A. Zorpas, D. Arapoglou, K. Panagiotis, Waste paper and clinoptilolite as a bulking material with dewatered anaerobically stabilized primary sewage sludge (DASPSS) for compost production, Waste Manag 23 (1) (2003) 27–35.
[143] A.A. Zorpas, M. Loizidou, V. Inglezakis, Heavy Metals fractionation before, during and after composting of sewage sludge with natural zeolite, Waste Manag 28 (2008) 2054–2060.

[144] A.A. Zorpas, Compost evaluation and utilization. In: Pereira, J.C., Bolin, J.L. (Eds.), Composting: Processing, Materials and Approaches, Nova Science Publishers, Hauppauge, NY, 2008a, pp. 31-68.
[145] A.A. Zorpas, Sewage sludge compost evaluation in oats, pepper and eggplant cultivation. Dyn. Soil Dyn. Plant 2 (2008b) 103-109.
[146] A.A. Zorpas, Compost evaluation and utilization, in: J.C. Pereira, J.L. Bolin (Eds.), Composting: Processing, Materials and Approaches, Nova Science Publishers, Hauppauge, NY, 2009, pp. 31–68.
[147] A.A. Zorpas, M. Loizidou, Heavy Metals Leachability before, during and after composting of sewage sludge with natural zeolite, Desalin. Water Treat. 8 (2009) 256–262.
[148] A.A. Zorpas, N.C. Costa, Combination of fenton oxidation and composting for the treatment of the olive solid residue and the olive mile wastewater from the olive oil industry in Cyprus, Biores. Technol. 101 (20) (2010) 7984–7987.
[149] A.A. Zorpas, Metals selectivity from natural zeolite in sewage sludge compost. a function of temperature and contact time, Dyn. Soil Dyn. Plant 5 (2011) 104–112.
[150] A.A. Zorpas, Sewage Sludge Compost Evaluation and Utilization, In: Zorpas, A.A., Inglezakis, V.J. (Eds.), Sewage Sludge Management; From the past to our Century. Nova Science Publishers, Hauppauge, NY, 2012a, pp. 173–216.
[151] A.A. Zorpas, Contribution of Zeolites in Sewage Sludge Composting. In: Inglezakis, V.J., Zorpas, A.A. (Eds.), Handbook on Natural Zeolite, Bentham Science Publishers Ltd., Bussum, The Netherlands, 2012b, pp. 182–199.
[152] A.A. Zorpas, K. Lasaridi, Measuring Waste Prevention, Waste Manag 33 (2013) 1047–1056.
[153] A.A. Zorpas, I. Voukkali, P. Loizia, Proposed treatment applicable scenario for the treatment of domestic sewage sludge which is produced from a sewage treatment plant under warm climates conditions, Desalin. Water Treat. 51 (13–15) (2013) 3081–3089.
[154] A.A. Zorpas, K. Lasaridi, K. Abeliotis, I. Voukkali, P. Loizia, L. Fitiri, C. Chroni, N. Bikaki, Waste prevention campaign regarding the Waste Framework Directive, Fresenius Environ. Bull. 23 (11a) (2014) 2876–2883.
[155] A.A. Zorpas, Recycle and reuse of natural zeolites from composting process: A 7 years project, Desalin. Water Treat. 52 (2014) 6847–6857.
[156] A.A. Zorpas, Sustainable waste management through end of waste criteria development, Environ. Sci. Pollut. Res. 23 (2016) 7376–7389.
[157] A.A. Zorpas, K. Lasaridi, I. Voukkali, P. Loizia, C. Chroni, Household waste compositional analysis variation from insular communities in the framework of waste prevention strategy plans. Waste Manag. 38 (2015a) 3–11.
[158] A.A. Zorpas, K. Lasaridi, I. Voukkali, P. Loizia, C. Chroni, Promoting sustainable waste prevention activities and plan in relation to the waste framework directive in insular communities. Environ. Process. 2 (2015b) 159–173.
[159] A.A. Zorpas, I. Voukkali, P. Loizia, Effectiveness of waste prevention program in primary student's schools. J. Environ. Sci. Pollut. Res. 24 (2017a) 14304–14311.
[160] A.A. Zorpas, I. Voukkali, P. Loizia, A prevention strategy plan concerning the waste framework directive in Cyprus. Fresenius Environ. Bull. 26 (2) (2017b) 1310–1317.
[161] A.A. Zorpas, K. Lasaridi, D.M. Pociovalisteanu, P. Loizia, Monitoring and evaluation of prevention activities regarding household organics waste from insular communities, J. Clean. Prod. 172 (2018) 3567–3577.
[162] A.A. Zorpas, Strategy Development in the Framework of Waste Management, Sci. Total Environ. 716 (2020) 137088.
[163] F. Zucconi, M. Forte, A. Monaco, M. De Bertoldi, Biological evaluation of compost maturity, BioCycle 22 (1981) 27–29.

CHAPTER EIGHT

# Environmental applications of tomato processing by-products

**Salah Jellali[a], Noureddine Hamdi[b,c], Khalifa Riahi[d], Helmi Hamdi[e], Mejdi Jeguirim[f]**

[a]PEIE Research Chair for the Development of Industrial Estates and Free Zones, Center for Environmental Studies and Research, Sultan Qaboos University, Al-Khoud, Muscat, Oman
[b]Higher Institute of Water Sciences and Techniques of Gabes, Zrig, Gabes, Tunisia
[c]National Center of Research in Materials Sciences (CNRSM), Borj Cedria Technopole, Tunisia
[d]High School of Engineers of Medjez El Bab (ESIM), UR-GDRES, University of Jendouba, Tunisia
[e]Center for Sustainable Development, College of Arts and Sciences, Qatar University, Doha, Qatar
[f]The institute of Materials Science of Mulhouse (IS2M), University of Haute Alsace, University of Strasbourg, CNRS, UMR, Mulhouse, France

## 8.1 Introduction

Freshwater has become scarce in various countries and its unavailability could induce major social and economic concerns. Furthermore, in various developing countries (in Africa and Asia), more than 80% of the produced wastewater is still discharged without appropriate treatment, which has negative effects on the environment and human health [29]. Therefore, sustainable water resources management (SWRM) has been pointed out as an urgent challenge to be seriously taken into account in order to preserve water resources and human health. One of the SWRM pillars concerns the adequate treatment and safe reuse of the produced wastewaters in a context of circular economy [103]. By the same way, this will permit the achievement of several sustainable development goals (SDGs) such as SDG3 concerning the "Ensuring of healthy lives and promotion of well-being for all at all ages," SDG6, regarding the "Ensuring of water and sanitation for all," SDG13 about "Taking urgent actions to combat climate change and its impacts," and finally SDG14 about "The conservation and the sustainably use the oceans, seas. and marine resources" [105].

The discharged urban and especially industrial wastewaters (even after classic treatment) could still contain various toxic compounds such as heavy metals and persistent organic pollutants at concentrations higher than the admissible levels fixed by national legislations or regional directives (for the European union water directive, please consult: www.ec.eurpa.eu/environment/water). Even at very low concentrations (i.e., μg/L to mg/L), these

*Tomato Processing by-Products*
DOI: https://doi.org/10.1016/B978-0-12-822866-1.00011-9

pollutants could significantly deteriorate the quality of surface water bodies, groundwater, and soil compartments. Some of these pollutants have a great potential to be accumulated in living organisms such as fish and crops. Indirect human exposure to these toxic pollutants via the food chain could cause serious health concerns [47].

In the last decades, various technologies have been tested for the removal of inorganic and organic pollutants from wastewaters. These techniques include coagulation-flocculation [119], low-cost chemical precipitation [36], biological purification [71], membrane filtration [42], zeolites [43], and advanced chemical oxidation [11]. These latter have serious drawbacks such as low efficiency end selectivity, production of huge amounts of sludge and relative high costs [81]. During the last decades, the use of low-cost materials, such as agricultural wastes, for the treatment of urban/industrial wastewaters has been proposed as a cost-effective and promising technology for both organic and inorganic pollutant removal from aqueous solutions [8,56,58,110].

Tomato is the most widely grown fresh vegetable crop in the world with a total production of about 243 million tons in 2018 (FAO, 2020). The fruit is often consumed fresh or processed (dried or canned). Besides, tomato plant residues (stem, leaves and roots) and 10%–18% of the processed tomato during the production of paste is lost as solid wastes [93]. These wastes have to be managed efficiently in order to avoid negative effects on the environment and human wellbeing. Turning these wastes into values could be, therefore, a great opportunity for industrials as well as stakeholders in charge of environment protection and water resource management. In this context, various biochemical and thermochemical conversion processes has been tested for their conversion into biogas and biofuels for energy purposes [14,55] and into a solid carbonaceous residue, named biochar. Besides their application as eco-friendly biofertilizers [34], raw and especially modified biochars have been used as efficient adsorbents for pollutants removal form aqueous and gaseous effluents [48].

This chapter summarizes and discusses the latest outcomes regarding the valorisation of raw as well as modified tomato wastes for wastewater treatment. It is divided in two sections. The first one is devoted to the synthesis and in-depth physicochemical characterisation of raw biochars and activated carbons derived from tomato wastes. The second section addresses the use of these modified materials for the removal of both inorganic and organic pollutants from aqueous solutions. At the end of the second section, mechanisms involved in the adsorption process onto adsorbents from tomato wastes is presented.

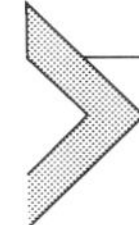

## 8.2 Synthesis and characterization of carbonaceous materials from tomato wastes

### 8.2.1 Nonactivated biochars

The thermochemical conversion of tomato wastes via the pyrolysis process was pointed out as a promising method [91,104]. This operation turns tomato wastes into biofuels for energetic purposes and a solid residue, which is a carbon-rich material, called biochar. Biochar has an appreciable mitigation potential of carbon sequestration and greenhouse gases emission [24]. It has also confirmed positive impacts on plant growth enhancement and wastewater quality improvement when used as eco-friendly fertilizer [89] and low cost adsorbent [110], respectively.

#### *8.2.1.1 Used raw feedstock properties and pyrolysis conditions*

In various countries such as Turkey, Poland, Pakistan, Tunisia, and New Zealand, tomato wastes have been thermochemically converted into biochars through pyrolysis process [25,53,68,73]. This process consists in the degradation of tomato wastes in the absence of oxygen at temperatures ranging between 300 and 900°C, heating rates of 0.1°C/min to 100°C/min and residence times of several seconds to few hours [68]. Before pyrolysis, tomato wastes are generally air-dried at least for 3 days, milled, and sieved to get particles size lower than 0.4 mm [68]. The pyrolysis process could be performed either by electric power or recently through solar energy [7].

Different raw tomato feedstock and pyrolysis conditions have been used for the production of biochars (Table 8.1). They include tomato crop green wastes [25], tomato harvest residues [26], tomato leaves and stem [68], tomato pruning [7], tomato peel [78], and tomato paste [53,73]. As for typical lignocellulosic materials, tomato wastes are mostly composed of cellulose and hemicellulose. For instance, tomato leaves and stems are constituted of 43.6% of cellulose, 25.7% of hemicellulose, 18.5% of ash, 11.6% of extractives, and 0.6% of lignin [68]. Raw tomato wastes contain relatively high carbon contents varying between 49.3% (leaves and stems) and 59.8% (pruning). Besides, high N contents were observed for tomato leaves and stems as well as tomato paste wastes with values of 4.2% and 3.8%, respectively (Table 8.1). Ash contents varied between 4.9% (tomato peels) and 22.0% (tomato green wastes) [25,65].

At the same time, a very wide range of experimental conditions has been tested during pyrolysis. Indeed, tomato wastes have been pyrolyzed at temperatures varying between 250 and 800°C [68], heating gradients from 8 and 30°C/min [7], and contact times from 40 min to 4 h [26]. Biochar yields from tomato wastes decrease with temperature as indicated in Table 8.1. For instance, increasing pyrolysis temperature from 250 and 800°C decreases biochar yields of tomato leaves and stems from 70% to 27% [68]. Besides, Prasad and Murugavelh [78] demonstrated that biochar production yields generated from the pyrolysis of tomato peels decreased from 44% to 26% when the pyrolysis temperature increased from 450 to 650°C. This behavior is due to the fact that higher is the used pyrolysis temperature higher is the decomposition efficiency of the contained volatile matter. For the same pyrolysis temperature, biochar yields seem to be very dependent on the used raw biomass as well. For instance, for a pyrolysis temperature of 600°C, biochar yields were determined to 35% for tomato leaves and stems [68] and to only 28% for tomato peels [78]. This is mainly attributed to differences in the physico-chemical characteristics of the used raw materials, especially the higher ash contents (Table 8.1) of tomato leaves and stems which is generally privileged at high nutrient content (e.g., Na, K, Ca, Mg, P) [35,99,118].

It is important to underline that the reported biochar yields generated from the pyrolysis of tomato wastes are comparable to the ones obtained from a range of lignocellulosic materials produced at similar temperatures like wood wastes, nut shells [16,54].

#### *8.2.1.2 Raw biochars characteristics*

Raw (non-activated) biochars generated from the pyrolysis of tomato wastes have been characterized using various analytical techniques in order to better assess their structural and textural properties as well as surface chemical characteristics [18,22,68,116]. These methods include the use of scanning electron microscope coupled with X ray (SEM/EDS), nuclear magnetic resonance (NMR), nitrogen physisorption, mercury porosimetry, X-ray fluorescence elemental analysis (XRF), Fourier transform infrared spectroscopy (FTIR), X-ray photoelectron spectroscopy (XPS), etc.

##### 8.2.1.2.1 Morphological and textural properties

Raw tomato wastes are generally arranged as platelet shape with very low porosity. The pyrolysis operation degrades volatile matter and generates biochar with more or less developed porosity depending on the used pyrolysis conditions (especially temperature). Fig. 8.1 compares the morphology of raw tomato wastes collected from a tomato processing factory (Marmara,

**Table 8.1** Effect of tomato waste types and pyrolysis conditions on biochar yields.

| | Raw feedstock properties | | | | | | | Pyrolysis conditions | | | | |
|---|---|---|---|---|---|---|---|---|---|---|---|---|
| Feedstock | C (%) | H (%) | O (%) | N (%) | Moisture | Volatile matter (%) | Ash (%) | T (°C) | G (°C/min) | t (h) | Biochar yield (%) | Reference |
| Tomato paste wastes | 49.7 | 7.4 | 39.10 | 3.8 | - | - | - | 350 | - | - | - | Önal et al. [73] |
| Tomato crop green waste | 55.0 | 1.6 | 43.2 | 0.2 | - | - | 22.0 | 550 | - | - | 38 | Dunlop et al. [25] |
| Leaves and stems of tomato | 49.3 | 7.4 | 31.8 | 4.2 | 10.4 | 86.4 | 18.1 | 250 | 10 | 1 | 70 | Mokrzycki et al. [68] |
| | | | | | | | | 300 | | | 56 | |
| | | | | | | | | 400 | | | 44 | |
| | | | | | | | | 500 | | | 37 | |
| | | | | | | | | 600 | | | 35 | |
| | | | | | | | | 800 | | | 27 | |
| Tomato Peel | 55 .0 | 7.9 | 34.0 | 2.8 | 4.7 | 78.1 | 4.9 | 450 | 20 | - | 44.0 | Prasad and Murugavelh [78] |
| | | | | | | | | 500 | | | 39.0 | |
| | | | | | | | | 550 | | | 26.4 | |
| | | | | | | | | 600 | | | 27.9 | |
| | | | | | | | | 650 | | | 26.0 | |

G, pyrolysis heating gradient; T, pyrolysis temperature; t, residence time.

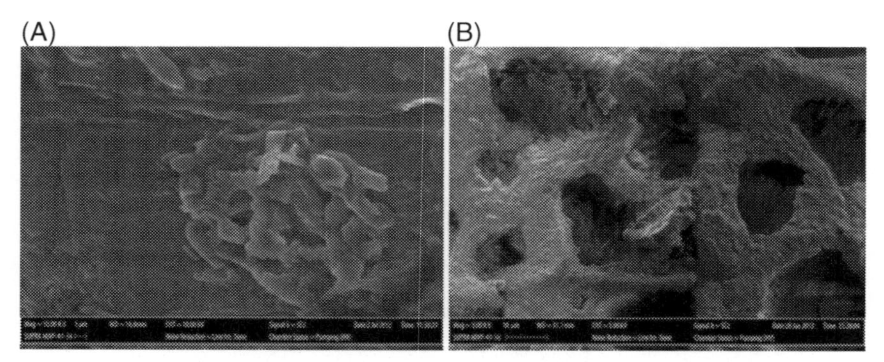


**Fig. 8.1** SEM images of raw tomato wastes (A) and its derived biochar at a temperature of 350°C (B). *Reproduced with permission from Önal et al.* [73].

Turkey) and its derived biochar at 350°C [73]. It can be clearly seen that the pyrolysis process has created a porous structure with different pore sizes. In addition, the biochar external surface has a sponge-like appearance with the presence of various cavities. The effect of the temperature on the structure of biochars generated from the pyrolysis of tomato pruning wastes is shown in Fig. 8.2. It appears that the increase of pyrolysis temperature from 450 to 600°C leads to more developed micro and meso-porosities. However, at an extreme pyrolysis temperature of 800°C, the microporous structure disappeared and bigger particles were formed leading to irregular morphologies.

Regarding the heating rate, Ayala-Cortés et al. [7] demonstrated that the use of low heating rate (8°C/min) instead of 30°C/min promotes the formation of microporous structure. However, a high heating rate showed the formation of macro and mesopores [7].

As for the structure of tomato wastes derived biochars, their textural properties depend not only on the nature of the raw feedstock but also on

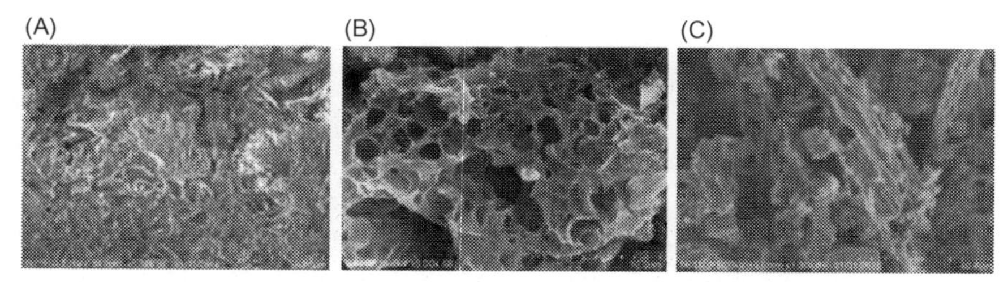


**Fig. 8.2** ***Effect of pyrolysis temperature and a constant heating rate of 30°C/min on tomato pruning biochar morphology.*** (A) raw biomass, (B) 450°C and (C) 600°C. *Reproduced with permission from Ayala-Cortés et al.* [7].

the used pyrolysis conditions [7]. Accordingly, for both tomato pruning and agave leaves, the BET surface areas increase with the increase of the solar pyrolysis temperature from 450 to 600°C. Compared to non-activated bio-chars from lignocellulosic materials, the tomato pruning derived biochars have important BET surface areas at a temperature of 600°C of 1100 $m^2/g$ [7]. These BET surface areas decrease to about 800 $m^2/g$ when increasing the pyrolysis temperature to 900°C. This behavior was attributed to the clogging of the micropores by precipitated salts.

However, when studying the pyrolysis of tomato leaves and stems at 250, 300, 400, 500, 600, and 800°C, Mokrzycki et al. [68] found that, on the basis of mercury porosimetry measurements, the total pore areas were relatively low for all biochars regardless of the pyrolysis temperature (13 to 19 $m^2/g$). Moreover, no significant impact of the pyrolysis temperature was observed on the total intrusion pore volume.

#### 8.2.1.2.2 Surface chemistry properties

The reactivity of biochars from non-activated tomato wastes depends mainly on its texture (especially specific surface area and porosity), mineral content, and surface functional groups nature and content. Fig. 8.3 shows the spectra of raw tomato wastes and its derived biochar produced at 350°C [73]. The spectra of the raw tomato wastes showed the presence of a broad and intense absorption band at 3280/cm related to the O-H stretching vibrations of absorbed water, cellulose and lignin [91]. This band disappeared after carbonization, which is most likely due to the vaporization of surface water and organic matter. Moreover, the peak observed at around 3000/cm was attributed to the asymmetric and

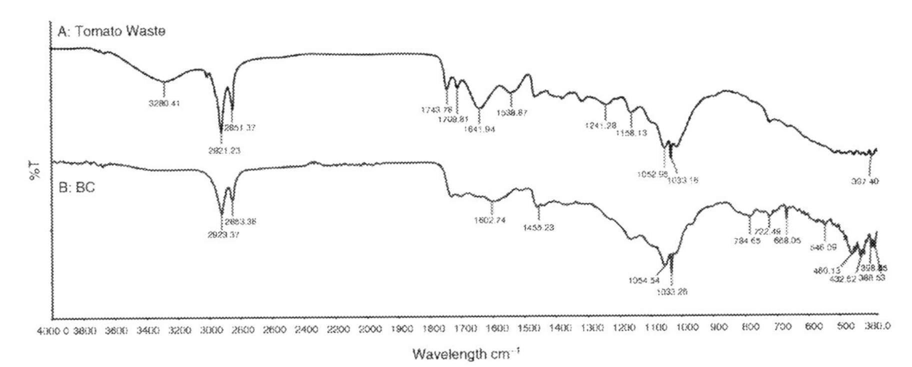


**Fig. 8.3** Infrared spectra of raw tomato wastes (A) and derived biochar at a temperature of 350°C (B). *Reproduced with permission from Önal et al.* [73].

symmetric of the aliphatic C–H, the intensities of these bands decreased after carbonization as well. The peaks observed in the raw tomato wastes at 2851 and 2921/cm were assigned to symmetric $-CH_2$ vibration and $-CH_2$ stretching vibration, respectively. For the case of biochars, these peaks were shifted to higher wavelengths. Bands between 1750 and 1640/cm were attributed to C = O stretching of carboxylic acids and aromatic C = C stretching [72]. Peaks located at 1158 and 1070/cm correspond to lactone groups and C-O stretching vibrations of lignin respectively [116]. This comparison study between the surface chemistry of the tomato waste and its derived biochar shows some significant differences especially a decrease in $OH^-$, $CH_2$ groups, and olefinic structure.

It is worth mentioning that the pH values of zero charge ($pH_{ZPC}$) of the produced biochars are generally much higher than those reported for the raw tomato wastes [68]. These $pH_{ZPC}$ generally increase with the temperature increase. For instance, Mokrzycki et al. [68] showed that the increase of the carbonization temperature of tomato leaves and stems from 250 to 800°C increased the $pH_{ZPC}$ value from about 8.5 to more than 13.2. This significant change was attributed to the increase of alkaline ash in the biochar at higher temperatures [99]. This result shows also that for acidic, neutral and slightly basic wastewaters, the biochars surface will be positively charged and therefore could efficiently remove anions such as fluorides, phosphates, nitrates [99,106].

## 8.2.2 Activated carbon

### *8.2.2.1 Activation procedures*

Activated carbons (ACs) can be produced from any natural or synthetic carbonaceous solid precursor and usually exhibit well-developed internal surface area and porosity [117]. In the last decade, the research on ACs synthesis has focused on the use of low-cost precursor from agricultural wastes or by-products [51,117]. The used precursor wastes include date pits waste [4] olive stone wastes [87], tea and coffee wastes [20,61], and tomato wastes [91]. The synthesis of ACs include three activation methods, namely physical, chemical and physicochemical as shown in Fig. 8.4. The physical activation consists in the carbonization of the biomass at relatively high temperatures (between 500 and 800°C) in presence or in absence of oxygen (pyrolysis). Then, the produced char is physically activated at high temperatures (700–1000°C) through various agents such as $O_2$, $CO_2$, stem [83]. The chemical activation consists in a simultaneous impregnation

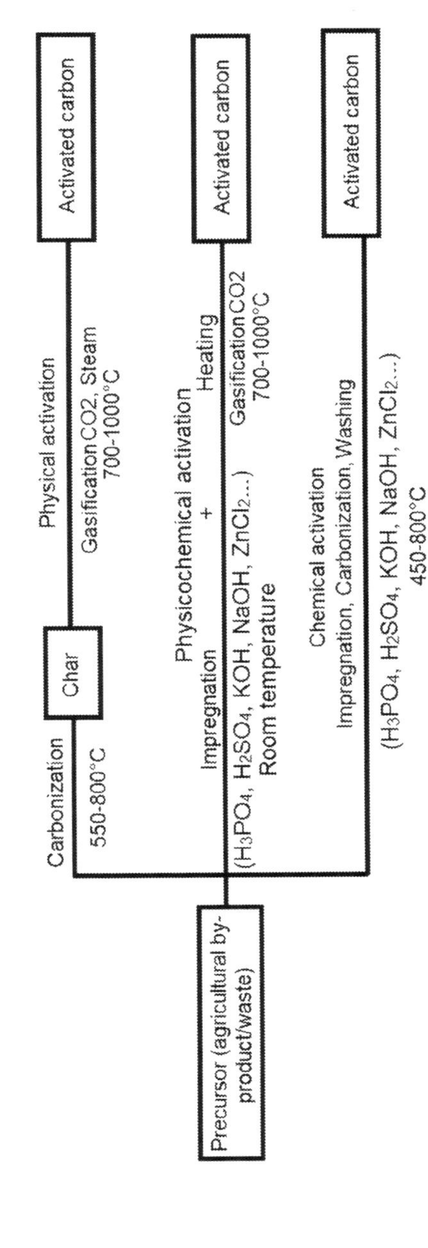


**Fig. 8.4** *Schematic representation of activated carbon manufacturing process.*

and carbonization at temperatures varying between 450 and 800°C [21]. The physicochemical activation consists in two steps process including impregnation followed by carbonization in presence of an oxidant flow [74]. The impregnation step is performed by an optimized mixing of the biomass with various chemicals (acids [$H_3PO_4$, $H_2SO_4$, HCl, $HNO_3$…], bases [KOH, NaOH…], salts [$ZnCl_2$, $FeCl_2$…], etc.) at ambient or low temperature. Then, the impregnated biomass is carbonized at temperatures between 700 and 1000°C. All these methods permit an important release of volatile compounds and the recovery of stable carbonaceous phase with a very well-developed surface area and porosity.

The physicochemical properties of the produced AC depend mainly on the raw biomass characteristics, the activation agent, the carbonization conditions especially the final temperature, the heating rate, and the residence time [57,67,117].

#### *8.2.2.2 Synthesis*

The preparation of ACs from tomato waste generally starts by an air-drying step at ambient temperature, crushing, and sieving in order to obtain particles with relatively similar sizes (~0.5 mm). The following step could be the carbonization which produces a stable structure, with an elementary and partially-developed pore structure [75]. As previously mentioned, the carbonized product is then activated by either physical, chemical or physico-chemical methods. Activation is necessary in order to remove the in-pore deposited decomposed products and tars. Different activation methods have been used for the synthesis of ACs from tomato wastes (Table 8.2). They include the use of $HNO_3$ [107], $H_3PO_4$ [112], KOH [76], NaOH [41], $ZnCl_2$ [18], and $FeCl_2$ [27]. As an example, Fu et al. [27] tested the feasibility of a novel activating agent ($FeCl_2$) in order to prepare activated carbons from tomato stems (TS). In this investigation, the impregnation ratios ($FeCl_2$/TS) ranged from 1/1 to 3/1. Several pyrolysis temperatures were also tested (from 500 to 800°C) at a heating rate and residence time of 10°C/min and 1 h in a conventional tubular furnace under a continuous nitrogen flow (100 $cm^3$/min). As an outcome, the optimum textural properties were obtained at 700°C.

Other studies on ACs have addressed the replacement of classic pyrolysis processes (electric ovens) by the use of microwave ovens in order to reduce the reaction time as well as energy consumption [112]. Results have been very promising regarding both the energy consumption reducing and also the activated carbon enhanced properties [72, 112].

**Table 8.2** Textural properties of different activated biochars prepared from tomato wastes.

| Precursor | Activating agent | Activation conditions | $S_{BET}$ ($m^2/g$) | $V_T$ ($cm^3/g$) | $V_{mic}$ ($cm^3/g$) | DP (nm) | Reference |
|---|---|---|---|---|---|---|---|
| Tomato paste waste | $ZnCl_2$ | T = 500°C; G = 10°C/min; t = 1h; IR = 1:1 | 722 | 0.476 | 0.201 | 2.64 | Güzel et al. [32] |
| Tomato waste | $ZnCl_2$ | T = 500°C; G = 10°C/min; t = 1h; IR = 1:1 | 617 | 0.437 | 0.313 | 2.64 | Sayğili and Güzel [91] |
| | | T = 500°C; G = 10°C/min; t = 1h; IR = 4:1 | 760 | 0.710 | 0.584 | 4.83 | |
| | | T = 500°C; G = 10°C/min; t = 1h; IR = 6:1 | 787 | 1.000 | 0.277 | 5.78 | |
| | | T = 400°C; G = 10°C/min; t = 1h; IR = 6:1 | 648 | 0.756 | 0.086 | 4.56 | |
| | | T = 600°C; G = 10°C/min; t = 1h; IR = 6:1 | 1093 | 1.569 | 0.129 | 5.92 | |
| | | T = 600°C; G = 10°C/min; t = 0.5h; IR = 6:1 | 522 | 0.662 | 0.191 | 5.02 | |
| | | T = 600°C; G = 10°C/min; t = 2h; IR = 6:1 | 1034 | 1.451 | 0.440 | 5.61 | |
| Tomato paste waste | KOH | T = 500°C; G = 15°C/min; t = 1h; IR = 0.25:1 | 157 | 0.092 | 0.043 | - | Ozbay and Yargic [76] |
| | | T = 500°C; G = 15°C/min; t = 1h; IR = 0.5:1 | 283 | 0.154 | 0.082 | - | |
| | | T = 500°C; G = 15°C/min; t = 1h; IR = 1:1 | 143 | 0.089 | 0.039 | - | |
| | $K_2CO_3$ | T = 500°C; G = 15°C/min; t = 1h; IR = 0.25:1 | 63 | 0.048 | 0.015 | - | |
| | | T = 500°C; G = 15°C/min; t = 1h; IR = 0.5:1 | 185 | 0.116 | 0.068 | - | |

*(continued)*

**Table 8.2** (Cont'd)

| Precursor | Activating agent | Activation conditions | $S_{BET}$ (m$^2$/g) | $V_T$ (cm$^3$/g) | $V_{mic}$ (cm$^3$/ g) | DP (nm) | Reference |
|---|---|---|---|---|---|---|---|
| | | T = 500°C; G = 15°C/min; t = 1h; IR = 1:1 | 221 | 0.135 | 0.075 | - | |
| | HCl | T = 500°C; G = 15°C/min; t = 1h; IR = 0.25:1 | 43 | 0.038 | 0.012 | - | |
| | | T = 500°C; G = 15°C/min; t = 1h; IR = 0.5:1 | 64 | 0.049 | 0.016 | - | |
| | | T = 500°C; G = 15°C/min; t = 1h; IR = 1:1 | 77 | 0.056 | 0.023 | - | |
| Tomato stems | $H_3PO_4$ | Microwave heating, 900 W; t = 3 min; IR = 2:1 | 813 | 0.991 | 0.1278 | - | Yagmur [112] |
| Tomato leaves | $H_3PO_4$ | Microwave heating, 900 W; t = 2 min; IR = 2:1 | 117 | 0.291 | 0.01 | - | |
| Tomato stem | $FeCl_2$ | T = 500°C; G = 10°C/min; t = 1h; IR = 2.5:1 | 474 | 0.265 | 0.229 | 2.186 | Fu et al. [27] |
| | | T = 600°C; G = 10°C/min; t = 1h; IR = 2.5:1 | 570 | 0.343 | 0.268 | 2.343 | |
| | | T = 700°C; G = 10°C/min; t = 1h; IR = 2.5:1 | 971 | 0.576 | 0.425 | 2.78 | |
| | | T = 700°C; G = 10°C/min; t = 1h; IR = 1:1 | 669 | 0.331 | 0.293 | 2.223 | |
| | | T = 700°C; G = 10°C/min; t = 1h; IR = 1.5:1 | 750 | 0.395 | 0.325 | 2.451 | |
| | | T = 700°C; G = 10°C/min; t = 1h; IR = 2:1 | 879 | 0.470 | 0.385 | 2.593 | |

DP, average pore diameter; IR, impregnation mass ratio (activating agent:biomass); $S_{BET}$, specific surface area; $V_T$, total pore volume; $V_{mic}$, micropore volume.

### 8.2.2.3 Characterization of activated carbons

#### 8.2.2.3.1 Morphological properties

The morphological properties of tomato waste-derived activated carbon depend not only on the nature of the raw feedstock but also on the type and conditions of the activation process. Figs. 8.5 and 8.6 represent SEM micrographs of tomato paste wastes and their activated carbon with $ZnCl_2$ and KOH, respectively. Fig. 8.5 clearly shows that the activation processes leads to the formation of more porous structure with the presence of various pores with different sizes.

As for $ZnCl_2$, the activation of tomato paste waste with KOH enhances the microporosity and surface area development. Fig. 8.6 shows SEM images of the used raw tomato wastes in comparison with the activated forms with KOH at two impregnation mass ratios (activating agent/biomass) of 25% and 50%. It can also be clearly seen that the activation process has formed large cavities and rough surfaces. At high impregnation ratio (50%), the activated carbon exhibited discontinuous and irregular surface which might be induced by the decomposition of the sample matrix by the KOH followed by the dehydrating action of them during heat treatment.

#### 8.2.2.3.2 Textural properties

The textural characteristics of the generated ACs from tomato wastes depend on both the nature of the tomato wastes precursor as well as the activation conditions (activation agent nature and concentration, pyrolysis temperature, residence time etc.) (Table 8.2). The most interesting textural properties with the highest surface area, total and microporous pore volume of 1093 $m^2/g$, 1.569 and 0.129 $cm^3/g$, respectively were found for an activation of tomato paste wastes with $ZnCl_2$ at a temperature of 600°C and

**Fig. 8.5** ***SEM micrographs of raw tomato paste waste (TW) and its derived activated carbon with $ZnCl_2$ (TAC).*** *Reproduced with permission from Sayğili and Güzel* [91].

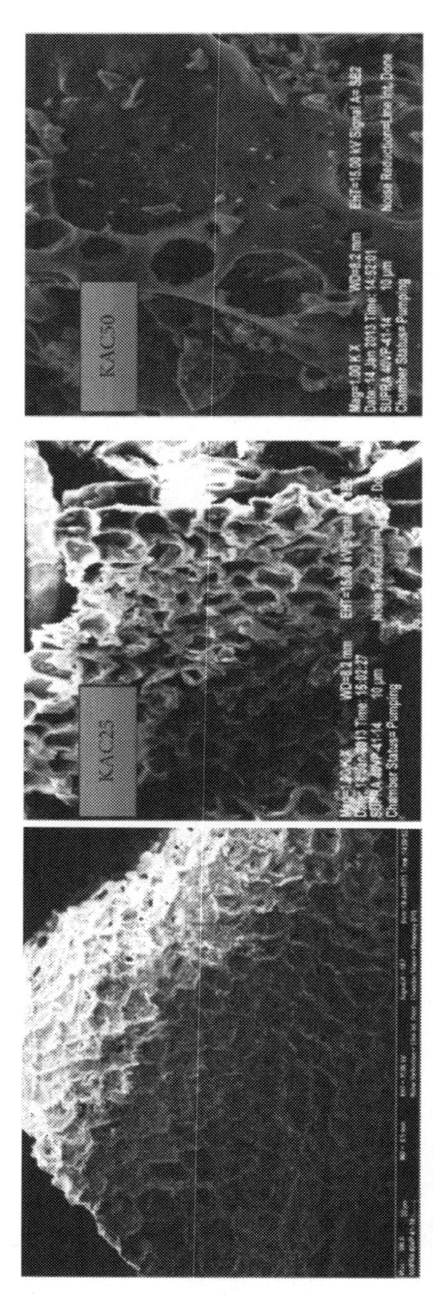


**Fig. 8.6** SEM images of tomato paste waste (left) and its activated carbon with KOH at two mass impregnation ratios (KOH/biomass) (ratio = 25%: KAC25 and ratio = 50%: KAC50). *Reproduced with permission from Ozbay and Yargic* [76].

an impregnation mass ratio (IR) ($ZnCl_2$:biomass) of 6:1 [91]. Impregnation of tomato stems with $FeCl_2$ showed also interesting textural properties with surface area, total pore and microporous pore volumes of 917 $m^2/g$, 0.576 and 0.425 $cm^3/g$, respectively [27]. Activation with $H_3PO_4$ [112] could be considered also as an attractive method that gives interesting textural properties (Table 8.2). Ozbay and Yargic [76] studied the activation of tomato paste wastes by three chemicals: KOH, $K_2CO_3$, and HCl at three IR (activating agent:biomass of 1:4; 1:2, and 1:1) at a fixed pyrolysis temperature and residence time of 500°C and 1 h, respectively. They found that KOH was the best activating agent since the AC for an IR of 1:2 had the highest surface area, total and micro pore volumes of 283 $m^2/g$, 0.154 and 0.082 $cm^3/g$, respectively (Table 8.2).

On the other hand, Sayğili and Güzel [91] studied the impact of the impregnation mass ratio, the pyrolysis temperature and the residence time on textural parameters of produced activated carbon. They showed that the higher the IR is, the more enhanced are the textural properties. Indeed, for constant temperature of 500°C and contact time of 1 h, the $S_{BET}$, $V_T$, and DP increase by about 28%, 129%, and 119% when the IR increased from 1:1 to 6:1. Globally similar behavior was observed by Ozbay and Yargic [76] when studying the impact of IR of KOH, $K_2CO_3$, and HCl on the textural properties of produced activated carbon (Table 8.2). Moreover, for a constant IR of 6:1 and a residence time of 1 h, when the temperature increases from 400 to 600°C, these parameters increase by about 69%, 108%, and 30%, respectively [91]. Similar behavior was observed by Fu et al. [27] who investigated the effect of pyrolysis temperature on activated tomato wastes by $FeCl_2$ at a fixed IR of 2.5:1 (Table 8.2). Finally, for a fixed pyrolysis temperature of 600°C and IR of 6:1, these three textural parameters increased by about 98%, 119%, and 12%, respectively when the residence time passed from 0.5 to 2 h. This improvement was mainly imputed to the formation of new pores because of the increase of volatile matter decomposition.

The textural characteristics of activated carbons were identified by the $N_2$ adsorption-desorption approach [27,32,91]. Fig. 8.7 shows clearly the importance of the used pyrolysis temperature (between 500 and 800°C) and the IR variation between 1:1 and 3:1 on the porosity and surface properties of activated carbons produced from tomato stems (TS) [27]. The impregnation step generally leads to the hydrolysis of the organic matter. Furthermore, the activating agent occupies a volume between biomass particles, which inhibits the contraction of the particles during the heat

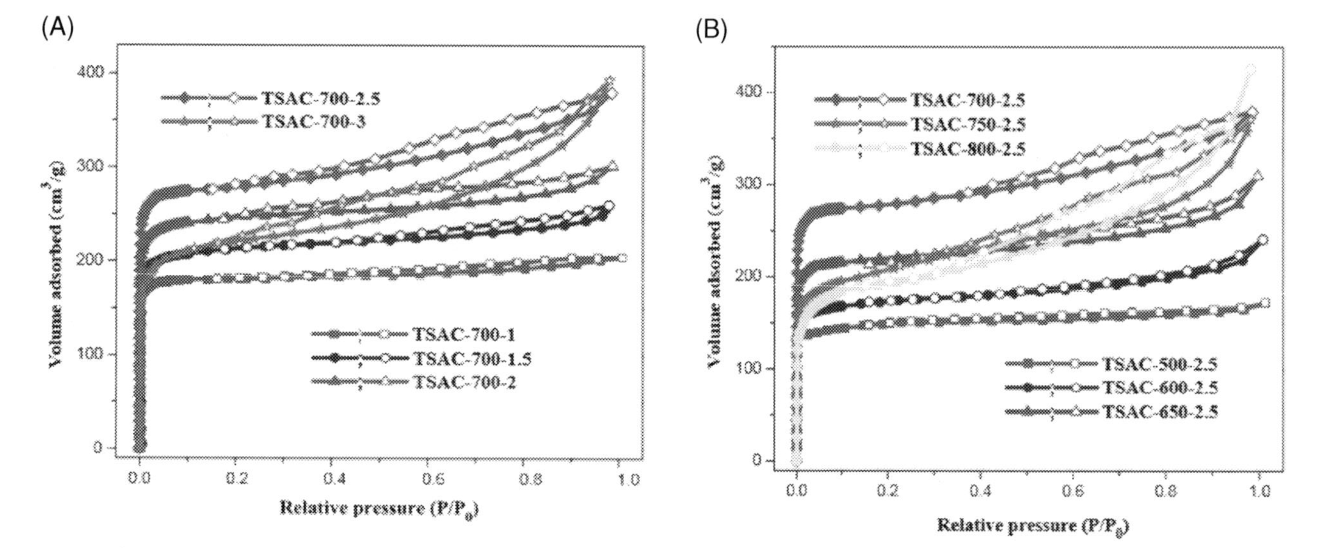


**Fig. 8.7** ***$N_2$ adsorption-desorption isotherms of tomato wastes activated biochars*** (A) prepared at different impregnation ratios with a fixed activation temperatures of 700°C and (B) prepared at different activation temperatures for a constant impregnation of 2.5:1. *Reproduced with permission from Fu et al.* [27].

treatment. Consequently, this treatment leaves a significant porosity after carbonization and washing with water.

As illustrated in Figs. 8.7A and B, the $N_2$ adsorption isotherms of all samples can be divided in two parts: at low pressure, the plots are of type I; at moderate and high pressure, the curve is of type IV. The adsorption isotherms exhibited steep increases in nitrogen uptake parallel to Y-axis at low relative pressures, which are due to the micropore filling, indicating the development of microporosity in the ACs materials. The hysteresis loops for all samples show the same form (type H4) and confirm the dominance of mesoporosity silt-shaped (IUPAC nomenclature, [96]). Besides, the same forms of $N_2$-isotherms curves were found by Sayğili and Güzel [91] when investigating tomato waste activation with $ZnCl_2$. However, for samples activated at a high temperature of 800°C (TSAC-800-2.5, Fig. 8.7B), the isotherm showed a steep $N_2$ adsorption at relative pressure close to one. This behavior was attributed to the widening of micropores and mesopores and the formation of macropores.

Fig. 8.8 presents pore size distributions (PSDs) of these tomato waste activated biochars [27]. Pore size distributions (width) seems to be significantly influenced by the activation temperature, as well as, the impregnation ratio. Indeed, larger pores were found for higher temperatures and IRs. In addition, a significant fraction of pores for all the produced activated carbon samples have dimensions below 1 nm confirming the dominance of microporosity. The ACs prepared by Sayğili and Güzel [91] using ZnCl2 as an activating agent, have higher surface area but larger pore width (Table 8.2).

As conclusion, the choice of the operating activation conditions, especially the impregnation ratios, the activation temperature and the residence time, is very decisive for the production of activated carbons with well-developed porosities and surface areas. For tomato wastes, the these parameters importance seems to be in the following order: carbonization temperature > residence time > impregnation ratio [27,91]. These results indicate that tomato wastes could be used as a promising alternative precursor for expensive commercial activated carbons.

#### 8.2.2.3.3 Surface chemistry properties

The surface chemistry of the activated carbons prepared from tomato waste has been characterized by several analytical techniques such as FTIR, XRD, XPS, Boehm titration, surface charge density, zeta potential. These surface properties are very influenced by the nature of tomato waste, as well as, the

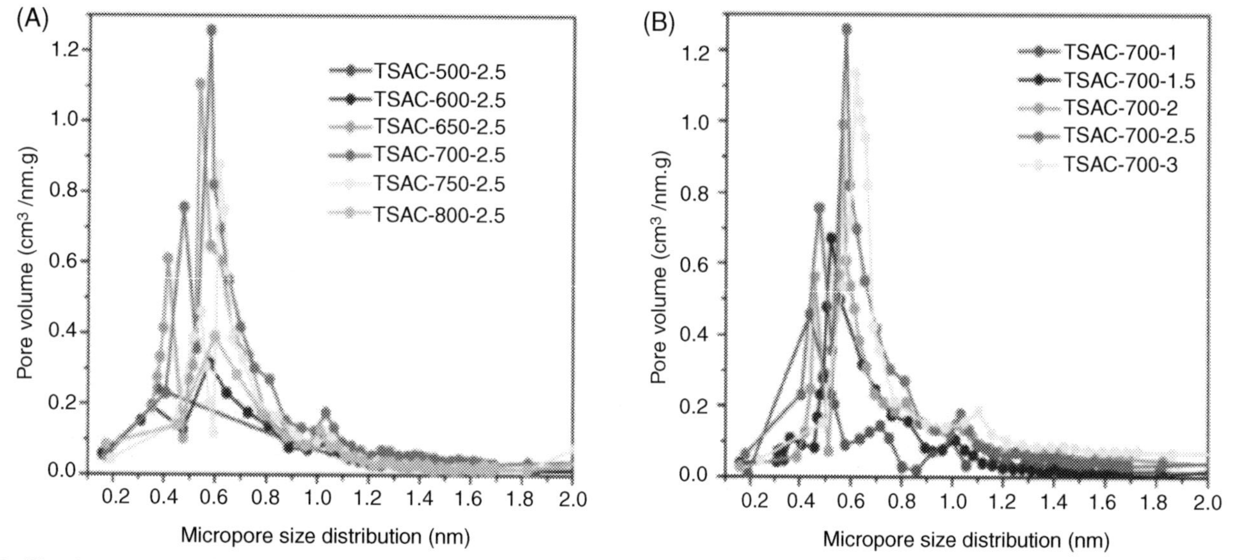


**Fig. 8.8** (A) micropore size distributions of activated carbons prepared at different activation temperatures with a fixed impregnation ratio of 2.5:1, (B) micropore size distributions of activated carbons prepared with different IR at a constant activation temperature of 700°C. *Reproduced with permission from Fu et al.* [27].

activation conditions (temperature, heating rate, residence time, and activation process).

The $pH_{ZPC}$ of the produced activated carbon depends mainly on the origin of the raw tomato waste in addition to the activating agent and pyrolysis experimental conditions. For instance, Ozbay and Yargic [76] showed that the activated carbon derived from the activation of tomato paste wastes by KOH at a temperature of 500°C, a residence time of 1 h, and an IR of 1:2 (KOH:biomass), has a $pH_{ZPC}$ of 2.8. This value is very low and indicates that its main surface functional groups are acidic. An acidic $pH_{ZPC}$ of 5.2 was also reported by Güzel et al. [32] for an activated carbon derived from tomato waste activation with $ZnCl_2$ at an impregnation ratio of 1:1, a pyrolysis temperature of 500°C and a residence time of 1 h. Boehm titration confirmed this acidic character since the contents of the acidic surface groups (phenolic, lactonic and carboxylic) and the basic groups were assessed to 1.33 and 0.95 meq/g, respectively. A slightly acidic $pH_{ZPC}$ (6.17) was found by Sayğili and Güzel [91] for an activated carbon derived from the pyrolysis at a temperature of 600°C and a residence time of 1 h of impregnated tomato wastes with $ZnCl_2$ at 6:1. The Boehm titration showed total acidic group content of 1.17 meq/g, which is about 12.5% higher than the basic groups (1.04 meq/g).

The FTIR analysis spectra of tomato waste-derived activated carbon compared to the raw feedstock demonstrates that the activation step contributes to the disappearance of several absorption bands due to the vaporization of organic matter at high temperatures [91,112]. Indeed, only few absorption bands remain after the activation process. They include oxygen-containing groups such as carboxylic and phenolic functional groups. In addition, XPS analysis indicates that significant differences existed on the percentage amounts of carbon and oxygen fractions after the carbonization procedure, which is in agreement with the FTIR and Boehm titration results.

On the other hand, the activation process of tomato wastes generally leads to amorphous to low crystalline structures. Indeed, the crystallization process would be inhibited by the functional groups existing on the surface of the ACs [32]. However, XRD analysis generally show that the activation process reduces the peak related to cellulose (observed at $2\theta = 21.8$). Other peaks linked to the used activating agent could be detected. For instance, when applying $ZnCl_2$, Sayğili and Güzel [91] observed new peaks related to zinc oxide and zinc carbides.

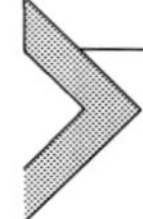

## 8.3 Use of tomato wastes for pollutant removal from aqueous solutions

### 8.3.1 Case of inorganic pollutants

The application of various types of raw (TW) and modified tomato wastes (MTW) (nonactivated biochar and activated carbon), as low-cost adsorbents for inorganic pollutant removal from aqueous solutions is currently of great interest. Indeed, various forms of raw and modified tomato wastes bio-sorbents are becoming the focus of several recent research studies. With relatively high carbon amounts and developed porous structure, modified tomato wastes have been explored in order to replace conventional commercial activated carbon and examined for the removal of various heavy metals such as Ni(II), Co(II), Cu(II), Pb(II), Fe(III), Mn(II), Cr(III), Cr(VI), As(III) [18,28,30,41,64,68,73,76,77,79,101,107,116], and inorganic phosphates [106,113]. This section summarizes and discusses results published data regarding the adsorption characteristics of several inorganic pollutants onto TW and MTW adsorbents under various operating conditions (pH, contact time, adsorbent dosage, temperature, and initial inorganic pollutants concentrations). It presents also the main physicochemical properties of the used adsorbents as well as the main involved mechanisms. Table 8.3 summarizes tomato wastes pretreatment, modification conditions, post-treatment, surface area analysis, adsorption conditions, the best fitting kinetic and isotherm models, and finally the estimated Langmuir adsorption capacities.

#### *8.3.1.1 Use of raw tomato wastes*

Gutha et al. [30] investigated the adsorption potential of raw tomato leaf powder as a low-cost agricultural waste biomass for the removal of Ni(II) ions from aqueous solution in batch mode. Batch experiments were performed for initial Ni(II) concentrations, adsorbent dosages, aqueous pH values and temperatures ranging respectively from 30 to 90 mg/L, 0.1 to 0.6 g/L, 2 to 8°C and 30 and 50°C. The contact time permitting to reach an equilibrium state was found to be about 3 h. Based on experimental data, the best fitting, was given by the Langmuir isotherm, and the estimated maximum adsorption capacities of Ni(II) was found to be very dependent on the used temperature. The Langmuir adsorption capacity of Ni(II) onto raw tomato leaves reached 47.6, 52.6, and 58.8 mg/g at temperatures of 30, 40, and 50°C, respectively. The kinetic experimental data were well fitted to the pseudo second order (PSO) model assuming a chemisorption adsorption process. The thermodynamic study indicated

**Table 8.3** Summary of inorganic pollutant removal by tomato wastes.

| Feedstock, origin | Pretreatment | Modification conditions | Post-treatment | Surface area ($m^2/g$) | Pore volume ($cm^3/g$) | Pore width (nm) | Adsorption conditions | Pollutant | Kinetic model | Isotherm | Langmuir $q_{max}$ (mg/g) / ARE (%) | Reference |
|---|---|---|---|---|---|---|---|---|---|---|---|---|
| Tomato leaves, Kammapalli, India | Sun-drying for 2 days, grinding, washing filtration, and drying for 6h at 60°C | - | - | 8.80 | 0.003 | 15.59 | $C_0$ = 30-90 mg/L; D = 0.4 g/L; pH = 5.5; t = 3 h; T = 50°C | Ni(II) | PSO | Langmuir | 58.82 | Gutha et al. [30] |
| Tomato peels from local market, Singapore[a] | Boiling for 10 min in water, Peels washing with 2-propanol, and drying | - | - | - | - | - | $C_0$ = 5-200 mg/L; D = 10 g/L; pH = 2 (As), pH = 4 (Cr), pH = 8 (Pb, Ni), t = 6 h; T = 30°C | Pb(II)<br>Ni(II)<br>As(III)<br>Cr (VI) | PSO | Freundlich | 0.2131<br>0.1101<br>0.0173<br>0.0442 | Mallampati and Valiyaveettil [64] |
| Tomato wastes from Marmara region, Turkey | Washing, air-drying, grinding, and sieving | Carbonization at 350°C | - | - | - | - | $C_0$ = 25-125 mg /L; D = 5 g/L; pH = 7; t = 1 h; T = 20°C | Co(II) | PSO | Langmuir | 166.67 | Önal et al., [73] |

*(continued)*

**Table 8.3** (Cont'd)

| Feedstock, origin | Pretreatment | Modification conditions | Post-treatment | Surface area ($m^2/g$) | Pore volume ($cm^3/g$) | Pore width (nm) | Adsorption conditions | Pollutant | Kinetic model | Isotherm | Langmuir $q_{max}$ (mg/g) / ARE (%) | Reference |
|---|---|---|---|---|---|---|---|---|---|---|---|---|
| Waste tomato leaves and Stems, Zgorzelec, Poland | Air drying for 3 days, grinding, and sieving | Pyrolysis at temperatures between 250–800°C, heating rate of 10°C/mi, residence time = 1h | - | 18 | 0.828 | 12.8 | $C_0$ = 25–300 mg/L; D = 0.16 g/L; pH = 5; t = 2 h | Cr (III) | PSO | Langmuir | 169.5 | Mokrzycki et al. [68] |
| Tomato tissues leaves, Florida, United States | Drying and grinding | Enrichment with Mg, and pyrolysis at a temperature of 600°C for 1 h | Washing and drying | - | - | - | $C_0$ = 30 mg/L; D = 2 g/L; t = 24 h; T = 22°C | Inorganic P | - | - | 88.5% | Yao et al. [113] |
| Tomato waste (mixture of peel, seeds, pulp) from a local shop, India | Washing and drying at 110°C for 24 h | Impregnation with $ZnCl_2$ at a mass ratio of 1:1 at a temperature of 40°C and 140 rpm for one day, filtration and drying at 110°C for 12 h, then pyrolysis at a temperature of 550°C with a residence time of 2 h: AC1 | Washing with 0.2 N HCl and distilled water, and drying at 110°C for 12 h | - | - | - | $C_0$ = 0.01 M; D = 0.6 g/L; pH = 7; t = 2 h | Co(II) | PSO | Langmuir | 170.06 | Changmai et al. [18] |

| | | | | | | | | | | | | |
|---|---|---|---|---|---|---|---|---|---|---|---|---|
| | | Mixture of 0.4 g of AC1 with 0.8 g of an activated FeCl3-polyethylene terephthalate at a temperature of 500°C for 2 h | Filtration and drying | | | | $C_0$ = 0.01 M; D = 0.8 g/L; pH = 7; t = 2 h | | | | 312.50 | |
| Tomato paste waste from food factory in Bursa, Turkey | Drying, grinding and sieving | Activation with KOH in N2 atmosphere at 500°C for 1 h and a heating gradient of 15°C/min | Washing and drying | 283 | 0.154 | - | $C_0$ = 100 mg/L; D = 2 g/L; pH = 8; t = 1 h; T = 20°C | Co(II) | - | - | 44.68[a] | Ozbay and Yargic [76] |
| Tomato processing waste pomace, Parvomay, Bulgaria | Grinding and sieving | Extraction of carotenoids: wrapping with aluminum foil, stirring in acetone at solid-liquid ratio of 1:35 for 90 min at 50°C | Filtration, washing and oven-drying at 60°C for 2 h | - | - | - | $C_0$ = 10-100 mg/L; D = 4g/L; pH = 6; t = 1 h; T = 25°C | Cu(II) | PSO | Langmuir Freundlich | 181.82 | Prokopov et al. [79] |

(continued)

**Table 8.3** (Cont'd)

| Feedstock, origin | Pretreatment | Modification conditions | Post-treatment | Surface area ($m^2/g$) | Pore volume ($cm^3/g$) | Pore width (nm) | Adsorption conditions | Pollutant | Kinetic model | Isotherm | Langmuir $q_{max}$ (mg/g) / ARE (%) | Reference |
|---|---|---|---|---|---|---|---|---|---|---|---|---|
| Tomato wastes, Shiraz, Iran | Grinding, sieving, washing and drying | Oxidation with HNO3 (1M) for 2 h at a temperature of 50°C | - | - | - | - | $C_0$ = 25–100 mg/L; D = 10 g/L; pH = 6; t = 1 h; T = 25°C | Cu(II) | PSO | Langmuir Freundlich | 25 | Vafakhah et al. [107] |
| Tomato wastes, Marmara, Turkey | Drying, grinding and sieving | Treatment with 3% v/v HCl solution, washing and drying | - | - | - | - | $C_0$ = 25–125 mg/L; D = 4 g/L; pH = 8; t = 1 h; T = 20°C | Cu(II) | PSO | Langmuir | 34.48 | Yargiç et al. [116] |
| Tomato wastes, local market, Indonesia | Grinding, washing, drying, and sieving | Dried-tomato waste poured in NaOH at 1:10 (m/v) and stirred for 24 h at room temperature | Washing, filtration, drying and sieving | 45.00 | $4.47.10^{-2}$ | 1.985 | $C_0$ = 20-100 mg/L; D = 1 g/L; pH = 4; t = 1.5 h; room temperature | Pb(II) | PSO | Langmuir Freundlich | 152.0 | Heraldy et al. [41] |
| Green tomato husk from a local market, Mexico[b] | Washing, drying, grinding, and sieving | Treatment with 0.2% of formaldehyde solution | - | 0.68 | 0.0015 | - | $C_0$ = 2.5–300 mg/L; D = 10 g/L; pH = 6; t = 24 h; T = 20°C | Mn (II) Fe(III) | PSO | Freundlich Mn (II) Langmuir Freundlich Fe (III) | 15.22 Mn (II) 19.83 Fe (III) | García-Mendieta et al. [28] |

[a] experimental $q_{max}$
[b] Freundlich $K_F$.
ARE, average removal efficiency.

that the adsorption of Ni(II) onto raw tomato leaves was a spontaneous and endothermic process [30].

A second study by Mallampati and Valiyaveettil [64] investigated the use of raw tomato peels for the removal of various heavy metals (Pb(II), Ni(II), As(III), and Cr(VI) under static conditions. They showed that tomato peels exhibited efficient adsorption towards these cationic inorganic pollutants (Table 8.3). Indeed, the adsorbed amounts significantly increased with the increase of the initial concentrations and time. An equilibrium state characterized by approximately constant adsorbed amounts was observed after a contact time of 6 h. These kinetic data were well fitted with the second-order model indicating that the adsorption of these metals might be mainly chemical. Moreover, Freundlich isotherm was found to be better than Langmuir model in fitting the experimental data suggesting a heterogeneous and multilayer adsorption process. Other experimental operating factors such as the aqueous pH and temperature had also an influence on metal removal efficiencies. As such, the removal of Ni(II) and Pb(II) was found to increase as the pH increases. Such behavior could be explained by the ionization of the surface acidic functional groups enhancing therefore the adsorption of positively charged cations through electrostatic attraction. Low pH values resulted in a higher adsorption capacity of the negatively charged pollutants (Cr(VI) and As(III)) due to the electrostatic attraction with the positively charged peels surface for pH lower than the $pH_{ZPC}$ [108]. Metal removal efficiencies were in the following order Pb(II) > Ni(II) > Cr(VI) > As(III) (Table 8.3). The presence of high percentages of acid and alcoholic functional groups on the surface of the tomato peels may explain such behavior. Desorption and regeneration process were also investigated for Ni(II) and Pb(II). Results showed that 96% of these pollutants were desorbed at pH 4 within only 10 min. Such results may be due to the cationic exchange process with $H^+$ that it is favored in acidic media. Furthermore, experimental results indicated that tomato peels can be reused for five consecutive adsorption-desorption cycles of Ni(II) and Pb(II) without significant loss of their adsorption capacities [64].

#### *8.3.1.2 Use of tomato waste biochars*

Various studies [68,73,76,113] have tested the use of tomato-derived biochars under different pyrolysis conditions for the removal of Co(II), Cr(III) and inorganic phosphates. In this context, Önal et al. [73], evaluated the performance of a biochar produced from the pyrolysis of tomato paste waste at a temperature of 350°C for the removal of cobalt from synthetic aqueous solutions. They tested the effect of different experimental conditions (pH, amount of adsorbent, initial concentration of the aqueous solution, contact time, and temperature). The experimental results showed that, the Langmuir maximum adsorption capacity was evaluated to about 167 mg/g for initial cobalt concentration range of 25–125 mg/L, a contact time of 60 min, a temperature of 20°C and a neutral pH. Adsorption of Co(II) increased with the increase of pH, adsorbent dosage, initial Co(II) concentration, and temperature. Furthermore, they pointed out that Co(II) adsorption process occurs mainly through complexation with the various functional groups formed during the pyrolysis process. The carbonization process seems to significantly enlarge pores width and increase its surface area. When the pH values were increased ($pH > pH_{ZPC}$), adsorbent surface became more negatively charged and therefore adsorption of metal ions was increased due to electrostatic attraction. Furthermore, the increase of the used adsorbent dosage has significantly increased the Co(II) adsorption efficiency. This behavior was attributed to the increase of the available adsorption sites on the adsorbent surface. Moreover, higher initial concentration provides increased driving force to overcome mass transfer resistance of metal ions from the aqueous to the solid phase resulting in higher probability of collision between Co(II) ions and biochars particles [9,49]. The increase of adsorption capacity with temperature may be related to the increase in the number of active surface sites available for adsorption on biochars. Finally, kinetic and equilibrium experimental data showed that they were well fitted by PSO and Langmuir models (Table 8.3).

Besides, biochars generated from tomato waste green leaves and stems through pyrolysis at various temperatures ranging between 250 and 800°C were tested for Cr(III) ion removal from synthetic aqueous solutions [68]. Experimental batch results proved that increasing pyrolysis temperature positively affected sorption of Cr(III) ions. The related sorption capacities of the used biochars increased from 62 mg/g at 250°C to more than 169 mg/g at 800°C due to increases of total surface area, average pore diameter and $pH_{ZPC}$ from 13 to 18 $m^2/g$, 8 to 12.8 nm and 8.45 to 13.23 respectively. Besides precipitation as $Cr(OH)_3$, Cr(III) ions were removed through cation exchange with some mineral present on the surface of the used adsorbent. Cr(III) sorption kinetic was examined using several models, while the best correlation with experimental data was achieved for PSO model. A desorption study with 0.1 mol/L HCl was conducted for a regeneration purpose of biochars. It showed that the sorption capacity significantly decreased from 160 mg/g in first cycle to about 39 mg/g in the second one. This behavior was imputed to ash content decrease as well as the net decrease of the precipitation process as pH decreases.

An innovative method was developed by Yao et al. [113] to produce engineered biochars from magnesium-calcium (Mg/Ca) enriched tomato tissues through slow pyrolysis. The efficiency of these biochars in removing inorganic phosphates from synthetic aqueous solutions was assessed in batch mode. The Mg-enriched biochar composites showed better removal of inorganic phosphates in aqueous solutions (88.5%) compared to the Ca-enriched biochar, laboratory control biochar and farm control biochar. This important removal efficiency of inorganic phosphates from aqueous solutions was observed for an initial $PO_4^{3-}$ concentration and adsorbent dosage of 30 mg/L and 2 g/L, respectively. These outcomes were in agreement with previous studies by Yao et al. [115] suggesting that Mg in biochars may form large amount of colloidal or nanosized oxide particles on carbon surface. Consequently, inorganic phosphate ions might be attracted electrostatically by these fine particles in aqueous solutions. Besides, phosphates complexes could be formed at the biochar surface through deposition mechanisms [115].

#### *8.3.1.3 Use of activated carbons*

Several research studies suggested the activation of tomato wastes through chemical treatments using agents such as $ZnCl_2$, polyethylene terephthalate (PET), NaOH, KOH, HCl, $HNO_3$, and formaldehyde for the removal inorganic pollutants (Co(II), Cu(II), Pb(II), Mn(II), Fe(III), inorganic phosphates) from aqueous solutions [18,28,40,77,107,116].

In this context, Co(II) removal by two ACs using tomato wastes (peel, seeds and pulp) and carrot wastes (peel and pulp) was investigated under static conditions (batch mode) [18]. The first AC was generated through the heating of impregnated tomato/carrot wastes with $ZnCl_2$ at a mass ratio ($ZnCl_2$:biomass) of 1:1. The heating temperature and residence time were fixed to 550°C and 2 h, respectively. The second AC consists in a mixture of 0.4 g of AC1 with 0.8 g of an activated $FeCl_3$-polyethylene terephthalate at a temperature of 500°C for 2 hours. The determined adsorption capacities of Co(II) were evaluated to about 170 and 313 mg/g for AC1 and AC2, respectively. These Co(II) adsorption capacities were much higher than the one determined by Ozbay and Yargic [76] with an AC generated from tomato paste wastes (44.8 mg/g for an initial pH of 8, an aqueous concentration of 100 mg/L, and a dosage of 1 g/L). This AC was generated from the pyrolysis at a temperature of 500°C and a residence time of 1 h of the impregnated tomato waste with KOH at an impregnation mass ratio of 1:2. These results prove that under the used experimental conditions, tomato waste activation with $ZnCl_2$ achieves ACs with better physicochemical properties than KOH.

The Cu(II) removal form aqueous solutions by modified tomato wastes was investigated by several authors [107,116].Vafakhah et al. [107] oxidized with a mixture of corn stalk with tomato wastes with 1 M $HNO_3$ at a temperature of 50°C for 2 h. They reported a Cu(II) maximum adsorption capacity of 25 mg/g, which is higher than the oxidized corn stalk alone (20.8 mg/g). A higher adsorption capacity (34.5 mg/g) was found by Yargiç et al. [116] for tomato wastes treated with 3% v/v HCl solution. Cu(II) removal by this HCl treated tomato wastes was found to be mainly controlled by chemical reactions (kinetic experimental data well fitted with PSO). Furthermore, the thermodynamic study proved that Cu(II) adsorption is a spontaneous and exothermic process. It is important to underline that these adsorption capacities are much lower than the one determined for carotenoid extracted from tomato pomace wastes (181.8 mg/g) [79]. This interesting ability was linked to the adsorbent richness in alcohol, phenolic and carboxylic groups assigned with stretching vibrations of intramolecular and intermolecular H-bridge between the OH groups at 3400/cm. The experimental isothermal and kinetic data were well fitted with Langmuir and PSO models, respectively.

Pb(II) removal from aqueous solutions by chemically modified tomato wastes was studied by Heraldy et al. [41]. Tomato wastes were modified by mixing with NaOH aqueous solution at a ratio of 1:10 (m/v) for 24 h at room temperature. This chemical modification increased the surface area, pore volume and pore size from 8.8 $m^2/g$, 0.008 $cm^3/g$, and 18.4 nm (before activation) to 45.0 $m^2/g$, 0.045 $m^2/g$, and 19.9 nm (after activation). These pore characteristics remain relatively low compared to other ACs activated with $ZnCl_2$, $H_3PO_4$, and $FeCl_2$ (see Table 8.2). Adsorption experiments were achieved for different adsorbent dosages (0.5–5 g $L^{-1}$); pH (2–10), contact times (15–90 min) and initial Pb(II) concentrations (20–100 mg/L). Results showed that the NaOH-treated tomato wastes could be considered as an interesting low cost material for Pb(II) removal under a wide experimental conditions [41]. Its Langmuir Pb(II) adsorption capacity was determined to 152 mg/g, which is considered much higher than other NaOH-modified materials including olive tree pruning [15] or alfalfa biomass [102].

Iron (Fe[III]) and manganese (Mn[II]) removal from aqueous systems by an activated carbon generated from green tomato husk treated with formaldehyde were also investigated [28]. Experimental results showed that, the modification with formaldehyde slightly improved the iron and manganese sorption (in comparison with the non-treated biomass). The sorption capacities of the formaldehyde modified green tomato husk were assessed to 15.2 and 19.8 mg/g for Mn(II) and Fe(III), respectively. Langmuir-Freundlich model for Fe(III) and Freundlich model for Mn(II) was found more suitable in describing the experimental data. In addition, the adsorption kinetic of Fe(III) and Mn(II) onto treated formaldehyde green tomato husk followed the PSO model. Ion exchange-complexation and precipitation-ion exchange are the main sorption mechanisms responsible for manganese and iron removal, respectively.

Finally, locally available tomato plant roots as cellulose-based biopolymers were used to create a low-cost filtration system for the remediation, recovery and potential reuse of inorganic phosphate from agricultural wastewater [106]. In this regard, carboxymethyl cellulose (CMC) was used in order to enhance inorganic phosphate precipitation in lab scale and in on-site tests. These authors noted that the addition of CMC enhances the precipitation of inorganic phosphate with an average removal rates of 97% and 71% for lab scale and on-site tests after 23 d, respectively. The use of CMC-tomato plant roots significantly enhances the precipitation of phosphates as Ca-P and their binding with some specific functional groups.

#### *8.3.1.4 Involved mechanisms*

The qualitative assessment of the involved mechanisms during inorganic pollutant removal from aqueous solutions by raw and modified tomato wastes has been reported in several studies through the fitting of the experimental isothermal data to Langmuir and Freundlich models and the kinetic data to pseudo-first-order and pseudo-second-order models. Most of studies related to inorganic pollutant removal by raw and modified tomato wastes have pointed out that Langmuir isotherm was the most suitable for the fitting of the experimental data at equilibrium [18,30,116]. This model assumes that the adsorption takes place homogeneously on monolayer of the adsorbent surface. Moreover, most of experimental investigations showed that the removal kinetic data of inorganic pollutants by raw or modified tomato wastes were best fitted with the pseudo second-order model, which indicates that chemisorption mechanism is the rate-controlling step [18,30,68,73,116]. Heraldy et al. [41] reported that

Pb(II) adsorption onto NaOH-activated biochar occurs mainly through the precipitation for high alkaline media. Moreover, García-Mendieta et al. [28] attributed Mn(II) and Fe(III) adsorption onto green tomato husk treated with formaldehyde to ion exchange and complexation for Mn(II), and precipitation and ion exchange for Fe(III). In addition, nanoscale $Mg(OH)_2$ and MgO particles were present onto the surface of the Mg-enriched carbon, which serve as the main adsorption sites for aqueous inorganic phosphates [113]. Finally, Ure et al. [106] demonstrated that CMC elution with tomato plant roots as a capture matrix enhances the precipitation of inorganic phosphates as Ca-P. The involved mechanisms during mineral adsorption by raw and modified biomasses (including tomato wastes) have been investigated by various studies [98,100]. The main cited mechanisms are: (1) electrostatic attraction, (2) ion exchange, (3) complexation with specific functional groups, and (4) precipitation at high pH values (Fig. 8.9).

### 8.3.2 Case of organic pollutants

The use of activated carbon/biochars produced from lignocellulosic materials for organic pollutants removal has been widely developed in scientific literature [6,17,52,85]. However, only a few publications have dealt with the use of tomato wastes for the removal of organics from aqueous solutions. In this context, tomato wastes (seeds) have been used as raw materials for dyes removal from aqueous solutions [69]. Besides, tomato peels were used for the adsorption of various organic pollutants [64]. However, in order to get better efficiencies in removing organics, tomato wastes have been either pyrolyzed or impregnated with various chemical agents such as $FeCl_2$ [27] or $ZnCl_2$ [5,32,33,90,92–95] before a controlled pyrolysis step. As for inorganic pollutants, these activated carbons have been tested as adsorbents for the removal of dyes, pharmaceuticals and insecticides. Besides, the hydrothermal carbonization of tomato wastes in order to produce efficient hydrochars for pollutants removal has recently been applied for the removal of a mixture of several persistent organic pollutants [109].

#### *8.3.2.1 Use of raw tomato wastes*

As mentioned above, Najafi et al. [69] investigated the use of raw tomato seeds (without activation) for the removal of two acid dyes: Acid Red 14 (AR14) and Acid Blue 92 (AB92). The $pH_{ZPC}$ of the used adsorbent was assessed to 4.5. For that reason, AR14 and AB92, as anionic dyes, were

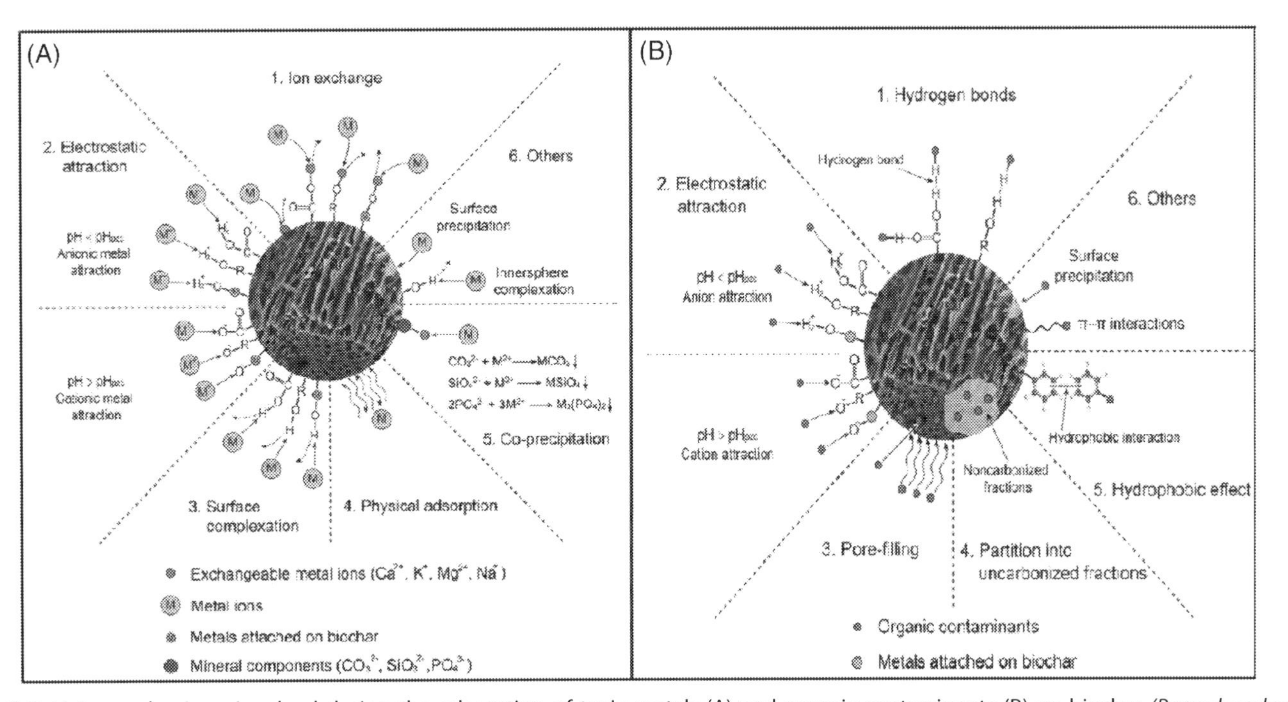


**Fig. 8.9** Main mechanisms involved during the adsorption of toxic metals (A) and organic contaminants (B) on biochar *(Reproduced with permission from* [100].

found to be highly adsorbed at an initial aqueous pH of 3 (due to electrostatic attraction), then significantly decreased with the increase of the used aqueous pH values. The maximum adsorption capacities of AR14 and AB92 by raw tomato seeds were 125 and 36.23 mg/g for an initial pH of 3 (Table 8.4). Moreover, the thermodynamic study showed that AR14 and AB92 adsorption was an exothermic and spontaneous process. These dyes desorption efficiency by distilled water solutions with pH values varying between 3 and 11 was relatively low. Indeed, they were only about 20% to 22% for pH 9 and reached 57% and 63% for AR14 and AB92 at pH 11, respectively. The cost analysis study showed that tomato seeds could be considered as a promising material for dye removal with a total cost of about 0.118$/g.

On the other hand, Mallampati and Valiyaveettil [64] applied raw tomato peels (without any modification) for the removal of pesticides (2,4,6-Trichlorophenol (TCP), p-chlorophenol (PCP), phenol, and p-nitrophenol (PNP)) and dyes (Alcian blue (AB), Coomassie brilliant blue G-250 (BB), neutral red (NR), and methylene blue (MB)) form aqueous synthetic solutions under different experimental conditions (Table 8.4). The kinetic study showed that a contact time of 12 hours was sufficient to attain an equilibrium state between the organic pollutants and tomato peel particles. Furthermore, dyes were more retained than pesticides. This finding was imputed to the weak $\pi$-$\pi$ and H bonding interactions. Cationic dyes are more efficiently adsorbed than anionic ones due to the presence of important acidic and alcoholic functional groups on tomato peels surface. Kinetic data were well fitted by the pseudo second order (PSO) model indicating that adsorption onto tomato peels is a chemical process. In addition, the used aqueous pH values seem to have an important effect on both the pesticides and dyes removal efficiencies. When the used aqueous pH values were higher than their pKa, pesticide adsorption was significantly reduced due mainly to electrostatic repulsion with the negatively charged adsorbent particles. In contrast, the increase of aqueous pH favored cationic dyes adsorption. This behavior is linked to the ionization of acidic functional groups, which enhances the adsorption of the positively charged cations through electrostatic attraction [10,66].

#### *8.3.2.2 Use of raw tomato wastes derived biochars*

Silvani et al. [95] investigated the use of non-activated biochar derived from the slow pyrolysis of greenhouse tomato wastes at a temperature of 550°C for the removal of hexachlorocyclohexane (HCH) isomers ($\alpha$, $\beta$, $\gamma$, and $\delta$),

**Table 8.4** Removal of organic contaminants by tomato wastes.

| Feedstock, origin | Pre-treatment | Modification conditions | Post-treatment | Surface area ($m^2/g$) | Pore volume/(micro-pore volume) ($cm^3/g$) | Average pore width (nm) | Adsorption conditions | pollutant | Kinetic model | Isotherm | Langmuir qmax (mg/g) | Reference |
|---|---|---|---|---|---|---|---|---|---|---|---|---|
| Tomato seeds from the by-products of a tomato paste factory, Iran | Washing with distilled water and then drying in an oven at 50°C for 12 h. Grinding and sieving for particles size lower than 250 μm | - | - | - | - | - | C0 = 50–150 mg/L; D = 0.6 g/L (for AR14) and 1 g/L (for AB92); pH = 3; t = 1 h; T = 30°C | Acid red 14 (AR14) Blue 92 (AB92) | PSO | Langmuir | 125.00<br>36.23 | Najafi et al. [69] |
| Tomato peels from local market, Singapore[a] | Washing with 2-propanol to remove anthocyanins, followed by water, then drying | - | - | - | - | - | C0 = 5–200 mg/L; D = 10 g/L; pH = 3; t = 24 h; T = 30°C | Alcian blue<br>Brilliant blue<br>Methylene blue<br>Natural red | PSO | Freundlich | 0.1012<br>0.0651<br>0.0683<br>0.0112 | Mallampati and Valiyaveettil [64] |
| Greenhouse tomato wastes, Norway[a,b] | - | - | - | 5.4 | 5.1 | 58.5 | C0 = 1–500 μg/L; D = 1.25 g/L; pH = -; t = 28 d; T = 25°C | Hexachloro-cyclo-hexane<br>α CHC<br>β CHC<br>γ CHC<br>δ CHC | - | Freundlich | 4.39<br>4.01<br>4.22<br>4.99 | (ilvani et al. (2019) |

*(continued)*

**Table 8.4** (Cont'd)

| Feedstock, origin | Pre-treatment | Modification conditions | Post-treatment | Surface area ($m^2/g$) | Pore volume/(micro-pore volume) ($cm^3/g$) | Average pore width (nm) | Adsorption conditions | pollutant | Kinetic model | Isotherm | Langmuir qmax (mg/g) | Reference |
|---|---|---|---|---|---|---|---|---|---|---|---|---|
| Tomato stem, Jinan, China | Washing with deionized water. Then, drying at 110°C for 24 h. Milling and sieving to sizes less than 0.178 mm. | Impregnation for 24 h with FeCl2 solutions at mass ratio (FeCl2: TS) of 2.5:1. Then, drying at 110°C. Afterwards, pyrolysis at a temperature of 700°C at a rate of 10°C/min and a residence time of 1 h. | Soaking in 0.1 M HCl solutions. Then washing with deionized water until a constant pH of 6.5. Crushing at 0.074 – 0.15 mm | 971 | 0.576/ 0.425 | 2.782 | C0 = 100 – 500 mg/L; D = 1 g/L; pH = 6.5; t = 24 h; T = 25°C | Congo Red | - | Langmuir | 158.730 | Fu et al. [27] |
| Waste of tomato juice and paste factories, Adana, Turkey | Drying in an air oven at 70°C for 24 h. Then, crushing and sieving particle size between 0.177 and 0.4 mm | Impregnation with ZnCl2 at mass ratio (ZnCl2: TW) of 1:1. Then, drying at 105°C for 12 h. Afterwards, pyrolysis at a temperature of 500°C at a rate of 10°C/min and a residence time of 1 h. | Washing with distilled water. drying at 105°C for 12 h. Crushing and sieving particle size between 0.177 and 0.4 mm | 722.17 | 0.476/ 0.201 | 2.64 | C0 = 75 – 200 mg/L; D = 0.1 g/L; pH = 2; t = 3 h; T: variable | Acid orange II | PSO | Langmuir | 185.00 (T = 20°C) 222.22 (T = 30°C) 277.78 (T = 40°C) 312.50 (T = 50°C) | Güzel et al. [31] |

| | | | | | | | | | | |
|---|---|---|---|---|---|---|---|---|---|---|
| | | | | | C0 = 20–350 mg/L; D = 2 g/L; pH = 8; t = 2.5 h; T: variable | Crystal violet | PSO | Freundlich | 51.55 (T = 20°C) 63.29 (T = 30°C) 64.52 (T = 40°C) 68.97 (T = 50°C) | Güzel et al. [32] |
| | Washing with distilled water. Then drying at 105°C for 12 h. Then, impregnation with Maghemite (ϒ-Fe2O3) nanoparticles | 154.90 | 0.264/0.027 | 6.80 | C0 = 20–350 mg/L; D = 2 g/L; pH = 4; t = 4.5 h; T: variable | Tetracycline | PSO | Langmuir | 39.06 (T = 20°C) 42.55 (T = 30°C) 51.28 (T = 40°C) 60.60 (T = 50°C) | Akkaya Sayıli et al. [5] |
| Impregnation for with ZnCl2 at mass ratio (ZnCl2: TW) of 6:1. Then, drying at 105°C for 12 h. Afterwards, pyrolysis at a temperature of 600°C at a rate of 10°C/min and a residence time of 1 h. | Impregnation with 0.2 N HCl solution. Then, washing with distilled water. Drying at 105°C for 12 h. Crushing and sieving particle size between 0.177 and 0.4 mm | 1093 | 1.569/ 0.129 | 5.92 | C0 = 200 – 400 mg/L; D = 0.2 g/L; pH = 5.7; t = 5 h; T: variable | Tetracycline | PSO | Langmuir | 370.4 (T = 15°C) 416.7 (T = 25°C) 500 (T = 35°C) | Sayğili and Güzel [92] |

*(continued)*

**Table 8.4** (Cont'd)

| Feedstock, origin | Pre-treatment | Modification conditions | Post-treatment | Surface area ($m^2/g$) | Pore volume/(micro-pore volume) ($cm^3/g$) | Average pore width (nm) | Adsorption conditions | pollutant | Kinetic model | Isotherm | Langmuir qmax (mg/g) | Reference |
|---|---|---|---|---|---|---|---|---|---|---|---|---|
| | | | | | | | C0 = 200 – 900 mg/L; D = 0.6 g/L; pH = 6.33 for MB and 6.23 for MY; t = 24 h; T = 30°C | Methylene blue (MB) Metanil yellow (MY) | - | Langmuir | 400<br>385 | Sayğili and Güzel [91] |
| | | | | | | | C0 = 100 – 900 mg/L; D = 0.2 g/L; pH = 6.95 t = 5 h; T: variable | Congo Red | PFO | Langmuir | 312.50 (T = 25°C)<br>384.62 (T = 35°C)<br>416.67 (T = 45°C)<br>434.78 (T = 55°C) | Saygl and Guzel [94] |
| | | | | | | | C0 = 200 – 400 mg/L; D = 0.2 g/L; pH = 4.0 t = 3 h; T: variable | Malachite green | PSO | Langmuir | 384.62 (T = 25°C)<br>434.78 (T = 35°C)<br>454.55 (T = 45°C)<br>526.32 (T = 55°C) | Sayğılı and Güzel [93] |

| | | | | | | | | | | | | |
|---|---|---|---|---|---|---|---|---|---|---|---|---|
| Skin and seeds of tomato, Sweden | Covering with ultra-pure water. Dry matter content of about 12% | Hydrothermal carbonization of 600 mL at a temperature of 600°C for 2 hours. | Drying overnight at 105°C. Then, washing with 0.1 M HCl and rinsing with ultra-pure water. Finally, drying at 105°C for 12 h and grinding. | 0.74 | - | - | C0 = 10 μg/L; D = 5 g/L; t = 25 min; T = 20°C | Fluconazole | - | - | ~0.26 | Weidemann et al. [109] |
| | | | | | | | | Ciprofloxacin | | | ~0.14 | |
| | | | | | | | | Paracetamol | | | ~0.0 | |
| | | | | | | | | Sulfamethoxazole | | | ~0.49 | |
| | | | | | | | | Trimethoprim | | | ~0.24 | |
| | | | | | | | | Octhilinone | | | ~0.65 | |
| | | | | | | | | Diphenhydramine | | | ~0.31 | |
| | | | | | | | | Bisphenol A | | | ~2.0 | |
| | | | | | | | | Diclofenac | | | ~2.0 | |
| | | | | | | | | Triclosan | | | ~0.72 | |

[a] qmax corresponds to log (Freundlich constant: $K_F$) (mg $kg^{-1}$) (mg $L^{-1}$)-n).
[b] textural analyses carried out by $CO_2$ adsorption at 0°C.

as halogenated compounds, from aqueous solutions. For initial concentrations of HCH isomers varying between 1–500 μg/L (single components) and between 4–2000 μg/L (multicomponent system), they showed that no significant sorption competition between HCH isomers on tomato derived biochars was observed. This behavior was mainly attributed to both the relatively low used concentrations and the abundance of available sorption sites. The adsorption strength of the tomato-derived biochar showed the following order: δ-CHC > α-CHC > γ-CHC > β-CHC since their Freundlich partition coefficients (log [KF]) were assessed to 4.99, 4.39, 4.22, and 4.01 (mg/kg) $(mg/L)^{-n}$. On the other hand, their study showed that tomato waste-derived biochars were more efficient in removing HCH isomers than durian shell derived biochars. This was due to its higher iron contents.

#### *8.3.2.3 Use of activated carbons*

Concerning tomato waste activation, Fu et al. [27] studied Congo Red dye removal by activated biochars generated from tomato stem wastes (TSW) under various conditions. The TSW were first impregnated for 24 h with $FeCl_2$ at mass ratios ($FeCl_2$:TSW) ranging from 1:1 to 3:1. Then, these resulting materials were pyrolyzed at temperatures in the range of 500–800°C at a rate of 10°C and a residence time of 1 h. It was shown that for a constant $FeCl_2$:TSW ratio of 2.5:1, increasing pyrolysis temperature from 500 to 700°C significantly promoted the creation of micropores and the increase of the overall surface area. In contrast, a further increase in the pyrolysis temperature resulted in a significant widening of the microporosity and a rapid increase of the mesoporosity or macroporosity. Moreover, at a given pyrolysis temperature (700°C), increasing $FeCl_2$:TS ratio from 1:1 to 2.5:1 did not only create new micropores but also enlarged existing pores. When increasing $FeCl_2$:TSW up to 3.0, an important enlargement of existing micropores into mesopores and even macropores was observed. The best characteristics were obtained for a $FeCl_2$:TS of 2.5:1 and a pyrolysis temperature of 700°C. The corresponding surface area, total pore volume, micropore volume and average pore width were respectively 971 $m^2$/g, 0.576 and 0.425 $cm^3$/g, and 2.782 nm. This material permitted a significant Congo red removal efficiency with a Langmuir adsorption capacity of about 158.73 mg/g. This capacity is about 782-, 24- and 3-fold higher than that observed for commercial activated carbons [63], activated carbon generated from coir pith wastes [70], and bituminous coal [60], respectively. A regeneration study of the dye-loaded tomato stem activated carbon showed that NaOH could be considered as a promising desorbent. In fact, after three cycles of Congo red adsorption-desorption, the removal and regeneration efficiencies were still relatively high: 73.9% and 88.1%, respectively.

The second type of activation addresses the use of $ZnCl_2$ as an effective impregnating agent before pyrolysis. In this context, various studies originating from Turkey were published between 2014 and 2018 related to the production of $ZnCl_2$-activated biochars from tomato wastes as well as their reuse for the removal of dyes, pharmaceuticals and insecticides from synthetic aqueous solution [31,92–94]. In this regard, Güzel et al. [31] and Güzel et al. [32] studied Acid Orange (II) (AOII, anionic dye) and crystal violet (CV, cationic dye) removal from aqueous solutions by $ZnCl_2$ activated biochar generated from the slow pyrolysis of tomato juice and paste wastes. For the pyrolysis process, the final temperature, the heating gradient and the residence time were fixed to 500°C, 10°C/min and 1 h, respectively. The derived biochar has an important surface area (722.11 $m^2/g$) and contains simultaneous micropores (0.201 $cm^3/g$: 42%) and mesopores (58%). Furthermore, this biochar exhibited an acidic behavior, with a $pH_{ZPC}$ of 5.3 and a total surface acidity of 1.33 meq/g distributed between phenolic, lactonic, and carboxylic groups with contents of 0.26, 0.53, and 0.54 meq/g, respectively. The surface basicity corresponds to a total content of 0.95 meq/g. All these interesting characteristics make this biochar very efficient in removing AOII from aqueous solutions especially for pH lower than $pH_{ZPC}$. Indeed, its adsorption capacity for a pH of 2 was reached in 3 hours and was assessed to 185 mg/g at a temperature of 20°C. It increased to 312.50 mg/g when the used temperature increased to 50°C (Table 8.4). The thermodynamic study demonstrated that AOII adsorption was a spontaneous and endothermic process. Regarding CV, it seems that, contrarily to AOII, its adsorption globally increased with the increase of the aqueous pH and reached a quasi-plateau from a pH of 8. The equilibrium time was reached in about 2.5 h and the adsorbed CV amounts increased from 51.55 to 68.97 mg/g when the temperature was increased from 20 to 50°C, respectively.

This same produced activated biochar was impregnated with Maghemite ($\Upsilon$-$Fe_2O_3$) nanoparticles and used for the removal of tetracycline from aqueous solutions [5]. This chemical modification resulted in a significant reduction in volatile matter, fixed carbon and carbon contents. At contrary, ash content increased significantly (from 1.47% to 73.53%). Moreover, this modification induced a significant decrease of surface area, total and micropore volumes from 722.17, 0.476 and 0.201 by about 78.5%, 44.5%, and 86.6%, respectively. FTIR and XRD analyses displayed the appearance of specific peaks that are assigned to FeO iron oxide and indicates that $\Upsilon$-$Fe_2O_3$ nanoparticles are effectively deposited on tomato waste-derived biochar. The kinetic adsorption of TC by the nanocomposite-based tomato wastes biochar was well fitted by the pseudo-second order model. The sorption isotherm was well described by Langmuir model. For an initial aqueous pH of 4, the adsorption capacity increased with the increase of the used temperature from 39.1 to 60.6 mg/g at temperatures of 20 and 60°C, respectively. Finally, the thermodynamic study proved that tetracycline adsorption by the used adsorbent was a spontaneous and endothermic process.

Sayğili and Güzel [91] studied the effect of tomato processing waste activation by $ZnCl_2$ under various experimental conditions (mass ratio of $ZnCl_2$/TW between 1:1 and 8:1; pyrolysis temperature between 400 and 800°C, and residence time between 0.5 and 4 h). The best activated biochars characteristics were obtained for an impregnation ratio of 6:1, a pyrolysis temperature of 600°C and a residence time of 1 h. Under these experimental conditions, the surface area, total pore volume mesoporosity and microporosity percentages and the average pore diameter were assessed to 1093 $m^2/g$, 1.569 $cm^{3/}g$, 92%, 8%, and 5.92 nm, respectively. In addition, the surface analysis showed that the $pH_{ZPC}$ of the activated biochar was 6.17. Furthermore, this biochar has globally similar contents of acidic and basic groups, which implies that this adsorbent will be effective in removing both cationic and anionic pollutants. The application of this biochar for the removal of tetracycline from aqueous solutions proved that it could be considered as a very promising material. In this regard, the equilibrium time was assessed to less than 5 h and its Langmuir adsorption capacity of tetracycline increased with the increase of temperature from 370.4 (at 15°C) to 500.0 (at 35°C) [92]. These adsorption capacities are considerably higher than the one observed for an activated tomato waste derived biochar impregnated with $\Upsilon$-$Fe_2O_3$ nanoparticles [5] and even other commercial activated carbons [84].

This same activated biochar exhibited relatively high adsorption capacities for methylene blue (MB: cationic dye) and metanil yellow (MY: anionic dye). For natural pH (between 6.2 and 6.3), the related adsorption capacities of MB and MY were 400 and 385 mg/g, respectively [91]. It is important to underline that this tomato waste derived biochar is about 2.2 and 1.9 times more efficient in removing MB from aqueous solutions than activated carbon derived from $H_3PO_4$-activated coffee grounds [82] and $CO_2$-activated cocoa shell [1], respectively. Besides, it has a MY adsorption capacity of 4.4 and 4.3 times higher than those observed for steam-activated rice husk [62] and steam-activated coconut shell [88]. This same activated biochar was tested as an adsorbent for Congo red (CR) [94] and malachite green (MG) [93] removal from aqueous solutions. They found that a contact time of 4 to 5 hours was sufficient to reach an equilibrium state between the dyes and the activated tomato waste derived biochars. The Langmuir adsorption capacity of CR and MG increased from 312 to 434 mg/g and from 384 to 526 mg/g as the temperature increased from 25 to 55°C (Table 8.4). Congo red removal by the tomato waste-activated biochar is more efficient than carbonized bael shell at a temperature of 450°C [2], a $H_3PO_4$-activated *Martynia annua* L. seeds [97] and even commercial activated carbons [80]. Besides, MG elimination by the used tomato derived biochar is better than activated carbon derived from a KOH-activated banana stalk [13], a KOH-activated coconut shell [12], and bamboo waste derived biochars that were activated with $K_2CO_3$ and subsequently gasified with $CO_2$ activated bamboo biochar [37]. Moreover, thermodynamic investigations showed that CR and MG adsorption processes were spontaneous and endothermic. Finally, the regeneration studies of their loaded activated tomato waste biochar through ethanol solutions (50% V/V) showed that the original adsorption capacity of CR and MG significantly dropped with the number of adsorption/desorption cycles. For five cycles, decreases by about 52% and 34.3% were observed for CR and MG, respectively.

Tomato waste-derived hydrochars was also used for the removal of a mixture containing ten organic persistent pollutants (octhilinone, triclosan, trimethoprim, sulfamethoxasole, ciprofloxacin, diclofenac, paracetamol, diphenhydramine, fluconazole, and bisphenol A) [109]. Their efficiencies were compared to hydrochars generated from the hydrothermal carbonization of horse manure, rice husks and olive press wastes. For an initial concentration of 10 μg/L and an adsorbent dosage of 5 g/L, they showed that rice husk and horse manure hydrochars exhibited the highest average of removal efficiencies for the ten pollutants (66% and 49%, respectively). These values were assessed to only 39% and 32% for olive and tomato waste derived hydrochars. Surface analysis including BET surface area measurements, diffuse reflectance infrared spectroscopy (DRIFT) and X-ray photoelectron spectroscopy (XPS) analyses showed that this adsorptive behavior was a result of a combination of the following factors: (1) the surface area of rice husk hydrochars (16.92 $m^2/g$) was higher than the values determined for horse manure, tomato, and olive press wastes hydrochars (4.62, 0.74, and 0.65 $m^2/g$, respectively), (2) the rice husk and manure hydrochars contain higher fractions of carbon species with a single bond to oxygen, (3) the rice husk and horse manure hydrochars are more polar than the tomato and olive waste derived hydrochars. The formers exhibited higher contents of $SiO_2$ and OH, and lower aliphatic carbon concentrations which will promote the removal of more polar molecules via H-bonding process. It is important to underline that the tomato waste derived hydrochars showed low adsorption capacities of paracetamol, ciprofloxacin, trimethoprim and fluconazole. However, they completely removed diclofenac and bisphenol A (table 8.4).

#### *8.3.2.4 Involved mechanisms [45]*

The adequate assessment of the involved mechanisms during pollutant removal by tomato wastes-based materials is a key parameter for a better understanding of the interactions at the adsorbent/adsorbate interface as well as the choice of the optimal desorption procedure [17,39,98]. However, in previous published papers regarding the use of tomato wastes for the removal of organics from aqueous solutions, there has been no sufficient attention paid to the accurate assessment of the involved mechanisms. In the majority of these papers, mechanisms are qualitatively determined through the interpretation of the kinetic and isotherms studies as well as some classic analyses such as FTIR and SEM/EDS analyses of the adsorbent before and after the adsorption process [27,92,93].

Generally, the involved mechanisms depend mainly on the nature and characteristics of the adsorbent, the physicochemical properties of the organic pollutant, and finally the adsorption conditions [23,44,45,59,86]. Hereafter, since the majority of organic removal experiments by tomato wastes have been performed with biochars, we will discuss the main involved mechanisms when using biochars of lignocellulosic origin. These mechanisms include mainly electrostatic attraction, pore-filling, π-π electron-donor acceptor interaction, H bonding, hydrophobic interactions, and partition into uncarbonized fraction [23,48] (Fig. 8.9).

Electrostatic attraction occurs between positively/negatively charged biochar particles (for pH lower/higher than $pH_{ZPC}$) and anionic/cationic organic pollutants. In case of anionic organic compounds, it is preferred to adjust the liquid effluents to pH lower than the $pH_{ZPC}$ and the opposite action applies for cationic organic pollutants. In this context, Inyang et al.

[46] suggested that electrostatic attraction was the main mechanism that controls MB molecules (positively charged) and carbon nanotube-biochar composite particles. The same finding was reported by Yao et al. [114] when studying MB adsorption by biochar-montmorillonite composite. Concerning the pore filling mechanism, it is mainly a function of biochar total micropore and mesopore volumes as well as surface areas [100]. It was reported as an important adsorption mechanism by Han et al. [38] and [46] when investigating phenol and MB removal from aqueous solutions by lignocellulosic material-derived biochars. Regarding the $\pi$-$\pi$ electron-donor acceptor interaction, it depends mainly on the intensity of the presence of the $\pi$-electron systems. Generally, the carboxylic acid, nitro, ketonic, hydroxyl and amine groups on the surface of biochar act as electron acceptors, forming $\pi$-$\pi$ electron-donor acceptor interaction with aromatic molecules [3,23]. This mechanism was evoked by Xie et al. [111] when investigating sulfonamides removal by pine wood biochars. Concerning the hydrogen bonding mechanism (H bonding), it is generally related to polar organic pollutants. It was reported as an important adsorption mechanism for carbamazepine removal by peanut shells biochar [19] as well as for dibutyl phthalate removal by rice straw and swine manure biochars [50]. It is important to underline that the surface of biochar is generally very heterogeneous due to the co-presence of both carbonized and non-carbonized fractions. Organics could be therefore, either uptaken through partition into the noncarbonized fraction or adsorbed onto the carbonized matter [100]. Finally, during the removal of organics by biochars, different mechanisms could take place simultaneously. The importance of each one of the involved mechanisms depends not only on the pollutant characteristics but also on the properties of the produced biochars (nature of the feedstock and pyrolysis conditions) [23,86].

## Conclusions

This chapter represents an extended scientific literature review on the recent developments related to the environmental valorisation of raw and modified tomato wastes as efficient adsorbents for the removal of both organic and mineral pollutants from aqueous solutions. The adsorptive properties of tomato-derived materials depend mainly on the raw precursor nature, and especially on the modification method and experimental conditions. Among various modification options (i.e., physical, chemical, and thermal treatment), production of activated carbons through chemical impregnation followed by a thermal treatment appears to be one of the most attractive methods. Indeed, these activated carbons could exhibit a well-developed porosity, specific surface areas, and richness in some specific functional groups, allowing for an efficient removal of pollutants from effluents. These characteristics are, however, very dependent on the nature of the used chemical reagent (acid, base, salt...), its mass ratio (impregnating agent/biomass) and thermal conditions (final temperature, residence time, and heating gradient). $ZnCl_2$, $FeCl_2$, and $H_3PO_4$ with adapted impregnating mass ratios (chemical/biomass ratios higher than 2:1) are among the best activating chemicals. The heating temperature should be sufficiently high but should not exceed 650–700°C. The residence time should be at least 60 min.

The efficiency of these produced activated carbons in removing pollutants from aqueous solutions depends on several parameters such as initial contaminant concentration, contact time, aqueous pH, adsorbent dosage, presence of other competitive pollutants, and temperature. The choice of the best adsorption conditions is specific to each adsorbate/adsorbent and could be carried out through optimizing methods such as response surface methodology (RSM).

The adsorption of pollutants onto tomato waste-derived materials (TWDM) occurs through various mechanisms. They depend on the nature of the pollutant (mineral or organic) and the properties of the adsorbent. Generally, several mechanisms can take place simultaneously during the adsorption tests. The main involved mechanisms in mineral removal include ion exchange, electrostatic attraction, complexation, precipitation, and physical adsorption. Organic pollutants can be removed through hydrogen bonds, electrostatic attraction, pore filling, hydrophobic interactions, precipitation, and $\pi$-$\pi$ interactions. The assessment of the quantitative importance of each of the above-cited mechanisms in the overall adsorption process is an important future research topic.

Different research avenues regarding TWDM valorization as effective and low-cost adsorbents for wastewater treatment should be explored in the future. These perspectives can address:

1. The optimization of the activation process permitting to produce adsorbents with promising physicochemical properties for an efficient removal of pollutants from wastewater. This optimization step should be carried out by using confirmed approaches such as RSM. It should take into account technical aspects but also economic and environmental concerns.
2. The study of TWDM embedment with specific nanocomposites for a better adsorption efficiency.

3. The promotion of the use of TWDM for the treatment of urban wastewater and especially for nutrient recovery. These nutrient-loaded materials could be susequently valorized in agriculture as eco-friendly biofertilizers in respect to the circular economy concept. In this context, the determination of the effect of such amendments on the physico-chemical, biological, and hydrodynamic properties of agricultural soils as well as plant growth is an urgent task.
4. The use of TWDM for pollutant removal from real effluents (instead of synthetic solutions) under dynamic conditions using laboratory columns and continuous renewed tank reactors (instead of batch assays). The numerical modelling of the pollutant breakthrough curves by adapted existing models is an important task for an easy up scaling to field conditions.
5. The use of sustainable management strategies of the pollutant-loaded TWDM that permit a regeneration and reuse of these materials in other adsorption cycles as well as the recovery of adsorbed chemicals and their reuse in various applications.

## References

[1] F. Ahmad, W.M.A.W. Daud, M.A. Ahmad, R. Radzi, Cocoa (Theobroma cacao) shell-based activated carbon by CO 2 activation in removing of cationic dye from aqueous solution: kinetics and equilibrium studies, Chem. Eng. Res. Des. (2012). https://doi.org/10.1016/j.cherd.2012.01.017.

[2] R. Ahmad, R. Kumar, Adsorptive removal of congo red dye from aqueous solution using bael shell carbon, Appl. Surf. Sci. (2010). https://doi.org/10.1016/j.apsusc.2010.08.111.

[3] M.B. Ahmed, J.L. Zhou, H.H. Ngo, M.A.H. Johir, L. Sun, M. Asadullah, D. Belhaj, Sorption of hydrophobic organic contaminants on functionalized biochar: protagonist role of Π-Π electron-donor-acceptor interactions and hydrogen bonds, J. Hazard. Mater. (2018). https://doi.org/10.1016/j.jhazmat.2018.08.005.

[4] M.J. Ahmed, Preparation of activated carbons from date (Phoenix dactylifera L.) palm stones and application for wastewater treatments: review, Process Saf. Environ. Prot. 102 (2016) 168–182. https://doi.org/10.1016/j.psep.2016.03.010.

[5] G. Akkaya Sayılı, H. Sayılı, F. Koyuncu, F. Güzel, Development and physicochemical characterization of a new magnetic nanocomposite as an economic antibiotic remover, Process Saf. Environ. Prot. (2015). https://doi.org/10.1016/j.psep.2014.10.005.

[6] A.M. Awad, R. Jalab, A. Benamor, M.S. Nasser, M.M. Ba-Abbad, M. El-Naas, A.W. Mohammad, Adsorption of organic pollutants by nanomaterial-based adsorbents: an overview, J. Mol. Liq. 301 (2020) 112335. https://doi.org/10.1016/j.molliq.2019.112335.

[7] A. Ayala-Cortés, C.A. Arancibia-Bulnes, H.I. Villafán-Vidales, D.R. Lobato-Peralta, D.C. Martínez-Casillas, A.K. Cuentas-Gallegos, Solar pyrolysis of agave and tomato pruning wastes: insights of the effect of pyrolysis operation parameters on the physicochemical properties of biochar, AIP Conference Proceedings, 2019. https://doi.org/10.1063/1.5117681.

[8] A.A. Azzaz, S. Jellali, H. Akrout, A.A. Assadi, L. Bousselmi, AC SC. J. Clean. Prod. (2018). https://doi.org/10.1016/j.jclepro.2018.08.023.

[9] A.A. Azzaz, S. Jellali, H. Akrout, A.A. Assadi, L. Bousselmi, Optimization of a cationic dye removal by a chemically modified agriculture by-product using response surface methodology: biomasses characterization and adsorption properties, Environ. Sci. Pollut. Res. 24 (2017). https://doi.org/10.1007/s11356-016-7698-6.
[10] A.A. Azzaz, S. Jellali, A.A. Assadi, L. Bousselmi, Chemical treatment of orange tree sawdust for a cationic dye enhancement removal from aqueous solutions: kinetic, equilibrium and thermodynamic studies, Desalin. Water Treat. 57 (2016). https://doi.org/10.1080/19443994.2015.1103313.
[11] M. Bargaoui, S. Jellali, A.A. Azzaz, M. Jeguirim, H. Akrout, Optimization of hybrid treatment of olive mill wastewaters through impregnation onto raw cypress sawdust and electrocoagulation, Environ. Sci. Pollut. Res. (2020). https://doi.org/10.1007/s11356-020-08907-w.
[12] O.S. Bello, M.A. Ahmad, Coconut (Cocos nucifera) shell based activated carbon for the removal of malachite green dye from aqueous solutions, Sep. Sci. Technol. (2012). https://doi.org/10.1080/01496395.2011.630335.
[13] O.S. Bello, M.A. Ahmad, N. Ahmad, Adsorptive features of banana (Musa paradisiaca) stalk-based activated carbon for malachite green dye removal, Chem. Ecol. (2012). https://doi.org/10.1080/02757540.2011.628318.
[14] D. Bilalis, P.E. Kamariari, A. Karkanis, A. Efthimiadou, A. Zorpas, I. KAkabouki, Energy inputs, output and productivity in organic and conventional maize and tomato production, under mediterranean conditions, Not. Bot. Horti Agrobot. Cluj-Napoca. (2013). https://doi.org/10.15835/nbha4119081.
[15] M. Calero, A. Pérez, G. Blázquez, A. Ronda, M.A. Martín-Lara, Characterization of chemically modified biosorbents from olive tree pruning for the biosorption of lead, Ecol. Eng. (2013). https://doi.org/10.1016/j.ecoleng.2013.07.012.
[16] R. Calvelo Pereira, J. Kaal, M. Camps Arbestain, R. Pardo Lorenzo, W. Aitkenhead, M. Hedley, F. Macías, J. Hindmarsh, J.A. Maciá-Agulló, Contribution to characterisation of biochar to estimate the labile fraction of carbon, Org. Geochem. (2011). https://doi.org/10.1016/j.orggeochem.2011.09.002.
[17] J.S. Cha, S.H. Park, S.C. Jung, C. Ryu, J.K. Jeon, M.C. Shin, Y.K. Park, Production and utilization of biochar: A review, J. Ind. Eng. Chem. 40 (2016) 1–15. https://doi.org/10.1016/j.jiec.2016.06.002.
[18] M. Changmai, P. Banerjee, K. Nahar, M.K. Purkait, A novel adsorbent from carrot, tomato and polyethylene terephthalate waste as a potential adsorbent for Co (II) from aqueous solution: Kinetic and equilibrium studies, J. Environ. Chem. Eng. (2018). https://doi.org/10.1016/j.jece.2017.12.009.
[19] J. Chen, D. Zhang, H. Zhang, S. Ghosh, B. Pan, Fast and slow adsorption of carbamazepine on biochar as affected by carbon structure and mineral composition, Sci. Total Environ. (2017). https://doi.org/10.1016/j.scitotenv.2016.11.052.
[20] C.H. Chiang, J. Chen, J.H. Lin, Preparation of pore-size tunable activated carbon derived from waste coffee grounds for high adsorption capacities of organic dyes, J. Environ. Chem. Eng. (2020). https://doi.org/10.1016/j.jece.2020.103929.
[21] Z.Z. Chowdhury, S.B.A. Hamid, R. Das, M.R. Hasan, S.M. Zain, K. Khalid, M.N. Uddin, Preparation of carbonaceous adsorbents from lignocellulosic biomass and their use in removal of contaminants from aqueous solution, BioResources (2013). https://doi.org/10.15376/biores.8.4.6523-6555.
[22] M. Chyli, M. Szyma, A. Zdunek, Imaging of polysaccharides in the tomato cell wall with Raman microspectroscopy, Plant methods 10 (2014) 14.
[23] Y. Dai, N. Zhang, C. Xing, Q. Cui, Q. Sun, The adsorption, regeneration and engineering applications of biochar for removal organic pollutants: A review, Chemosphere 223 (2019) 12–27. https://doi.org/10.1016/j.chemosphere.2019.01.161.

[24] P.D. Dissanayake, S. You, A.D. Igalavithana, Y. Xia, A. Bhatnagar, S. Gupta, H.W. Kua, S. Kim, J.H. Kwon, D.C.W. Tsang, Y.S. Ok, Biochar-based adsorbents for carbon dioxide capture: a critical review, Renew. Sustain. Energy Rev. (2020). https://doi.org/10.1016/j.rser.2019.109582.
[25] S.J. Dunlop, M.C. Arbestain, P.A. Bishop, Closing the loop : use of biochar produced from tomato crop green waste as a substrate for soilless, hydroponic tomato production, HortScience. 50 (10) (2015) 1572–1581.
[26] I. Erdal, M. Memici, A. Dogan, C. Yaylaci, K. Ekinci, Effects of tomato harvest residue derived biochars obtained from different pyrolysis temperature and duration on plant growth and nutrient concentrations of corn, in, Engineering for Rural Development (2018). https://doi.org/10.22616/ERDev2018.17.N137.
[27] K. Fu, Q. Yue, B. Gao, Y. Wang, Q. Li, Activated carbon from tomato stem by chemical activation with FeCl 2, Colloids Surf. A 529 (2017) 842–849. https://doi.org/10.1016/j.colsurfa.2017.06.064.
[28] A. García-Mendieta, M.T. Olguín, M. Solache-Ríos, Biosorption properties of green tomato husk (Physalis philadelphica Lam) for iron, manganese and iron-manganese from aqueous systems, Desalination (2012). https://doi.org/10.1016/j.desal.2011.08.052.
[29] L. Guppy, K. Anderson, Water crisis report - the facts, United Nations University Institude for Water, Environment and Health, 2017.
[30] Y. Gutha, V.S. Munagapati, M. Naushad, K. Abburi, Removal of Ni(II) from aqueous solution by Lycopersicum esculentum (Tomato) leaf powder as a low-cost biosorbent, Desalin. Water Treat. (2015). https://doi.org/10.1080/19443994.2014.880160.
[31] F. Güzel, H. Say, G. Akkaya, F. Koyuncu, Elimination of anionic dye by using nanoporous carbon prepared from an industrial biowaste. J. Mol. Liq. 194 (2014a) 130–140. https://doi.org/10.1016/j.molliq.2014.01.018.
[32] F. Güzel, H. Sayğili, G.A. Sayğili, F. Koyuncu, Decolorisation of aqueous crystal violet solution by a new nanoporous carbon: Equilibrium and kinetic approach. J. Ind. Eng. Chem. (2014b) https://doi.org/10.1016/j.jiec.2013.12.023.
[33] H.S.F. Güzel, Usability of activated carbon with well-developed mesoporous structure for the decontamination of malachite green from aquatic environments : kinetic, equilibrium and regeneration studies, J. Porous Mater. (2017). https://doi.org/10.1007/s10934-017-0459-1.
[34] K. Haddad, M. Jeguirim, S. Jellali, N. Thevenin, L. Ruidavets, L. Limousy, Biochar production from Cypress sawdust and olive mill wastewater: agronomic approach, Sci. Total Environ. (2020). https://doi.org/10.1016/j.scitotenv.2020.141713.
[35] K. Haddad, M. Jeguirim, B. Jerbi, A. Chouchene, P. Dutournié, N. Thevenin, L. Ruidavets, S. Jellali, L. Limousy, Olive mill wastewater: from a pollutant to green fuels, agricultural water source and biofertilizer, ACS Sustain. Chem. Eng. 5 (2017). https://doi.org/10.1021/acssuschemeng.7b01786.
[36] K. Haddad, S. Jellali, S. Jaouadi, M. Benltifa, A. Mlayah, A.H. Hamzaoui, Raw and treated marble wastes reuse as low cost materials for phosphorus removal from aqueous solutions: Efficiencies and mechanisms, Comptes. Rendus. Chim 18 (2015). https://doi.org/10.1016/j.crci.2014.07.006.
[37] B.H. Hameed, M.I. El-Khaiary, Equilibrium, kinetics and mechanism of malachite green adsorption on activated carbon prepared from bamboo by K2CO3 activation and subsequent gasification with CO2, J. Hazard. Mater. (2008) https://doi.org/10.1016/j.jhazmat.2007.12.105.
[38] Y. Han, A.A. Boateng, P.X. Qi, I.M. Lima, J. Chang, Heavy metal and phenol adsorptive properties of biochars from pyrolyzed switchgrass and woody biomass in correlation with surface properties, J. Environ. Manage. (2013). https://doi.org/10.1016/j.jenvman.2013.01.001.

[39] L. He, H. Zhong, G. Liu, Z. Dai, P.C. Brookes, J. Xu, Remediation of heavy metal contaminated soils by biochar: Mechanisms, potential risks and applications in China, Environ. Pollut. 252 (2019) 846–855. https://doi.org/10.1016/j.envpol.2019.05.151.
[40] E. Heraldy, Journal of Environmental Chemical Engineering Biosorbent from tomato waste and apple juice residue for lead removal, J. Environ. Chem. Eng. 6 (2018) 1201–1208. https://doi.org/10.1016/j.jece.2017.12.026.
[41] E. Heraldy, W.W. Lestari, D. Permatasari, D.D. Arimurti, Biosorbent from tomato waste and apple juice residue for lead removal, J. Environ. Chem. Eng. (2018). https://doi.org/10.1016/j.jece.2017.12.026.
[42] S. Hube, M. Eskafi, K.F. Hrafnkelsdóttir, B. Bjarnadóttir, M.Á. Bjarnadóttir, S. Axelsdóttir, B. Wu, Direct membrane filtration for wastewater treatment and resource recovery: a review, Sci. Total Environ. (2020). https://doi.org/10.1016/j.scitotenv.2019.136375.
[43] V.J. Inglezakis, A.A. Zorpas, Handbook of Natural Zeolites (2012). https://doi.org/10.2174/97816080526151120101.
[44] M. Inyang, E. Dickenson, The potential role of biochar in the removal of organic and microbial contaminants from potable and reuse water: A review. Chemosphere 134 (2015a) 232–240. https://doi.org/10.1016/j.chemosphere.2015.03.072.
[45] Inyang, M., Dickenson, E., 2015b. Chemosphere the potential role of biochar in the removal of organic and microbial contaminants from potable and reuse water : a review. Chemosphere 134, 232–240. https://doi.org/10.1016/j.chemosphere.2015.03.072.
[46] M. Inyang, B. Gao, A. Zimmerman, M. Zhang, H. Chen, Synthesis, characterization, and dye sorption ability of carbon nanotube-biochar nanocomposites, Chem. Eng. J. (2014). https://doi.org/10.1016/j.cej.2013.09.074.
[47] R. Islam, S. Kumar, J. Karmoker, M. Kamruzzaman, M.A. Rahman, N. Biswas, T.K.A. Tran, M.M. Rahman, Bioaccumulation and adverse effects of persistent organic pollutants (POPs) on ecosystems and human exposure: A review study on Bangladesh perspectives, Environ. Technol. Innov. (2018). https://doi.org/10.1016/j.eti.2018.08.002.
[48] S. Jellali, M. Labaki, A.A. Azzaz, H. Akrout, L. Limousy, M. Jeguirim, Biomass-derived chars used as adsorbents for liquid and gaseous effluents treatment, in: char and carbon materials derived from biomass: production, characterization and applications (2019). https://doi.org/10.1016/B978-0-12-814893-8.00007-9.
[49] S. Jellali, M.A. Wahab, M. Anane, K. Riahi, L. Bousselmi, Phosphate mine wastes reuse for phosphorus removal from aqueous solutions under dynamic conditions, J. Hazard. Mater. 184 (2010). https://doi.org/10.1016/j.jhazmat.2010.08.026.
[50] J. Jin, K. Sun, F. Wu, B. Gao, Z. Wang, M. Kang, Y. Bai, Y. Zhao, X. Liu, B. Xing, Single-solute and bi-solute sorption of phenanthrene and dibutyl phthalate by plant- and manure-derived biochars, Sci. Total Environ. (2014). https://doi.org/10.1016/j.scitotenv.2013.12.033.
[51] L. Joseph, B.M. Jun, J.R.V. Flora, C.M. Park, Y. Yoon, Removal of heavy metals from water sources in the developing world using low-cost materials: a review, Chemosphere (2019). https://doi.org/10.1016/j.chemosphere.2019.04.198.
[52] M. Kadhom, N. Albayati, H. Alalwan, M. Al-Furaiji, Removal of dyes by agricultural waste, Sustain. Chem. Pharm. 16 (2020) 100259. https://doi.org/10.1016/j.scp.2020.100259.
[53] B. Khiari, M. Massoudi, M. Jeguirim, Tunisian tomato waste pyrolysis: thermogravimetry analysis and kinetic study, Environ. Sci. Pollut. Res. (2019). https://doi.org/10.1007/s11356-019-04675-4.

[54] T.J. Kinney, C.A. Masiello, B. Dugan, W.C. Hockaday, M.R. Dean, K. Zygourakis, R.T. Barnes, Hydrologic properties of biochars produced at different temperatures, Biomass Bioenergy (2012). https://doi.org/10.1016/j.biombioe.2012.01.033.
[55] N. Kraiem, M. Lajili, L. Limousy, R. Said, M. Jeguirim, Energy recovery from Tunisian agri-food wastes: Evaluation of combustion performance and emissions characteristics of green pellets prepared from tomato residues and grape marc, Energy 107 (2016) 409–418. https://doi.org/10.1016/j.energy.2016.04.037.
[56] G.Z. Kyzas, M. Kostoglou, Green adsorbents for wastewaters: a critical review, Materials (Basel) (2014) https://doi.org/10.3390/ma7010333.
[57] A. Li, H.L. Liu, H. Wang, H.Bin Xu, L.F. Jin, J.L. Liu, J.H. Hu, Effects of temperature and heating rate on the characteristics of molded Bio-Char, BioResources (2016). https://doi.org/10.15376/biores.11.2.3259-3274.
[58] W. Li, B. Mu, Y. Yang, Feasibility of industrial-scale treatment of dye wastewater via bio-adsorption technology, Bioresour. Technol. (2019). https://doi.org/10.1016/j.biortech.2019.01.002.
[59] Z. Liu, S. Singer, Y. Tong, L. Kimbell, E. Anderson, M. Hughes, D. Zitomer, P. McNamara, Characteristics and applications of biochars derived from wastewater solids, Renew. Sustain. Energy Rev. 90 (2018) 650–664. https://doi.org/10.1016/j.rser.2018.02.040.
[60] Lorenc-grabowska, E. Adsorption characteristics of Congo Red on coal-based mesoporous activated carbon 74 (2007) 34–40. https://doi.org/10.1016/j.dyepig.2006.01.027.
[61] M. Malhotra, S. Suresh, A. Garg, Tea waste derived activated carbon for the adsorption of sodium diclofenac from wastewater: adsorbent characteristics, adsorption isotherms, kinetics, and thermodynamics, Environ. Sci. Pollut. Res. (2018). https://doi.org/10.1007/s11356-018-3148-y.
[62] P.K. Malik, Use of activated carbons prepared from sawdust and rice-husk for adsoprtion of acid dyes: a case study of acid yellow 36, Dye. Pigment. (2003). https://doi.org/10.1016/S0143-7208(02)00159-6.
[63] Mall, I.D., Srivastava, V.C., Agarwal, N.K., Mishra, I.M., 2005. Removal of congo red from aqueous solution by bagasse fly ash and activated carbon : kinetic study and equilibrium isotherm analyses. Chemosphere 61, 492–501. https://doi.org/10.1016/j.chemosphere.2005.03.065.
[64] R. Mallampati, S. Valiyaveettil, Application of tomato peel as an efficient adsorbent for water purification - alternative biotechnology? RSC Adv (2012). https://doi.org/10.1039/c2ra21108d.
[65] K. Midhun Prasad, S. Murugavelh, Experimental investigation and kinetics of tomato peel pyrolysis: performance, combustion and emission characteristics of bio-oil blends in diesel engine, J. Clean. Prod. (2020). https://doi.org/10.1016/j.jclepro.2020.120115.
[66] A. Mlayah, S. Jellali, Study of continuous lead removal from aqueous solutions by marble wastes: efficiencies and mechanisms, Int. J. Environ. Sci. Technol. 12 (2015). https://doi.org/10.1007/s13762-014-0715-8.
[67] N. Mohamad Nor, L.C. Lau, K.T. Lee, A.R. Mohamed, Synthesis of activated carbon from lignocellulosic biomass and its applications in air pollution control - A review, J. Environ. Chem. Eng. (2013). https://doi.org/10.1016/j.jece.2013.09.017.
[68] J. Mokrzycki, I. Michalak, P. Rutkowski, Tomato green waste biochars as sustainable trivalent chromium sorbents, Environ. Sci. Pollut. Res. Int. (2019).
[69] H. Najafi, E. Pajootan, A. Ebrahimi, M. Arami, The potential application of tomato seeds as low-cost industrial waste in the adsorption of organic dye molecules from colored effluents 3994. Desalin Water Treat. (2016) https://doi.org/10.1080/19443994.2015.1072060.

[70] C. Namasivayam, D. Kavitha, Removal of Congo Red from water by adsorption onto activated carbon prepared from coir pith, an agricultural solid waste, Dyes Pigm. 54 (54) (2002) 47–58.

[71] Y.V. Nancharaiah, M. Sarvajith, Aerobic granular sludge process: a fast growing biological treatment for sustainable wastewater treatment, Curr. Opin. Environ. Sci. Heal. (2019). https://doi.org/10.1016/j.coesh.2019.09.011.

[72] V.O. Njoku, K.Y. Foo, B.H. Hameed, Microwave-assisted preparation of pumpkin seed hull activated carbon and its application for the adsorptive removal of 2,4-dichlorophenoxyacetic acid, Chem. Eng. J. (2013) https://doi.org/10.1016/j.cej.2012.10.068.

[73] E. Önal, N. Özbay, A.Ş. Yargıç, R.Z.Y. Şahin, Ö. Gök, Performance evaluation of the bio-char heavy metal removal produced from tomato factory waste, Progress in Exergy, Energy, and the Environment, Springer International Publishing, Switzerland, (2014). https://doi.org/10.1007/978-3-319-04681-5_70.

[74] C.H. Ooi, W.K. Cheah, Y.L. Sim, S.Y. Pung, F.Y. Yeoh, Conversion and characterization of activated carbon fiber derived from palm empty fruit bunch waste and its kinetic study on urea adsorption, J. Environ. Manage. (2017). https://doi.org/10.1016/j.jenvman.2017.03.083.

[75] N.B. Osman, N. Shamsuddin, Y. Uemura, Activated carbon of oil palm empty fruit bunch (EFB); core and shaggy, in, Procedia Engineering (2016). https://doi.org/10.1016/j.proeng.2016.06.610.

[76] N. Ozbay, A.S. Yargic, Comparison of surface and structural properties of carbonaceous materials prepared by chemical activation of tomato paste waste: the effects of activator type and impregnation ratio, J. Appl. Chem. (2016). https://doi.org/10.1155/2016/8236238.

[77] D. Permatasari, E. Heraldy, W.W. Lestari, Biosorption of toxic lead (II) ions using tomato waste (Solanum lycopersicum) activated by NaOH 030022. (2016) https://doi.org/10.1063/1.4941488.

[78] K.M. Prasad, S. Murugavelh, Experimental investigation and kinetics of tomato peel pyrolysis : performance, combustion and emission characteristics of bio-oil blends in diesel engine, J. Clean. Prod. 254 (2020) 120115. https://doi.org/10.1016/j.jclepro.2020.120115.

[79] T.V. Prokopov, M.I. Nikolova, D.S. Taneva, N.T. Petkova, Use of carotenoid extracted tomato residue for removal of cu (II) ions from aqueous solution, Sci. Study Res. Chem. Chem. Eng. Biotechnol. Food Ind. 20 (2) (2019) 135–150.

[80] M.K. Purkait, A. Maiti, S. DasGupta, S. De, Removal of congo red using activated carbon and its regeneration, J. Hazard. Mater. (2007). https://doi.org/10.1016/j.jhazmat.2006.11.021.

[81] V.V. Ranade, V.M. Bhandari, Industrial wastewater treatment, recycling, and reuse: an overview, in: Industrial Wastewater Treatment, Recycling and Reuse. (2014). https://doi.org/10.1016/B978-0-08-099968-5.00001-5.

[82] A. Reffas, V. Bernardet, B. David, L. Reinert, M.B. Lehocine, M. Dubois, N. Batisse, L. Duclaux, Carbons prepared from coffee grounds by H3PO4 activation: characterization and adsorption of methylene blue and Nylosan Red N-2RBL, J. Hazard. Mater. (2010). https://doi.org/10.1016/j.jhazmat.2009.10.076.

[83] S. Rezma, M. Birot, A. Hafiane, H. Deleuze, Physically activated microporous carbon from a new biomass source: Date palm petioles, Comptes Rendus Chim (2017). https://doi.org/10.1016/j.crci.2017.05.003.

[84] J. Rivera-Utrilla, C.V. Gómez-Pacheco, M. Sánchez-Polo, J.J. López-Peñalver, R. Ocampo-Pérez, Tetracycline removal from water by adsorption/bioadsorption on activated carbons and sludge-derived adsorbents, J. Environ. Manage. 131 (2013) 16–24. https://doi.org/10.1016/j.jenvman.2013.09.024.

[85] L.S. Rocha, D. Pereira, É. Sousa, M. Otero, V.I. Esteves, V. Calisto, Recent advances on the development and application of magnetic activated carbon and char for the removal of pharmaceutical compounds from waters: a review, Sci. Total Environ. 718 (2020) 137272. https://doi.org/10.1016/j.scitotenv.2020.137272.
[86] E. Rosales, J. Meijide, M. Pazos, M.A. Sanromán, Challenges and recent advances in biochar as low-cost biosorbent: From batch assays to continuous-flow systems, Bioresour. Technol. 246 (2017) 176–192. https://doi.org/10.1016/j.biortech.2017.06.084.
[87] J. Saleem, U.Bin Shahid, M. Hijab, H. Mackey, G. McKay, Production and applications of activated carbons as adsorbents from olive stones, Biomass Convers. Biorefinery. (2019). https://doi.org/10.1007/s13399-019-00473-7.
[88] A.K. Santra, T.K. Pal, S. Datta, Removal of metanil yellow from its aqueous solution by fly ash and activated carbon produced from different sources, Sep. Sci. Technol. (2008). https://doi.org/10.1080/01496390701885729.
[89] P. Sashidhar, M. Kochar, B. Singh, M. Gupta, D. Cahill, A. Adholeya, M. Dubey, Biochar for delivery of agri-inputs: current status and future perspectives, Sci. Total Environ. (2020). https://doi.org/10.1016/j.scitotenv.2019.134892.
[90] H. Say, F. Güzel, Ecotoxicology and Environmental Safety Effective removal of tetracycline from aqueous solution using activated carbon prepared from tomato (Lycopersicon esculentum Mill.) industrial processing waste. 131, (2016) 22–29. https://doi.org/10.1016/j.ecoenv.2016.05.001.
[91] H. Sayğili, F. Güzel, High surface area mesoporous activated carbon from tomato processing solid waste by zinc chloride activation: process optimization, characterization and dyes adsorption. J. Clean. Prod. (2016a). https://doi.org/10.1016/j.jclepro.2015.12.055.
[92] H. Sayğili, F. Güzel, Effective removal of tetracycline from aqueous solution using activated carbon prepared from tomato (Lycopersicon esculentum Mill.) industrial processing waste. Ecotoxicol. Environ. Saf. (2016b). https://doi.org/10.1016/j.ecoenv.2016.05.001.
[93] H. Sayğılı, F. Güzel, Usability of activated carbon with well-developed mesoporous structure for the decontamination of malachite green from aquatic environments: kinetic, equilibrium and regeneration studies, J. Porous Mater. (2018). https://doi.org/10.1007/s10934-017-0459-1.
[94] H. Saygl, F. Guzel, Behavior of mesoporous activated carbon used as a remover in Congo red adsorption process, Water Sci. Technol. (2018). https://doi.org/10.2166/wst.2018.100.
[95] L.Silvani, G.Cornelissen, S.E.Hale, Sorption of Α-, Β-, Γ- and Δ-hexachlorocyclohexane isomers to three widely different biochars: sorption mechanisms and application, Chemosphere (2019). https://doi.org/10.1016/j.chemosphere.2018.12.070.
[96] K.S.W. Sing, D.H. Everett, R.A.W. Haul, L. Moscou, R.A. Pierotti, J. Rouquerol, T. Siemieniewska, Reporting physisorption data for gas/solid systems with special reference to the determination of surface area and porosity, Pure Appl. Chem. (1985). https://doi.org/10.1351/pac198557040603.
[97] V. Sivakumar, Removal of Congo red dye using an adsorbent prepared from Martynia annua, L. Seeds. Am. Chem. Sci. J. (2014). https://doi.org/10.9734/acsj/2014/6680.
[98] T. Sizmur, T. Fresno, G. Akgül, H. Frost, E. Moreno-Jiménez, Biochar modification to enhance sorption of inorganics from water, Bioresour. Technol. 246 (2017) 34–47. https://doi.org/10.1016/j.biortech.2017.07.082.
[99] B. Smider, B. Singh, Agriculture, Ecosystems and Environment Agronomic performance of a high ash biochar in two contrasting soils, Agric. Ecosyst. Environ 191 (2014) 99–107. https://doi.org/10.1016/j.agee.2014.01.024.

[100] X. Tan, Y. Liu, G. Zeng, X. Wang, X. Hu, Y. Gu, Z. Yang, Application of biochar for the removal of pollutants from aqueous solutions, Chemosphere 125 (2015) 70–85. https://doi.org/10.1016/j.chemosphere.2014.12.058.
[101] P. Taylor, Y. Gutha, V.S. Munagapati, M. Naushad, K. Abburi, Desalination and water treatment removal of Ni (II) from aqueous solution by Lycopersicum esculentum (tomato) leaf powder as a low-cost biosorbent, Desalin Water Treat (2014) 37–41. https://doi.org/10.1080/19443994.2014.880160.
[102] K.J. Tiemann, G. Gamez, K. Dokken, J.G. Parsons, J.L. Gardea-Torresdey, Chemical modification and X-ray absorption studies for lead(II) binding by Medicago sativa (alfalfa) biomass, Microchem. J. (2002). https://doi.org/10.1016/S0026-265X(02)00021-8.
[103] UN-Water, Waterwater - The Untapped Resources, The United Nations World Water Development Report. Wastewater. The Untapped Resource, United Nations Education, Scientific & Cultural Organization, Paris, France, 2017. https://doi.org/10.1017/CBO9781107415324.004.
[104] B. Üner, K. Kömbeci, M. Akgül, The utilization of tomato stalk in fiber production: NaOH and CaO pulping process, Wood Res. 61 (6) (2016) 927-936.
[105] United Nations, 2019. The sustainable development goals report 2019. United Nations Publ. issued by Dep. Econ. Soc. Aff.
[106] D. Ure, A. Awada, N. Frowley, N. Munk, A. Stanger, B. Mutus, Greenhouse tomato plant roots/carboxymethyl cellulose method for the efficient removal and recovery of inorganic phosphate from agricultural wastewater, J. Environ. Manage. (2019). https://doi.org/10.1016/j.jenvman.2018.12.053.
[107] S. Vafakhah, M.E. Bahrololoom, M. Saeedikhani, Adsorption kinetics of cupric ions on mixture of modified corn stalk and modified tomato waste, J. Water Resour. Prot. (2016). https://doi.org/10.4236/jwarp.2016.813095.
[108] W.S. Wan Ngah, M.A.K.M. Hanafiah, Removal of heavy metal ions from wastewater by chemically modified plant wastes as adsorbents: a review, Bioresour. Technol. (2008). https://doi.org/10.1016/j.biortech.2007.06.011.
[109] E. Weidemann, M. Niinipuu, J. Fick, S. Jansson, Using carbonized low-cost materials for removal of chemicals of environmental concern from water, Environ. Sci. Pollut. Res. 25 (2018) 15793–15801.
[110] W. Xiang, X. Zhang, J. Chen, W. Zou, F. He, X. Hu, D.C.W. Tsang, Y.S. Ok, B. Gao, Biochar technology in wastewater treatment: a critical review, Chemosphere (2020). https://doi.org/10.1016/j.chemosphere.2020.126539.
[111] M. Xie, W. Chen, Z. Xu, S. Zheng, D. Zhu, Adsorption of sulfonamides to demineralized pine wood biochars prepared under different thermochemical conditions, Environ. Pollut. (2014). https://doi.org/10.1016/j.envpol.2013.11.022.
[112] E. Yagmur, Preparation of low cost activated carbons from various biomasses with microwave energy, J. Porous Mater. (2012). https://doi.org/10.1007/s10934-011-9557-7.
[113] Y. Yao, B. Gao, J. Chen, M. Zhang, M. Inyang, Y. Li, A. Alva, L. Yang, Engineered carbon (biochar) prepared by direct pyrolysis of Mg-accumulated tomato tissues: Characterization and phosphate removal potential, Bioresour. Technol. (2013). https://doi.org/10.1016/j.biortech.2013.03.057.
[114] Y. Yao, B. Gao, J. Fang, M. Zhang, H. Chen, Y. Zhou, A.E. Creamer, Y. Sun, L. Yang, Characterization and environmental applications of clay-biochar composites, Chem. Eng. J. (2014). https://doi.org/10.1016/j.cej.2013.12.062.
[115] Y. Yao, B. Gao, M. Inyang, A.R. Zimmerman, X. Cao, P. Pullammanappallil, L. Yang, Removal of phosphate from aqueous solution by biochar derived from anaerobically digested sugar beet tailings, J. Hazard. Mater. (2011). https://doi.org/10.1016/j.jhazmat.2011.03.083.

[116] A.S. Yargiç, R.Z. Yarbay Şahin, N. Özbay, E. Önal, Assessment of toxic copper(II) biosorption from aqueous solution by chemically-treated tomato waste, J. Clean. Prod. (2015). https://doi.org/10.1016/j.jclepro.2014.05.087.
[117] Z.M. Yunus, Y. G, A. Al-Gheethi, N. Othman, R. Hamdan, N.N. Ruslan, Advanced methods for activated carbon from agriculture wastes; a comprehensive review, Int. J. Environ. Anal. Chem. (2020). https://doi.org/10.1080/03067319.2020.1717477.
[118] H. Zheng, Z. Wang, J. Zhao, S. Herbert, B. Xing, Sorption of antibiotic sulfamethoxazole varies with biochars produced at different temperatures, Environ. Pollut. 181 (2013) 60–67. https://doi.org/10.1016/j.envpol.2013.05.056.
[119] A.A. Zorpas, I. Voukalli, P. Loizia, Chemical treatment of polluted waste using different coagulants, Desalin. Water Treat. (2012). https://doi.org/10.1080/19443994.2012.692043.

CHAPTER NINE

# Thermochemical conversion of tomato wastes

**Mejdi Jeguirim[a], Besma Khiari[b]**
[a]The institute of Materials Science of Mulhouse (IS2M), University of Haute Alsace, University of Strasbourg, CNRS, UMR, Mulhouse, France
[b]Laboratory of Wastewaters and Environment, Centre of Water Researches and Technologies (CERTE), Technopark Borj Cedria, Tunisia

## 9.1 Introduction

Tomato is one of the most consumed vegetables in the world either by direct consumption (70%) or after industrial transformation (30%) [1]. The post-harvest circuits as well as the processing by-products generate important losses and bring low returns to growers, processors and traders. More precisely, one third of the whole production quantity is converted to undesirable wastes (immature tomatoes, lesioned tomatoes for mechanical or microbial action, processing residues, pomace, wastes of refining or cleaning, skins, and seeds). Moreover, the refuse might be an additional cost for companies due to the disposal processes, which have to be rapid and adequate. Indeed, the accumulation of these residues promotes uncontrolled anaerobic fermentations, especially in warm periods, leading to environmental problems [2]. Besides, as these residues have high moisture contents, storage and transport are costly steps impeding a reasonable discharge way. The treatment processes are also themselves difficult to master as the physical and the chemical properties vary significantly within and between the raw tomato wastes (climatic conditions, soil properties, etc.) and the industrial by-product ones (ranging from simple preservation such as sun drying or freezing, to modern, complex and capital-intensive methods such as formulating a frozen meal or a make-up with the right balance of ingredients).

For all these reasons, tomato manufacturing companies used to give their production residues for free to other companies that generally use them for feeding livestock or as soil amendment. Recently, energetic use of the tomato plant waste was proposed as an alternative recovery issue. Many researchers found out that tomato wastes could be suitable for different thermal conversion processes (combustion, pyrolysis, gasification and hydrothermal conversion). Indeed, the high volatile matter

*Tomato Processing by-Products*
DOI: https://doi.org/10.1016/B978-0-12-822866-1.00008-9

and carbon suggests the high potential of energy recovery. Besides, the increasing competition for solid biomass, such as wood pellets, creates space for relatively novel biomass sources to enter the market, among which tomato wastes have a great potential. The elemental composition reported in many papers confirmed the applicability of a thermal degradation process to these residues. For example, the H/C and O/C ratios varied respectively from 12.9% to 14.9% and from 37% to 78.6%, which is in the same range of many other hydrocarbons [2–4].

However, low mass density has been reported to be limiting factors for direct use as biofuels. This was circumvented by pelleting the TW in order to optimize energy production [5,6].

Still, other physical properties, such as moisture content, particle size and distribution, durability, etc., are critical parameters in the choice of the process and the equipment. Chemical properties are also of great importance for the energy efficiency, environmental pollution, and ash-related operating problems. Depending on the type of application, the above-mentioned variations may also affect the systems performance.

## 9.2 Combustion

Tomatoes are one of the most widely produced vegetables on the planet. After harvest campaigns or at the end of a processing unit, tomato plants and/or wastes are often burnt on the field.

This open burning as well as non-ideal combustion conditions, may lead to environmental issues. In addition, the cost and the non-benefit from the generated waste, making tomato processing industries vulnerable.

For a the suitable recovery through combustion process, several parameters must be taken into account during the plant design, namely the fuel characteristics (water content, particle size, density, temperature, etc.), the fuel injection and its frequency, minimum storage autonomy, etc. An effective way to control these aspects is through laboratory simulations, using either balances for thermalgravimetric analysis (TGA) or laboratory/pilot scale reactors for more accurate calculations.

### 9.2.1 Thermogravimetric and kinetic analyses

#### *9.2.1.1 Thermal degradation behavior*

The thermal degradation behavior that is, mass loss evolution with increasing temperature, can be followed using thermogravimetric analysis. Few works have investigated thermal degradation of tomato wastes under oxidative atmosphere in TG conditions at different heating rates.

Font et al. [7] as well as Khiari et al. [8] carried out such experiments. During these studies, an initial small endothermic peak was due to the evaporation of the humidity while the subsequent small exothermic peaks were attributed to devolatilization step. The main exothermic phase of decomposition was related to the combustion of char [9]. In the above mentioned investigations, whatever is the heating rate, the mass loss profiles were reported very similar and close to each other. For instance, thermal degradation of TW is composed of three consecutive and overlapping steps (Fig. 9.1). First, the particle heats up and the moisture in the sample is removed causing a small loss weight stage in the TG profile. With temperature rising, pyrolysis takes place, which can be seen in the large weight loss phase in the curve (up to 85%) where a cloud of volatile matter, containing probably oxidizable materials, is formed around the particle. Once the volatiles are exhausted, the oxygen of the oxidizing gas (air) can reach the surface of the solid residue [10]. The heterogeneous combustion phase then begins and lasts a quite long time during the combustion process until ultimately only ash remains.

It is important to notice that increasing the heating rate resulted in higher temperatures, broader combustion ranges and increasing maximum mass loss rates. The discrepancies in the above mentioned works may be attributed to the differences in texture and climatic conditions of plantation (Tunisia and Spain).

#### 9.2.1.2 Kinetic parameters extraction

The thermal degradation under oxidative atmosphere occurs through multiple physical and chemical events including crystallographic structure destruction, chemical bonds breaking, solid product recrystallization, and gas desorption [9,11]. The reaction rate is then formalized by taking into consideration the geometry and the global kinetics of progression of the reaction interface [7].

Several kinetic models have been proposed, according to the mechanism described by the decomposition reactions. Kinetics can then be determined in isothermal or in non-isothermal conditions. Three variants of this last method are usually used: Friedman [12], Flynn–Wall–Ozawa (FWO) [13], and Kissinger–Akahira–Sunose (KAS) [14] (Table 9.1).

Khiari et al. have applied these methods at four different heating rates (5, 10, 20 and 30°C/min) to extract kinetic parameters and construct the kinetic model [8]. The mean activation energy values for tomato waste using the KAS, FWO and Friedman methods were, respectively,

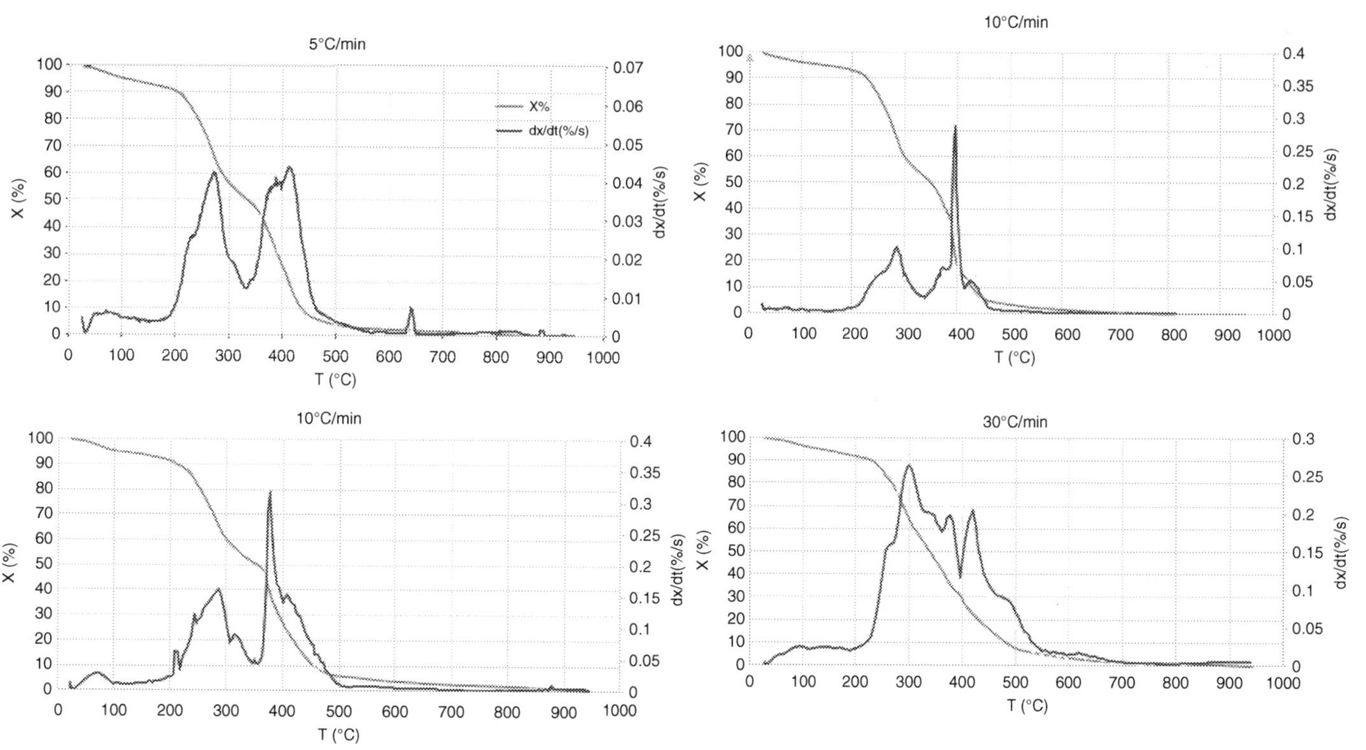


**Fig. 9.1** ***TG and DTG profiles of Tomato wastes combustion at different heating rates*** [8].

**Table 9.1** Isoconversional kinetic models.

| Method | Expression | Plots |
|---|---|---|
| Friedman | $\text{Ln}\frac{dX}{dt} = \text{Ln}\left(\beta\frac{dX}{dT}\right) = \text{Ln}\left[Af(X)\right] - \frac{Ea}{RT}$ | $\text{Ln}\left(\beta\frac{dX}{dT}\right)$ vs $\frac{1}{T}$ |
| FWO | $\text{Ln}\beta = \text{Ln}\frac{AEa}{g(X)R} - 2.315 - \left(\frac{1.0516Ea}{RT}\right)$ | $\text{Ln}\beta$ vs $\frac{1}{T}$ |
| KAS | $\text{Ln}\frac{\beta}{T^2} = \text{Ln}\frac{AR}{Eag(X)} - \frac{Ea}{RT}$ | $\text{Ln}\frac{\beta}{T^2}$ vs $\frac{1}{T}$ |

231, 224 and 199 kJ/mol. The activation energy increased, reached maximum values of 495.96, 481 and 500kJ/mol at 55-60% conversion rate then dropped until the end of the degradation process denoting the decomposition of different macro-components contained in the tomato waste (Fig. 9.2). At low conversion levels (10–30%), activation energies between 170 and 186 kJ/mol are probably related to hemicelluloses degradation while those recorded afterward (190–207 kJ/mol) are attributed to the cracking of cellulose as they are in the range of 150–250 kJ/mol met in literature [11,15,16]. Lignin is the most stable compound; consequently, it degrades over a wide interval between 55% and 75% conversion rates with maximum activation energy. Finally, energy values tend to decrease until the end of the combustion process, corresponding to the phase of char combustion. The fittings between theoretical and experimental masses were good and the models validated by first-order or contracted sphere models according to the applied heating rate (Fig. 9.3).

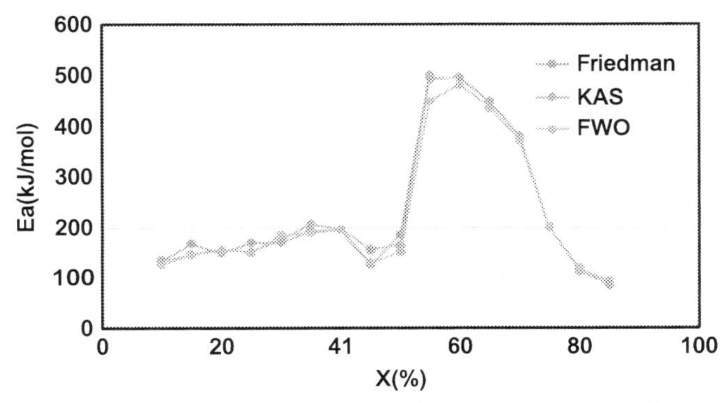


**Fig. 9.2** ***Activation energy values calculated by Friedman, KAS and FWO models*** [8].

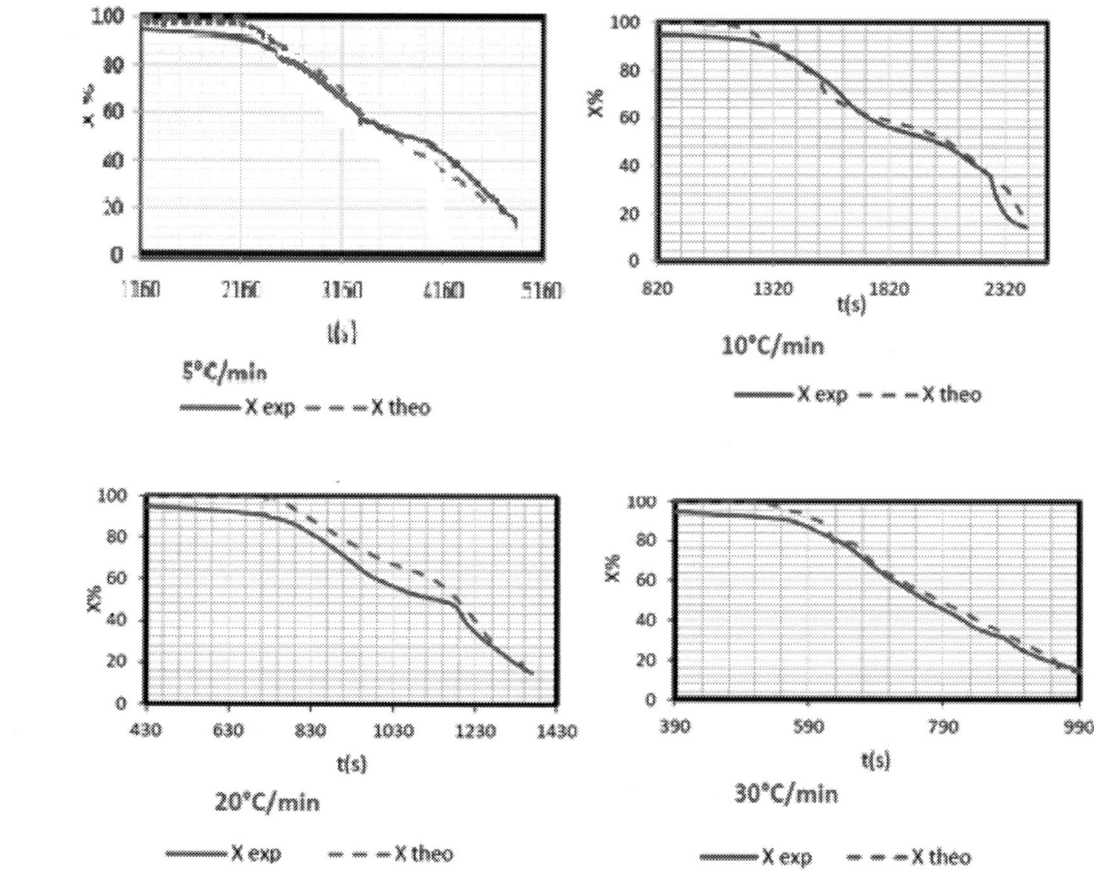


**Fig. 9.3** ***Model validation for different heating rates*** *(adaptef from* [8]*).*

### 9.2.2 Combustion

Several investigations have made an effort to minimize some of the uncertainty regarding the open burning or the uncontrolled combustion of tomato wastes. Therefore, several reactors have been used for the optimization of tomato waste combustion. For example, Molto et al. have used a residential stove in order to obtain a combustion chamber with poor oxygen-fuel mixture, similar to the one obtained in an open-burning fire, but with the possibility to reproduce the operating conditions in the greenhouses. The same authors carried out combustion tests in a batch laboratory scale horizontal tubular reactor for continuous simulations (Fig. 9.4).

González et al., have evaluated the performance of a mural boiler for TW combustion (Fig. 9.5) [6]. They found that tomato residues give higher boiler efficiency than other biomasses they have tested such as forest residues, sorghum and, almond pruning and reed [18].

In another study, Gonzalez et al. studied the performance of the same reactor by varying two operating parameters (fuel mass flow and chimney draught) applied to mixtures of tomato residues with olive stones (50%–50%) and with forest residues (75%–25%) as well as other agricultural residues [6].

The mass flow effect on the main combustion parameters was similar for all studied residues: An increase in mass flow produces a decrease in the $O_2$ content in the fumes, in $\lambda$ and in $q_A$, and an increase in $CO_2$ content in the fumes, in the fumes volumetric flow, in air outlet temperature and in the combustion efficiency. Similarly, the chimney draught effect on the main combustion parameters was the same for all pellets. An increase in draught produces an increase in the $O_2$ content in the fumes, in $\lambda$, in qA and in the fumes volumetric flow, and a decrease in CO content and the combustion efficiency.

The optimal experimental conditions and the main combustion parameters for tomato-based pellets are given in Table 9.2. The sequence of the combustion process efficiency was tomato pellet 75% - forest pellet 25% > tomato pellet 50% - olive stone 50% > 100% tomato pellet. This sequence is the same as for the fuel HHV and C content. The authors also stated that the influence of the fuel mixture mass flows on the different combustion parameters was similar to that observed for individual fuel combustion, i.e. an increase in mass flow causes a decrease in the $O_2$ content in the fumes and in the air excess coefficient, and an increase in the $CO_2$ content in the fumes and in the air outlet temperature [6].

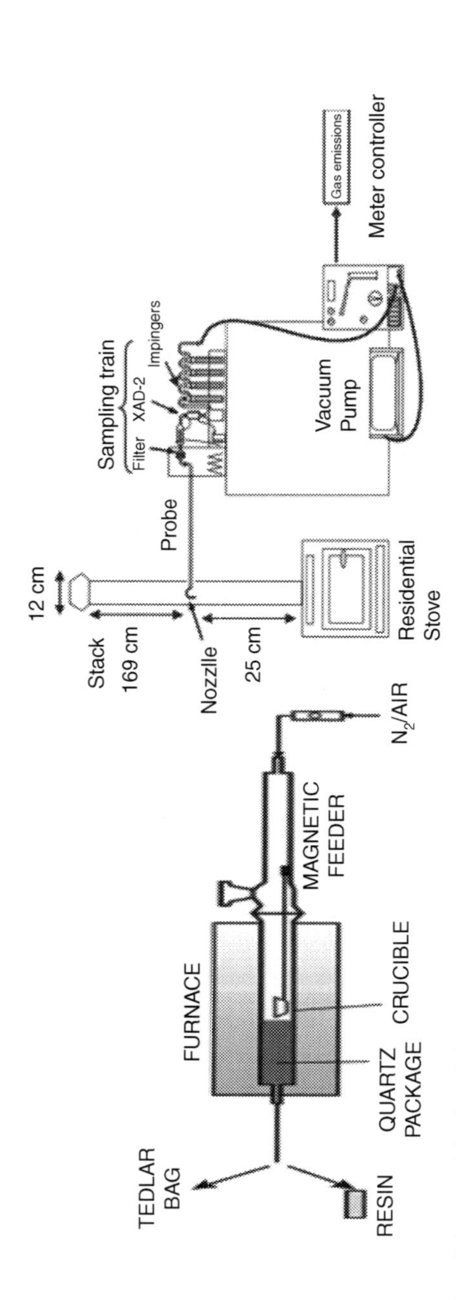


**Fig. 9.4** Scheme of the batch laboratory scale tubular reactor (left) and the experimental set-up used to simulate open burning (right) [17].

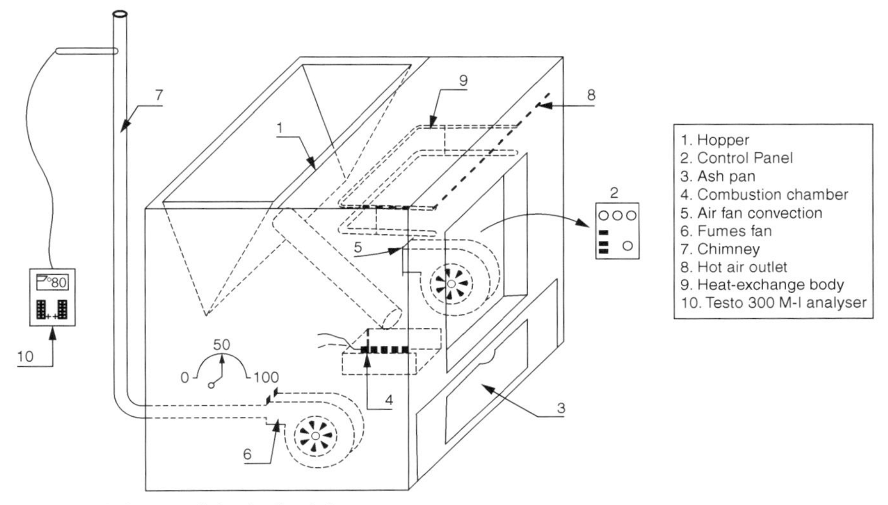


**Fig. 9.5** ***Scheme of the boiler*** [6].

Gonzales et al. concluded their works by confirming that tomato waste based pellets were excellent substitutes for the forest pellet recommended by the boiler manufacturer.

However, as it is going to be detailed in the next sections, the combustion of TW can liberate quite important quantities of gas, dust, and aerosols and, lead to deposit formation (slagging, fouling) and corrosion of the combustion systems due to high ash, N, K, S and Cl increased by the use of fertilizers, pesticides and herbicides in agriculture compared to wood. More precisely, these elements lead to quite important NOx, SOx, and HCl emissions compared to wood pellets. K plays a role in both particulate

**Table 9.2** The main combustion parameters for different tomato-based pellets.

| | Tomato pellet | Tomato 75% – forest 25% | Tomato 50% – olive stone 50% |
|---|---|---|---|
| Mass flow (kg/h) | 2.42 | 1.30 | 1.27 |
| $O_2$ (%) | 10.4 | 7.9 | 13 |
| CO (ppm) | 2414 | 5159 | 2013 |
| $CO_2$ (%) | 7.2 | 8.9 | 5.5 |
| Air outlet temperature AOT (°C) | 102 | 98 | 66 |
| Fumes temperature FT (°C) | 107 | 101 | 79 |
| Sensitive heat loss of the fumes qA (%) | 7 | 4.4 | 5.9 |
| Air excess coefficient $\lambda$ | 1.9 | 1.6 | 2.6 |
| Efficiency $\eta$ (%) | 91 | 92.4 | 92 |

emissions and slagging of an increased ash volume, while lowering the softening temperature of the fuel. A high Cl content brings out corrosion problems on the boiler surface and contributes into the dioxins and furans formation, higher than during the combustion of textile wastes, waste lube oil, meat and bone meals and waste paper [19].

Finally, as well as large scale are concerned, the high ash and the low melting point, play in favor of straw pellets in grate combustion or fluidized bed systems as stated by Kraiem et al. [5]. The authors confirmed that some of these obstacles could partially be overcome by using multi-fuel boilers ranging between 10 to 60 kW. Such arrangement could be more suitable for agripellets burning or for agricultural residue co-firing with fossil fuels or with sawdust. This also could be beneficial when cleaning the agricultural residues before pelletizing by lowering their ash contents.

Besides, low mass density has been reported to be limiting factors for direct use as biofuels, suggesting pelletizing as a mean to optimize energy production. These aspects are detailed in the following sections.

#### *9.2.2.1 Densification*

The production of thermal energy by tomato wastes has shown a clear trend toward densified biofuels (called tomato waste agripellets). This is due to the consistent size and shape that can be more easily delivered to homes, businesses, and power plants and that can be automatically fed into advanced pellet boilers in a controlled and calibrated way. The use of densified TW also reduces the costs associated with handling and transportation, due to the increase in density involved by the densification processes. However, they do have certain differences compared to conventional wood pellets.

Indeed, pellets made from TW, just like any other agricultural residue, are more suitable for large-scale combustion plants than for wood pellets that should be preferred for small-scale systems, both in terms of technical and environmental aspects. As mentioned earlier, quantities are sometimes not sufficient and the main challenge regarding TW-pellets is to improve small-scale combustion devices and to produce high quality fuels; the largest ones being equipped with sophisticated combustion control systems and flue gas cleaning systems.

As a matter of fact, Ruiz-Celma et al. reported a high heating value of tomato seeds and peels pellets and an energy density around 8 $GJm^{-3}$, very close to other agricultural waste pellets, regardless their low bulk densities [20]. Gonzalez et al. also studied a number of pellets made of different agricultural residues and revealed that the tomato pellet is the fuel with the highest contents of C and H and, therefore, the highest HHV (22.7 MJ/kg) [6]. This potential fuel had very low S content (0.074%) and the possibility to form NOx is practically negligible, due to the fact that the temperature reached in the combustion chamber used was relatively low (less than 600°C). The Cl content is low too (0.12%), eliminating the risk of chlorine dioxide in the combustion, which attacks the metallic surfaces of the boiler. Indeed, the use of fuels with high Cl content in heat generation installations is not advisable, especially if the stainless steel of the boiler is not of good quality. The tomato pellets prepared by Gonzales et al. had very high volatile matter contents (80.1%) and quite high ash content (3.5%). This last is problematic as, if the ashes melt in the combustion chamber at low temperatures, this would obstruct the orifice for inlet air in the heat generation plant.

Nevertheless, the high nitrogen and ash contents strongly limit the certification of pure TW-pellets as a widely used energy source. In this regard, several studies have shown that blending tomato wastes with sawdust before pelleting could help to meet the required certifications. For instance, Kraiem et al. could, with their four agropellet types: 100% tomato waste, 50-50% tomato waste-sawdust, 25-75% tomato waste-sawdust and 100% sawdust, meet French certifications such as AQHP (Agro Quality High Performance) and AQI (Agro Industrial Quality), and the European standard EN14961-1-6 [5]. Indeed, the authors found that the characteristics of the prepared pellets played in favor of the tomato wastes densification. For instance, the length to diameter (L/D) ratios varied between 2.2 and 4.5, which is very important to insure a good combustion [21]. The apparent densities also were good (473 $kg/m^3$ to 522 $kg/m^3$), not far from that of the pure beech sawdust pellet samples (601 $kg/m^3$). As for the mineral contents, potassium, calcium, magnesium and sodium were the major elements (more than 70% wt. of the total composition of the inorganic elements). However, the high nitrogen, sulfur, chlorine and ash contents have limited the marketing of pure tomato waste pellets. For example, the ash content was up to 11 wt.% for the TW pellets, while chlorine and Sulfur were both higher than 0.2 wt.%. More characteristics for 100% TW pellets and 50% TW – 50% PS pellets are gathered in Table 9.3; In this table, fuel indexes in the form of molar ratios (K + Na)/(2S + Cl) and (Si + P + K)/(Ca + Mg) are also logged. The first may be an indicator for HCl and $SO_2$ emission levels while the second ratio may help to analyze the slagging tendency in the bottom ash. Compared to sawdust, these two ratios are

**Table 9.3** Chemical and physical parameters of the prepared pellets.

| Parameter | Unit | TW | TW-PS | PS |
|---|---|---|---|---|
| Moisture | (wt %, ar) | 10 | 11 | 13 |
| Ash | (wt %, db) | 11 | 4 | 0.6 |
| Bulk density | ($kg/m^3$) | 522 | 473 | 601 |
| $LHV^{wb}$ | ($MJ/kg^1$) | 19.5 | 17.6 | 16.4 |
| $ED_{pellets}$ | ($GJ/m^3$) | 10.2 | 8.3 | 9.8 |
| N | $(\%)_{db}$ | 1.5 | 0.8 | 0.2 |
| S | $(g/kg^1)_{db}$ | 2.96 | 2.08 | 0.12 |
| K | | 30.48 | 16.61 | 0.36 |
| Cl | | 5.75 | 3.42 | 0.31 |
| Ca | | 1.45 | 1.22 | 0.36 |
| Si | | 0.19 | 0.18 | 0.01 |
| Na | | 0.35 | 0.29 | 0.01 |
| P | | 0.93 | 0.64 | 0.08 |
| Mg | | 0.59 | 0.45 | 0.02 |
| Al | | 0.12 | 0.11 | 0.01 |
| Fe | | 0.10 | 0.10 | 0.02 |
| Mn | | 0.09 | 0.01 | 0.05 |
| K + Na + S + Cl | | 39.54 | 22.40 | 0.8 |
| (K +Na)/ (2S+ Cl) | mol/mol | 2.29 | 1.93 | 0.59 |
| (Si +P +K)/ (Ca + Mg) | mol/mol | 13.51 | 9.23 | 1.24 |

Ar, as received; db, dry basis.

high, which indicated that higher concentrations of HCl and $SO_2$ would be found in the flue gas on one hand and that the risk of slagging in bottom ash is important on the other hand.

Still, the combustion tests of the same authors indicated that the prepared pellets had no significant negative effects on the combustion efficiency nor on the boiler performances (from 80.6% to 82.8% versus 83.4% for wood pellets). Moreover, the least value of combustion efficiency was still higher than the minimum value required by the European Standard EN 303-5 (77%). For the other samples, tomato waste, as an example, and even though that it had a higher low heating value (19.5 MJ/kg) than wood pellets (16.4 MJ/kg), registered a slightly lower efficiency. Authors explained the fact that the combustion efficiency may be low for a sample that has a high LHV by several reasons such as low air supply and high ash and fine contents [22]. To improve boiler and combustion efficiencies, they recommended manual adjustment of combustion by providing primary and secondary air, checking the air factor which value must be between 1.5 and 2.5 as well as the air excess for which emission of CO could be minimum.

#### 9.2.2.2 *Emission issues*

During combustion, tomato residues produce soot, particulate matter, carbon monoxide (CO), light hydrocarbons, volatile organic compounds (VOC) and semi-volatile compounds including several persistent bio-accumulative and toxic pollutants mainly polycyclic aromatic hydrocarbons (PAH). The relatively high nitrogen content can generate environmental problems in terms of NOx emissions.

When combustion is incomplete (often because of low combustion temperatures, short residence times or/and oxygen shortage), TW agripellets results in CO, VOC, particles, tars, and PAH.

Emission factors of several compounds were determined in the combustion of tomato plant in a residential stove and in a laboratory scale reactor. In all the runs, nearly the same PAHs were identified, naphthalene being the main one. Combustion of tomato plant at 500°C in the laboratory scale reactor produced the highest emission factors for all compounds analyzed, even if the $CO/(CO + CO_2)$ ratios were close to 0.8–0.9, indicating similar combustion conditions concerning the carbon oxides formation [23].

Combustion of tomato plant produces higher PCDD/F amount than the combustion of textile wastes, waste lube oil, meat and bone meals and wastepaper; otherwise in combustion of PVC and some samples of sewage sludge much higher amounts of PCDD/Fs are obtained.

Blending tomato wastes with woody biomass was again proposed as a solution to reduce the emissions, but the reduction of NOx, $SO_2$, HCl, and total particulate matter was substantial only for a ratio of at least 50 wt % wood [24]. For Kraiem et al., who combusted different pellets made of tomato wastes and sawdust in a commercial boiler that was initially meant for woody biomass, noted that the CO and, to a lesser extent VOC, emitted from the different TW based pellets present significant peaks on the graph of gaseous emissions (Fig. 9.6 and Table 9.4) in comparison with woody pellets which were more stable. Therefore, the authors suggested to adjust the boiler operation parameters since the air excess for the woody pellets might be sufficient for the combustion process while for tomato waste based pellets, the distribution between the primary and secondary air might be not optimal. They also didn't rule out the idea that it might be due to the ash low melting temperature, which causes their agglomeration on the plate, leading to the partial blockage of the air inlet and the unburned gaseous emissions. These are also the reasons given to explain the increased emissions of NOx (higher

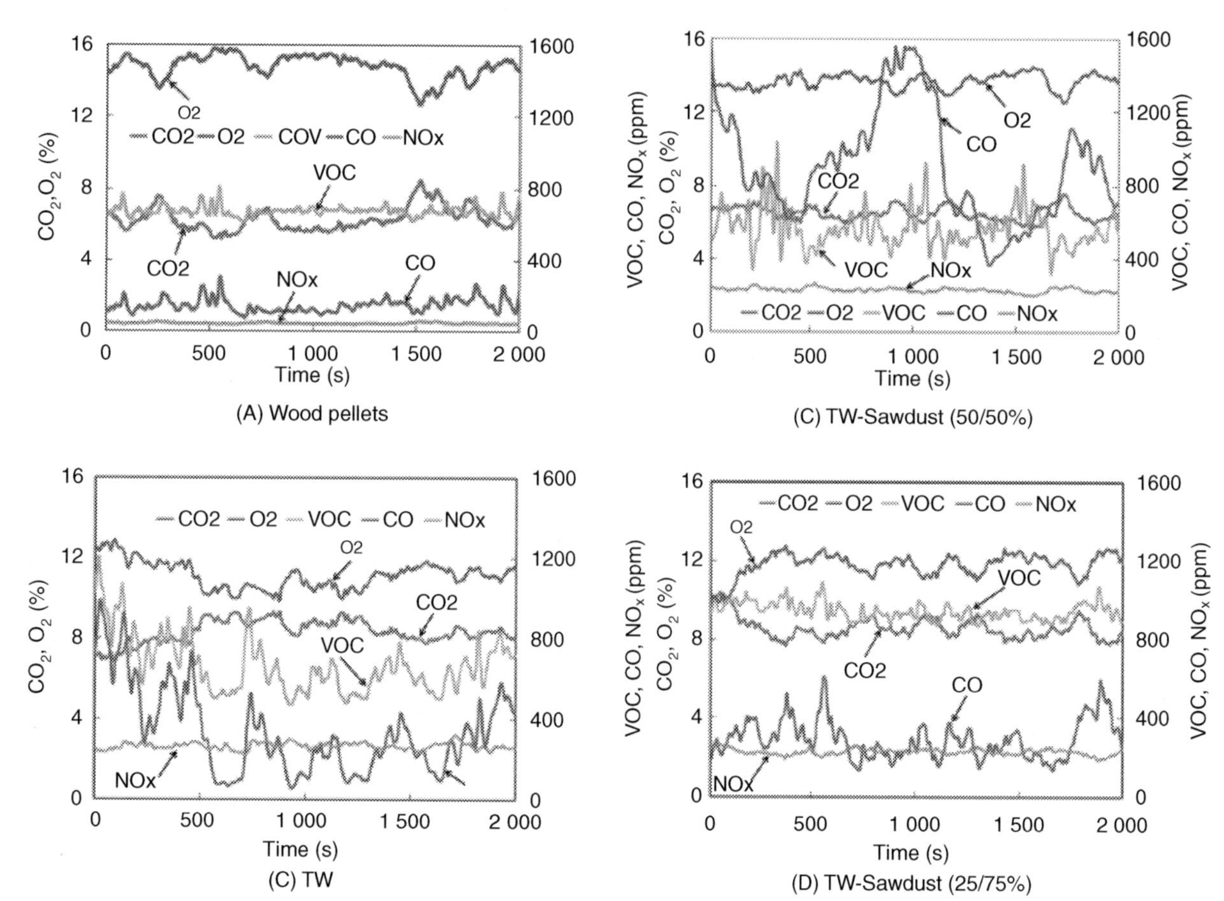


**Fig. 9.6** ***Gaseous emissions during combustion tests of the different pellets*** *(adapted from [5])*.

**Table 9.4** Mean values of the raw gas emissions (adapted from [5]).

| | | | CO (mg/Nm³) | $NO_x$ (mg/Nm³) | VOC (mg/Nm³) | $SO_2$ (mg/Nm³) |
|---|---|---|---|---|---|---|
| **Sample** | **$CO_2$ (%)** | **$O_2$ (%)** | **At 10% of $O_2$** | | | |
| Wood pellets | 6 | 15 | 346 | 116 | 914 | 16 |
| TW | 8 | 11 | 495 | 428 | 566 | 231 |
| TW-sawdust (25%–75%) | 8 | 12 | 465 | 399 | 861 | 225 |
| TW-sawdust (50%–50%) | 7 | 13 | 809 | 537 | 515 | 235 |

than 400 and up to 537 mg/Nm$^3$ vs 116 mg/Nm$^3$ for wood pellets). Besides, the increase in CO was correlated with the increase of ash content. Indeed, the addition of tomato waste to sawdust increased the ash contents as follows: CO wood pellets (346 mg/Nm$^3$) < CO TW-sawdust (465 mg/Nm$^3$) <CO TW (495 mg/Nm$^3$). CO fluctuations were also explained by the fact that the emission depends on the temperature conditions, the gas residence time (ratio of volume of the combustion chamber/gas flow), turbulence, and excess air, which are decisive for the optimization of combustion process [5, 22].

The VOC emissions followed the opposite trend of the CO emissions for the different samples, suggesting that significant amounts of VOCs were not completely oxidized and converted into CO. More precisely, TW-sawdust have the lowest VOC emissions (566 mg/Nm$^3$) while TW (50 wt %)-Sawdust, TW (25 wt %)-Sawdust and wood pellets presented the highest VOC emissions values (675, 861, and 914 mg/Nm$^3$, respectively).

The NOx emissions for wood pellets were less than those for tomato waste (428 mg/Nm$^3$), TW-sawdust 25%–75% (399 mg/Nm$^3$) and TW-sawdust 50%–50% (359 mg/Nm$^3$), which was expected as tomato residues had higher nitrogen contents in comparison with wood pellets. The same observations were made for $SO_2$, to be correlated with the sulfur contents higher for tomato wastes than for wood.

The mean values of total PM emissions observed during the combustion at 10% and 13% $O_2$ of the four previously mentioned pellets are seen in Table 9.5 and Fig. 9.7.

The addition of tomato wastes to wood pellets led to an increase of the PM concentration, mainly due to their ash contents. The main part of the particles generated during the combustion step corresponded to submicrometric ones, due probably to the presence of K and Ca, which might

**Table 9.5** Comparison of PM emissions obtained with different pellets (adapted from [5]).

| Sample | $O_2$ | Particles (mg/Nm$^3$) |
|---|---|---|
| Sawdust | 10 | 143 |
| | 13 | 104 |
| TW | 10 | 2719 |
| | 13 | 1978 |
| TW-Sawdust (25%–75%) | 10 | 1006 |
| | 13 | 731 |
| TW-Sawdust (50%–50%) | 10 | 2279 |
| | 13 | 1658 |

have catalytic effects and led to the fractionation of larger particles of more than 1 mm into smaller ones. These fine particles include soot agglomerates and condensable organic vapors. As for the bottom ashes, the more silicon was present in the ash, the fewer particles were emitted even if the biomass contained high amounts of minerals. These results were explained by the partial encapsulation of light minerals such as K and Na by silicon. The authors also noted that the presence of higher amounts of minerals in the pellets induced an increase of the mean particle diameter emitted during the combustion. They attributed this to the sintering of minerals when ash content is high.

The chemical composition of the bottom ash showed that the major oxides were CaO, $K_2O$, and $SiO_2$. The high concentrations of $P_2O_5$ and MgO were attributed to significant P and Mg contents in the initial TW (Table 9.6).

#### *9.2.2.3 Deposit formation*

Tomato waste boiler issues regarding slagging, fouling, and corrosion are related to alkali species, mainly potassium and sodium compounds. Combined with silica, these alkali species condense on cold boiler surfaces causing fouling and corrosion. This is referred to as slagging when the deposits are in a molten or highly viscous state, or fouling when the deposits are built up largely by species that have vaporized and then condensed. Slagging is often found in the radiant section of the furnace, while fouling occurs in the cooler furnace regions where the heat exchanger equipment is located [25]. The negative effects of slagging and fouling are high furnace material wear, heat transfer efficiency reduction with pressure drop, and increased corrosion of the boiler.

In order to overcome these problems, special boilers have been designed with lower furnace exit temperatures or low operation temperature.

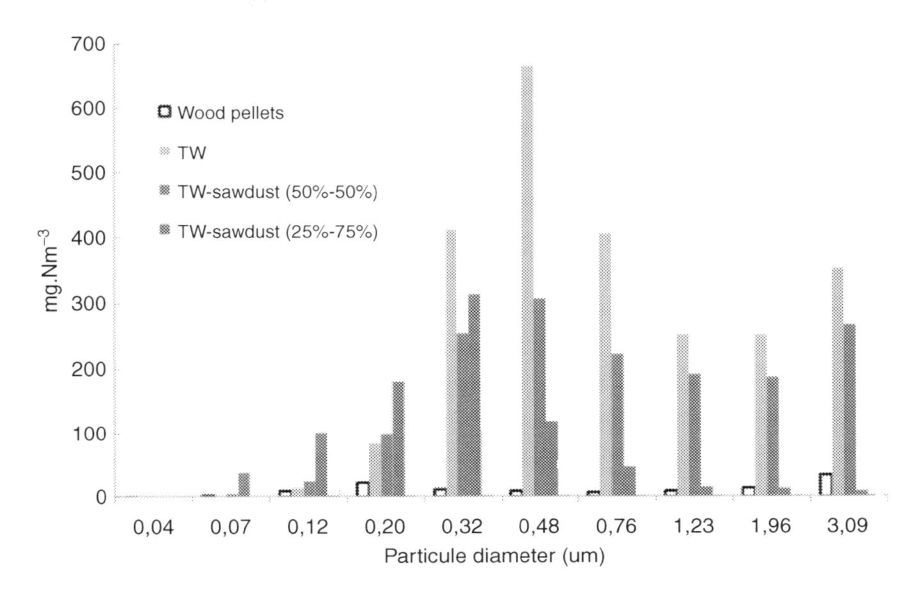


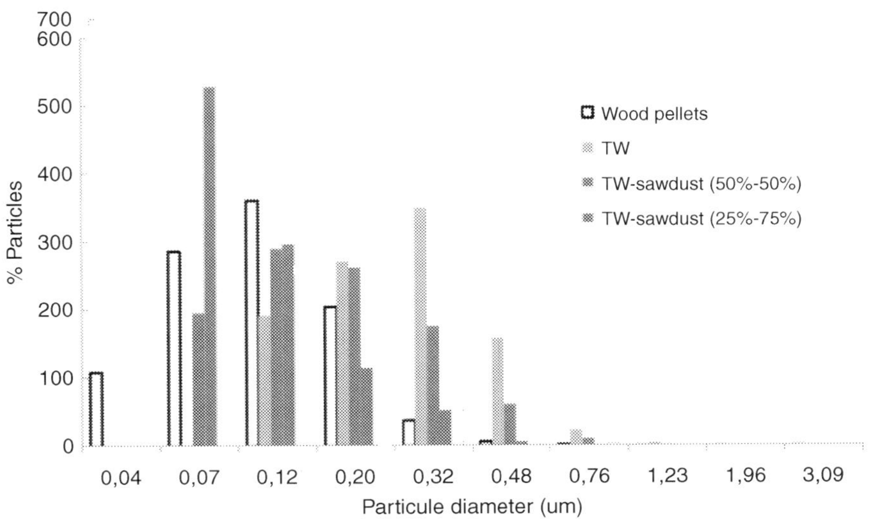


**Fig. 9.7** ***PM emissions concentrations of tomato residues based pellets depending on particle size distribution*** *(adapted from* [5]*).*

### 9.2.2.4 Corrosion

Corrosion of the combustors (the superheater tubes) by the chlorinated elements (direct corrosion) or by the alkali (molten phase corrosion) in the TW can be avoided by co-firing TW with aluminum silicates containing fuel, such as coal [26]. Keeping superheater metal temperature below

**Table 9.6** Bottom ash analysis (wt.%) normalized to 100% (adapted from [5]).

| Element | PS | TW | TW-PS |
|---|---|---|---|
| $K_2O$ | 20.82 | 26.84 | 26.00 |
| $SiO_2$ | 9.61 | 9.42 | 12.64 |
| CaO | 44.62 | 17.88 | 22.19 |
| $Fe_2O_3$ | 8.38 | 1.86 | 2.11 |
| $SO_3$ | 2.22 | 3.56 | 3.64 |
| $P_2O_5$ | 3.59 | 16.22 | 16.03 |
| $Na_2O$ | 2.27 | 2.32 | 2.50 |
| MgO | 3.69 | 10.10 | 11.08 |
| Cl | 1.32 | 1.51 | 0.75 |
| $Al_2O_3$ | 2.05 | 1.88 | 2.54 |
| $TiO_2$ | 0.22 | 0.06 | 0.07 |
| SrO | 0.12 | 0.13 | 0.14 |
| CuO | - | - | - |
| $Cr_2O_3$ | - | - | - |
| $ZrO_2$ | 0.01 | - | - |
| ZnO | 0.04 | - | 0.03 |
| $Rb_2O$ | 0.04 | - | 0.01 |
| $MnO_2$ | 0.98 | 0.22 | 0.23 |
| Br | - | - | - |
| Other oxides | - | - | - |
| Sum | 100 | 100 | 100 |

the first melting temperature of deposits (less than 500°C for TW-pellets) would also prevent the molten phase corrosion. Besides, it was proved that if the sulfur-to-chlorine atomic ratio (S/Cl) in a given fuel is less than two, then the risk for superheater corrosion is high. For a ratio of at least four, the blend could be regarded as noncorrosive.

### *9.2.2.5 Ash recycling for agricultural applications*

The rapidly growing number of pellet heating installations illustrates an increased interest in environmentally friendly heating systems. The problems associated with the use of agripellets are essentially linked to the ash management. Thus, the recycling or storage of TW-pellets ash deserves a special attention.

Ashes from TW pellets are produced in high quantities compared to woody biomass (5%–10% vs 2%), making the emptying of the ash storage under the furnace in small scale stoves and boilers more frequent. More embarrassing, James et al. suggest that inefficiencies in boilers and furnaces result in high percentages of unburned organic matter in ash [27]. This carbon content may be recycled to the boiler or furnace to improve energy output and increase the process efficiency.

However, the contents in heavy metals for each fraction (bottom, coarse ashes and even fly ashes which usually accumulate the largest part of heavy metals) seem to be lower. Besides, they contain nutrients, primarily potassium, and other soil-fertilizing elements like magnesium, phosphorus, and calcium and can therefore be applied in agriculture as fertilizer. It seems that ash is strongly alkaline (pH of 11–12) and could cause sharp increase of pH and ion concentration in the soil after spreading. Thus, ash should not be used unless a soil pH test has been done. Such a phenomenon would be harmful with respect to plant growth. Consequently, ash could be treated (e.g., granulated) in some way to reduce impact on soil. In regard to acidic soil correction, agripellets ashes as a garden amendment are a much more convenient means than the traditionally used ground limestone, bearing in mind that it is an absolutely costless resource.

According to Gomez-Barea et al., the utilization of ash has also seen its application in the construction industry [28]. Fly ash can be used as a cement replacement in concrete, for soil stabilization, as a road base, structural filler in asphalt and asphalt base products, lightweight bricks and synthetic aggregate.

## 9.3 Pyrolysis, torrefaction, and hydrothermal carbonisation processes

### 9.3.1 Thermogravimetric and kinetic analyses

#### *9.3.1.1 Thermal degradation behavior*

Few investigations have examined the thermal decomposition of tomato wastes using thermogravimetric analysis. Font et al. have evaluated the pyrolysis of tomato plant using thermogravimetric (TG) analysis [7]. Tomato plant is the agriculture residue that is usually decomposed and combusted on the field. The TG analysis was performed under nitrogen atmosphere from room temperature to 800°C at different heating rates. The TG and DTG curves are shown in Fig. 9.8 for 10°C/min.

Fig. 9.8 indicates that the thermal degradation of tomato plant follows similar trends as the major lignocellulosic biomass resources [29]. In particulier the DTG curve presents a shoulder at 550 K, a sharp peak at 600k and small peak at 720 K. In general, the shoulder is attributed to the hemicellulose degradation, the intense peak to cellulose degradation and the small peak to ligning degradation. Font et al. mentionned that the wide central band is a result of the overlapping of the three main polymers [7].

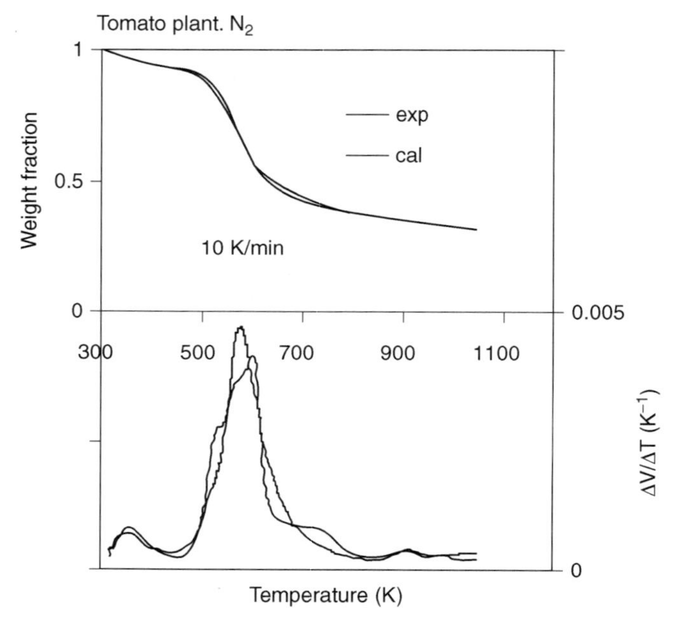


**Fig. 9.8** ***Weight fraction and its derivative during pyrolysis tests.***

Khiari et al. have evaluated the thermal degradation behaviour of tomato processing by-products under inert atmosphere using different heating rates (5,10, 20, and 30°C/min) [30]. Fig. 9.9 shows weight losses (TG curves) and their derivatives (DTG) curves versus temperature during experiments performed at 5°C.min$^{-1}$.

Fig. 9.9 indicates that the thermal degradation of tomato wastes occurred in four main zones. A first slow drying step continues until 160°C. The second degradation step occurs between 180°C and 310°C. This step is attributed to the hemicellulose degradation with a maximum degradation rate of 0.03%/s obtained at 280°C. The third decomposition steps occurs between 310°C and 390°C. This phase corresponds to the cellulose degradation with a maximum decomposition rate of 0.026 %/s $^{1}$ obtained at 370°C. The fourth degradation zone occurs between 390°C and 510°C with a maximum rate of 0.032 %/s obtained at 425°C. This final stage corresponds to the lignin structure decomposition and represents the main contributor to the final mass of char. This final char yield obtained at 600°C is 24 wt % which is the same order of magnitude as several food processing residues [29].

Similar recent investigation was performed by [31]. Authors have studied tomato peel pyrolysis using thermogravimetic analysis at different

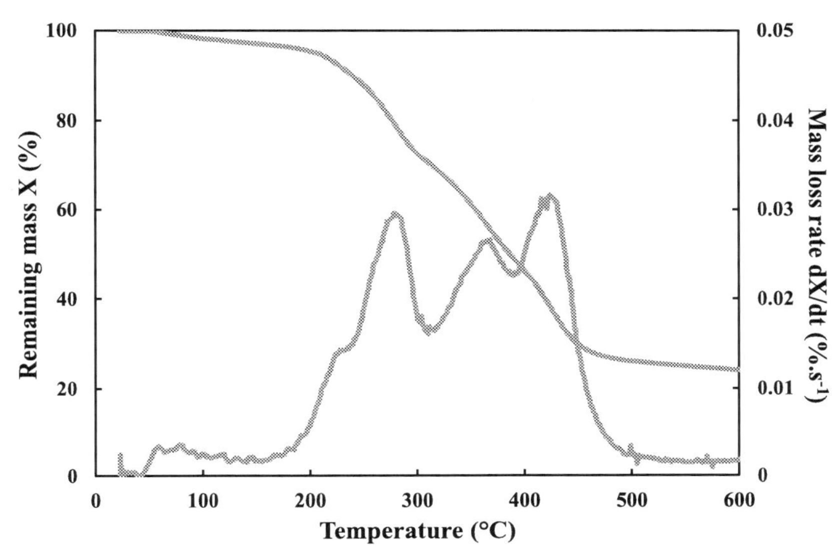


***Fig. 9.9*** *TG and DTG profiles of the thermal degradation behavior of tomato processing residues under nitrogen atmosphere at 5°C.min$^{-1}$.*

heating rates (5, 10, 15, 20 and 25°C/min$^{1}$). The DTG curves have similar shapes as the ones obtained by Khiari et al. (2019). In particular, authors defined three degradation zones corresponding to hemicellulose, cellulose and lignin decomposition (Fig. 9.10) [31].

Furthermore, Fig. 9.10 shows that increasing the heating rate from 5 to 25°C/min$^{1}$ leads to the expected increase in the degradation rate and the classical shift of the different maximum rates to higher temperatures [32].

### 9.3.1.2 Kinetic parameters extraction

The thermogravimetric analyses performed in the previous investigations were used for the kinetic modelling of tomato wastes pyrolysis [8,31]. Different isoconversional models including Friedman, Flynn-Wall-Ozawa (FWO), and Kissinger-Akahira-Sunose (KAS) were applied to extract the kinetic parameters. The corresponding equations are given in Table 9.2.

Linear forms of KAS, FWO, and Friedman models allow getting the values of activation energies $E_a$ presented in Fig. 9.11. The activation energies values varied with the conversion rates since it is attributed to the degradation of macro-components of the tomato processing residues. KAS and Friedman give similar average values but lower than that calculated by FWO (respectively 135.64, 125.91, and 157.35 kJ/mol). At low conversion

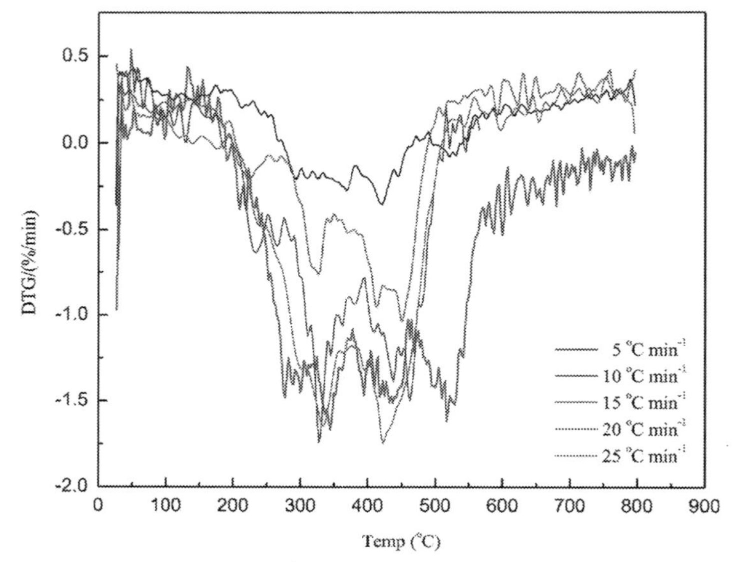


**Fig. 9.10** ***Derivative thermogravimetric analysis of tomato peel under different heating rates*** [31].

levels (from 10 to 20%), activation energies between 132 and 172 kJ/mol can be linked to hemicelluloses degradation. At higher conversion rates, activation energy values increase to 268.1 and 263.1 kJ/mol for FWO and

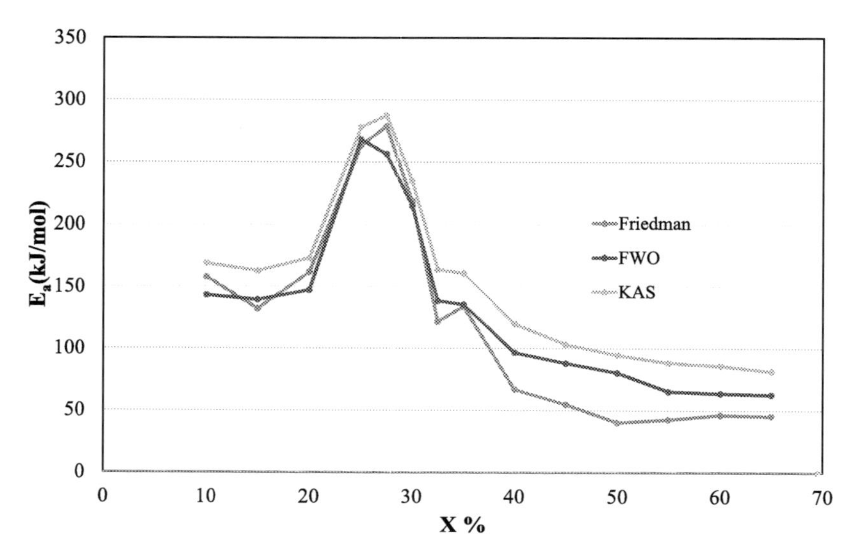


**Fig. 9.11** ***Activation energies versus conversion rate according to KAS, FWO and Friedman models.***

Friedman and to 277.7 kJ/mol for KAS model, which could be attributed to cellulose cracking. Until the conversion rate of X = 30%, the activation energy increases then at more elevated temperatures, the activation energy decreases for all models. The lipid contents, mostly in the tomato seeds, decomposes at 32.5% through an activation energy of 138.5 and 121.6 kJ /mol respectively for FWO and Friedman and 163.9 kJ/mol. Beyond this rate, the energy decreases again, which is the consequence of the continuous and ongoing degradation of cellulose and lignin as well as the char rearrangement through secondary and more complex reactions.

Similar trends were obtained by Midhun Prasad and Murugavelh during the kinetic modelling of thermogravimetric analysis data obtained during the tomato peels thermal degradation [31]. Authors found that the activation energy of tomato peel pyrolysis were reported to be 112.7 kJ/ mol, 113.85 kJ/mol, and 234.47 kJ/mol, respectively by Kissinger, KAS, and FWO models.

The extraction of activation energies values are not sufficient to describe the thermal degradation process of tomato processing residues. Khiari et al., proposed to combine the isoconversional models with several kinetic models described according to the decomposition reaction mechanism [8]. The latter kinetic models use a conversion rate function f(X), the form of which depends on the geometric progression of the reaction interface, the process nucleation and growth and/or the diffusion mechanism.

During the fitting between experimental and calculated values, the first-order kinetic model and the FWO method was the most appropriate to describe the thermal degradation process for applied for 10-20°C.min$^{-1}$ heating rates (Fig. 9.12).

For 5°C/min heating rate, a good fit is obtained the first hour. Afterward, conversion rates are under-estimated by the first order model. Therefore, n$^{th}$ order models with n = 1/3 were applied to fit calculated and experimental values above 3500 s (Fig. 9.13) which corresponds to lignin decomposition. The latter is better simulated by a contracted sphere model with n = 1/3.

### 9.3.2 Pyrolysis

Pyrolysis is the thermal degradation of organic materials in absence or low concentration of oxygen. Generally, the process takes place between 400 and 600°C, but higher or lower temperatures could be applied in order to target one pyrolytic fraction and/or one specific propriety of the pyrolysis product. In fact, pyrolysis produces three different fractions, liquid (bio-oil), gas (syngas), and solid fractions (biochar) that could be valorised

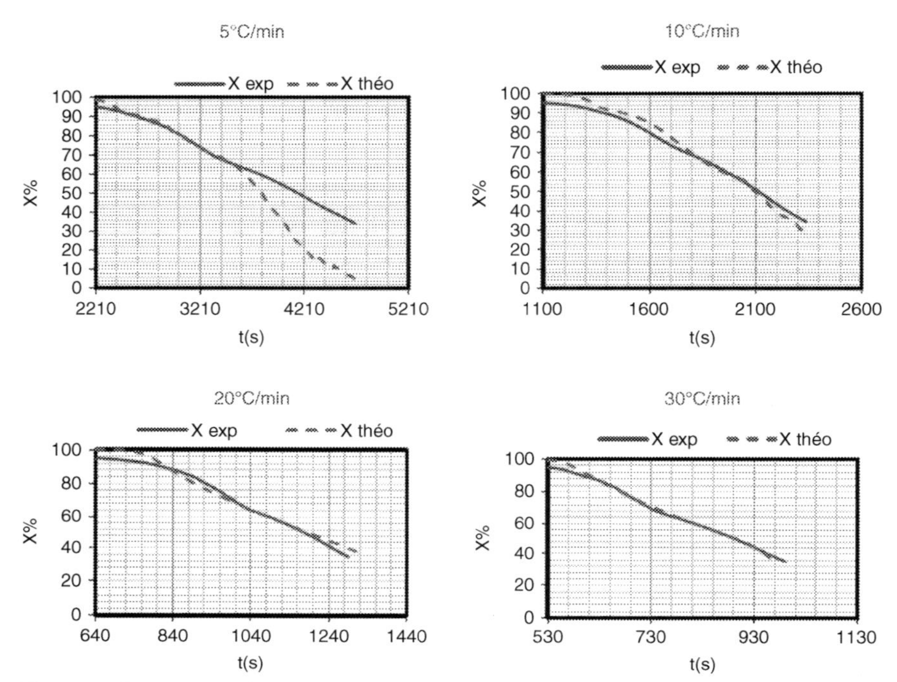


**Fig. 9.12** ***Experimental and simulated conversion rates of tomato residues pyrolysis for different heating rates.***

in energy, environmental and agronomic applications. Pyrolysis products yields depend strongly on the pyrolysis process, the operating conditions, the biomass type, etc.

Several investigations have performed for tomato wastes conversion using the pyrolysis technique. The final objective differs from biochar production for soil amendment to bio-oil for energy production. Therefore, different operating conditions including the heating rate variability, the addition of catalyst or other feedstocks, were applied. These investigations are detailed below.

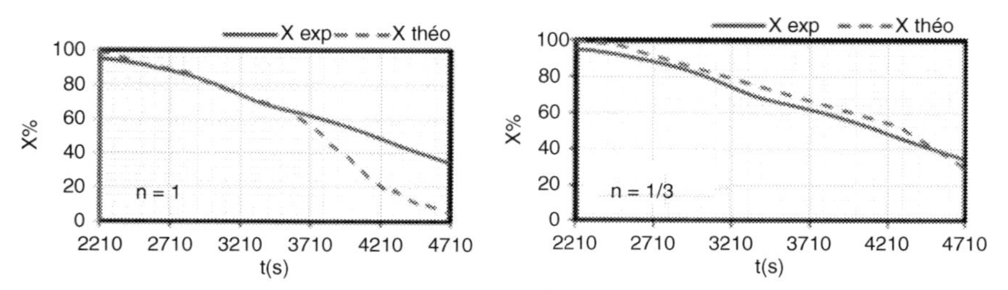


**Fig. 9.13** ***Experimental and calculated conversion rates for 5°C/min pyrolysis.***

#### 9.3.2.1 Biochar and carbon materials production

Biochar production through the slow pyrolysis of tomato waste is the most common recovery route in the literature. According to Mohan et al. the slow pyrolysis could be considered when the process occurs for as a heating rate <10°C $min^{-1}$ between 350 and 800°C for a residence times ranging between 10 min and 2 days [33].

Khiari et al. have evaluated the physical and the chemical properties of the biochar produced from the slow pyrolysis (3–6°C/min heating rate) of tomato harvest waste (THW: stalks and leaves) at various temperature (300–700°C) and duration (40–240 min). The pyrolysis trials were realized by placing 4 kg of tomato waste in a cylindrical reactor ($D_i$ = 48 cm; $D_e$= 60 cm) [34,35].

Memici and Ekinci have found that the biochar yield fallen when increasing temperature while no clear trend detected as function of residence time. In particular, the biochar yield increases from 30.65 to 53.03% at 400°C when residence time varies from 40 to 240 min. In addition, the biochar yields were approximately 53, 17.7, 15, 8.8, and 11.5% at 300, 400, 500, 600, and 700°C, respectively, for a residence time of 240 min. The decrease in biochar yields were already observed for several agricultural and agrifood processing residues [34–36]. The reduction in the biochar yield can be attributed to the organic matter removal with increasing temperature.

Several properties including pH, volatile and ash contents, C and N concentrations, were determined to assess the suitability of the THW biochar for soil applications. Memici and Ekinci found that the mean pH values were ranging between 5.41 and 9.97. The European Biochar Certificate (EBC) indicates that the standard pH value of soil-improving biochar varies between 8 and 10 [37]. The alkalinity of biochar is increasing with temperature and residence time as observed for several biomass derived biochars. This alkalinity is attributed to the degradation of organic acid and carbonate during the pyrolysis. Such effects are confirmed through the decrease of volatiles contents with temperature. In particular, volatiles decreased from 67.58% to 30.82% when temperature increased from 300 to 700°C at the 40-min residence time.

The C content ofTHW derived biochars were analyzed. The C content is around 40 w.t% a and varies with temperature and residence time. In particular, the biochar was enriched in carbon when the temperature and residence time increases. However, the high C-rich biochar (44 wt%) was obtained at 500°C and 240°C min. Similar behavior was obtained for pine-derived biochars when the temperature increased from 300 to 500°C [38].

The N content analysis shows that THW derived biochar contains between 1.5 and 2.5 wt%. Memici et Ekinci found that the pyrolysis temperature increase at the lower residence times (40-80 min) increased the N contents, whereas the higher residence times (240 min) at the same temperature decreased them.

In order to assess the biochar suitability for agronomic applications, Memici and Ekinci have performed field trial to investigate the influence of different biochars on soil $CO_2$ emissions in the field conditions. Authors found that the lowest cumulative soil $CO_2$ emission was obtained for the THW biochar produced at 500°C and the residence time of 240 min. In contrast, the highest cumulative soil $CO_2$ emission happened with the biochars obtained at 300°C for a holding time of 40 min, the difference between the two cases was 38.6 %. Therefore, the soil $CO_2$ emissions increased with the biochars produced at the low temperatures and decreased with the biochars produced at the high tones. Authors pointed to 500°C as the cut-off point to produce the THW derived biochars, baring in mind the associated energy consumption and the soil $CO_2$ emission

Ayala-Cortés et al. have converted tomato pruning biomass into carbon materials using solar pyrolysis. Pyrolysis experiments were carried by placing 27–29 g of biomass in the solar reactor shown in Fig. 9.14 and technically detailed in [39]. Samples were pyrolyzed at 450, 600, and 900°C with a heating

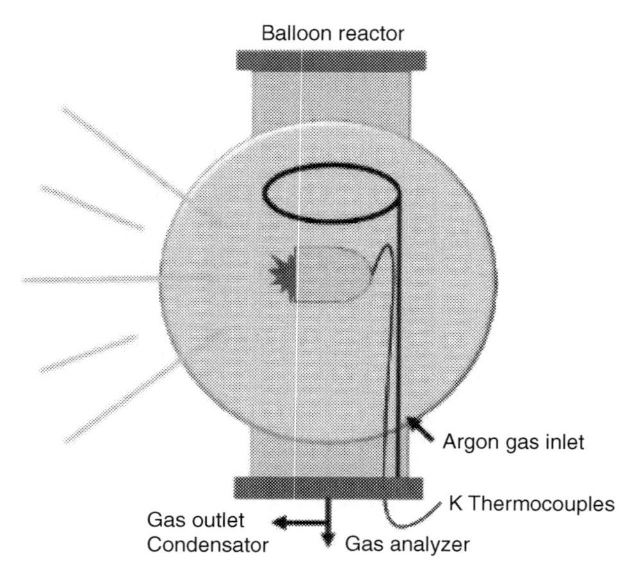


**Fig. 9.14** ***Sketch of the solar pyrolysis reactor.***

rate of 30°C/min and a residence time of 2 hours. The obtained biochar was, then, treated to eliminate the remaining ashes and to obtain a clean carbon material. The latter was characterized using various analytical techniques in order to determine elementary analysis (CHONS), specific surface area (nitrogen physisorption) and capacitance values (cyclic voltammetry).

The elemental analysis of carbon materials produced from tomato pruning waste at different temperature shows that the carbon content increases from 59.8 to 67.1 wt% when temperature increases from 450°C to 900°C. Such behaviour is attributed to graphitization of carbon structure. In contrast, oxygen and hydrogen content decreases from 29.3 to 23.3 wt% and from 2.0 to 1.5 wt%, respectively with increasing temperature from 450°C. The decrease in oxygen and hydrogen content is attributed to the cracking of weaker bonds into the structure of biomass.

The morphology of tomato pruning chars prepared at different temperature was analysed using scanning electron microscopy. Fig. 9.15 shows that at low pyrolysis temperatures a porous structure like a sponge was obtained. This structure has numerous porous sizes ranging between 1μm and 10μm. When pyrolysis temperature increases, the macroporous structure disappear,

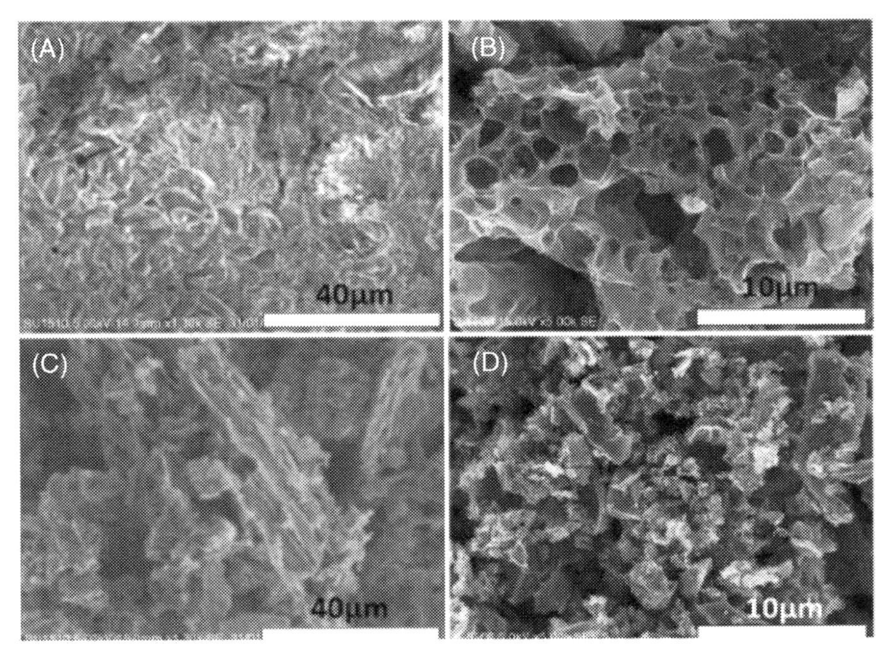


**Fig. 9.15** ***Effect of pyrolysis temperature on tomato pruning biochar morphology.*** (A) raw biomass, (B) 450°C (C) 600°C, and (D) 900°C.

forming micro and mesoporous surface. This observation is confirmed by high surface area (above 400 $m^2/g$) and capacitance values of tomato pruning carbons. Such results were attributed to the inorganic compounds in the tomato pruning wastes (ash contents: 36%) that produced an autoactivation process during solar pyrolysis. These inorganic compounds has also an effect on the capacitive behaviour since heteroatoms such as nitrogen and sulfur are electrochemically active and contributing to the pseudo-capacitance processes which promote the energy storage properties. In addition, nitrogen elements improve the contact between the carbon materials and the electrolytes.

The physicochemical properties of tomato pruning biochar shows an efficient behavior as supercapacitor. In particular, there is no need to addition chemical and/or physical activation process since the structural, chemical and electrochemical properties are already suitable for the biochar to be used as a competitive electrode material for supercapacitors.

#### *9.3.2.2 Biofuel (energy, alternative fuel) production*

Several investigations have examined the tomato waste pyrolysis and characterized the solid, liquid and gaseous products for their possible use in energy production. Encinar et al. have studied different operating parameters in terms of temperature (400–800°C), the initial sample mass (2.5–10 g of tomato plant waste) and the particle size (0.63–2.00 mm) in order to optimize the pyrolysis process [9].

The effect of temperature on the different yields is shown in Table 9.8. It is clearly seen that an increase of the temperature leads to a decrease in the solid and the liquid yields and to an increase in the yield of the gas phase. The decrease in the liquid yield is more pronounced from 55.8 wt% at 400°C to 24.8 wt% at 800°C. In the opposite, a significant increase in the gas phase yield was observed with temperature (17.2 wt% at 400°C, 58.4 wt% at 800°C). The main components of this gas phase detected were $H_2$, CO, $CH_4$, $CO_2$ and traces of some hydrocarbons such as ethane and ethylene. The increase of the gas phase yield is attributed to the strong cracking reactions of the liquid phase as the temperature increases. Such observation is already mentioned during the pyrolysis of other lignocellulosic residues [34–36, 40].

For the solid fraction, as expected, an increase in temperature leads to a decrease in volatile matter and an increase in fixed carbon and ash contents. However, one may note that Encinar et al. have observed no significant effect of the initial sample mass and its particle size on the different products yield [9].

**Table 9.8** Influence of temperature on yields of different fractions, proximate analysis and gas production (particle size: 0.63–1 mm, initial sample mass: 5 g, nitrogen flow rate: 150 mL/min).

| Parameter | Temperature (°C) | | | | |
|---|---|---|---|---|---|
| | 400 | 500 | 600 | 700 | 800 |
| Fraction yields (wt %) | | | | | |
| Solids | 27 | 25;4 | 21.8 | 23 | 16.8 |
| Liquids | 55.8 | 54 | 41 | 36.6 | 24.8 |
| Gases | 17.2 | 20.6 | 37.2 | 40.4 | 58.4 |
| Proximate analysis of charcoal (wt %) | | | | | |
| Fixed Carbon | 37.61 | 54.27 | 53.06 | 51.02 | 58.82 |
| Volatile matter | 52.92 | 35.51 | 34.55 | 36.07 | 27.15 |
| Ash | 9.48 | 10.22 | 12.40 | 12.91 | 14.03 |
| Gas production (mol/kg residue) | | | | | |
| $H_2$ | 0.099 | 0.487 | 2.268 | 5.313 | 7.030 |
| CO | 0.988 | 2.176 | 3.784 | 5.147 | 8.414 |
| $CH_4$ | 0.145 | 0.493 | 1.746 | 2.874 | 3.975 |
| $CO_2$ | 2.364 | 2.472 | 4.714 | 4.499 | 6.071 |

In order to assess the potential of the pyrolysis products for energy production, their higher heating values (HHV) were determined. The solid char has an average HHV of 26 MJ/kg, the bio-oil (liquid phase) has a HHV of 7.8 MJ/kg at 400°C that decreases with and the gas fraction has an HHV between 0.5 and 8.0 MJ/kg of raw material.

According to these different values as well as their physicochemical properties, the pyrolysis products could be recovered in different manners. In particular, the char can be used as fuel or precursor for the activated carbons production. The bio-oil (liquid phase) could be used as liquid fuel or as organic-compounds source. The gas fraction could be used to heat the pyrolysis reactor or to generate heat and electricity in a gas-turbine/vapor-turbine combined cycle.

#### 9.3.2.2.1 Bio-oil production

The thermochemical conversion of tomato waste into liquid fuel (bio-oil) using pyrolysis technique has received particular attention. Midhun Prasad and Murugavelh have examined the bio-oil production from the pyrolysis of tomato peel in an auger reactor [31]. This pyrolytic reactor made in stainless steel is shown in Fig. 9.16. This reactor is designed for operating in batch and continuous mode. The technical details are found in [31].

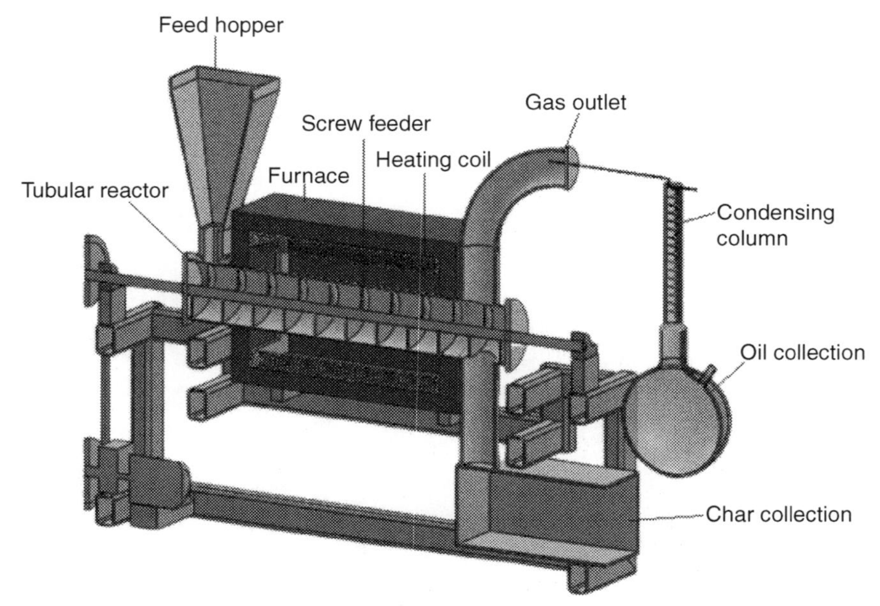


**Fig. 9.16** ***Reactor for pyrolysis of tomato peels.***

The bio-oil collected during pyrolysis tests was characterized using various analytical techniques. The elementary analysis, density, flash point, kinematic viscosity and calorific of the tomato peel oil are presented in Table 9.9.

Density of the bio-oil (973 kg/m$^3$) was higher than diesel fuel (830 kg/m$^3$). Also, Kinematic viscosity (11.82 cSt) and flash point (94°C) of the bio-oil were higher that diesel fuel. Elementary analysis of the bio-oil indicated a higher amount of oxygen 11.2% comparing to fuel diesel. Such oxygen content may explain the lower calorific value of tomato peel bio-oil (33.04 MJ/kg$^1$) comparing to diesel (44 MJ/kg$^1$). Such results encourage the blend of tomato bio-oil with diesel to be used as an alternative fuel for engines.

**Table 9.9** Physical and chemical properties of tomato peel bio-oil and diesel.

| Parameters | TPO | Diesel |
|---|---|---|
| Kinematic viscosity at 40°C (cSt) | 11.82 | 3.36 |
| Flash point (°C) | 94 | 60 |
| Density (kg/m$^3$) | 973 | 830 |
| Calorific value (MJ/kg) | 33.04 | 44 |
| Elements of C (wt %) | 75 | 85.72 |
| Elements of H (wt %) | 9.28 | 13.2 |
| Elements of N (wt %) | 4.4 | 0.18 |
| Elements of S (wt %) | 0.12 | 0.3 |
| Elements of O (wt %) | 11.2 | 0.6 |

Furthermore, the functional groups and the major compounds of the tomato peel bio-oil was assessed using FTIR and GC-MS analysis, respectively. The bio-oil FTIR spectra reported peaks at the range of 3200-3550 $cm^{-1}$ showing the presence of O-H stretch of phenol group. The peak around 2850–2950/cm corresponds to C-H stretch of asymmetric alkanes. A parallel peak in the same interval helps detecting the alkyl C-H bond of alkyl group. The alkenyl groups can be identified from the peak at 1630–1690/ cm while the carboxylic acid was determined from the peak in the range of 2500–3000/cm. The set of the small peaks at 1500–1700/ cm indicated C= O stretch of aldehyde, ketone, ester, carboxylic acid and amide in tomato peel bio-oils. It can also be observed from Fig. 9.17 that the tomato peel oil contains more peaks at the range of 1000-1300 $cm^{-1}$ indicating the presence of alcohol.

GC-MS of the tomato peel bio-oil shows numerous peaks at different retention time. The major peaks were attributed to hydrocarbons. In particular, GC-MS showed the presence of petroleum based components indane, indene, nonane, benzene, which are highly inflammable.

In order to improve the bio-oil quality via deoxygenation, several researchers have used solid acid catalysts during biomass pyrolysis. In general, catalytic pyrolysis occurs in two stages. The first step is the biomass degradation while and the second step is the conversion of pyrolysis vapors to smaller molecules with the catalyst.

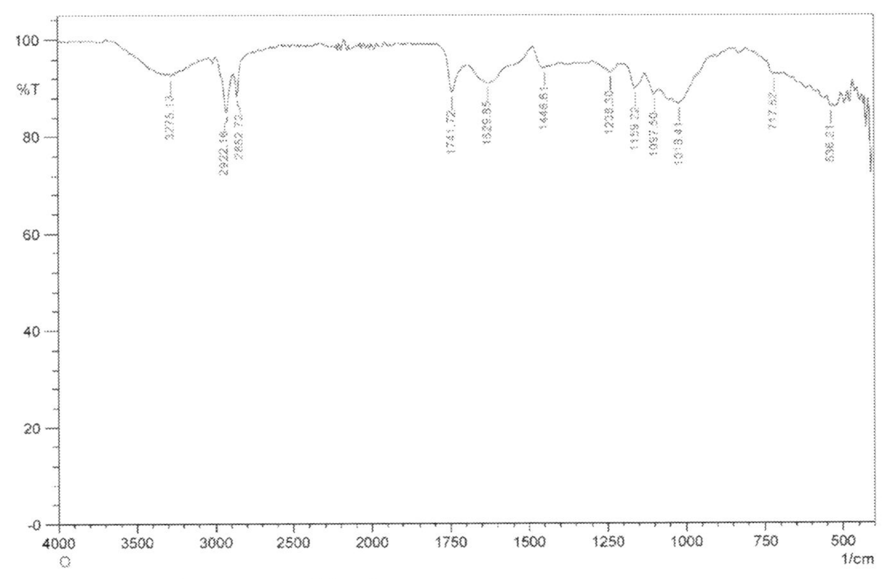


**Fig. 9.17** ***FTIR of Tomato peel bio-oil.***

Ozbay et al. have studied the effect of different copper-loaded alumina catalysts on the pyrolysis products yields and bio-oil properties during the pyrolysis of tomato wastes [41]. In particular authors have produced oils using two compositions (5 and 10 wt.% Cu loading) and two methods (sol-gel and coprecipitation). The 5 and 10 wt% copper-loaded catalysts prepared by sol-gel method were designated as SG1 and SG2, respectively. Similarly, the 10 wt.% copper-loaded alumina catalysts prepared by co-precipitation method were assigned as CP1 and CP2, respectively. Pyrolysis experiments were performed in absence and in presence at 40 and 100°C/min heating rates for 500°C final temperature and 15 min holding time.

Figs. 9.18 and 9.19 show the pyrolysis product yields for different heating rates (40 and 100, catalyst preparation methods (solegel and co-precipitation) and catalyst loading ratios (5% and 10%), respectively.

It is clearly seen that in absence of catalyst, the bio-oil yield is around 26 wt% for both heating rates. The increasing in heating rates does not significantly affect the bio-oil yield but favour gasification to the detriment of char production. The comparison between both catalyst preparation methods show that the higher bio-oil yields were obtained with co-precipitated catalysts. Authors attributed such behavior to the stronger metal support interaction between Cu and $Al_2O_3$ support. However, Figs. 9.18 and 9.19 shows a reduction of bio-oil yield when increasing the metal loading ratio due to limited selectivity. This limited selectivity is explained by the agglomeration of the copper particles at higher loading and the increase of activation energy for pyrolytic reactions increases with the additional copper loading.

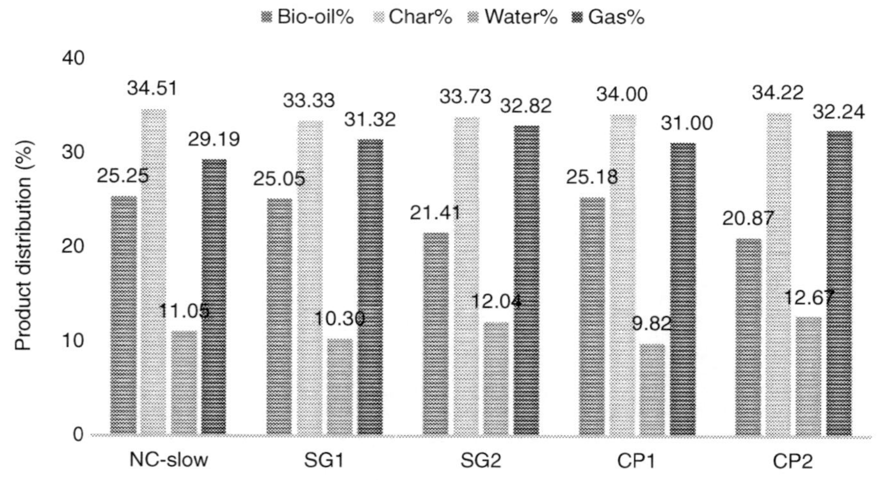


**Fig. 9.18** ***Effect of catalyst on tomato waste pyrolytic yield at heating rate of 40°C/min.***

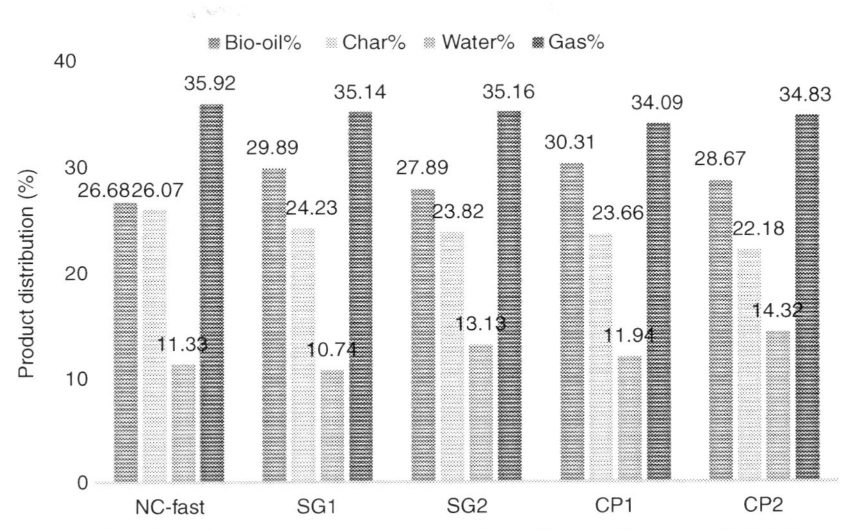


**Fig. 9.19** ***Effect of catalyst on tomato waste pyrolytic yield at heating rate of 100°C/min.***

The analysis of the different preparation and operating conditions shows that the maximum bio-yield was 30.31% by using 5% copper loaded co-precipitated catalyst for pyrolysis tests performed at 100°C/min.

The physicochemical properties of the tomato waste bio-oil were assessed through elementary analysis and higher heating values (HHV) determination. It was found that the bio-oils carbon and hydrogen contents increase when pyrolysis occurs in presence of catalysts while the nitrogen content seems to not be significantly affected. In contrast, the oxygen contents decrease with the presence of catalysts and also significantly with increasing heating rates to 100°C/min. The low oxygen content of bio-oils obtained from catalytic fast pyrolysis (around 20 wt%) may explain the promising high heating values (35.4 MJ/kg) that was reached in presence of SG1 catalyst.

In order to identify the catalysts role, the functional groups and the composition of the tomato of bio-oils were analyzed using FTIR and GC-MS techniques, respectively. For the FTIR spectra, it is seen that the O-H stretching vibrations between 3500 and 3100 $cm^{-1}$ attributed to the presence of phenols disappeared for the catalytic bio-oils. This behavior confirms the oxygen removal during catalytic pyrolysis which is in agreement with the elementary analysis and HHV values. However, for the functional groups, all the pyrolytic oil in absence or in presence of catalyst, presented similar peaks. The main difference is observed in their intensity.

GC-MS analysis shows also similar compounds in the catalytic and non-catalytic pyrolysis. These compounds include alkanes and alkenes (dodecane, pentadecane, hexadecane, tridecane, 1-dodecene, 1-decene, 1-tetradecene, 1-tridecene, 3-heptadecene, heneicosane, and hexacosane), phenolic compounds (phenol, 2-methyl phenol, 4-methyl phenol, 2,3-dimethyl phenol, 2,4-dimethyl phenol, and 3-ethyl phenol), polycyclic aromatic hydrocarbons (naphthalene, alkylated naphthalene), aromatics (toluene, benzene), Carboxylic acids and carbonyl compounds (pentadecanoic acid, 9,12-octadecadienoic acid, methyl ester-9-octadecenoic acid, and cis-9-hexadecenal)

From these results including yields, physicochemical properties, functional groups and composition, it could conclude that catalytic bio-oil is suitable for the production of many chemical feedstock and fuels.

#### 9.3.2.2.2 Combustion and emission characteristics of bio-oil in diesel engine

The recovery of tomato waste bio-oil as an alternatative fuel for diesel engine was examined. Midhun Prasad and Murugavelh have evaluated the performance and emission characteristics analysis of different blends of tomato peel oil and diesel in a 4-stroke water cooled DI open chamber diesel engine of KIRLOSKAR TV1 with a power output capacity of 5.2 kW at the speed of 1500 rpm. Engine block diagram and diesel engine specification are shown in Fig. 9.20 and Table 9.10, respectively [31].

During combustion tests, different performance parameters including brake thermal efficiency, brake specific fuel consumption and exhaust gas emission were evaluated at different loading conditions. In addition, heat release rate, mass burnt fraction and ignition delay were also analyzed.

The Brake thermal efficiency (BTE) under full load condition of 5%, 15%, and 25% blends of tomato peel bio-oil was estimated to be 31.5%, 31.7% and 30%, respectively. This value for diesel fuel was slightly higher (32.5%). The presence of oxygen in the tomato peel bio-oil might have improved the combustion resulting in BTE similar to that of diesel fuel. However, the slight reduction for tomato peel bio-oil comparing to diesel fuel canbe attributed to the tomato peel oil properties including viscosity, spray characteristics and calorific value.

The brake specific fuel consumption (BSFC) for diesel fuel is lower than tomato peel bio-oil for all the load condition. As an example, at At engine load of 25% the BSFC of diesel fuel, blend 5%, blend 15% and blend 25% was found to be 400, 445, 452 and 460 g k/Wh, respectively. Higher viscosity and lower calorific value can explain the higher BSFC of tomato peel oil comparing to diesel fuel.

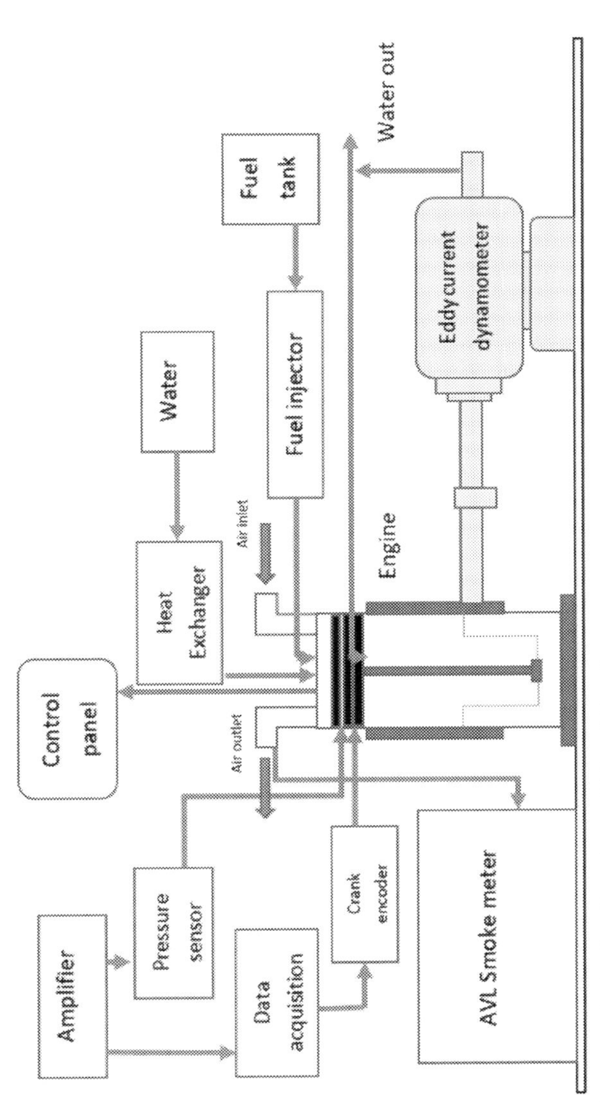


**Fig. 9.20** *Engine setup for testing Tomato peel bio-oil blends.*

**Table 9.10** Engine specification.

| Model | Kiloskar TV1 |
|---|---|
| Type | Single cylinder, four stroke |
| Fuel type | Diesel engine |
| Injection type | Direct injection |
| Bore diameter | 87.5 mm |
| Stroke length | 110 mm |
| Capacity | 661 $cm^3$ |
| Compression ratio | 17.5: 1 |
| Engine speed | 1500 rpm |
| Rated power | 5.2 Kw |
| Dynamometer | Eddy Current |
| Cooling type | Water cooling |
| Displacement | 661 $cm^3$ |
| Injection time | 23° BTDC |
| Injection pressure | 23 bar |

The emission characteristics of the different blends show that unburnt HC, CO emissions decreased for the tomato peel bio-oil blends, whereas NOx emission is higher. Such behaviour is attributed to the oxygen content of the tomato peel oil.

## 9.4 Torrefaction application to tomato wastes

Torrefaction is a thermochemical process, called sometime mild pyrolysis since it occurs, like pyrolysis in absence or in presence of low concentration of oxygen, but at temperatures of 200-350°C. In general, torrefaction purpose is different from pyrolysis since this technique is applied to decrease the water and volatiles contents from the biomass, and therefore to improve some fuel properties including higher energy density, hydrophobic behavior, biological activity elimination, grindability,....

Brachi et al. have studied the torrefaction of tomato peel residues using fluidized bed technology [42]. In particular, the influence of temperature and residence time on the mass and energy yields as well as on and the main phyiscochemical properties (i.e., elemental composition, ash content, calorific value and equilibrium moisture content) was examined.

Torrefaction tests were performed in the laboratory-scale fluidized-bed apparatus schematized in Fig. 9.21.

Fluidized-bed torrefaction tests were realized in a silica sand bed (fine SS: particle density = 2800 kg/$m^3$; bulk density = 1475 kg/$m^3$;

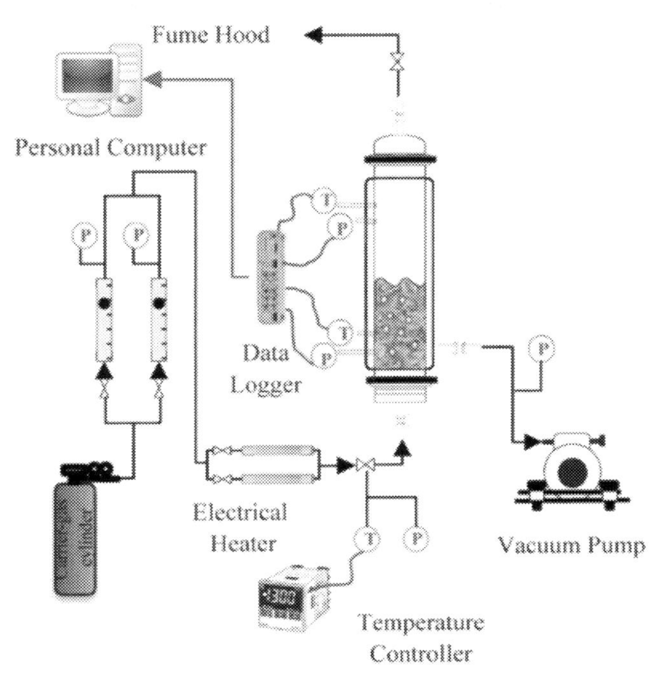


**Fig. 9.21** ***Engine setup for testing Tomato peel bio-oil blends.***

D= 140 μm) at three different temperatures (i.e., 200, 240, and 285°C) and torrefaction times (i.e., 5, 15, and 30 min).

The main results of the batch fluidized bed torrefaction tests are shown in Fig. 9.22A and D in order to show the effect of the torrefaction temperature and time on mass and energy yields, energy densification index and torrefied samples composition. It is seen that higher that 285°C and 30 min holding time led to an increase in the calorific value of the torrefied tomato by a factor of 1.2 peels with respect to the parent one. Under the same experimental conditions, a reduction in the O/C elemental ratio of 40 % and an improvement in the hydrophobicity of the torrefied tomato peels were also observed. These positive effects of the torrefaction treatment took place while maintaining the mass yields (between 75 and 94 %, dry and ash free (daf)) and the energy yields (between 90 and 96 %, daf) at satisfactory levels.

Furthermore, an uniform and consistent quality of the torrefied tomato peels is observed during the different tests (Fig. 9.23). This homogeneity shows that fluidized-bed technology is particularly suitable to cope with the exothermicity associated with the thermal degradation of non-woody biomasses, which easily tend to ignite or to be carbonized during torrefaction.

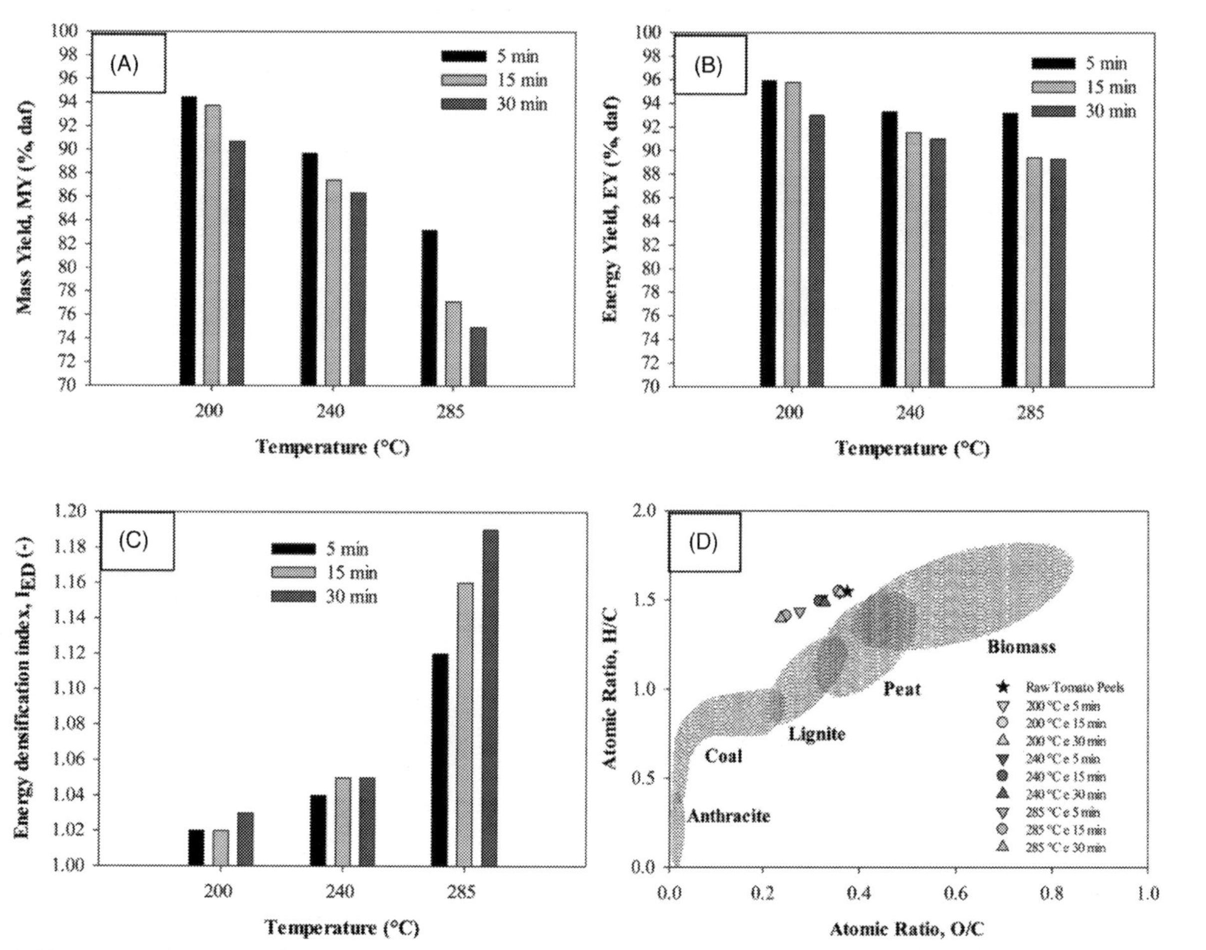


**Fig. 9.22** (A) Mass yield, (B) energy yield, (C) energy densification index, and (D) Van Krevelen diagram of torrefied tomato peels obtained in a fluidized bed reactor.

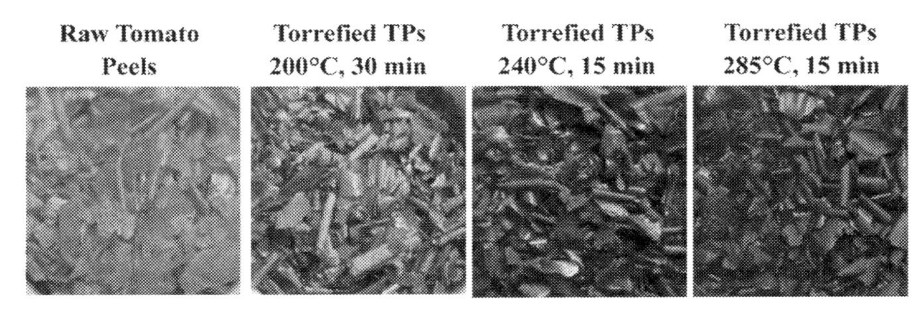


**Fig. 9.23** *Solid product homogeneity and color change during torrefaction.*

## 9.5 Hydrothermal carbonisation

As torrefaction process, other pre-treatment techniques, including drying [43], pelleting [5,44], are usually applied in order to improve the biomass physicochemical properties for its further use in environmental and energy applications. Recently, hydrothermal carbonization (HTC) has been selected as a promising technique for the energy densification by of biomass resources or the production of sustainable carbonaceous materials.

During hydrothermal carbonisation, biomass is heated in water under autogenous conditions, to generate a carbonaceous fraction called hydrochar. The latter is more stable and has higher C content comparing to the raw materials. HTC has the advantage to process biomass without a preliminary drying step. Therefore, the HTC energy balance is very interesting and a large amount of energy which would be necessarily supplied in a pyrolysis process is saved.

Sabio et al. have studied the HTC of tomato peels and seeds generated by the Spanish tomato processing industry [45]. The HTC experiments were performed by placing a mass of tomato peel (1.6–18.4 g) and 150 mL of deionised water (biomass/water ratio, R: 1.6–15% by weight) in a 0.2 L stainless steel autoclave. Then, the system, after remaining an overnight at room temperature, was heated up in an electric furnace at chosen temperatures (150–250°C), during a selected processing time (1.6–18.4 h). At the end of the experiment and after cooling, the solid phase was separated from liquid by vacuum filtration and dried at 80°C. The produced hydrochars were characterized in terms of their solid yield (% mass weight), high heating values (MJ $kg^{-1}$ dry basis), elemental analysis, FTIR and scanning electronic microscopy (SEM) analysis.

Table 9.11 shows the experimental conditions of each test and the corresponding values of solid yield (SY) and HHV. It is clearly seen that

**Table 9.11** Solid yield (SY, %) and higher heating value (HHV, MJ/kg) for hydrochars.

| Sample | R (%) | Temperature (°C) | Time (h) | SY (%) | HHV (MJ/kg) |
|---|---|---|---|---|---|
| 1 | 6.7 | 150 | 10 | 65 | 27.2 |
| 2 | 3.3 | 170 | 5 | 64.6 | 26.4 |
| 3 | 10 | 170 | 5 | 68 | 26.7 |
| 4 | 3.3 | 170 | 15 | 60.8 | 27.4 |
| 5 | 10 | 170 | 15 | 61.3 | 29 |
| 6 | 6.7 | 200 | 1.6 | 87.7 | 23.6 |
| 7 | 12.3 | 200 | 10 | 61.5 | 31.1 |
| 8 | 6.7 | 200 | 10 | 62 | 29.2 |
| 9 | 6.7 | 200 | 10 | 61.1 | 29.1 |
| 10 | 6.7 | 200 | 10 | 62 | 29.2 |
| 11 | 6.7 | 200 | 10 | 61.9 | 28.9 |
| 12 | 1.1 | 200 | 10 | 49.8 | 29.1 |
| 13 | 6.7 | 200 | 18.4 | 47.9 | 32.2 |
| 14 | 3.3 | 230 | 5 | 49.6 | 31.2 |
| 15 | 10 | 230 | 5 | 62.2 | 28.3 |
| 16 | 3.3 | 230 | 15 | 27.6 | 32.9 |
| 17 | 10 | 230 | 15 | 35.4 | 34.8 |
| 18 | 6.7 | 250 | 10 | 29.4 | 34.6 |

the experimental conditions affect strongly the SY that ranged between 27.6% (3.3%–230°C–15 h) and 87.7% (10%–200°C–1.6 h). Similarly, the HHV is ranging from 23.6 MJ $kg^{-1}$ (6.7%–200°C–1.6 h) and 34.8 MJ $kg^{-1}$ (10%–200°C–15 h). The latter HHV value is in the same order of magnitude of charcoal.

The elemental composition of tomato peel and the elaborated hydrochars were determined. Fig. 9.24 shows the Van Krevelen diagram plotted from the obtained H/C and O/C ratios. It is clearly seen that HTC process leads to the decrease of both ratios, moving in some cases the biomass composition towards a more densified material, such as peat. Sabio et al. have related the HTC process to the decomposition products of lignin, in the form of monomeric radicals [45].

The structural and chemical properties of hydrochars were evaluated by FTIR analysis. The same spectrum is obtained for the different hydrochars no matter the experimental conditions. The spectrum is also very similar to the one fin in the litterature for different lignin samples. The different peaks and their assignments are well detailed in [45]. This FTIR analysis shows that the tomato peel chars obtained consisted mainly of barely degraded lignin, which confirm hypothesis of the monomer removal mentioned previously.

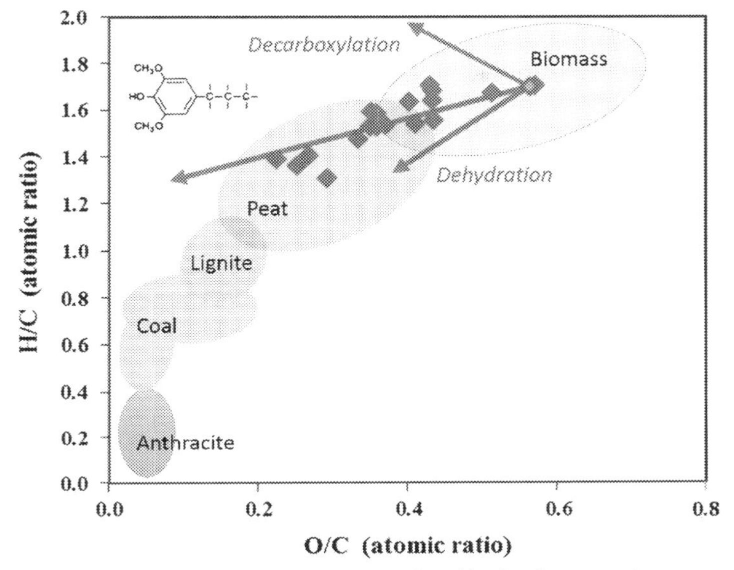


**Fig. 9.24** *Van Krevelen diagram for tomato peel and hydrochar samples.*

Morphological properties of the produced hydrochars were analyzed using SEM (Fig. 9.25). The surface morphology of the HCs maintains the original scaffold and did not show remarkable differences as a consequence of the change of the experimental conditions used. A slight erosion of the tomato peel layers for increased HTC times can be mentioned.

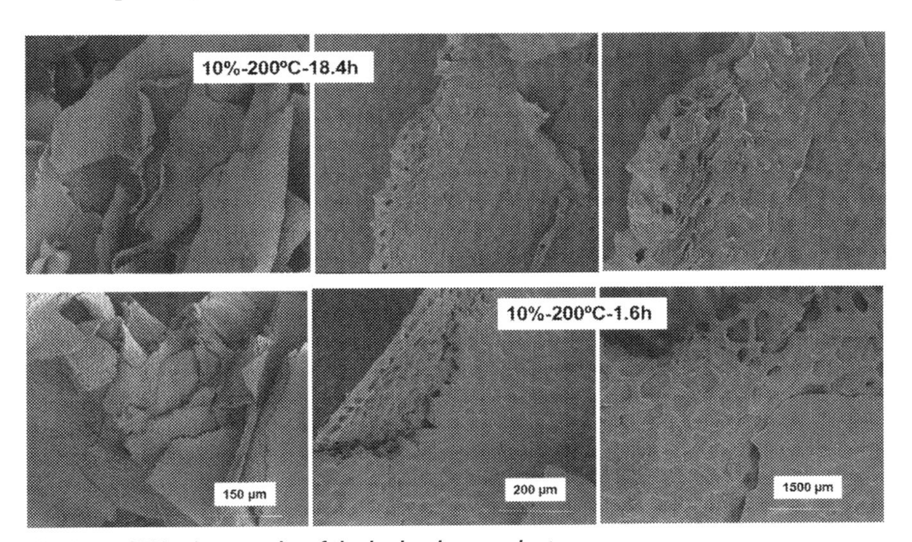


**Fig. 9.25** *SEM micrographs of the hydrochar products.*

## 9.6 Gasification

Gasification is a process where the organic wastes or biomasses are heated, generally between 500 and 1000°C, in a partially oxidative atmosphere, using gasifying agents such as steam, air or carbon dioxide. The output is small amounts of tars, chars, and ashes and mainly a valuable gaseous product, called syngas composed of hydrocarbons, hydrogen, carbon monoxide, and carbon dioxide. The syngas, also called producer gas, can be burnt directly to give energy, heat or power, or converted into other compounds that will be used as fuels or chemical products, such as hydrogen and second-generation biofuels.

Although gasification is flexible to convert any type of biomass or waste into a variety of fuels and chemicals in addition to energy, works dealing with gasification of tomato residues are very scarce.

Madenoglu et al. investigated supercritical water gasification in a continuous flow reactor (Temperature = 600°C, pressure = 35 MPa, residence time = 0.3 minutes) of five selected biomasses, including tomatoes residue [46]. Compared to other tested biomasses, alkali salts in the tomato wastes increased the $H_2$ and $CO_2$ yields and decreased the CO yield (Fig. 9.26). The carbon gasification efficiency was improved with two catalysts (0.86% for Potassium carbonate and 0.73% for Trona versus 0.7% without any catalyst).

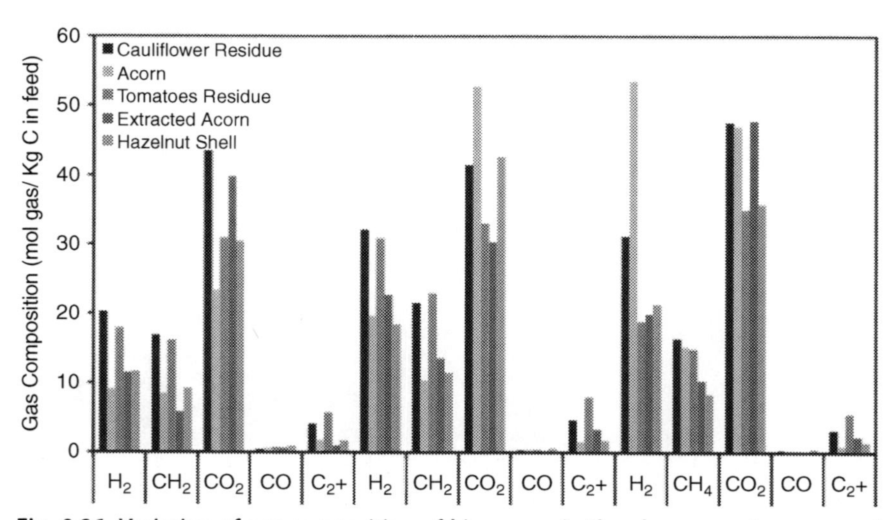


**Fig. 9.26** ***Variation of gas composition of biomasses in the absence and presence of catalyst (mol gas/kg C in feed)*** [46].

In another study, Brachi et al. aimed to quantify the impact of torrefaction pretreatment on the quality of the producer gas arising from the gasification with steam and steam-oxygen mixtures of previously-torrefied tomato peels in high temperature entrained flow reactors [42]. To this aim, they developed a chemical equilibrium model which allowed predicting the product gas composition as a function of process temperature, equivalence ratio, steam-to-biomass ratio and biomass elemental composition. A global sensitivity analysis with respect to the model input parameters was performed to assess the impact of torrefaction and gasification operating conditions on the quality of the product gas in terms of heating value and composition metrics typically adopted in the process industry ($H_2/CO$ ratio, stoichiometric module, etc.). Their results highlighted that the quality of product gas arising from the oxygen-steam gasification of torrefied and of untreated tomato peels did provide only a marginal improvement in the product gas quality, although torrefied feedstocks produced more $H_2$ and CO and less $CO_2$ than the parent one (Fig. 9.27).

The goal of Kocer et al. was to evaluate the hydrogen production potential of greenhouse residues to tomato-pepper blending in different rates (0%, 25%, 50%, 75% and 100%) by air-steam gasification using developed biomass gasification model [47]. Steam and air were used as the gasification agents. Air to fuel rate and steam to fuel rates were 0.05 due to high content of $O_2$ in the greenhouse residues. The gasifier temperature was set at 877°C. The main result of this study was that increasing tomato residues blending rate increases $H_2$ production (Fig. 9.28). This was attributed to higher $O_2$ in tomatoe wastes compared to that of pepper residues, being therefore an important factor in the syngas production and composition, seen in Table 9.12.

Pinna-Hernández et al. considered the utilization of a fluidized bed gasifier incorporated into a thermal electric system to improve power plant production both thermally and electrically [48]. They tested different biomasses: almond shells, olive tree prunings, holm oak prunings and greenhouse tomato and pepper plant residues. Technical and economic and criteria were applied to determine the most appropriate biomass to use in the gasification process. They found that all biomasses are eligible for such process except greenhouse vegetable residues (tomato and pepper) which did not have suitable technical parameters (moisture content = 82.6%–29.6%, dry base ash content = 35.5%–6.4%, volatile matter content = 75.1%–59.1%, and gross net calorific value = 17277–11529 kJ/kg). Indeed, even if they did have the best cost per kilogram

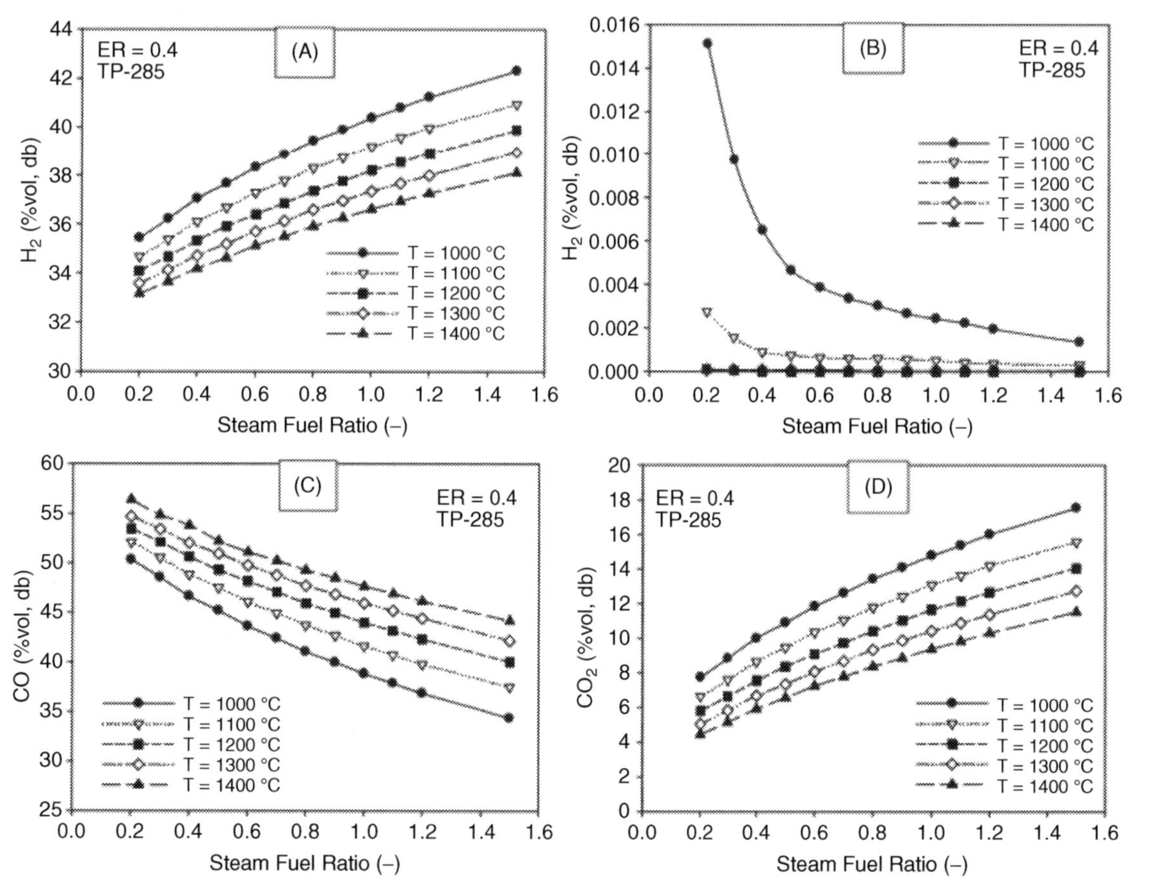


**Fig. 9.27** ***Effect of the gasification temperature and the steam-fuel ratio on the composition of the producer gas arising from the gasification of tomato peels torrefied at 285°C with equivalence ratio 0.4*** [42].

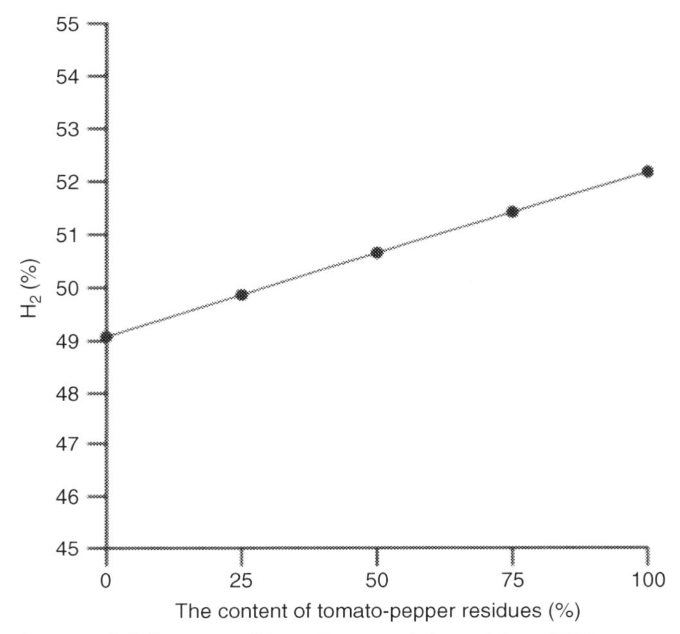


**Fig. 9.28** ***Content of $H_2$ in composition of syngas*** *(adapted from* [47]*).*

(0.01€/kg), which is the price paid by the farmer to the waste manager, greenhouse vegetable residues from tomato and pepper plants could have negative consequences when developing the gasification process, such as corrosion and fouling of the different components. Another parameter that has an undesirable value is the calorific value which is low for its energy use. In addition, the heterogeneity of this biomass must be taken into account, assuming an added level of complexity.

**Table 9.12** The composition of syngas from gasifier (air to fuel = 0.05, steam to fuel = 0.05, P = 1 atm, T = 1150 K) (adapted from [47]).

| | The content of tomato residues (%) | | | | |
|---|---|---|---|---|---|
| **Syngas composition** | **0** | **25** | **50** | **75** | **100** |
| $H_2$ (%) | 49.08 | 49.87 | 50.66 | 51.43 | 52.19 |
| CO (%) | 31.55 | 31.52 | 31.50 | 31.47 | 31.45 |
| $CH_4$ (%) | 2.82 | 2.20 | 1.59 | 0.99 | 0.39 |
| $CO_2$ (%) | 12.80 | 12.80 | 12.80 | 12.80 | 12.80 |
| $H_2O$ (%) | 0.03 | 0.03 | 0.03 | 0.03 | 0.03 |
| $N_2$ (%) | 3.72 | 3.57 | 3.42 | 3.28 | 3.13 |
| $H_2$/CO | 1.56 | 1.58 | 1.61 | 1.63 | 1.66 |
| HHV (MJ/kg) | 17.67 | 17.56 | 17.44 | 17.33 | 17.22 |
| Efficiency (%) | 83.45 | 83.60 | 83.74 | 83.89 | 84.03 |

## 9.7 Conclusion

The investigations about thermal conversion performed and synthesized in this chapter unveiled the technologic and economic interest in the valorization of tomato by-products. A typical feature of tomato products that affects the whole sector is their high perishability as well as their scattering over large areas, making the aggregation challenging and economically unattractive. In an increasing competitive context for the global agricultural sector, where the circular economy has become a frontline issue, the energy recovery from TW can be an advantage in the tomato processing industries. Besides, a better environmental management of tomato by-products would require a better understanding of the economics of these wastes processing for bio-fuel production and for resolving issues related to these processes in practical applications. In this vein, the tomato waste may need to be processed with other wastes (for example blend and pelleting with sawdust).

## References

[1] D. Singh, D. Sharma, S.L. Soni, S. Sharma, P. Kumar Sharma, A. Jhalani, A review on feedstocks, production processes, and yield for different generations of biodiesel, Fuel 262 (082208) (2020) 116553–116568.
[2] G. Rossini, G. Toscano, D. Duca, F. Corinaldesi, E. Foppa Pedretti, G. Riva, Analysis of the characteristics of the tomato manufacturing residues finalized to the energy recovery, Biomass Bioenergy 51 (2013) 177–182.
[3] M. Hijosa-Valsero, J. Garita-Cambronero, A.I. Paniagua-García, R. Díez-Antolínez, Tomato waste from processing industries as a feedstock for biofuel production, Bioenergy Res 12 (2019) 1000–1011.
[4] D. Shao, C. Venkitasamy, X. Li, Z. Pan, J. Shi, B. Wang, et al., Thermal and storage characteristics of tomato seed oil, LWT - Food Sci. Technol 63 (1) (2015) 191–197.
[5] N. Kraiem, M. Lajili, L. Limousy, R. Said, M. Jeguirim, Energy recovery from Tunisian agri-food wastes: Evaluation of combustion performance and emissions characteristics of green pellets prepared from tomato residues and grape marc, Energy 107 (2016) 409–418.
[6] J.F. González, C.M. González-García, A. Ramiro, J. González, E. Sabio, J. Gañán, et al., Combustion optimisation of biomass residue pellets for domestic heating with a mural boiler, Biomass Bioenergy 27 (2) (2004) 145–154.
[7] R. Font, J. Moltó, A. Gálvez, M.D. Rey, Kinetic study of the pyrolysis and combustion of tomato plant, J. Anal. Appl. Pyrolysis 85 (1–2) (2009) 268–275.
[8] B. Khiari, M. Moussaoui, M. Jeguirim, Tomato-processing by-product combustion: thermal and kinetic analyses, Materials (Basel) 12 (4) (2019) 553–563.
[9] J.M. Encinar, J.F. González, G. Martínez, Energetic use of the tomato plant waste, Fuel Process. Technol 89 (11) (2008) 1193–1200.
[10] A. Chouchene, M. Jeguirim, B. Khiari, G. Trouvé, F. Zagrouba, Study on the emission mechanism during devolatilization/char oxidation and direct oxidation of olive solid waste in a fixed bed reactor, J. Anal. Appl. Pyrolysis 87 (1) (2010) 168–174.
[11] A. Chouchene, M. Jeguirim, B. Khiari, F. Zagrouba, G. Trouvé, Thermal degradation of olive solid waste: Influence of particle size and oxygen concentration, Resour. Conserv. Recycl 54 (5) (2010) 271–277.

[12] H.L. Friedman, Kinetics of thermal degradation of char-forming plastics from thermogravimetry. Application to a phenolic plastic, J. Polym. Sci. Part C: Polym. Symp 6 (1) (1964) 183–195.
[13] J.H. Flynn, L.A. Wall, General treatment of thermogravimetry of polymers, J. Res. Natl. Bur. Stand 70A (6) (1966) 487–523.
[14] H.E. Kissinger, Variation of peak temperature with heating rate in differential thermal analysis, J. Res. Natl. Bur. Stand 57 (4) (1956) 217–221.
[15] A. Aires, R. Carvalho, M.J. Saavedra, Reuse potential of vegetable wastes (broccoli, green bean and tomato) for the recovery of antioxidant phenolic acids and flavonoids, Int J. Food Sci. Technol 52 (1) (2017) 98–107.
[16] D.A. Tillman, Biomass cofiring: the technology, the experience, the combustion consequences, Biomass Bioenergy 19 (6) (2000) 365–384.
[17] J. Moltó, R. Font, A. Gálvez, M.D. Rey, A. Pequenín, Analysis of dioxin-like compounds formed in the combustion of tomato plant, Chemosphere 78 (2) (2010) 121–126.
[18] J.F. González, C.M. González-García, A. Ramiro, J. Gañán, A. Ayuso, J. Turegano, Use of energy crops for domestic heating with a mural boiler, Fuel Process. Technol 87 (8) (2006) 717–726.
[19] J.A. Conesa, R. Font, A. Fullana, I. Martín-Gullón, I. Aracil, A. Gálvez, et al., Comparison between emissions from the pyrolysis and combustion of different wastes, J. Anal. Appl. Pyrolysis 84 (1) (2009) 95–102.
[20] A. Ruiz Celma, F. Cuadros, F. López-Rodríguez, Characterization of pellets from industrial tomato residues, Food and Bioproducts Processing 90 (4) (2012) 700–706.
[21] I. Obernberger, G. Thek, Physical characterisation and chemical composition of densified biomass fuels with regard to their combustion behaviour, Biomass Bioenergy 27 (6) (2004) 653–669.
[22] M. Jeguirim, N. Kraiem, M. Lajili, C. Guizani, A. Zorpas, Y. Leva, et al., The relationship between mineral contents, particle matter and bottom ash distribution during pellet combustion: molar balance and chemometric analysis, Environ. Sci. Pollut. Res. Int 24 (11) (2017) 9927–9939.
[23] J. Moltó, J.A. Conesa, R. Font, I. Martín-Gullón, Organic compounds produced during the thermal decomposition of cotton fabrics, Environ. Sci. Technol 39 (14) (2005) 5141–5147.
[24] T. Zeng, N. Weller, A. Pollex, V. Lenz, Blended biomass pellets as fuel for small scale combustion appliances: Influence on gaseous and total particulate matter emissions and applicability of fuel indices, Fuel 184 (2016) 689–700.
[25] J. Werkelin, B.-J. Skrifvars, M. Zevenhoven, B. Holmbom, M. Hupa, Chemical forms of ash-forming elements in woody biomass fuels, Fuel 89 (2) (2010) 481–493.
[26] H.P. Nielsen, F.J. Frandsen, K. Dam-Johansen, L.L. Baxter, The implications of chlorine-associated corrosion on the operation of biomass-fired boilers, Progr. Energy Combust. Sci. 26 (3) (2000) 283–298.
[27] A.K. James, R.W. Thring, S. Helle, H.S. Ghuman, Ash management review—applications of biomass bottom ash, . Energies. 5 (10) (2012) 3856–3873.
[28] A. Gómez-Barea, L.F. Vilches, C. Leiva, M. Campoy, C. Fernández-Pereira, Plant optimisation and ash recycling in fluidised bed waste gasification, Chem. Eng. J 146 (2) (2009) 227–236.
[29] M. Jeguirim, Y. Elmay, L. Limousy, M. Lajili, R. Said, Devolatilization behavior and pyrolysis kinetics of potential Tunisian biomass fuels, Environmental Progress & Sustainable Energy 33 (4) (2014) 1452–1458.
[30] B. Khiari, M. Massoudi, M. Jeguirim, Tunisian tomato waste pyrolysis: thermogravimetry analysis and kinetic study, Environ. Sci. Pollut. Res 26 (2019) 35435–35444.
[31] K. Midhun Prasad, S. Murugavelh, Experimental investigation and kinetics of tomato peel pyrolysis: Performance, combustion and emission characteristics of bio-oil blends in diesel engine, J. Clean. Prod 254 (2020) 120115.

[32] M. Jeguirim, S. Dorge, A. Loth, G. Trouvé, Devolatilization Kinetics of Miscanthus Straw from Thermogravimetric Analysis, Int. J. Green Energy 7 (2) (2010) 164–173.
[33] D. Mohan, C.U. Pittman, P.H. Steele, Pyrolysis of Wood/Biomass for Bio-oil: A Critical Review, Energy Fuels 20 (3) (2006) 848–889.
[34] B. Khiari, M. Jeguirim, Pyrolysis of grape marc from tunisian wine industry: feedstock characterization, thermal degradation and kinetic analysis, Energies 11 (4) (2018) 730–744.
[35] B. Khiari, I. Ghouma, A.I. Ferjani, A.A. Azzaz, S. Jellali, L. Limousy, et al., Kenaf stems: thermal characterization and conversion for biofuel and biochar production, Fuel 262 (2020) 116654.
[36] B. Khiari, A. Ibn Ferjani, A.A. Azzaz, S. Jellali, L. Limousy, M. Jeguirim, Thermal conversion of flax shives through slow pyrolysis process: in-depth biochar characterization and future potential use, Biomass Conversion and Biorefinery 11 (2021) 325–337.
[37] EBC, European Biochar Certificate -Guidelines for a Sustainable Production of Biochar, Version 6.1, European Biochar Foundation (EBC), Arbaz. Switzerland, 2012.
[38] K.H. Kim, J.-Y. Kim, T.-S. Cho, J.W. Choi, Influence of pyrolysis temperature on physicochemical properties of biochar obtained from the fast pyrolysis of pitch pine (Pinus rigida), Bioresour. Technol 118 (2012) 158–162.
[39] A. Ayala-Cortés, C.A. Arancibia-Bulnes, H.I. Villafán-Vidales, D.R. Lobato-Peralta, D.C. Martínez-Casillas, A.K. Cuentas-Gallegos, Solar pyrolysis of agave and tomato pruning wastes: insights of the effect of pyrolysis operation parameters on the physicochemical properties of biochar, AIP Conf. Proc 2126 (2019) 180001.
[40] A. Gagliano, F. Nocera, F. Patania, M. Bruno, S. Scirè, Kinetic of the pyrolysis process of peach and apricot pits by TGA and DTGA analysis, Int. J. Heat Technol 34 (2) (2016) S553–SS60.
[41] N. Ozbay, A.S. Yargic, R.Z. Yarbay Sahin, Tailoring Cu/Al2O3 catalysts for the catalytic pyrolysis of tomato waste, J. Energy Ins 91 (3) (2018) 424–433.
[42] P. Brachi, R. Chirone, F. Miccio, M. Miccio, G. Ruoppolo, Entrained-flow gasification of torrefied tomato peels: combining torrefaction experiments with chemical equilibrium modeling for gasification, Fuel 220 (2018) 744–753.
[43] A.A. Azzaz, M. Jeguirim, V. Kinigopoulou, C. Doulgeris, M.-L. Goddard, S. Jellali, et al., Olive mill wastewater: from a pollutant to green fuels, agricultural and water source and bio-fertilizer – Hydrothermal carbonization, Sci. Total Environ 733 (2020) 139314.
[44] S. Dorge, M. Jeguirim, G. Trouvé, Thermal degradation of Miscanthus pellets: kinetics and aerosols characterization, Waste Biomass Valori 2 (2011) 149–155.
[45] E. Sabio, A. Álvarez-Murillo, S. Román, B. Ledesma, Conversion of tomato-peel waste into solid fuel by hydrothermal carbonization: Influence of the processing variables, Waste Manag 47 (Part A) (2016) 122–132.
[46] T. Güngören Madenoğlu, N. Boukis, M. Sağlam, M. Yüksel, Supercritical water gasification of real biomass feedstocks in continuous flow system, Int. J. Hydrog. Energy 36 (22) (2011) 14408–14415.
[47] A. Kocer, I.F. Yaka, A. Gungor, Evaluation of greenhouse residues gasification performance in hydrogen production, Int. J. Hydrog. Energy 42 (36) (2017) 23244–23249.
[48] M.G. Pinna-Hernández, I. Martínez-Soler, M.J. Díaz Villanueva, F.G. Acien Fernández, J.L.C. López, Selection of biomass supply for a gasification process in a solar thermal hybrid plant for the production of electricity, Ind. Crop. Prod 137 (2019) 339–346.

CHAPTER TEN

# Biofuels production: Biogas, biodiesel and bioethanol from tomato wastes

**Andrius Tamošiūnas[a], Besma Khiari[b], Mejdi Jeguirim[c]**
[a]Lithuanian Energy Institute, Plasma Processing Laboratory, Kaunas, Lithuania
[b]Laboratory of Wastewaters and Environment, Centre of Water Researches and Technologies (CERTE), Technopark Borj Cedria, Tunisia
[c]The institute of Materials Science of Mulhouse (IS2M), University of Haute Alsace, University of Strasbourg, CNRS, UMR, Mulhouse, France

## Acronyms

| | |
|---|---|
| AD | Anaerobic digestion |
| AP | Acidification |
| AV | Acid value |
| CN | Cetane number |
| CHP | Combined heat and power |
| COD | Chemical oxygen demand |
| FA | Fatty acids |
| FAME | Fatty acids methyl ester |
| FE | Fresh water eutrophication |
| GHG | Greenhouse gases |
| HHV | Higher heating value |
| HRT | Hydraulic retention time |
| HSS-AD | Hemi-solid state anaerobic digestion |
| IPSW | Iberian pig slaughterhouse wastes |
| IRR | Internal rate of return |
| KV | Kinematic viscosity |
| L-AD | Liquid anaerobic digestion |
| LR | Loading rate |
| ME | Marine eutrophication |
| MFRD | mineral, fossil and renewable depletion |
| NPV | Net present value |
| OSI | Oxidative stability index |
| PB | Payback period |
| RD | Relative density |
| RI | Refractive index |
| SS-AD | Solid state anaerobic digestion |
| TE | Terrestrial eutrophication |

*Tomato Processing by-Products*
DOI: https://doi.org/10.1016/B978-0-12-822866-1.00003-X

| | |
|---|---|
| TIP | Tomato industry processing waste |
| TPW | Tomato processing waste |
| TS | Total solids |
| VFA | Volatile fatty acids |
| VS | Volatile solids |
| TVSA | Total volatile solids |

## 10.1 Introduction

In accordance with sustainable development objectives, tomato wastes could be a reliable, available, environmental-friendly and economically feasible energy source. As other biomass sources, these wastes are considered as neutral carbon and could mitigate global warming issues. Indeed, the chemical energy content stored inside the plant cell walls in tomato wastes (carbohydrates and sugars) is suitable for renewable energy applications, despite the recalcitrant nature of the biomass (cellulose crystallinity and amorphous lignin non-reactivity) which may hinder the chemical and the biological processing methods. However, once applying the required pre-treatment technique, biomass derived fuels (called biofuels) present two major advantages compared to other renewable energy sources. These advantages are related to the diverse forms since the produced biofuel could be liquid, gaseous and solid as well as their various applications in heat recovery, electricity generation and high value chemicals production [29,52,55].

Tomato wastes generated from crops cultivation and food processing industries are considered to be cellulose based second generation biofuel, which could be converted to biogas, biodiesel or bioethanol through different conversion techniques. In particular, biogas is obtained through anaerobic digestion [41]. However, bioethanol production is ensured through tomato sugars fermentation while biodiesel production is performed through tomato fats extraction and transesterification [25,58].

The main objective of this chapter is to provide a detailed presentation of all works that were published during the last decades in the field of tomato wastes valorization for biofuel production through biochemical and extraction process. The details for the thermochemical conversion process investigations (mainly combustion and pyrolysis) are already provided in chapter 9.

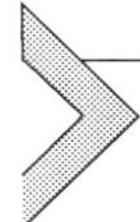

## 10.2 Biogas production

### 10.2.1 Overview

Anaerobic digestion (AD) is a biological degradation process by which microorganisms break down biodegradable organic material in the absence of oxygen (air). The resultant products of such process are a mixture of gases known as "biogas" and a nutrient-rich solid part called "digestate", which can be used as organic fertilizer for soil amendment [21,33,35,36]. The biogas mixture predominantly consists of the following compounds:

1. 50%–75% methane ($CH_4$)
2. 25%–45% carbon dioxide ($CO_2$)
3. 2%–7% water vapour ($H_2O$)
4. <2% oxygen ($O_2$)
5. <2% nitrogen ($N_2$)
6. <1% ammonia ($NH_3$)
7. <1% hydrogen sulphide ($H_2S$)
8. <2% trace gases

The AD process involves four steps: hydrolysis, acidogenesis, acetogenesis, and methanogenesis, which occur simultaneously (Fig. 10.1). Each process involves different groups of microorganism, i.e. in hydrolysis step hydrolytic bacteria, in acidogenesis acidogenic (fermentative) bacteria, in acetogenesis acetogens, and in methanogenesis methanogens. Such process needs to take place in wet and warm environment in order to achieve successful decomposition of organic matter and biogas production rates. The AD process is sensitive to disturbances and inhibition, which can affect the microorganisms making their effort less efficient while producing biogas. Therefore, various concepts are used for biogas production. They differ according to the process characteristics (dry matter content, way of feeding, number of phases, temperature, etc.) and are summarized in Table 10.1.

Biogas offers a variety of alternative use, for instance heat and electricity production in a combined heat and power (CHP) process, the feed-in of upgraded biogas to biomethane to natural gas grid, and the use as fuel or in the chemical industry (Fig. 10.2).

Generally, AD provides triple benefits of decreasing the levels of organic wastes, the use of fossil fuels (production of biogas/biomethane) and chemical fertilizers (production of digestate as organic fertilizer) [14]. Moreover, AD contributes to reduction of greenhouse gas (GHG) emissions and promotion of using renewable resources for green energy production.

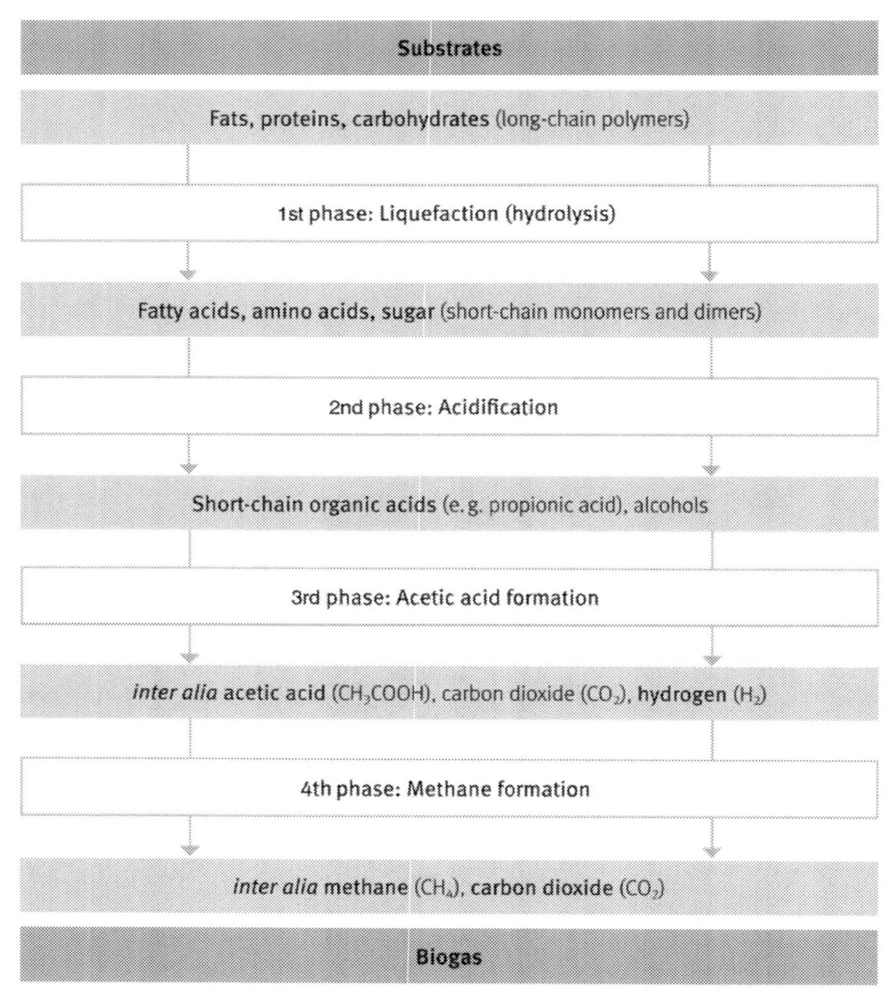


**Fig. 10.1** ***Simplified diagram of the decomposition of organic matter during biogas production*** [14] *(Reproduced with the permission of FNR).*

## 10.2.2 Digestion and codigestion of tomato wastes

According to the statistics of Food and Agriculture Organization of the United Nations (FAOSTAT) [13], tomato is one of the most popular vegetable crop in the world, which annual production in 2017 exceeded about 180 million tons. This means that a large quantity of tomato residues (stalks, leaves and spoiled tomato fruits) are generated annually. Therefore, the disposal of tomato wastes becomes a critical problem of the tomato production industry. One of the possible ways for tomato waste utilization is biogas production via

**Table 10.1** Classification of the AD process for generating biogas according to different criteria.

| Criterion | Features |
|---|---|
| Dry matter content of the substrates | Wet digestion |
| | Dry digestion |
| Type of feeding | Discontinuous |
| | Quasi-continuous |
| | Continuous |
| Number of process phases | Single-phase |
| | Two-phase |
| Process temperature | Psychrophilic (4–10°C) |
| | Mesophilic (30–38°C) |
| | Thermophilic (49–57°C) |

AD process. Tomato wastes could be successfully digested or co-digested mixing with other organic waste residues in order to produce biogas. Numerous investigations have examined tomato waste digestion and co-digestion to recover energy and other valuable products [9,21,33,34,42,45], Sarada and Joseph, 1994 [3,32,38,40,41,43,51], Sarada and Joseph, 1994 [24,44]. These research studies are detailed in the following sections.

#### *10.2.2.1 Digestion*

Sarada et al. [42,45], Sarada and Joseph, 1994 [43], Sarada and Joseph, 1994 [44] performed various experimental studies regarding the possibility to produce methane from tomato processing waste. Various factors, such as type of microorganisms, hydraulic retention time, loading rate and temperature as well as process phases and biochemical changes, influencing biogas/methane production from tomato waste were examined. In one of the earliest studies, the start-up of AD of tomato processing waste (TPW] for biogas/methane production was investigated [41]. Tomato waste collected from several local companies was shredded into fine pieces, air dried and stored at 4°C for further use. A 5.5 litre four laboratory digesters operating in batch/semi-continuous mode at 30°C were used to carry out the experiments. The inoculum for starting the process was taken from commercial digester regularly fed with animal house wastes. The laboratory digesters were filled with 6% (dry mass) cow manure slurry and inoculated with 10% of an active starter culture. One digester was fed only with TPW, which was used as a control. The process proceeded for eight weeks in a batch mode. A total process time for start-up studies was 12 weeks. Experimental conditions and main results are summarized in Table 10.2.

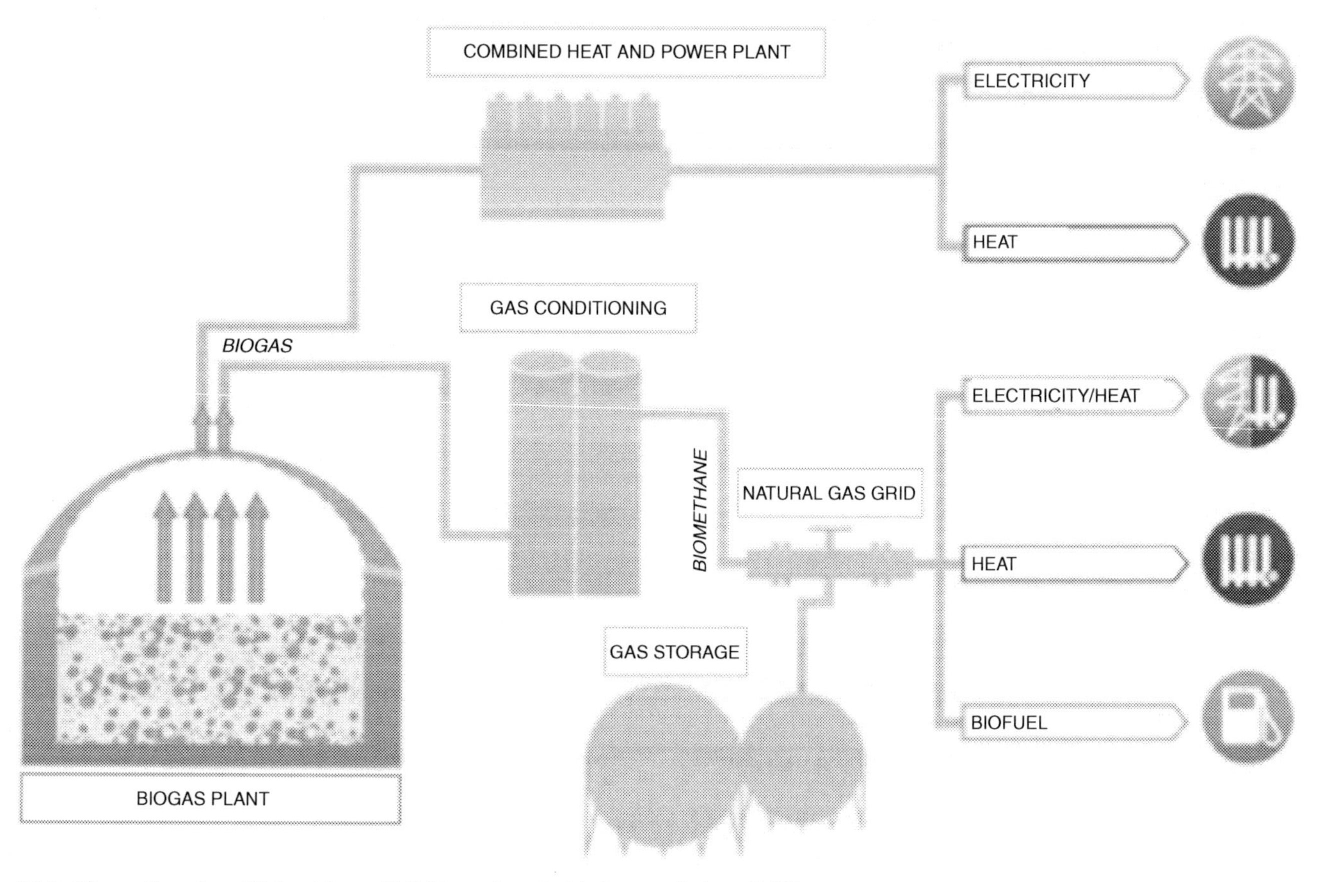


**Fig. 10.2** ***Alternatives for utilizing biogas*** [14] *(Reproduced with the permission of FNR).*

**Table 10.2** Experimental conditions and main results [41].

| Parameter | Digester 1 | Digester 2 | Digester 3 | Digester 4 |
|---|---|---|---|---|
| Feedstock | Cow manure only[a] | Cow manure[a] | Cow manure[a] | TPW only[a] |
| | Cow manure only[b] | Cow manure + TPW[b,c] | Cow manure + TPW[b,d] | TPW only[b,d] |
| Average biogas production, $m^3$/kg VS[e] | 0.2 | 0.6 | 0.54 | 0.06–0.07 |
| Methane content, %[e] | 50 | 72 | 60–70 | 40 |
| Volatile fatty acids | low | low | high | high |
| pH | 6.8–7 | 6.8–7 | 6.8–7 | 5 |

[a]Operation in a batch mode for 8 weeks.
[b]Operation in a semi-continuous mode of feeding by removing and adding daily 156 mL of slurry at a hydraulic retention time (HRT) of 32 days.
[c]Cow manure-TPW feeding proportion: 90:10, 80:20, 70:30, 60:40, 50:50, 40:60, 30:70, 20:80, 10:90, 0:100. Feedstock in the feed 6% total solids (TS).
[d]Start of feeding of TPW (6% TS) after 8 weeks of operation.
[e]Average value after 8 weeks of operation.

According to Table 10.2, the best results were obtained in the Digester 2. A stepped feeding of TPW was reported to be an optimum choice for a stable AD and yield of biogas production with the highest methane content. A stable operation of such process was reached within 8–10 weeks from the start-up. The acid content was low and the pH value did not require special adjustment. The processing of tomato waste alone for biogas production is very limited due to high acidity and low pH value as well as process stability and low yield of biogas production. Generally, the operation mode of gradual increase of TPW in the feedstock may offer an advantage and stability at large scale anaerobic digestion process too.

Moreover, Sarada and Joseph [43,44] have investigated the influence of hydraulic retention time (HRT), loading rate (LR) and temperature on the yield of biogas/methane production during AD of tomato waste. The tomato waste feedstock used for the process was received from local fruit-processing companies in India. The composition of the TPW was as follows: 95.45% volatile solids, 24% lipid, 17% protein, 25% cellulose, 14% hemicellulose. The laboratory digesters used for the tests were the same as described above and were monitored for 4-5 months. Total gas production was monitor daily, whereas gas composition, pH, total volatile fatty acids (TVFA) and ammoniacal nitrogen were analysed every week. The chosen HRT values were 4, 8, 16, 24 and 32 days at solids loading rate of 4.5 kg TS/$m^3$ day and temperature of 33±2 °C. This corresponds to daily replacement of effluent with fresh TPW slurry of 1250, 625, 313, 208 and 156 ml, respectively. The effect of loading rate was investigated at 3.0, 4.5, 6.0 and 7.5 kg TS/$m^3$ day at constant temperature and HRT. The impact of temperature was studied at mesophilic conditions at a temperature range of 25, 30, 35, 40, and 45°C, at the HRT (24 days) and LR (4.5 kg TS/$m^3$ day) constant. The main experimental results are summarized in Tables 10.3, 10.4 and 10.5.

It was stated that taking into account the obtained results, the optimum HRT was 24 days. Therefore, HRT of 16 and 24 days were further used to examine the effects of loading rate and temperature on biogas/methane production.

**Table 10.3** Summarized results on the effect of HRT on biogas production from TPW at feedstock loading rate of 4.5 kg/$m^3$ TS day and temperature of 33 ± 2°C (Sarada and Jospeh, 1994).

| Parameter | | | | | |
|---|---|---|---|---|---|
| HRT, days | 4 | 8 | 16 | 24 | 32 |
| Added slurry, ml/day | 1250 | 625 | 313 | 208 | 156 |
| Biogas yield, $m^3$/kg VS | 0.035 | 0.12 | 0.5 | 0.7 | 0.7 |
| $CH_4$ yield, $m^3$/kg VS | 0.015 | 0.02 | 0.35 | 0.42 | 0.42 |
| TVFA, g/L | 5.1 | 4.0 | 0.5 | 0.2 | 0.2 |
| pH | 5.0 | 5.5 | 6.8 | 7.0 | 7.0 |

**Table 10.4** Summarized results on the effect of LR on biogas production from TPW at constant temperature of 35 °C and HRT of 24 days (Sarada and Jospeh, 1994).

| Parameter | | | | |
|---|---|---|---|---|
| LR, kg TS/$m^3$ day | 3.0 | 4.5 | 6.0 | 7.5 |
| Biogas yield, $m^3$/kg VS | 0.59 | 0.65 | 0.58 | 0.55 |
| $CH_4$ yield, $m^3$/kg VS | 0.4 | 0.43 | 0.4 | 0.38 |
| TVFA, g/L | 0.1 | 0.1 | 0.1 | 0.2 |
| pH | 7.0 | 7.0 | 7.0 | 7.0 |

**Table 10.5** Summarized results on the effect of temperature on biogas production from TPW at feedstock loading rate of 4.5 kg/$m^3$ TS day and HRT of 24 days (Sarada and Jospeh, 1994).

| Parameter | | | | | |
|---|---|---|---|---|---|
| Temperature, °C | 25 | 30 | 35 | 40 | 45 |
| Biogas yield, $m^3$/kg VS | 0.6 | 0.63 | 0.7 | 0.7 | 0.1 |
| $CH_4$ yield, $m^3$/kg VS | 0.3 | 0.35 | 0.43 | 0.4 | 0.05 |
| TVFA, g/L | 0.9 | 0.6 | 0.5 | 0.2 | 0.2 |
| pH | 6.8 | 6.9 | 7.0 | 7.0 | 5.5 |

It was reported that the yield of biogas/methane increased with increase of the LR. Also, the yields of biogas (~16%) and methane (~20%) were greater at 24 days than that at 16 days HRT for a given LR of 4.5 kg TS/$m^3$ day. At the highest LR of 7.5 kg TS/$m^3$ day, scum formation and choking of the digester was observed determining a decrease in biogas yield. Generally, the optimum loading rate of 4.5 kg TS /$m^3$ day was reached at 24 days HRT giving the highest biogas and methane concentrations of ~0.65 and 0.43 $m^3$/kgVS, pH value of ~7.0, and TVFA value close to zero, respectively.

In terms of temperature effect on biogas/methane production, it was obtained that at lower temperatures of 25–30°C, the production of biogas and methane was lower accompanied by higher concentrations of TVFA (0.6–0.9 g/L). The best results were found at the maximum temperatures of 3°C up to 40°C, while the digestion of TPW at 45°C was unstable with rapid decrease of biogas production to almost zero.

To sum up, the overall results of HRT, LR and temperature show that the optimal process conditions for processing tomato waste to biogas/methane via anaerobic digestion were obtained at 24 days hydraulic retention time, 4.5 kg TS/$m^3$ day loading rate and 35°C temperature.

Since the major constituents of tomato waste are cellulose (25% dry weight basis) and lipid (25% dry weight basis) followed by protein (17% dry weight basis) and hemicellulose (14% dry weight basis), Sarada and Joseph [42] have performed a comparison of the digestibility of the major components of tomato processing waste between the closed batch system and the semi-continuous system of anaerobic digestion process. Authors found that in the batch process proteins were most efficiently degraded compared to other compounds. Around 70% of proteins, 65% of hemicellulose and 50% of cellulose were utilized in the AD process, whereas lipids were least converted and only 35% was degraded. The average production of biogas for the entire 100 days period of digestion of TPW was in the range of 46–56 l/kg VS. Only during the first four days the cumulative biogas production was high and decreased rapidly thereafter possibly caused by biohydrogenation of proteins. During the start-up stages up to 36 hours, hydrogen was the major component of the gases formed constituting 80 vol.% followed by carbon dioxide, which comprised 20 vol.%. Later, $CO_2$ was found to be the basic gas with traces of $CH_4$ (1–2 vol.%). Only after 10 days, the content of methane increased significantly to 10%–18%, but the total biogas production drastically decreased. Despite the other compounds, such as acetate, propionate, isobutyrate, valerate, isovalerate and caproate, constituting the total volatile fatty acids, butyrate was found to be the major component with concentration of ~4 g/L. The total concentration of VFA increased from 0.1 to 8.2 g/L over the whole 100 days digestion period with concomitant decrease in pH value from 6.5 to 4.5.

Unlike in the batch process, lipids (~95%) were mostly utilised while proteins (30%–35%) were least degraded in the semi-continuous process at the hydraulic retention time of 32 days. Meanwhile the utilisation of hemicellulose and cellulose was the same as in the batch mode process at the same HRT. It was also found that the higher the HRT, the higher the level of degradation of lipid. The concentration of volatile fatty acids of 5.2 g/L was only high at 8 day HRT, which had an impact on decrease of pH of the slurry from 7.0 to 5.0. Subsequently the level of VFA decreased to 0.5 g/L and the pH value remained at 7.0 at the HRT of 32 days. Propionate was the most dominant compound found from 8 to 16 days and at 32 days HRT in the VFA composition. Very high concentrations of isobutyrate and isovalerate (50% of TVFA) were determined in 16 day HRT slurry. In the meantime, acetate and propionate were the only fatty acids present in the slurry at the HRT of 32 days. The highest biogas/methane yields of 0.52 and 0.31 $m^3$/kg VS were observed at the highest HRT of 32 days, respectively. The produced biogas from TPW consisted of more than 60% methane with rest being $CO_2$.

In another study, Sarada and Joseph [43] monitored predominant physiological groups of microorganisms which degrade cellulose, hemicellulose, pectin, protein and lipid at various hydraulic retention times of anaerobic digestion of TPW to biogas. The main microorganisms identified were cellulolytics, xylanolytics, pectinolytics, proteolytics, lipolytics, and methanogens. The experiments were performed in both batch (1 l Buchner flasks containing 6% TPW slurry and 10% inoculum at 33±2 °C) and semi-continuous (5.5 l capacity bottles each containing 5 l of slurry with 4.5% solids at 8, 16, 24 and 32 HRT) modes. In the batch mode lasting up to 110 days, the number of cellulolytics, xylanolytics, pectinolytics, proteolytics and lipolytics increased steadily up to 40 days, but declined thereafter possibly due to the decrease in pH.. In the semi-continuous digestion mode, the number of cellulolytics, proteolytics, and lipolytics was found to be greater at 24 and 32 days HRT than at 8 and 16 days HRT. Whereas the number of xylanolytics and pectinolytics varied very little at different retention time. It was also observed that the higher the HRT, the higher the number of methanogens was present. Moreover, the predominant microorganisms in the batch process mode belonged to *Bacteroides, Propionibacterium,* and *Selenomonas,* while in the semi-continuous process mode they belonged to *Eubacterium, Fusobacterium,* and *Bacteroides. Lactobacilli* were found in both studied processes.

Sarada and Joseph [45] also performed a comparative study of single and two stage AD of TPW to biogas/methane. In this regard, the experiments have been done in such a way to separate acidogenic and methanogenic stages of anaerobic digestion process with final scope to evaluate the efficiency of biogas production. In acidogenic stage (first stage) the process was designed to perform hydrolysis leading to high concentrations of volatile fatty acids in the slurry and low methane production. This was therefore done using a digester containing 1 liter of active slurry at 4 and 8 days hydraulic retention time. The TPW was added daily at 4.5 kg/m$^3$ loading rate. The process temperature was in the range of 28–31°C.

A 5.5 litre flasks were used for methanogenic process (second stage), which was operated at a HRT of 20 days or 250 ml of digester effluent was daily replaced with an equivalent volume of slurry from the acidogenic stage digester at 4 days HRT. The pH of the acidogenic digester was ~5.0 and was not specially adjusted before slurry addition to the methanogenic process. The data was collected for a period of 3–4 months under steady state conditions. The results of both single and two stage AD of TPW to biogas are shown in Table 10.6.

Comparison of the results shows that a 20%–34.4% and a 14%–24% increase in the yields of both biogas and methane was observed in the two stage AD process of TPW, respectively. The overall improved performance of the latter process can be explained as follows. In the acidogenic process stage, the easily biodegradable portion of organic matter is hydrolysed and fermented to volatile fatty acids at a much faster rate. This results in the decrease of pH which is favourable for acidogens, but inhibitory to methanogens resulting in less biogas production with a low methane content. It was also ascertained that proteins and fats were degraded to a lesser extent under these conditions. As the slurry of the single acidogenic stage was added to the methanogenic stage, more time was available for the degradation of organic matter and volatile fatty acids, as well and conversion to methane.

A 95% of the theoretical biogas yield was achieved in the two stage process and only 83% in the single stage process of tomato waste processing in the digesters under optimal conditions. This is considered under the condition that a theoretical yield of biogas is 0.81–0.83 m$^3$/kg VS at the typical feedstock (plants, manures) composition of carbohydrates (40%–47%), protein (17.7%–23.3%), and fat (24.3%–17%). Theoretically, if all carbon is converted to biogas then a 0.1 m$^3$/kg VS yield of biogas can be produced. Therefore, in terms of total carbon, which content in the TPW was estimated to be 54.9%, a 78% and 68% conversion to biogas in the two stage and single stage process was determined, respectively.

**Table 10.6** Main results of the AD process efficiency of single and two stage processing of TPW [45].

| Parameter | Single stage[a] | Two stage[b] |
|---|---|---|
| Total gas production rate, vol/vol/day | 1.25 | 1.68 |
| $CH_4$ production rate, vol/vol/day | 0.8 | 0.96 |
| Biogas yield, $m^3$/kg VS | 0.7 | 0.8 |
| $CH_4$ yield, $m^3$/kg VS | 0.42 | 0.52 |
| $CH_4$ concentration, % | 60–65 | 65 |
| Total volatile fatty acids, g/L | 0.36–0.39 | 0.36–0.4 |
| Ammoniacal nitrogen, g/L | 0.644–0.784 | 0.34–0.625 |
| pH | 7.0–7.2 | 7.0 |
| Theoretical/Actual biogas yield, $m^3$/kg VS: | | |
| Calculated for carbohydrates (40%–47%), protein (17.7%–23.3%), fat (24.3%–17%) | $0.81–0.83^T$/ $0.66–0.7^A$ | $0.81–0.83^T/0.76–0.8^A$ |
| Calculated for total carbon | $1000^T/0.66–0.7^A$ | $1000^T/0.76–0.8^A$ |
| Efficiency, %: | | |
| Calculated for carbohydrates (40%–47%), protein (17.7%–23.3%), fat (24.3%–17%) | 83 | 95 |
| Calculated for total carbon | 68 | 78 |

A, actual; T, theoretical.
[a]experimental conditions: HRT – 24 days, temperature 35 °C, loading rate 4.5 kg VS/$m^3$ day.
[b]HRT – 20 days, temperature 28–31°C.

To sum up, the segregation of the acidogenic and methanogenic stages resulted in a more efficient conversion of TPW to biogas/methane without the need to adjust the pH by adding any alkali in the digesters.

Hills and Nakano [23] investigated the effect of particle size on methane production via AD of tomato solid wastes at laboratory scale and mesophilic conditions (35 ± 1°C). Tomato solid waste obtained from a cannery was chopped to average particle sizes of 1.3, 2.4, 3.2, 12.7, and 20 mm. These were fed at 3 g VS/liter to a 4 liters digester with hydraulic retention time of 18 days. It was observed that the greatest biogas production and volatile solids reduction occurred using the most finely chopped tomato substrate of 1.3 mm. With this fraction of substrate it was possible to produce 0.81 volume of methane per digester volume per day (vol/vol/day) with a VS reduction of 60.3%. Meanwhile the 20 mm substrate yielded only 0.25 vol/vol/day of $CH_4$ and VS reduction was 21.1%. The methane content in the produced biogas decreased slightly from 62.3% to 59.9% with finer substance used. One of the possible explanation is that a chopping process for particle size reduction greatly increases the available surface are of the substance accessible for enzymatic attack. As a general conclusion it was stated that the production rate of $CH_4$ appeared to be inversely linear to the substrate's average particle diameter.

Saghouri et al. [40] modelled and experimentally evaluated anaerobic digestion process of tomato processing waste for biogas production in a single phase laboratory digester at mesophilic conditions (35°C). Biogas production from TPW was approximately 142 liters equivalent to 0.14 $m^3$/kg VS. This was much lower as reported in [41] (~0.6 $m^3$/kg VS) probably due to different inoculum used. The methane content in the produced biogas was around 60.5% corresponding to 86 liters in the biogas.

The results of modelling of one of the most important parameter pH during the AD process showed that the process proceeded in two stages and four phases (hydrolysis, acidification, acedification, and methanogenesis) (Fig. 10.3). The first stage is called self-compatibility stage. During this stage, there is a decreasing trend in the pH value due to production of the VFA. During 12 days period, the pH decreases from 7.2 to 5.75 (Fig. 10.3A), and then increases gradually. In the second phase of the first stage (Fig. 10.3B), the methane-forming bacteria starts their activity thus consuming VFA and converting to $CH_4$ and $CO_2$. The first stage is completed in the 24th day of the process. After, the second stage, namely the stabilization stage begins. The first phase (Fig. 10.3C) shows a slow increase trend of pH. Finally,

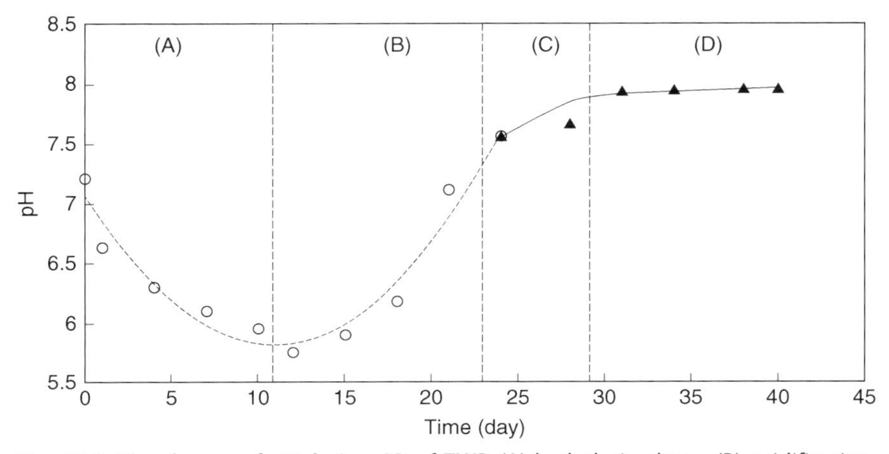


**Fig. 10.3** ***The change of pH during AD of TWP,*** (A) hydrolysis phase, (B) acidification, (C) acedification, (D) methanogenesis Solid-dashed line – model results, symbols – experimental data. *(Reproduced with permission from* [40]*).*

in the second phase of the second stage (Fig. 10.3D) the pH reaches its constant value to approximately 8 with final conversion of the remaining VFA. It could be stated that the Gaussian and the quadratic models were the most appropriate models for variation of the pH, biogas/methane and $CO_2$ versus retention time.

Bacenetti et al. [3] evaluated, from the environmental point of view, two different tomato processing waste management systems of tomato industry in Italy. The evaluation was based on a life cycle assessment (LCA) method of the tomato by-products (discarded tomatoes, skins and seeds) formed after the production of tomato puree utilization for energy production via AD thus reducing environmental load of tomato puree. Two scenarios were taken into consideration: Baseline Scenario (BS), when tomato by-products are taken back to the fields as organic fertilizers; alternative scenario (AS) – tomato processing waste is used in a nearby biogas plant for energy production. In order to properly compose the LCA model the following subsystems (SS) were included: (1) crop cultivation (SS1), (2) tomato processing (SS2), and (3) by-product management (SS3). The subsystems involve the subdivision:

1. SS1: soil tillage and transplanting; crop growth; harvesting; transportation.
2. SS2: tomatoes unloading; selection and washing; chopping; blenching; concentration and refinement; pasteurization.
3. SS3: management of by-products arisen from SS2 in the AD plant with CHP production. The electric capacity of the engine is 300 kW. Produced electric energy is sold to grid, while the heat is used for the AD process (self-consumption: around 15% of the produced thermal energy). The remaining thermal energy is consumed in SS2.

The following nine impact potentials were chosen for evaluation and comparison between the BS and AS according to the selected method: climate change (CC), ozone depletion (OD), particulate matter (PM); photochemical oxidant formation (POF); acidification (AP), freshwater eutrophication (FE), terrestrial eutrophication (TE), marine eutrophication (ME), and mineral, fossil and renewable resource depletion (MFRD). The comparison between the BS and AS in terms of environmental impact of 1 kg of tomato puree production and its by-product utilisation is shown in Fig. 10.4.

For all the nine studied impact measures, the AS showed better environmental performances over the BS. Anaerobic digestion of tomato processing waste allows a reduction of the overall environmental load of tomato puree to around 13% instead of its utilisation as organic fertilizer directly spread on the fields. For instance, for the climate change (CC) category,

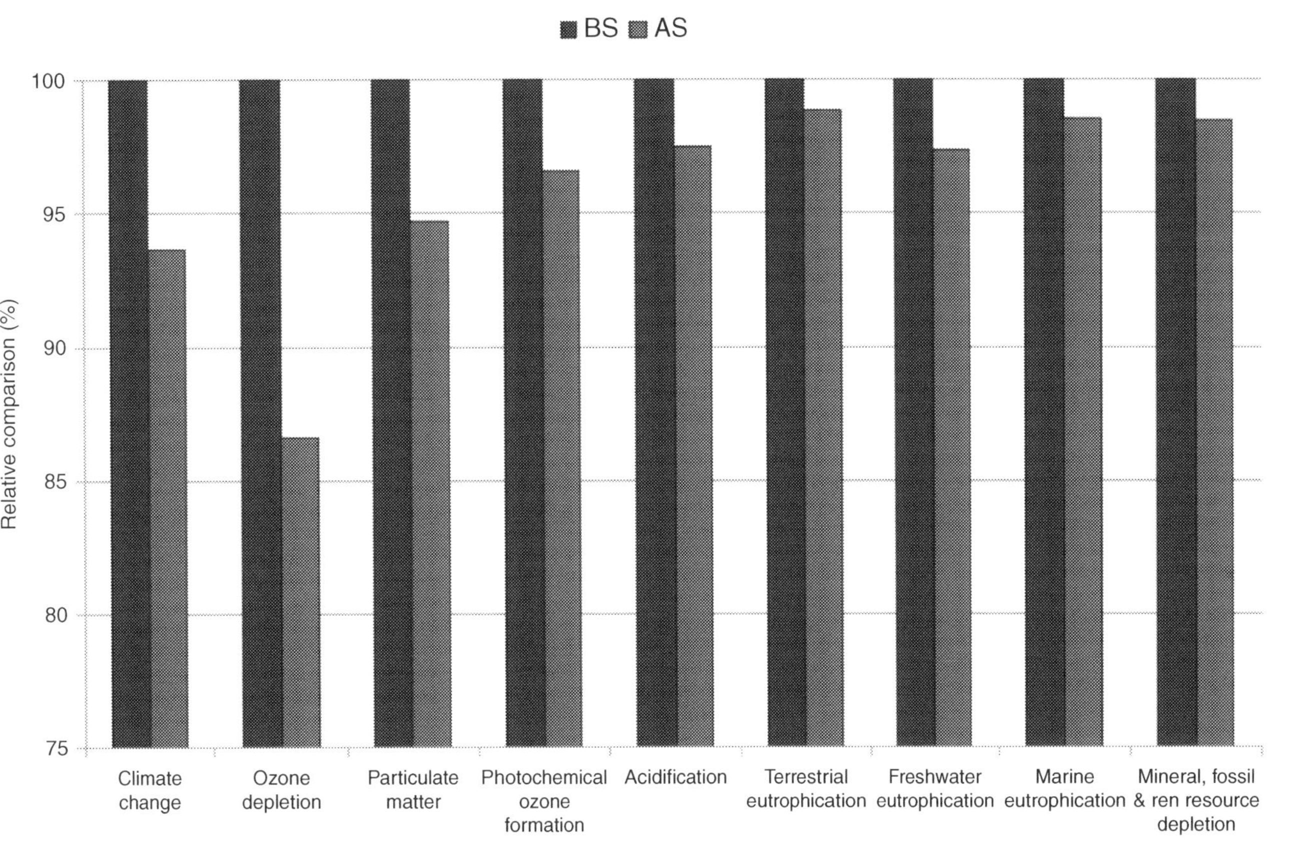


**Fig. 10.4 *Comparison among the environmental impact of 1 kg of tomato puree considering the two different management solutions for tomato by-products*** *(Reproduced with permission from* [3]*).*

the AD of tomato processing waste permits a savings of GHG emission of about 7%. Moreover, the environmental benefits due to processing of tomato by-products in the AD plant enables the credits of "green" energy production; electricity from renewable source thus avoiding generation from fossil fuels and, secondary, heat generated in the CHP of the AD plant and its use in tomato processing (SS2). Allison et al. [1,2] studied ionic liquid pre-treatment of tomato pomace and lycopene extraction from tomato pomace, and methane production from the extracted tomato pomace during anaerobic digestion. Due to superior efficiency of ionic liquid in dissolving lignocellulosic biomass, it was stated that ionic liquid (1-ethyl-3-methylimidazolium acetate ([C2mim][OAc]) pretreatment of tomato pomace could potentially boost sugar release during enzymatic digestion of pomace cell wall polysaccharides with subsequent improvement of the yield of biofuels. Anaerobic digestion was performed using both pre-treated and untreated pomace samples to investigate the potential of ionic liquid pre-treatment to enhance biomethane production.

It was reported that no significant difference in methane yield for AD of the pre-treated pomace versus the untreated pomace was found, suggesting that loss of nutrients from the pomace during pre-treatment was not the sole factor preventing enhanced AD of the pre-treated pomace.

#### *10.2.2.2 Codigestion*

Co-digestion is usually a beneficial way to improve biogas/methane production yield in AD [7]. Li et al. [32] investigated a co-digestion of tomato residues with dairy manure and corn stover for biogas production at 20% TS under mesophilic temperature of 35°C for 45 days in 1 liter reactor. Firstly, a co-digestion of tomato wastes (stalks, leaves, residual tomatoes) with dairy manure and corn stover on methane yield and system stability was evaluated. Secondly, the effect of feedstock (tomato waste, manure, corn stover) mixing ratio on methane yields and process stability was studied. The percentage of tomato residues in the mixed feedstock tested was 13%, 27%, 40%, 54% based on wet weight.

During an individual feedstock digestion in the reactor with 100% of tomato residues methane production failed. In the first 21 days the methane yield was less than 5 $L/d/VS_{feed}$ and dropped to zero after day 21. The daily yield of methane in the reactor loaded with 100% dairy manure increased gradually in the first 7 days and subsequently rapidly increased reaching a peak of 18.8 $L/d/VS_{feed}$ at day 13. The peak of methane daily yield of 100% corn stover was reached at day 29 and was 11 $L/d/VS_{feed}$.

The experiments performed with a ternary feedstock mixture showed the improved methane yields, except the case when the content of tomato residues in the mixture was 54% (33% dairy manure and 13% corn stover). In this case almost zero methane production was observed. Only by adding $NaHCO_3$ to adjust the pH value to 7.0, methane yield increased reaching a daily peak production of 23.2 $L/d/VS_{feed}$ on day 35. With less amount of diary manure and fixed amount of corn stover (33%) in the multiple feedstock, the daily methane yield was lower. For instance, at 13% of tomato by-products content in the feedstock (33% corn stover and 54% manure), the daily methane yield was 33% (27.2 $L/d/VS_{feed}$) higher than that of 54% tomato by-products (33% corn stover and 13% manure), 20.4 $L/d/VS_{feed}$. The average methane content in the produced biogas ranged from 53 to 57% until the stop of the digestion process. It was higher than that of the individual feedstocks digestion, 48% and 47% for pure dairy manure and pure corn stover, respectively.

The cumulative methane yield obtained in the 100% tomato residues anaerobic digestion was around 37.3 $L/kgVS_{feed}$ which was inhibited. While the highest cumulative methane yields of 100% manure and 100% corn stover digestion were 279 $L/kgVS_{feed}$ and 223.2 $L/kgVS_{feed}$, respectively.

The cumulative methane yields of ternary mixture were also varied depending on the ratio of feeds. The highest cumulative yield of methane was obtained at the tomato residues ratio in the substrate ranging from 13% to 27%, thus giving a 25% increase at 33% fixed dairy manure content. Further increase of tomato residues in the substrate from 27% to 54%, yielded a decrease of the cumulative methane yield by 10%. Moreover, at a fixed corn stover content of 33% (dairy manure drop from 54% to 13%), the increased content of tomato residues from 13% to 54% caused a reduction of cumulative methane yield by 76%.

Co-digestion of tomato waste with manure and corn stover substantially improved methane yields (Fig. 10.5). Tomato waste and dairy manure played an important role in increasing the methane yield.

Among the all ternary mixture studied, the highest methane yield was determined from the substrate mixture consisting of 33% corn stover, 54% dairy manure and 13% tomato waste. This lead to a 0.5-, 0.9-, and 10.2-fold higher compared to anaerobic digestion of dairy manure, corn stover and tomato waste, respectively. Generally, the use of this feedstock mixture enabled to achieve the highest volatile solids reduction of 46.2% and cumulative methane yield of 415.4 $L/kgVS_{feed}$.

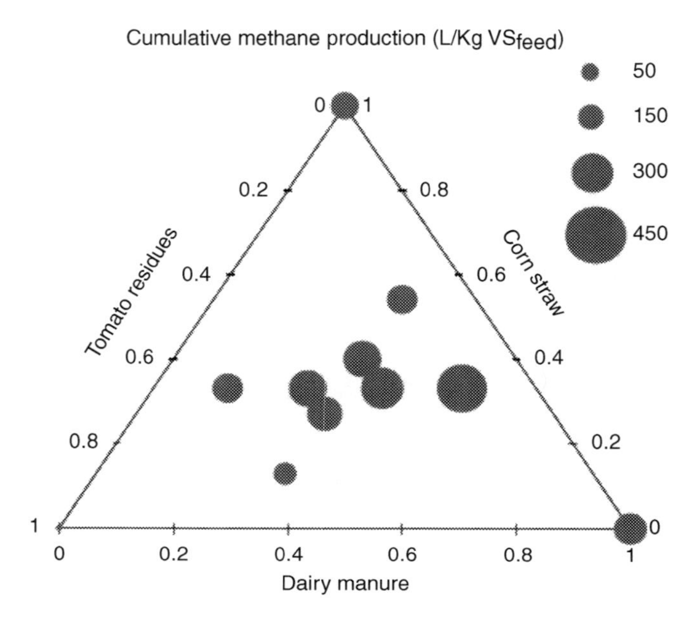


**Fig. 10.5** ***Accumulative $CH_4$ yields from anaerobic digestion of mono- and ternary feed-stocks*** *(Reproduced with permission from* [32]).

Moreover, inhibition of VFA to biogas production took place when more than 40% tomato waste was added to the mixture. Using ternary mixtures enabled to dilute the inhibitors which caused inhibition in the AD of pure tomato residues used.

In another research studies performed by Li et al. [33], a similar co-digestion tests with same ternary feedstock (tomato residues, dairy manure and corn stover) were carried out including energy and economic analysis of the AD process. It was concluded that adding 20%–40% of tomato waste to a substrate of corn stover and dairy manure was beneficial to the AD process, while more than 60% of tomato residues addition inhibited biogas production due to the excess of VFAs and pH drop. More methane was produced using ternary mixtures even with a low C/N ratio of 10–15. Although the addition of tomato waste to the mixture increased the energy input in chopping and transportation, but supplement of 40% tomato residues had the highest net energy production. This had a 135% increase in net energy production compared to AD of dairy manure used as mono-feedstock, and 88% higher compared to binary mixture composed of 60% manure and 40% corn stover. The extra energy input required for adding tomato waste for co-digestion could be compensated by the increase of methane yield.

The energy required for heating co-digested substrates had the prevailing impact on the total energy consumption. The highest volatile solids reduction of 57.0%, methane yield of 379.1 L/kg $VS_{feed}$, and net energy production were obtained using the co-digested ternary mixture of 24% corn stover, 36% dairy manure, and 40% tomato waste. The anaerobic co-digestion process can be improved by locating AD plant near greenhouses and utilising $CO_2$ produced in the digestor thus promoting photosynthesis of plants, which will have energy benefits in transportation and tomato productivity.

Li et al. [33] also performed economic evaluation of AD codigestion of diary manure with corn stover and tomato residues under liquid (L-AD), hemi-solid (HSS-AD), and solid state (SS-AD) conditions. It was assumed that the total volume of theoretical AD digester is 300 $m^3$ with operational volume of 240 $m^3$ (80% of total volume). The AD reactor was assumed to be operated in a batch mode for 35 days at the process temperature of 35°C. The treatment capacity was 200 tons of total agricultural solid waste (dairy manure, corn stover and/or tomato residues). The following assumptions were also applied in this study:

1. The lifetime of the AD plant is 20 years, with 7560 annual operating hours
2. Discount rate was fixed at 8%
3. The inoculum was used from the L-AD system
4. The produced electricity by CHP is fed into the grid
5. Part of the produced heat is used to sustain the internal technological process (keeping digester's temperature at 35 °C), whereas excess heat is wasted to the atmosphere
6. Subsidies from government were not taken into consideration

Economic analysis of the biogas plant based on net present value (NPV), which enables to evaluate economic feasibility of the biogas production scenarios, was used [8]. An NPV > 0 indicates profitability of the investment. Three scenarios based on the different total solids (TS) of the feedstock used for co-digestion were selected:

1. Scenario 1: liquid AD system (L-AD), TS = 8%
2. Scenario 2: hemi-solid-state AD (HSS-AD), TS = 15%
3. Scenario 3: solid state AD (SS-AD), TS = 20%

The composition of ternary feedstock: 48% dairy manure, 32% corn stover; 20% tomato waste (VS based). These proportions were selected based on recommendations of a previous study [34]. The results of economic analysis of the biogas production from ternary feedstock are summarized in Table 10.7.

**Table 10.7** Results of economic analysis of the biogas plant under different scenarios [33].

| Parameter | Scenario 1 | Scenario 2 | Scenario 3 |
|---|---|---|---|
| Input cost | | | |
| Raw materials, $/year | | | |
| Dairy manure | 0.0 | 0.0 | 0.0 |
| Corn stover | 3028 | 5678 | 7570 |
| Tomato waste | 0.0 | 0.0 | 0.0 |
| Fresh water | 628 | 309 | 81 |
| Transport costs for substrate, $/year | 421 | 790 | 1054 |
| Capital cost, $/20 years | 192,462 | 181,362 | 181,362 |
| Labor, $/year | 16,065 | 16,065 | 16,065 |
| Operating and maintenance costs, $/year | 11,753 | 11,192.5 | 10,435.3 |
| Liquid digestate treatment, $/year | 7906 | 6318 | 5184 |
| Annual assets, $/year | | | |
| Electricity | 22,893 | 35,571 | 35,669 |
| Solid digestate | 12,096 | 22,680 | 30,240 |
| Economic results | | | |
| NPV, $ | -26,341 | 196,589 | 263,986 |
| IRR, % | – | 8.1 | 11.7 |
| PB | – | 17.6 | 10.9 |

IRR, internal rate of return; NPV, net present value; PB, payback period.

SS-AD of tomato residues, dairy manure, and corn stover at ratios of 20:48:32 (VS basis) had the highest NPV (2.6 million US$) and IRR (11.7%), and the shortest payback period (10.1 year), which indicated Scenario 3 the most profitable of all scenarios studied. Moreover, economic analysis results indicated that capital and labor costs had the dominant effect on total investment. The highest revenue from selling electricity and solid digestate as organic fertilizer was achieved in Scenario 3 as well. It was also reported that the sales of solid digestate was more economically profitable than electricity based on the selected market price. The support subsidies from government were not considered in this study and therefore the payback period was relatively longer compared to the other works [37].

González-González et al. [21] studied environmental, energetic and economic feasibility of anaerobic co-digestion of Iberian pig slaughterhouse wastes (IPSW) mixed with tomato industry processing waste (TIP) in a semi-continuous laboratory digester at mesophilic temperature. The research mainly focused on the chemical oxygen demand (COD) removal efficiency, quality of effluents, assessment of optimal values of methane production using different ratios of substrate IPSW:TIP (100:0, 80:20, 60:40, 40:60, 20:80, 0:100), and economic feasibility. It was reported that a 54–80% reduction in COD and 6–19 $Nm^3/m^3$ substrate methane production was achieved. Furthermore, 0.79–0.88 $m^3$ water/$m^3$ substrate was recovered, which might be utilized as irrigation water, and 0.12–0.21 $m^3$ agricultural amendment/$m^3$ substrate with 91–98% moisture content. The optimal ratio of IPSW and TIP for the highest biogas production rate was reported to be at 60:40, respectively. Economic analysis showed the feasibility of both anaerobic co-digestion plant operating with 60% IPSW/40% TIP and AD plant using only Iberian pig slaughterhouse waste. The payback times are reported to be 14.86 and 3.73 years, respectively. However, the NPV and the IRR were reported to be negative for the AD plant operating on 60% IPSW/40% TIP.

Tomato residues can also be mixed and codigested for biogas production with various feedstocks other than manure or maize producing similar volumes of methane. Jagadabhi et al. [24] reported tomato, cucumber, common reed and grass silage digestion to biogas in a two-stage anaerobic process. Trujillo and Cebreros [51] performed AD of a substrate composed of tomato and rabbit wastes. However, in all the studied cases it was reported that no matter what kind of feedstocks used, the co-digestion of tomato residues with other feedstocks is only possible to some extent. The digestion of pure tomato wastes is inhibitory for biogas production. It's limited due to dominance of acidification thus forming excess VFAs and decreasing pH value, which is a crucial parameter for a stable AD plant operation and biogas/methane production. Therefore, depending on the type of inoculum, organic feedstock origin, process conditions, and mixing ratios of tomato waste with other substrates, it is possible to obtain higher methane yields compared to biogas production from mono-feedstocks, such as manure or vegetable crops.

#### *10.2.2.3 Digestate*

The residual solid digestate left after biogas production has a big value to be used as an organic fertilizer in farming and other sectors. Digestate can replace appropriate amounts of synthetic mineral fertilisers. Synthetic mineral fertiliser adds nutrients to the cycle while digestate uses the nutrients already available in the cycle thus closing the nutrient loop. The carbon-nitrogen rate (C/N-rate of 8–10) of the digestate means that nitrogen is easily available for the crops. Therefore, the use of digestate is beneficial in terms of environmental and economic aspects.

## 10.3 Biodiesel production

### 10.3.1 Biodiesel production and utilization in general

Biodiesel is a form of diesel fuel consisting of long-chain fatty acid esters made through a process called transesterification, carried out by several methods including common batch process, heterogeneous catalysts/enzymes, supercritical, ultrasonic and microwave methods. During this operation, crude glycerine is separated from the fat, using an alcohol, usually methanol (Fig. 10.6). The process leaves behind two products: Glycerol and methyl esters (biodiesel). This latter can be used, alone or blended with petroleum diesel, directly and without modification in both diesel engines and in distribution infrastructure, despite some differences. For instance, biodiesel has large Cetane index and consequently a reduced ignition time but also a lower heating value and consequently lower generated power [49]. In addition to renewability, biodiesel is cleaner than diesel as its oxygen content (up to 15% versus zero in the fossil diesel) induces a more complete combustion with less pollutant emissions (mainly reduced amounts of suspended particles and lower $SO_2$, CO, and nonburnt hydrocarbons in the exhaust gases).

Different biomasses have been examined as possible sources of biodiesel production by considering developing engine technologies, decreasing harmful emissions, ameliorating the fuel efficiency, optimizing biodiesel use in combustion chambers, etc. However, few attempts have evaluated this potential production from tomato wastes [17,19]. These investigations examined mainly the oil extraction from the fatty acids which are predominant in the tomato seeds [26].

### 10.3.2 Oil extraction and characterization

Usual agro-food products obtained from tomatoes are juice, puree, sauce, ketchup, salsa and paste. The remaining residues are made mainly of skin, peels, seeds and poor-quality tomatoes eliminated during the sorting process [27]. Seeds are the major wastes of the tomato processing industries (about 60% of the pomace), with an average of 20%–40% fat and up to 44% on dry basis [39], making them excellent candidate for biodiesel production [18].

To this end, seeds are separated from pulp, cleaned in water, dried naturally or in an oven and finally ground. The following step may be carried out by different methods, including mechanical pressing, solvent refluxing, supercritical fluid extraction, etc.

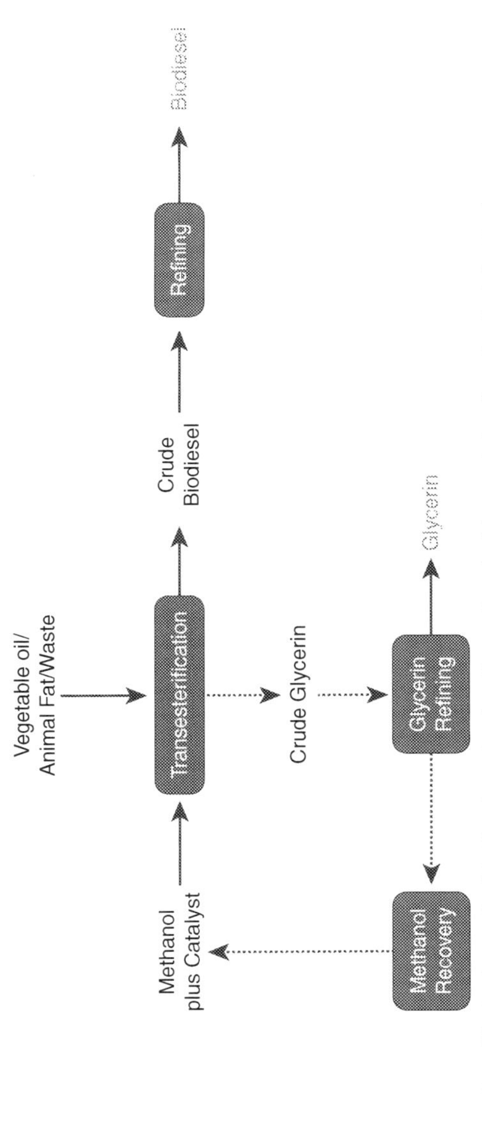


**Fig. 10.6** ***Simplified schema for biodiesel production*** *(Reproduced with permission from Bagheri, S., 2017).*

In the mechanical system, the dried seeds are simply pressed by a (screw) pressing system which mechanically squeezes the oil from seeds. The output of the pressing might be lower than the solvent extraction but it is also more ecofriendly. Indeed, during the solvent-extraction process, even though the solvent is removed later (using a rotary vacuum evaporator), gaseous losses and solvent waste are very likely.

Non-polar solvent (such as petroleum ether or n-hexane) are often used during the oil extraction in a batch reactor, in which the following succeeding and overlapping steps occur: (1) solvent diffusion into the seed powder, (2) oil dissolution in the solvent, and (3) oil diffusion from the seed powder to the surrounding solvent, leading to a mixture of oil and solvent.

Traditional solvent extraction requires solvent and oil warming during and at the end of the extractive process while stirring solvent extraction, which is carried out at room temperature, has the additional advantage of lower costs (equipment and energy). In laboratories, it is the Soxhlet apparatus which is commonly used for oil extraction from tomato pomace as it is the most similar to the industrial extractive plants [18].

The potential use of the tomato seed oil as biodiesel depends on its physicochemical properties, that in turn, are function of the particle size of ground seeds, the extraction system, the solvent type, the solvent/seed ratio, the extraction duration, temperature, etc. More precisely, it is well established that the small particle size facilitates the solvent diffusivity in the seed powder and that temperature affects considerably the process. Some authors reported also that the higher the temperature, the higher the oil yield for petroleum ether (itself higher than diethyl ether [5]) but, for n-hexane and acetone, exceeding these solvents' boiling points, the oil recovery decreased due to low solvents' densities in the samples.

Polarity, matrix and toxicity of the solvent are also fundamental. For example, Giuffrè et al. [20] showed that chloroform and stirring petroleum ether produced good tomato seed oils and, the high toxicity of the chlorinated solvent, in the first case was a minor element. The highest seed oil recovery is generally obtained by the Soxhlet extraction, due probably to the purity of the solvent that is in contact with the seed powder. In fact, after refluxing, only pure solvent drops into the extraction tube, whereas in the case of the stirrer apparatus, after the initial phase, when the largest oil quantity is extracted, the solvent is always in a mixture with the extracted oil and the extractive power is reduced Giuffrè et al. [20].

In comparison with many other seeds oils such as cotton and rapeseed, the extracted oil from tomato seeds has higher cetane rating which reduces the occurrence of incomplete combustion. It has also lower ashes, which can reduce the possibility of plugging of filters, and injector problems. It is also reported that sulphur content is lower in tomato seeds oil, which reduces levels of corrosive sulphuric acid accumulating in the engine crankcase oil [49].

Other parameters characterize the extracted oils for biodiesel production purpose. One can name the refractive index (RI), the relative density (RD), the fatty acids (FA), the fatty acid methyl esters (FAMEs), the acid value (AV), the oxidative stability index (OSI), the kinematic viscosity (KV), the high heating value (HHV) and the cetane number (CN).

Each vegetable oil has a fatty acid composition, which influences its molecular weight. Besides, a high unsaturated FAME content facilitates oil polymerization whereas a high saturated FAME content facilitates oil solidification at low temperatures Consequently, it is necessary to know the composition of these FA to calculate the molar ratio of the alcohol needed for esterification. The lower are the FA, the more efficient is the extraction operation. In biodiesel production, low linolenic acid content is appreciated and the saturated fatty acid content has to be considered. In tomato seed oil, about 15 FAMEs are detected and identified, including linoleic acid (around 55%) oleic acid (around 20%), palmitic acid (around 15%). In general, polar solvents extract more free fatty acids than non-polar ones, from tomato seeds oil, in which linoleic acid (two double bonds) was 55–56% of the total FAME content [31]. According to Giuffrè et al. [20], the combinations Soxhlet-acetone, Soxhlet-petroleum ether and stirring- petroleum ether produce the highest FAMEs.

The RD is also an important property of biodiesel fuel. This parameter is usually greater than for conventional diesel, which means that volumetrically operating fuel pumps will inject greater mass of biodiesel than petrodiesel [16]. This is the reason why different standards give a limit value, not to be exceeded for bio-oils. As density decreases linearly with temperature for vegetable oils, heating at a temperature of 120°C or higher could ensure this condition. The Soxhlet extraction systems generally produce the densest oils, but supercritical acetone extraction and mechanical cold pressing [17] could be in the range (for example between 910 and 925 $kg/m^3$ in the German standard for rapeseed oil).

The significance of AV (mg KOH/g of oil) lies on the fact that low values would reduce the loss yield due to the saponification reaction. The standards therefore expect a maximum limit, beyond which the extracted oils have to be de-acidified. In general, extraction by stirring processes give oils with adequate AV values, contrarily to solvents. This is was also true for tomato pomace oils [47].

The determination of the OSI is correlated to the induction time (h) as it indicates the resistance of oil to the oxidation, in other words, to high temperatures and reduced polymerization. Stirr-extraction gives lower OSI than solvent-extraction, probably due to the higher PV of tomato oils [18].

The KV ($mm^2/s$) indicates the fluidity of a given oil, which affects the operation of fuel injection, especially at low temperatures. In particular, high viscosity leads to poorer atomization of the fuel spray [28]. As KV is sensitive to temperature (the higher is the latter, the lower is the former), international standards impose a maximum limit value at 40°C. Tomato seed oils KV are below or in the borderline of the threshold limit value according to the chosen standard (Table 10.8). It is however worthy note that KV varies also significantly with the extraction method and with the solvent. In fact, for all temperatures (30, 40 and 50°C), the Soxhlet extraction gives always higher values of KV than stir-extraction for all solvents, except for ethyl acetate (Table 10.8).

The RI measures the light speed through the oil and it is mainly sensitive to the temperature. In fact, the lower is the temperature, the denser is the oil, the higher is its viscosity and the lower is its RI (Table 10.9).

The HHV (MJ/kg) is the maximum energy that can be obtained by the complete combustion, in standard conditions, of the tomato seeds oil. This latter yields interesting quantity of energy, above the minimum required by the different standards (36 MJ/kg for example in the German standard for rapessed oils), whatever the extraction process (Table 10.10). This might be correlated to the great saturations of the tomato seeds oils. Indeed, fuels with highest saturations generally have higher energy content [28].

The CN, one of the most important properties of a diesel fuel, is a measure of the ignition quality or the auto ignition tendency of a fuel under

**Table 10.8** KV rounded values ($mm^2/s$) for different temperatures and different extraction methods [20].

| | **Acetone** | | **Ethyl acetate** | | **Chloroform** | | **Petroleum ether** | |
|---|---|---|---|---|---|---|---|---|
| | ***Stirring*** | ***Soxhlet*** | ***Stirring*** | ***Soxhlet*** | ***Stirring*** | ***Soxhlet*** | ***Stirring*** | ***Soxhlet*** |
| 30°C | 34 | 45 | 47 | 40 | 27 | 38 | 41 | 48 |
| 40°C | 26 | 34 | 37 | 31 | 20 | 28 | 30 | 37 |
| 50°C | 21 | 25 | 27 | 25 | 16 | 22 | 24 | 26 |

**Table 10.9** RI values for different extraction methods and at different temperatures.

| Extraction method | Solvent | Temperature | RI value | Reference |
|---|---|---|---|---|
| Soxhlet | Ethyl acetate | 20°C | 1.4680 | [20] |
| Stirring | Ethyl acetate | 20°C | 1.4692 | |
| Stirring | Acetone | 20°C | 1.4693 | |
| Pressing | - | 20°C | 1.4708 | [6] |
| Soxhlet | Hexane Chloroform:methanol (2:1) | 25°C | 1.4700 | [30] |
| Soxhlet | Hexane | 25°C | 1.472 | [12] |
| Stirring | Hexane | 25°C | 1.471 | [47] |

**Table 10.10** HHV values for different extraction methods and at different temperatures.

| Extraction method | Solvent | HHV (MJ/kg) | Reference |
|---|---|---|---|
| Soxhlet | Acetone | 38.85 | [20] |
| | Ethyl acetate | 39.32 | |
| | Chloroform | 38.88 | |
| | Petroleum ether | 38.98 | |
| Stirring | Acetone | 38.08 | |
| | Ethyl acetate | 38.57 | |
| | Chloroform | 39.02 | |
| | Petroleum ether | 38.66 | |
| Soxhlet | Hexane | 39.4 | [17] |

the compression ignition process. A high CN allows a short fuel ignition delay whereas a low CN results in poor combustion characteristics and leads to excessive emissions of smoke and particulates [4]. For tomato seeds oil, this number is higher than the minimum required in the different standards (51 in the EN 14214 Europe standard; 47 in the ASTM 6751 (B100) USA standard and 36 in the German standard for rapeseed oil as fuel (DIN 51605)). Indeed, CN was reported to be more than 52 for both solvent and stirring extraction systems [20]. Some authors correlated such good CN to the saturated and long carbon chain FAMEs in tomato seeds oils.

### 10.3.3 Biodiesel production

The same procedure of biodiesel production from the crude oil extracted from fruits or vegetables was applied to tomato residues oils in the literature [16,17,20,49,54]. The biodiesel is obtained by transesterification, which is a chemical reaction where glycerides of the oil are transformed into their esters. This reaction involves an alcohol (usually methanol or ethanol) which is mixed with the oil in the presence of a catalyst (enzymes, acid-base catalysts, and heterogeneous (or surface) catalysts).

The functional group of the alcohol substitutes the functional group of the fatty acids and produces different water molecules. The higher the molar ratio alcohol/oil, the higher the reaction yields.

A catalyst is a substance that can be added to a reaction to increase the reaction rate without getting consumed in the process. Although the removal of homogeneous catalyst (such as KOH) after reaction is more difficult (and therefore causes a lower yield) than heterogeneous catalysts, the former are gaining attention because of their good catalytic activity, faster reaction rate, more complete conversion of the oils, and moderate reaction temperature conditions, between 50 and 70°C [18]. Enzyme catalysts (lipases), recently used in transesterification of lipids from tomato pomace oils [46], are advantageous as they are "greener", can decrease the inhibition rate and can assure an easy separation of product from enzymes. Besides, the oil does not require a pre-treatment and, moisture content in the raw oil does not hinder the reaction [53].

Reaction temperature, reaction time, amount of alcohol and amount of reaction catalyst are the most influential parameters [48]. These variables impact mainly the yield and the chemical and physical parameters of the final product. Finally, the obtained mixture is raw biodiesel, that contains soap and glycerol (formed during the transesterification), that have to be removed from the esters by decantation and/or water washing until a neutral pH is reached. Any excess catalyst or water is also to be removed from the resulting biodiesel. Heating and clarification are the ultimate steps before storage and analysis.

### 10.3.4 Biodiesel characterization

Biodiesel quality is characterized by a large number of parameters, the threshold values, the calculation and the measurement procedure of which are imposed by international standards such as ASTM and EN standards. The most relevant of these properties were exposed in section III.2., for the tomato pomace oils, specifically the density, the kinematic viscosity, the acid value and the cetane number. Another significant parameter is the iodine value (IV).

The density is important because it gives an indication of the delay between the injection and combustion of the fuel in a diesel engine (ignition quality) and the energy delivered per unit mass (specific energy). This can influence the efficiency of the fuel atomization for airless combustion systems. Density of tomato wastes diesel ranged from 883 to 915 $kg/m^3$, which is within the range established by European regulations (Table 10.11). This parameter varies significantly; in particular it is lower in the samples in

**Table 10.11** Characterization of the tomato biodiesel.

| Characteristics and standards | | | Results | | |
|---|---|---|---|---|---|
| **Parameter** | **Test method** | **Limit** | **Karami et al., 2019** | **[20]** | **[49]** |
| AV (mg KOH/g) | ASTM D 664 | <0.5 | nd | 0.06–0.17 | nd |
| IV (g $I_2$/100 g) | EN 14214 | <120 | nd | 116–126 | nd |
| KV at 40°C ($mm^2$/s) | EN 14214 | 3.5–5.0 | 5 | 4.92–8.66 | 2.8 |
| RD at 15°C (kg/$m^3$) | EN 14214 | 860–900 | 883 | 880–891 | 915.1 |
| CN | EN 5165 | >51 | 47.7 | 53.1–54.5 | 54.71 |

nd, not determined.

which a higher yield was obtained. This is due to the fact that the oil has a higher density than the esters of its fatty acids.

Kinematic viscosity refers to the thickness of the oil and describes the biodiesel internal resistance to flow through an orifice. It range from 2.8 to 8.7 $mm^2$/s, which outboards the limits of the European standard, but, except for the sample with 8.7 $mm^2$/s, KVs are in the range accepted by the ASTM D 445 standard which extends from 1 to 6 $mm^2$/s.

The acid value is used to evaluate fuel acidity over time. Diesel from tomato oils always presented AV values below the maximum legal limit.

The IV (g $I_2$ /100 g of biodiesel) is related to unsaturated fatty acid composition. This quantitative parameter finds significance because it does affect (directly or not) almost all the other qualitative parameters. For example, the higher the saturation, i.e. less double bonds, the higher the stability and consequently the OSI. Transesterification significantly affects this parameter. Tomato seed oil has a high degree of un-saturation, which is then reproduced in biodiesel. Actually, linolenic acid increases the instability of biodiesel and thus worsens the biofuel quality, since its three double bonds are easily oxidizable. The legal limit is set at 12% of the total FAME composition, considerably higher than the values found in tomato biodiesel (1.98 % in [56], 1.88 % in [20]).

Globally, results confirm the possibility that biodiesel could be technically processed from tomato wastes with characteristics compliant with the different regulations. However, to date, there is a lack of essential research and substantial knowledge that must be addressed before these residues can be chosen as a commercial fuel.

## 10.4 Bioethanol production

Alternative liquid fuels such as bioethanol are indispensable to decrease the reliance on fossil fuels and to encourage an economic growth with respect to the environment. In fact, bioethanol can be generated from sustainable biomass sources and adapted to the current fuel supply systems for a cleaner combustion. Bioethanol is generally produced from crops such as sugar cane and corn following different stages including pretreatment, hydrolysis and fermentation stages. However, the great interest to these food crops were followed by a rapid increase in their prices leading to economic concern. Therefore, cheap and abundant biomass feedstocks such as food processing residues are suitable for bioethanol production at low cost [50].

In this context, tomato derived wastes including pomaces, peels, seeds and stalks, could be a promising low cost and sustainable feedstock for bioethanol production [15]. However, the characteristics and the composition of these by-products are variable and specific procedures should be adapted for each waste.

### 10.4.1 Pretreatment techniques

Lignocellulosic biomasses generated by the food processing industries are mainly composed of hemicellulose, cellulose and lignin. These polymers could not be metabolized directly by conventional fermentation microorganisms. Therefore, pretreatment techniques are required to remove lignin and to deliver the sugars contained in hemicellulose and cellulose.

Several procedures including, drying and milling [11,57] dilute acid pretreatment [11], alkali pretreatment [10], hydrothermal processes [11,22], steam explosion [10] and supercritical $CO_2$ extraction [25,57] have been applied to tomato wastes to increase the total yield of fermentable sugars during the hydrolysis step. Moreover, combinations of successive pretreatment techniques have been also applied in the literature [10,11].

Del Campo et al. [11] have examined the pretreatment of tomato processing wastes though drying, milling, and hydrothermal process. Samples were homogenized, oven-dried at 45°C and milled prior to pretreatment step (particle size $< 1$ mm). In order to determine the optimum pretreatment conditions, authors have tested temperatures from 100 to 130°C and reaction times from 5 to 30 min for a solid:liquid ratio of 5%. Some of the pretreatment tests were performed on a larger scale using a 2-L stirred tank reactor. The obtained results indicated that the best assay conditions were 110°C and 5 min. The amount of single sugars recovered was $35.46 \pm 2.23$ (% w/w) [11].

Cöpür et al., [10] have compared different chemical pretreatments ($NaOH$, $H_2SO_4$, $NaBH_4$) for steam-exploded and dry-milled vegetable stalks mixture (tomato, pepper and eggplants). Steam explosion was performed in a 20 L reaction vessel through placing 1 kg of stalks in presence of saturated steam at 200°C and 15 bar during 5 min. The obtained sample was filtered then dried and grinded (d <74 μm) using a laboratory mill. Both steam exploded and dry milled samples were exposed to 2% (w/v) NaOH, $H_2SO_4$, $NaBH_4$ concentrations. These tests were carried out in an autoclave for 10% (w/v) solid loading at 121°C during residence times of 30 and 90 min. The obtained results indicated that vegetable stalks are valuable feedstocks for bioethanol production due to their higher content of polysaccharides (65%). The comparison between different pretreatment techniques showed that steam explosion is efficient in lignin removal and preserves glucan in the structure. Furthermore, NaOH treatment lead to highest lignin removal (6.40%–6.74%) but with some undesirable glucan removal (4.54%–10.5%). In contrast, $NaBH_4$ preserved more glucan (1.15%–4.33% removal) and removed some lignin from the structure (3.85%–4.73%) [10].

Recently, the supercritical $CO_2$ has been used as a green technique for lycopene and carotenoids extraction from tomato wastes [25,57]. The remaining lignocellulosic residue represents an important polysaccharidic industrial waste. Lenucci et al. [57] have carried out $CO_2$ extraction by submitting freeze-dried and milled (500 μm) tomato matrix to a $CO_2$ flow rate of 18−20 kg/h at a temperature of 65−70°C and a pressure of 450 bar. The extraction was perfomed for 3 h. The supercritical $CO_2$ extracted tomato matrices were submitted to hydrolysis and fermentation steps for bioethanol production [57].

### 10.4.2 Hydrolysis step, saccharification

During the hydrolysis step, acid and enzymatic methods have been generally applied for tomato wastes. The dilute acid hydrolysis procedure is the simplest and oldest technique for generating bioethanol from lignocellulosic resources. Del Campo et al. [11] have submitted tomato processing wastes to hydrolysis reaction at temperatures from 100 to 130°C, low $H_2SO_4$ concentration of 0.5% and 1.0% (w/w), reaction times from 5 to 30 min and a solid:liquid ratio of 5%. The optimal conditions were obtained at 110°C, 5 min reaction time and 0.5 (%w/w) $H_2SO_4$ concentration. The highest amount of the recovered single sugars was 36.86±4.42 (% w/w) [11].

Gaspar et al. [15] have used the standard analytical procedure for the hydrolysis method (NREL/TP-510-42618 Technical Report) established by the National Renewable Energy Laboratory in order to determine the hemicellulose, cellulose and lignin contents. Authors aimed to calculate the potential of bioethanol production. During this method, 300 mg of dried tomato wastes were digested (72% $H_2SO_4$ solution, 30°C, 60 min), and the resulting hydrolysates were diluted and autoclaved (121°C, 60 min) [15].

The technical difficulties in recuperating sugar from the acid stimulate the enzymatic method application. This technique is more effective and occurred at room temperature without the generation of harmful waste. Hijosa-Valsero et al. [22] have achieved the enzymatic hydrolysis of tomato pomace by cellulases actions. Enzyme mixture of two commercially cellulase enzymes: Cellic CTec 2 (Novozymes, China) and Viscozyme L (Novozymes, Denmark) was used. The enzyme hydrolyses were performed at 50°C, with shaking at 180 rpm 48 h or 120 h, depending on the biomass load (20% or 30%, respectively). Hijosa-Valsero et al. [22] have found in the obtained broth 44.1 g/L and 85.1 g/L of total sugars for biomass load of 20% and 30%, respectively. These concentrations imply saccharification efficiencies of 61.0 and 66.8% for biomass load of 20% and 30%, respectively.

### 10.4.3 Fermentation step and bioethanol production

Yeast microorganisms are commonly used during biomass fermentation. More specifically yeast *Saccharomyces cerevisiae* has generally been used to ferment the glucose derived from coffee processing residues to bioethanol.

Lennuci et al. (2013) [57] have exposed various hydrolyzed tomato wastes (tomato pomace, tomato serum: by-products of $CO_2$ supercritical) to to fermentation with a selected Saccharomyces cerevisiae strain, "Cispa 161", immobilized in sodium alginate beads. The bioethanol produced from each waste was usually >50% of the calculated theoretical amount [57].

Hijosa-Valsero et al. [22] have compared twelve bacterial strains for their performance in tomato waste fermentation. For the tomato pomace sample hydrolyzed with a 20% solid-to-solvent ratio, fermentation experiments produced 0.22–5.95 g/L acetone, 5.82–7.00 g/L butanol, 0.17–0.39 g/L ethanol, 0.05–8.28 g/L isopropanol and a total of 11.03–15.07 g/L ABEI (acetone-butanol-ethanol-isopropanol). For the tomato pomace sample hydrolyzed with a 30% solid-to-solvent ratio, fermentation experiments produced 20.1–21.7 g/L ethanol [22]. The experimental conditions for each bioethanol production step and the mass balance are shown in the following figure. Fig. 10.7.

## 10.5 Conclusion

The conversion of tomato derived wastes into bioethanol, biodiesel and biogas was investigated in this literature. The investigation on biogas production from the tomato wastes firstly started since 1989 by Sarada et co-workers. Authors found that the optimal process conditions for

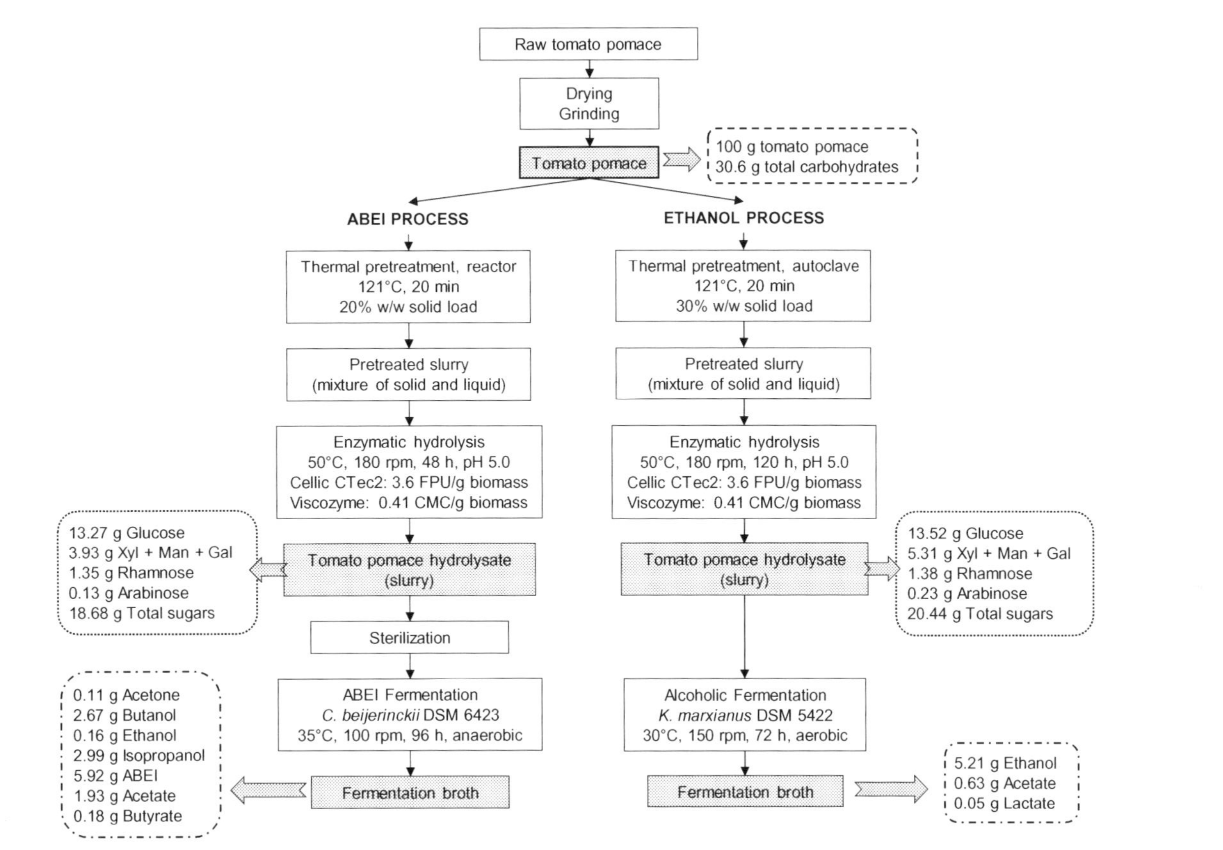


**Fig. 10.7** *Mass balance of ABEI and alcoholic fermentation (normalized to an initial amount of 100 g dried tomato pomace) for two model microorganisms.*
*Mass losses due to evaporation have been considered for the calculation. Note: Xyl, xylose; Man, mannose; Gal, galactose* [22].

processing tomato waste to biogas/methane via anaerobic digestion were obtained at 24 days hydraulic retention time, 4.5 kg TS/$m^3$ day loading rate and 35°C temperature. Furthermore, authors showed that the biogas yield increased when a two stage AD process of tomato waste was applied. Other investigations have confirmed the promising potential of TW for biogas production. They have proposed the co-digestion with other feedstocks such as animal manure and corn stover to improve the methane yield.

The lipids extraction from tomato wastes was examined in the literature using different solvent. This oily fraction has been converted into fatty acid methyl ester (FAME) and fatty acid ethyl ester (FAEE) via the biodiesel trans-esterification reaction. This latter process is influenced mainly by the reaction temperature and time and by the amount of the alcohol and the catalyst used. Most successful operations, in terms of compliance with the international regulations, were conducted at temperatures around 55°C, for a period of 1 hour, using 1:8 to 1:6 oil/methanol molar ratio and 1 to 2% wt. of catalyst.

The direct conversion of tomato wastes to bioethanol was poorly investigated in the literature due to the higher lignin and lipid contents. Therefore, pretreatment techniques such as diluted acid hydrolysis, alkali treatment, hydrothermal treatment, $CO_2$ supercritical and steam explosion and have been used to increase the production rate and the total yield of fermentable sugars during the saccharification step.

## References

[1] B.J. Allison, J.C. Cádiz, N. Karuna, T. Jeoh, C.W. Simmons, The effect of ionic liquid pretreatment on the bioconversion of tomato processing waste to fermentable sugars and biogas, Appl. Biochem. Biotechnol. 179 (2016) 1227–1247.

[2] B.J. Allison, C.W. Simmons, Valorization of tomato pomace by sequential lycopene extraction and anaerobic digestion, Biomass Bioenergy 105 (2017) 331–341.

[3] J. Bacenetti, D. Duca, M. Negri, A. Fusi, M. Fiala, Mitigation strategies in the agro-food sector: the anaerobic digestion of tomato purée by-products. An Italian case study, Sci. Total Environ. 526 (2015) 88–97.

[4] N. Bezaire, K. Wadumesthrige, Ng K.Y. Simon, S.O Salley, Limitations of the use of cetane index for alternative compression ignition engine fuels, Fuel 89 (2010) 3807–3813.

[5] C. Botineştean, A.T. Gruia, I. Jianu, Utilization of seeds from tomato processing wastes as raw material for oil production, J. Mater. Cycles. Waste Manag. 17 (2015) 118–124.

[6] C. Botineştean, N.G. Hădărugă, D.I Hădărugă, I. Jianu, Fatty Acids composition by gas chromatography –mass spectrometry (GC-MS] and most important physical chemicals parameters of tomato seed oil, J. Agroaliment. Processes Technol. 18 (2012) 89–94.

[7] D. Brown, Y. Li, Solid state anaerobic co-digestion of yard waste and food waste for biogas production, Bioresour. Technol. 127 (2013) 275–280.

[8] W.M. Budzianowski, D.A. Budzianowska, Economic analysis of biomethane and bioelectricity generation from biogas using different support schemes and plant configurations, Energy 88 (2015) 658–666.

[9] P.S. Calabrò, R. Greco, A. Evangelou, D. Komilis, Anaerobic digestion of tomato processing waste: Effect of alkaline pretreatment, J. Environ. Manage. 163 (2015) 49–52.

[10] Y. Cöpür, Ö. Özyürek, A. Tozluoglu, S. Kütük, Enzymatic digestibility of tomato, pepper, and eggplant stalks mixture, BioRes 7 (2012) 3252–3261.
[11] I. Del Campo, I. Alegría, M. Zazpe, M. Echeverría, I. Echeverría, Diluted acid hydrolysis pretreatment of agri-food wastes for bioethanol production, Ind. Crop. Prod. 24 (2006) 214–221.
[12] M. Fahimdanesh, M.E. Bahrami, Evaluation of physicochemical properties of iranian tomato seed oil, J. Nutr. Food Sci. 3 (2013) 1–4.
[13] Food and Agriculture Organization of the United Nations, Statistics Available online: http://www.fao.org/faostat/en/#data/QC (Accessed on 29 January 2020).
[14] Fachagentur Nachwachsende Rohstoffe e.V. (FNR). Biogas, an introduction. 2013, https://mediathek.fnr.de/media/downloadable/files/samples/b/r/brosch.biogas-2013-en-web-pdf.pdf.
[15] M.C. Gaspar, C.V.T. Mendes, S.R. Pinela, R. Moreira, M.G.V.S. Carvalho, M.J. Quina, et al., Assessment of agroforestry residues: their potential within the biorefinery context, ACS Sustain. Chem. Eng. 7 (2019) 17154–17165.
[16] E.G. Giakoumis, A statistical investigation of biodiesel physical and chemical properties, and their correlation with the degree of unsaturation, Renew. Energy 50 (2013) 858–878.
[17] P.N Giannelos, S. Sxizas, E. Lois, F. Zannikos, G. Anastopoulos, Physical, chemical and fuel related properties of tomato seed oil for evaluating its direct use in diesel engines, Ind. Crops Prod. 22 (2005) 193–199.
[18] A.M. Giuffrè, M. Capocasale, C. Zappia, M. Poiana, Biodiesel from tomato seed oil: transesterification and characterisation of chemical-physical properties. Agron. Res. 15, (2017a) 133–43.
[19] A.M. Giuffrè, V. Sicari, M. Capocasale, C. Zappia, T.M. Pellicanò, M. Poiana, Physicochemical properties of tomato seed oil (Solanum Lycopersicum L.) for biodiesel production, Acta Hortic. 1081 (2015) 237–244.
[20] A.M. Giuffrè, C. Zappia, M. Capocasale, Tomato seed oil: a comparison of extraction systems and solvents on its biodiesel and edible properties. La rivista italiana delle sostanze grasse, XCIV (2017b).
[21] A. González-González, F. Cuadros, A. Ruiz-Celma, F. López-Rodríguez, Energy-environmental benefits and economic feasibility of anaerobic codigestion of Iberian pig slaughterhouse and tomato industry wastes in Extremadura (Spain), Bioresour. Technol. 136 (2013) 109–116.
[22] M. Hijosa-Valsero, J. Garita-Cambronero, A.I. Paniagua-García, R. Díez-Antolínez, Tomato waste from processing industries as a feedstock for biofuel production, Bioenergy Res. 12 (2019) 1000–1011.
[23] D.J. Hills, K. Nakano, Effects of particle size on anaerobic digestion of tomato solid wastes, Agric. Wastes 10 (1984) 285–295.
[24] P.S. Jagadabhi, P. Kaparaju, J. Rintala, Two-stage anaerobic digestion of tomato, cucumber, common reed and grass silage in leach-bed reactors and upflow anaerobic sludge blanket reactors, Bioresour. Technol. 102 (2011) 4726–4733.
[25] M. Kehili, L.M. Schmidt, W. Reynolds, A. Zammel, C. Zetzl, I. Smirnova, et al., Biorefinery cascade processing for creating added value on tomato industrial by-products from Tunisia, Biotechnol. Biofuels 9 (2016) 261.
[26] B. Khiari, M. Moussaoui, M. Jeguirim, Tomato-processing by-product combustion: thermal and kinetic analyses. Materials 12, (2019a) 553.
[27] B. Khiari, M. Moussaoui, M. Jeguirim, Tunisian tomato waste pyrolysis: thermogravimetry analysis and kinetic study. Environ. Sci. Pollut. Res. 26, (2019b) 35435–35444.
[28] G. Knothe, K.R. Steidley, Kinematic viscosity of biodiesel fuel components and related compounds. Influence of compound structure and comparison to petrodiesel fuel components, Fuel 84 (2005) 1059–1065.
[29] N. Kraiem, M. Lajili, L. Limousy, R. Said, M. Jeguirim, Energy recovery from Tunisian agri-food wastes: evaluation of combustion performance and emissions characteristics

of green pellets prepared from tomato residues and grape marc, Energy 107 (2016) 409–418.
[30] A.S. Kulkarni, V.I. More, R.R. Khotpal, Composition and lipid classes of orange, tomato and pumpkin seed oils of vidarbha region of Maharashtra, J. Chem. Pharm. Res. 4 (2012) 751–753.
[31] E.S. Lazos, J. Tsaknis, S. Lalas, Characteristics and composition of tomato seed oil, Grasas y Aceites 49 (1998) 440–445.
[32] Y. Li, Y. Li, D. Zhang, G. Li, J. Lu, S. Li, Solid state anaerobic co-digestion of tomato residues with dairy manure and corn stover for biogas production, Bioresour. Technol. 217 (2016) 50–55.
[33] Li, Y., Lu, J., Xu, F., Li, Y., Li, D., Wang, G., Li, S., Zhang, H., Wu, Y., Shah, A., et al. 2018a. Reactor performance and economic evaluation of anaerobic co-digestion of dairy manure with corn stover and tomato residues under liquid, hemi-solid, and solid state conditions. Bioresour. Technol. 270, 103–112.
[34] Li, Y., Xu, F., Li, Y., Lu, J., Li, S., Shah, A., Zhang, X., Zhang, H., Gong, X., Li, G. 2018b. Reactor performance and energy analysis of solid state anaerobic co-digestion of dairy manure with corn stover and tomato residues. Waste Manag.73, 130–139.
[35] P. Loizia, N. Neofytou, A.A. Zorpas, The concept of circular economy in food waste management for the optimization of energy production through anaerobic digestion, Environ. Sci. and Pollut. R. 26 (2019) 14766–14773.
[36] F. Passos, V. Ortega, A. Donoso-Bravo, Thermochemical pretreatment and anaerobic digestion of dairy cow manure: experimental and economic evaluation, Bioresour. Technol. 227 (2017) 239–246.
[37] R.J. Patinvoh, O.A. Osadolor, I. Sárvári Horváth, M.J. Taherzadeh, Cost effective dry anaerobic digestion in textile bioreactors: Experimental and economic evaluation, Bioresour. Technol. 245 (2017) 549–559.
[38] M. Rasapoor, B. Young, R. Brar, A. Sarmah, W.Q. Zhuang, S. Baroutian, Recognizing the challenges of anaerobic digestion: critical steps toward improving biogas generation, Fuel 261 (2020) 116497.
[39] G. Rossini, G. Toscano, D. Duca, F. Corinaldesi, E. Foppa Pedretti, G. Riva, Analysis of the characteristics of the tomato manufacturing residues finalized to the energy recovery, Biomass Bioenergy 51 (2013) 177–182.
[40] M. Saghouri, Y. Mansoori, A. Rohani, M.H.H. Khodaparast, M.J. Sheikhdavoodi, Modelling and evaluation of anaerobic digestion process of tomato processing wastes for biogas generation, J. Mater. Cycles Waste Manag. 20 (2018) 561–567.
[41] R. Sarada, K. Nand, Start-up anaerobic digestion of tomato-processing wastes for methane generation, Biol. Wastes 30 (1989) 231–237.
[42] R. Sarada, R. Joseph, Biochemical changes during anaerobic digestion of tomato processing waste, Process Biochem 28 (1993) 461–466.
[43] R. Sarada, R. Joseph, Characterization and enumeration of microorganisms associated with anaerobic digestion of tomato-processing waste. Bioresour. Technol. 49 (1994a) 261–265.
[44] R. Sarada, R. Joseph, Studies on factors influencing methane production from tomato-processing waste. Bioresour. Technol. 47, (1994b) 55–57.
[45] R. Sarada, R. Joseph, A comparative study of single and two stage processes for methane production from tomato processing waste, Process Biochem 31 (1996) 337–340.
[46] M. Sarno, M. Iuliano, Highly active and stable $Fe_3O_4$/Au nanoparticles supporting lipase catalyst for biodiesel production from waste tomato, Appl. Surf. Sci. 474 (2019) 135–146.
[47] D. Shao, C. Venkitasamy, X. Li, Z. Pan, J. Shi, B. Wang, et al., Thermal and storage characteristics of tomato seed oil, LWT - Food Sci. Technol. 63 (2015) 191–197.

[48] D. Singh, D. Sharma, S.L. Soni, S. Sharma, P. Kumar Sharma, A. Jhalani, A review on feedstocks, production processes, and yield for different generations of biodiesel, Fuel 262 (2020) 116553.
[49] H. Sivasubramanian, V. Sundaresan, S.K. Ramasubramaniam, S.R. Shanmugaiah, S.K. Nagarajan, Investigation of biodiesel obtained from tomato seed as a potential fuel alternative in a CI engine, Biofuels 11 (2020) 57–65.
[50] S. Sukumara, J. Amundson, F. Badurdeen, J. Seay, A comprehensive techno-economic analysis tool to validate long-term viability of emerging biorefining processes, Clean Technol. Environ. Policy 17 (2015) 1793–1806.
[51] D. Trujillo, F.J. Cebreros, Energy recovery from wastes ad of tomato plant mixed with rabbit wastes, Bioresour. Technol. 45 (1993) 81–83.
[52] I. Vardopoulos, I. Konstantopoulos, A.A. Zorpas, L. Limousy, S. Bennici, V. Inglezakis, I. Voukkali, Sustainable metropolitan areas perspectives though the assessment of the existing waste management strategies, Environ. Sci. Pollut. Res. (2020). https://doi.org/10.1007/s11356-020-07930-1.
[53] W. Xie, N. Ma, Immobilized lipase on $Fe_3O_4$ nanoparticles as biocatalyst for biodiesel production, Energy Fuels 23 (2009) 1347–1353.
[54] A. Yousuf, F. Sannino, D. Pirozzi, Production of microbial lipids from tomato waste to be used as feedstock for biodiesel, Environ. Eng. Manag. J. 16 (2017) 7.
[55] A.A. Zorpas, Strategy development in the framework of waste management, Sci. Total Environ. 716 (2020) 137088.
[56] A. Zuorro, R. Lavecchia, F. Medici, L. Piga, Enzyme-assisted production of tomato seed oil enriched with lycopene from tomato pomace, Food Bioprocess Technol. 12 (2016) 3499–3509.
[57] M.S. Lenucci, et al., Journal of Agricultural and Food Chemistry (2013), doi:10.1021/jf4005059.
[58] Karami, et al., Journal of Food Biochemistry (2019), doi:10.1111/jfbc.12800.

CHAPTER ELEVEN

# Biorefinery concept for the industrial valorization of tomato processing by-products


Miguel Carmona-Cabello[a], Antonis A. Zorpas[b], M. Pilar Dorado[a]

[a]Department of Physical Chemistry and Applied Thermodynamics, EPS, Ed. Leonardo da Vinci, Campus of Rabanales, University of Cordoba, Campus of International Agrifood Excellence ceiA3, Spain

[b]Open University of Cyprus, Faculty of Pure and Applied Sciences, Laboratory of Chemical Engineering and Engineering Sustainability, Latsia, Nicosia, Cyprus


## 11.1 Introduction

Tomato is among the most popular foods, due to its versatility as a cooking ingredient and its high antioxidant nature. Nowadays, its market share is 8.7 billion US $ worldwide, with an annual production of around 180 million tons. However, between 5%–30% tomato of the global production is discarded, disposed in landfills or in the best scenario, used as animal feed [1,2]. Food waste (FW) generation is an issue which has been gaining attention in the last decades due to its magnitude. Globally, FW represents approximately one-third of total food production, which means that 1300 million tons of food are annually wasted [3–9]. In terms of environmental impact, Scialabba et al. [10] found that, FW produces 13.7 billion tons of $CO_2$ equivalent and is involved in eutrophication (0.74 M-eq) and acidification (1.66 Mt $SO_2$-eq) of soil. In sum, tomato waste (TW) is generated in the primary and postharvest sectors, which produce 9.1 million tons/year and represent 64% of greenhouse emissions [11].

TW is considered a promising industrial resource due to its composition. TW is rich in bioactive substances such as polysaccharides, proteins, fibers, and phytochemicals. Carotenoid composition is of special interest; it provides the characteristic tomato color and antioxidant properties. Moreover, other components, namely carotenoids, phenolic compounds, dietary fiber, and polyunsaturated fatty acids are also included in TW [12,13]. In fact, the bioactive compounds can only be synthetized by plants or microorganisms, but no by humans and need to be incorporated into the dietary habits. TW can be returned to food industry and substitute synthetic additives for natural ones. Lycopene and polyphenol have an important role in the nutraceutical industry, used to reduce the effect in chronic diseases, cardiovascular problems, and


*Tomato Processing by-Products*
DOI: https://doi.org/10.1016/B978-0-12-822866-1.00006-5

cancer disease [14]. Other components, like pectin, are used in encapsulation applications or as corrosion inhibitors [15]. Traditional lycopene extraction methods are not aligned with green chemistry principles [16]. Conventional extraction requires high amount of energy and toxic solvent consumption, that is, hexane. However, new recovery strategies such as ultrasound-assisted extraction (UAE) [17], supercritical fluid extraction (SFE) [18], electric field-assisted extraction [19], emulsion extraction [20] have been investigated with relevant results.

Previous set of substances may be recovered, and its application represents the spirit of circular economy. Biorefineries may play a significant role in TW recovery, enhancing new efficient recovery processes and transforming biomass into high-value added products, fine chemicals, biomaterials, power, or heat. This methodology increases the added value of TW, thus reducing the cost of tomato production [21–24]. On the other hand, TW valorization reduces the environmental impact, in line with EU policies for circular economy and United Nations [25–30]. Current EU policies include the reduction, to half, of food losses and waste production by 2030 [31–33]. Also, application of TW as raw material for biodiesel production or biogas through anaerobic digestion (AD) is in agreement with the frame line of policies for Energy and Climate Change 2021–2030 ('Framework 2030' from the Directive 2018/410/EC of the European Parliament) targeting a reduction of 40% of greenhouse gas emissions, a 27% increase in renewable energy and a 27% improvement in energy efficiency compared to values from 1990. Those reductions underlie the Kyoto agreement and further United Nations framework on climate change (Paris 2015 and Madrid 2019).

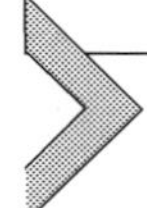

## 11.2 Tomato waste characterization and basic ideas about recovery

Tomato is a fruit botanically classified as berry, bright red colored and divided inside into two or more locules (the maximum division number being 18). Size is heterogeneous among different cultivars; in general, it ranges between one to 15 cm in diameter, approx. Fig. 11.1 shows tomato structure through across-section picture. Tomato fruit is divided in the following external components, namely exocarp, mesocarp and reproductive system remains (pedicel and sepal), and internal endocarp, that comprises placenta, seed locular cavity and gelatinous mass, plus seeds (rich in anthocyanins, flavones, and flavonoids) [34,35]. The mesocarp is a red layer, which is composed of parenchymal tissue. The exocarp, which is called skin or

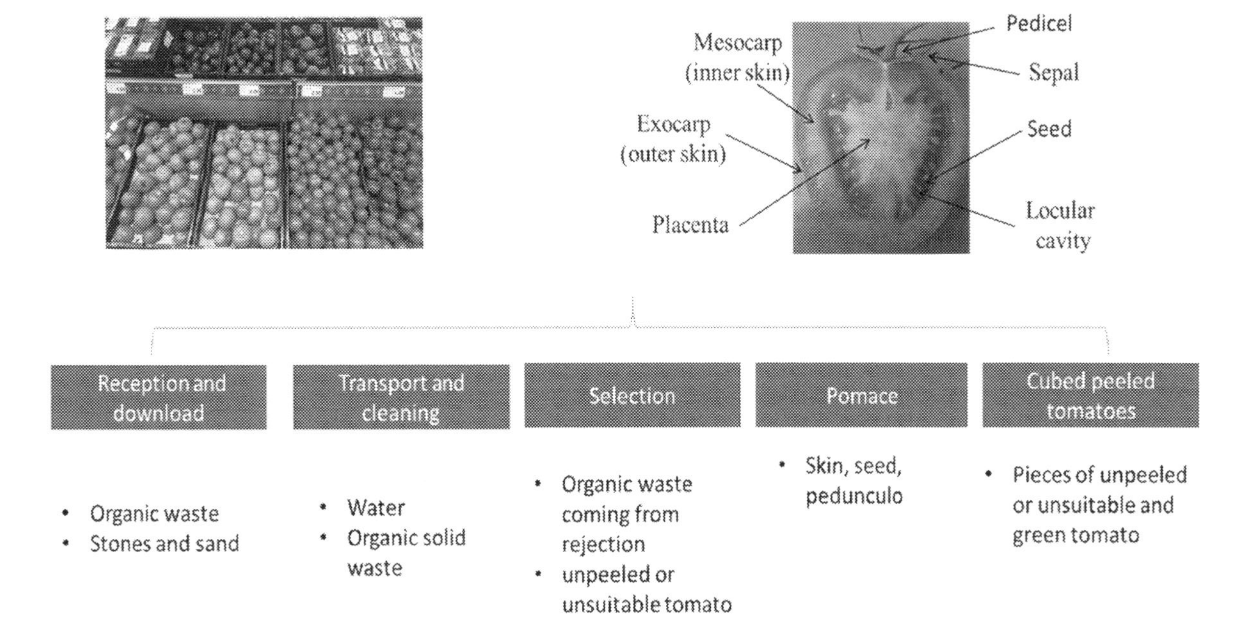


Fig. 11.1 *Tomato physiognomy and diagram of tomato waste generation.*

peel, is the outlier layer and is also composed of parenchymal cell, but of a smaller size compared to mesocarp. This layer includes cuticle, epidermal cell layer and hypodermal cell. Exocarp is discharged in tomato processing [16]. As may be inferred from Fig. 11.1, TW processing involves a wide range of steps like concentration, peeling, crushing, and powdering. The initial step is usually applied to both inorganic waste (i.e., sand or stone) and organic waste (i.e., pruning remains or unsuitable tomato). TW is a heterogeneous chemical matrix, which affect the recovery route and its processing depends on the initial product conditions. In this sense, peeling differs depending on the final use of the product. Also, the physicochemical treatment may affect TW composition. Hot lye and steam treatment affect exocarp and mesocarp layers. Moreover, it dissolves wax from cuticles and produces cellular degradation. Hot (boiling) treatments affect fiber, i.e. pectin, polysaccharide and cuticle [36].

Both fresh tomato and, therefore, its residues lack in uniformity between varieties. Shape may be globose, elongated, flattened, pear-shaped, oblate and with a smooth epicarp, while color may be red, yellow, and green. Seeds may appear either flattened or oval in shape. Considering TW, this variation provides a heterogeneous mix of peels, seen, and pomace. Table 11.1 summarizes the main components found in each tomato fraction. As may be seen, seeds show a lipid range from 17% to 25% (around 3% wax) and protein between 32% and 45%. However, fiber concentration is below 15%. On the contrary, skin and peel depict the higher content of fiber, with range between 70% and 80%, mainly including polysaccharides like cellulose, hemicellulose, lignin, or pectin. Their protein level is close to 10%. Pomace composition is similar to that of skin, but with higher protein content (19% to 25%). Table 11.1 also shows that the composition of a mixture of seeds, peels, and pomace is similar to that of pomace.

From the recovery point of view, tomato seed oil (TSO) is the most interesting macronutrient. The oil has a strong odor and color. The main fatty acid is linoleic acid (C18:2, 49.70%) followed by oleic acid (C18:1, 23.64%). Saturated fatty acids (C16:0, 17.08% and C18:0, 5.97) represent 25% of the total. Fig. 11.2, shows a principal component analysis (PCA), which provides information about the correlation between oils from different origin. PC1 (OX) shows saturated oils (mainly composed by C16:0 and C18:0) while PC2 represents the unsaturated fraction (C18:1 and C18:2). TSO shows a similar fatty profile to other seeds, namely almond, soybean and *Jatropha curcas*. Regarding oil properties, TSO presents a high iodine value (148 g $I_2$/kg of oil), which in due to its unsaturated fatty

**Table 11.1** Main composition of each tomato fruit fraction.

| Tomato fraction | Moisture | Ash | Oil | Fiber | Protein | Reference |
|---|---|---|---|---|---|---|
| | % | %, dry weight basis | | | | |
| Seed + peel + pomace | 7.90 | 7.10 | 8.80 | 64.50 | 20.00 | [37] |
| Seed + peel + pomace | 7.90 | 7.01 | 8.83 | 64.12 | 20.14 | [38] |
| Seed | – | – | – | – | 28.66 | [39] |
| Fiber | – | – | – | 15.00[a] | – | [13] |
| Seed | – | – | 26.00 | 19[b] | 29.00 | |
| Skin | – | – | 3[c] | 65[d]/32[a] | – | |
| Pomace[e] | – | 48.62 | 122.85 | 350.34 | 197.97 | [12] |
| Peel[f] | | 5.90 | 4.04 | 78.56 | 10.50 | [40] |
| | | 2.03–2.25 | – | 87.61–88.53 | 6.00–6.20 | [41] |
| | | 4.92–6.46 | 4.94–5.50 | 74.46–76.67 | 11.02–11.13 | [42] |
| | | 1.04–3.26 | 1.63–1.99 | 62.76–69.94 | 0.99–1.85 | [43] |
| Seed[f] | | 5.00 | 22.00 | 16.00 | 32.00 | [44] |
| | | 3.37 | 17.80 | – | 27.24 | [17] |
| | | 4.68–5.33 | 22.20–22.74 | – | 39.72 | [45] |
| | | 3.64 | 24.50 | – | 23.60 | [46] |
| | | – | 19.84-23.44 | – | 35.02–40.94 | [47] |
| Pomace[f] | | 3.92 | 5.85 | 59.03 | 19.27 | [12] |
| | | 5.29 | 9.87 | 39.11 | 24.67 | [48] |
| | | 2.88–4.40 | 8.37–16.24 | 58.53–68.04 | 15.08–22.70 | [49] |
| | | 4.00 | 2.00 | 46.00 | 16.00 | [44] |

[a]Cellulose, hemicellulose, and lignin.
[b]Cellulose.
[c]Wax.
[d]Cutin (mix of fatty acids, flavonoids, and phenolic compounds).
[e]Units for this reference are in g/kg.
[f]g/100 g dry weight (dw).

acid profile. Indeed, the high linoleic acid content is related to positive physiological effects against heart disease and cancer. The acid value is also high (12.8 g/kg) and similar to other oils, namely soybean and palm oil [16,37–39]. Those properties prevent the use of this oil as raw material for biodiesel production.

Considering TW recovery, trace elements play an important role. Ash represents the mineral content and ranges between 1%–9%. Azabou et al. [37] reported the main metal composition, regarding each tomato fraction (Table 11.2). It may be seen that potassium is the main component of all fractions, followed by calcium. Grassino et al. [43] reported the presence of elements like cadmium (0.16 mg/kg), mercury (0.14 mg/kg), and arsenic (4.35 mg/kg), which may play a toxic role. Table 11.2 also includes the amino acids present in seed, as an indicator of protein quality. According to Sarkar et al. [44], the percentage of essential amino acids is around 39.5%.

One important aspect of TW composition is the antioxidant activity, provided by the presence of lycopene, carotenoids, and polyphenol. Anwesha et al. [44], performed an antioxidant experiment (DPPH method, Blois 1958) finding out that tomato peel presents 37.5% radical inhibition per g of sample, while tomato seed meal had a 21.0%. According to Santangelo et al. [45], metabolic pathways play an important role in bioactive concentration. In fact, the different phenotypes of tomato fruit (different color) are due to the concentration of chlorophyll, carotenoid, flavonoid, and polyphenol (Table 11.3).

## 11.3 Bioactive compounds in tomato waste

TW shows a rich composition in carotenoids, antioxidants, fibers (i.e., pectin), fatty acids, proteins, etc. This profile has a large amount of bioactive compounds [46]. Lycopenes, phenolic compounds, and flavonoids are present and can be considered a natural source of exogenous antioxidants. The antioxidant activity is equivalent to a trap for free radicals and reactive oxygen species (ROS), which damage lipids, proteins and deoxyribonucleic acid (DNA) [47]. A promising pharmaceutical industry is based on the extraction of this components, based on low cost and healthy benefits compromise. In fact, supplements of these natural antioxidants may reduce the risk of cardiovascular disease and other chronic diseases, i.e. stomach, prostate and lung cancers. In addition to the antioxidant properties, TW has also a great potential in the food industry, due to its gelling, coloring

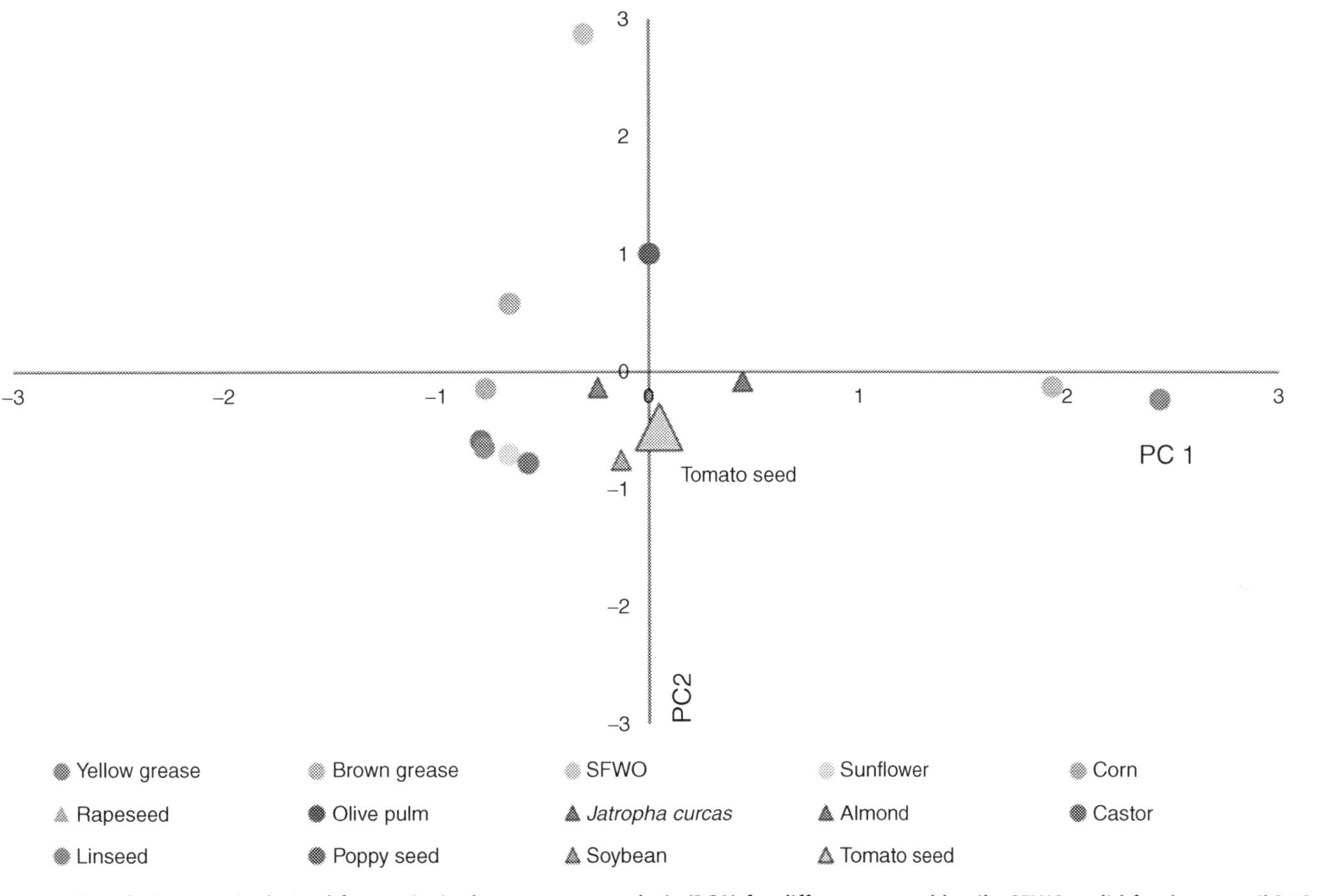


**Fig. 11.2** ***Correlation matrix derived from principal component analysis (PCA) for different vegetable oils. SFWO: solid food waste oil*** [40]. ***Vegetable oil composition*** [41]. ***Animal fat*** [42].

**Table 11.2** Metal distribution in tomato fractions and amino acid distribution seed protein.

| Metal composition (ppm) [37] | | | |
|---|---|---|---|
| | Tomato pomace | Tomato peel | Tomato seed |
| Potassium | 808.54 | 559.56 | 375.13 |
| Sodium | 191.68 | 128.53 | 812.80 |
| Calcium | 371.52 | 240.83 | 148.00 |
| Magnesium | 3.08 | 1.66 | 37.40 |
| Iron | 10.98 | 6.16 | 24.50 |
| Zinc | 1.83 | 1.21 | 19.40 |
| Copper | 0.09 | 0.09 | 0.80 |
| Amino acid (mg/g protein) [44] | | | |
| Histidine (HIS) | | | 25.01 |
| Isoleucine (Iso) | | | 49.30 |
| Leucine (Leu) | | | 77.90 |
| Lysine (Lys) | | | 59.63 |
| Sulfur-containing amino acids | | | 30.58 |
| Aromatic amino acid | | | 87.32 |
| Threonine (Thr) | | | 36.49 |
| Tryptophan (Trp) | | | 12.36 |
| Valine (Val) | | | 55.19 |

**Table 11.3** Bioactive content in function of color.

| | | Chlorophyll | Carotenoid | Polyphenol |
|---|---|---|---|---|
| Line | Color | g/kg | | |
| Yellow flesh | Yellow | >0.005 | >0.01 | 0.4 |
| Tangarine | Orange | >0.005 | 0.03 | 0.70 |
| Hight beta | Orange | >0.005 | 0.01 | 0.39 |
| Colorless epidermis | Pink | >0.005 | 0.025 | 0.39 |
| Anthocyanin fruit | Black red | >0.005 | 0.05 | 0.90 |
| Pigment diluter | Light red | >0.005 | 0.02 | 0.50 |
| Green flesh | Dark brown | 0.03 | 0.06 | 0.58 |
| Control | Red | >0.005 | 0.025 | 0.50 |

and stabilizing properties. Additionally, TW may be used as functional food ingredient, to enrich meat products, oils and in the canning industry. TW improves food nutritional quality, reducing lipid oxidation and increasing food shelf-life [48–50]. Table 11.4 summarizes some of the most frequent applications found in literature.

Bioactive extraction is a complex process, which establishes a delicate balance between extraction yield, functional activity, solvent toxicity, energy, and cost. The main step consists in passing through, and even

**Table 11.4** Carotenoid extraction methodologies.

**Solvent extraction**

| Residue | Pretreatment/operational parameters | Treatment | Extraction process | Extraction yield | Reference |
|---|---|---|---|---|---|
| Tomato waste (TW) | 0.8 mL of enzyme solution was added and the mixture was incubated in a shaking water bath for 1 h. The incubation temperature was 50°C (when P or C were used) or 45°C (for V-treated samples), based on enzyme optimum temperatures | Solvents: acetone, ethyl acetate, ethanol, their 1:1 mixture and 1:1 ethanol: water mixture. 4 g of ground TW was mixed with 20 mL of solvent. The mixture was incubated in a shaker for 30 min at room temperature | Enzymatic reaction temperature: 40°C reaction time: 5 h enzyme/substrate ratio: 0.2 mL/g, solvent/substrate ratio: 5 mL/g, extraction time:1 h enzyme/enzyme ratio: 1 | 11.5 mg lycopene/g TW | [102] |
| Dehydrated TW | – | Polar solvents: acetone, ethyl acetate, ethanol; nonpolar solvent: hexane | Factorial design of experiments: mixtures of hexane and ethyl acetate: 10/90 (v/v), mixtures of hexane-acetate 80/20 (v/v), solvent-to-waste ratio: 3/1, and 10/1 (v/w) particle size: 0.5 and 1.0 mm, temperature: 25°C, time: 30 min | 36.5 mg/kg (combination of ethyl acetate and hexane, 1: | [65] |

*(continued)*

**Table 11.4** (Cont'd)

**Solvent extraction**

| Residue | Pretreatment/operational parameters | Treatment | Extraction process | Extraction yield | Reference |
|---|---|---|---|---|---|
| TW | Step 1: disruption of cellular wall, solvent: ethanol/acetone | Step 2: lycopene extraction, hexane/ethanol/acetone. Step 3: separation of nonpolar from aqueous layer, 1 mL of saturated NaCl solution | Step 1: 1g FW + solvent, vortexed for 30 s, step 2: step 1+ solvent, vortexed in 15 s pulses with a 5 s rest between pulses, for a total of 60 s | 94.6% (1037.2 mg/100 TW dry matter) conditions: step 1 (12.5% ethanol + 12.5% acetone), step 2 (25% acetone + 25% ethanol), step 3 (25% water) | [51] |
| TW | 40–80°C temperature, 200–400 rpm agitation speed and 2.5–5.5 (%, w/v) tomato peels-to-refined olive oil (ROO) ratio | Maceration in ROO | 2.5% (w/v) biomass-to-oil ratio, 80°C of temperature, 400 rpm magnetic stirring, 45 min time | 1244.50 ± 30.41 mg/kg on dry basis | [47] |
| TW | | For peels: ethanol for seeds: hexane | Peels: dark flasks at 1:60 (w/v) sample-to-solvent ratio, 150 rpm, 25°C, 6 h extraction time Seeds: 25 g tomato seeds placed in dark flasks, homogenized with 250 mL of n-hexane, mixed 4 h in a shaker, at 180 rpm | 199.35 ± 0.35-mg GAE/g and 102.10 ± 0.03-mg QE/g, respectively; edible oil: 17.15% | [37] |

| | | | | | |
|---|---|---|---|---|---|
| Ultrasound (US)-assisted extraction | | | | | |
| Tomato pomace | Step 1 (US treatment): 37W, 90% amplitude, 15 min, step 2: 0.2 mL/kg enzyme concentration, 30 min time, 35°C | Surfactant: saponin co-surfactant: glycerol | | 409.68 μg/g | [20] |
| Tomato pomace | Step 1 (US treatment): 37W, 90% amplitude, 15 min, step 2: 0.2 mL/kg enzyme concentration, 30 min, 35°C | Surfactant: olive oil cosurfactants: lecithin, 1-propanol, water | 4 extraction cycles, 5 g microemulsion composed of lecithin/1-propanol/olive oil/water (53.33%/26.67%/10%/10%, w/w) | 328.82 μg/g | [73] |
| Tomato paste processing waste | | Solvent: sunflower oil: TW solid/oil ratio (3.18%–36.82%, w/v) | 70 W/m$^2$ US intensity, 10 min | 91.49 μg/g | [74] |
| Tomato pomace | | Eco-friendly solvent mixture (ethyl lactate and ethyl acetate) | 63.4°C, 30% (v/v) solvent, 100 ml/g solvent/ sample ratio, 20 min, 100 W | 1334.8 μg/g | [67] |
| Tomato processing waste | | (2:1:1) hexane/acetone/ methanol | 2.5 g/100 g dry matter cellulase, 0.8 g/100 g dry matter pectinase, 250 Hz US frequency, 60 min, pH: 5.3 | 94.3 μg/g | [60] |
| Tomato tissue waste | | (10:7:6:7, v/v) hexane/ acetone/methanol/toluene | Step1: freeze (-20 to -40°C) Step ep 2: UAE 45 min at 45 Hz | UAE: 45.51 μg/g FD: 104.10 μg/g UAE + FD: 138.82 μg/g | [103] |

*(continued)*

**Table 11.4** (Cont'd)

**Solvent extraction**

| Residue | Pretreatment/operational parameters | Treatment | Extraction process | Extraction yield | Reference |
|---|---|---|---|---|---|
| Tomato | | Hexane/ethanol | Pressure: 50 kPa Amplitude: 94 μm Temperature: 45°C Hexane: 50% | 140.8 μg/g | [71] |
| Supercritical carbon dioxide (SC-$CO_2$) | | | | | |
| TW | | | Temperature: 80°C Pressure: 380 bar Solvent flow ratio: 15 kg/h | 84.6% (w/w) | [104] |
| Tomato peels and seeds | | water-in-oil emulsion | Pressure: 380 bar Solvent flow ratio: 90 kg/h | 97% (w/w) | [86] |
| TW | | 40–80°C, 30–50 MPa, and 30/70 to 70/30 peel/seed ratio | 80°C, 50 MPa, 70/30 peel/seed ratio | 1.32 mg/kg TW | [105] |
| Industrial TW (skin and seeds) | | – | 60°C, 300 bar, 0.59 g/min solvent flow ratio, 0.36 mm particle size, 4.6% moisture content | 93% (w/w) | [81] |
| Industrial TW (skin and seeds) | | 37/63 (tomato peal/seed ratio) | 90°C, 40 MPa, 0.59 g/min solvent flow ratio, 1.05mm particle size, 4.6% moisture content: | 56% (w/w) | [84] |
| TW | | | 90°C, 500 bar, 25 kg/kg TW solvent flow ratio | Minimum cost: 1.8 k€/kg lycopene | [85] |

| Waste | Pretreatment | Solvent | Optimal conditions | Extraction yield | |
|---|---|---|---|---|---|
| TW | | | 333K, 30 MPa, 15 kg/h solvent flow ratio, 120 min | 1.58 mg/L | [59] |
| Enzyme treatment | | | | | |
| Waste | Pretreatment | Solvent | Optimal conditions | Extraction yield | |
| Ripened tomato | | | Total dosage: 25 U/g Time: 180 min Temperature: 45–55°C pH: 5.5 | 4.30 mg lycopene/kg TW | [106] |
| Tomato peel residues | | Span20 | Time: 1 h, Enzyme content: 250 U/mL Solvent-to-waste ratio: 9:1 (mol/mol lycopene) | 20%–25 % (w/w) | [107] |
| Whole tomato/tomato peel/lab/pulper waste/industrial waste | | | Pectinase: 2% (w/w tomato peel) Cellulase: 3% (w/w tomato peel) | Pectinase: 1104 μg/g Cellulase: 429 μg/g | [108] |
| Tomato processing industry | | | Pectinase + cellulase: 3% (w/w tomato peel) NaOH: 4% (w/w tomato peel) | 95% (w/w) | [109] |
| Tomato waste (skin/seed) | | Ethyl lactate | Pectinase: 70 U/g Cellulase: 122 U/g Pressure: 700 MPa Time: 10 min Solvent/feed ratio: 10:1 | Carotenoid: 165.27 mg/kg Lycopene: 82.48 mg/kg | [110] |
| Tomato peels | | Acetone/ethanol/hexane | Cellulase 1.5% (w/w tomato peel) Pectinase: 2% (w/w tomato peel) Time: 4 h | Cellulase: 847.33 μg/g Pectinase: 1262.56 μg/g | [111] |

*(continued)*

**Table 11.4** (Cont'd)

**Solvent extraction**

| Residue | Pretreatment/operational parameters | Treatment | Extraction process | Extraction yield | Reference |
|---|---|---|---|---|---|
| Tomato peels | | Hexane /ethanol/acetate | Step 1: Pectinase + cellulase + hemicellulose activities Time: 1 h Step 2: Solvent: 50 / 25 / 25 Time: 3 h Temperature: 40°C | 440 mg/100 g | [112] |
| Waste | Pretreatment | Solvent | Optimal conditions | Extraction yield | |
| Encapsulation by spray drying | | Solvent: gelatin and sucrose | 3/7 gelatin/sucrose ratio 1/4core/wall material Feed temperature: 55°C, Inlet temperature: 190 °C Homogenization pressure: 40 MPa | Encapsulation efficiency: 82.02% Lycopene purity: 90.3% | [113] |
| Encapsulation by spray drying | | Wall material: Arabic gum: 5%–35% Inulin: 5%–25% | Drying air flow: 43 $m^3/h$ Inlet temperature range: 110–120°C | Encapsulation efficiency: Arabic gum: 20.5% Inulin: 25.3% Antioxidant activity: 3.5 μmol trolox/mg | [57] |
| Encapsulation by spray drying (tomato oleoresin) | US pre-treatment to improve emulsifying properties 544.59 $W/cm^2$, 2 cm liquid height and 50% duty cycle (2 s, on; 2 s, off) | Wall material: protein from corn gluten meal | Emulsion spray-dried: Flow rate: 9 mL/min Inlet temperature:120°C Outlet air temperature: 55–60°C | Encapsulation efficiency: 91.57%, $T_{1/2}$: 10.90 days | [114] |

| | | | | | |
|---|---|---|---|---|---|
| Encapsulation by freeze drying (TW) | Solvent extraction: acetone/ethanol: 36:64 (v/v) solid ratio: 1:20 (w/v) oleoresin: extract/sunflower oil ratio: 26:1 (v/v) | Wall material: Proteins: soy (EPS), pea (EPP). Polysaccharides: inulin (EIN) and gum arabica (EGA) | Two steps: 1. Freeze overnight at -70°C. 2. Freeze at -40°C, 0.01 mbar, 72 h | Best encapsulation efficiency: β-carotenoid: 53.7% (EGA) lycopene: 51.44% (EPP) | [115] |
| Encapsulation by freeze drying (tomato peels) | US-assisted extraction | Wall material: 2:1 whey protein/acacia gum | Coacervates freeze-dried at -42°C, 10 Pa, 48 h | Lycopene concentration: 27.34 mg lycopene/g dry weight (DW) (72% lycopene), antioxidant activity of 2.15 mmol Trolox/g DW | [116] |
| Encapsulation by electrospinning | Extraction homogenized with acetone and hexane (50:50) Extraction solvent ratio: 1:10 (w/v) | Wall material: protein | 12 kV, 15 μlmin−1 of solution flow rate, 10 cm distance between needle tip and collector | 90% encapsulation efficiency Lycopene and antioxidant activity (AA) during 14-days storage | [117] |
| Encapsulation by spray drying | Osmosis and diafiltration | Wall material: maltodextrin | Drying air flow rate: 700 l/h, feed flow rate: 34 ml/min, inlet temperature: 160°C, outlet air temperature: 80°C | Lycopene concentration: 494.4 μg/g Antioxidant capacity: 25.07 μmol Trolox/g | [118] |

*(continued)*

**Table 11.4** (Cont'd)

**Solvent extraction**

| Residue | Pretreatment/operational parameters | Treatment | Extraction process | Extraction yield | Reference |
|---|---|---|---|---|---|
| | | Pulsed electric fields | | | |
| Residue | Pretreatment/operational parameters | solvent | Extraction process | Yield | |
| Tomato-processing peels | - | Acetone and ethyl lactate | Acetone: Field strengths: 5 kV/cm Energy input:5 kJ/kg Ethyl lactate: Field strengths 3 kV/cm Energy input:10 kJ/kg | Acetone: 17.53 µg/g Ethyl lactate: 10.14 µg/g | [19] |
| Peel and seed from juicing steps | – | Acetone | Field strengths: 1 kV/cm Pulses: 500 Energy input: 28.5 kJ/kg | Carotenoid: 177.70 µg/g Lycopene: 142.9 µg/g Protein: 143.4 µg/g Phenolic compound: 56.16 mg of gallic acid/kg Antioxidant capacity: 0.73 mmol Trolox equivalent/100 g | [101] |
| Red tomato waste | | Hexane / acetone / ethanol | Field strengths: 5kV/cm Temperature: 20°C Energy input:5 kJ/kg Solvent ratio: 50/25/25 | Carotenoid: 58.81 µg/g tomato peel | [2] |

| | | | | | |
|---|---|---|---|---|---|
| "Datterino" tomato | Peeling of steam blanched tomato | Acetone | Field strengths: 5 kV/cm Temperature: 20°C Energy input: 5 kJ/kg | Carotenoids: 804 μg/g tomato peel | [98] |
| Mature ripe red tomato pulp | | Acetone/ethyl alcohol/hexane | Field strengths: 16 kV/cm Capacity of discharge capacitor: 2μF Pulses: 50 Solvent ratio: 1/1/2 | Lycopene: 68.80% (w/w) | [100] |

DW: dry weight; EGA: gum arabica; EIN: inulin; EPP: pea protein; EPS: soy protein; FD: freeze drying; FW: food waste; GAE: gallic acid equivalents; QE quercetin equivalent; ROO: refined olive oil; TW: tomato waste; UAE: ultrasound-assisted extraction.

breaking down, the cell structure and solubilizing the target bioactive [51]. Conventional extraction is carried out using a solvent, which is selected depending on its solubility. The use of a polar or non-polar solvent is based on the extraction objective. However, the solvent may have an negative impact on the extraction yield, due to low solubility, thus reducing the antioxidant activity [52]. According to Azabou et al. [53], an inverse relationship between antioxidant extraction and lycopene extraction was found. While in distilled water the percentage of antioxidant yield was 15% (w/w), the lycopene yield was 1.1% (w/w). However, in hexane, the lycopene yield was close to 20% (w/w) while antioxidant yield was 8% (w/w). On the other hand, phenolic and flavonoid activity in the extract is an important factor in the evaluation of the solvent; the best yield of extraction was found in water. However, the best phenol (170 mg gallic acid-equivalents per gram (GAE)/g) and flavonoid (180 mg quercetin equivalent (QE)/g) activity was provided by ethanol. The high yield, but low antioxidant activity, in distilled water may be associated to the extraction of polar components, namely proteins, carbohydrates, organic acids and salts. The concentration of moisture is considered also a negative factor, which may reduce the concentration of extracted lycopene by 1.4 time [54]. Antioxidant compounds are also dependent on the type of extract. On the other hand, organic solvents, i.e. benzene, chloroform and methylene chloride, provide a good extraction yield. However, they are hazardous and toxic materials, and its residues may be introduced in the value chain. The biorefinery industry should be focused on the selection of the most appropriate technologies, to improve bioactive extraction yield, thus helping to develop TW recovery factory, in agreement to circular economy and green chemistry principles. Fig. 11.3, displays alternative extraction techniques, namely UAE [55], SFE [56], encapsulation technology [57], or oleoresin [58], which produce a disruption of cell structure with low environmental impact. The objective of these new process generation, is to improve the extraction yield, thus including TW in the circular economy concept and biorefinery process.

## 11.3.1 Carotenoids

Carotenoids are a secondary metabolite produced by plants and microorganisms, which are a natural pigment. Their chemical structure is based on monomers of isoprene with a 40-C skeleton. Carotenoid can show a linear chemical structure, such as lycopene or a cyclic one, as α-carotene or β-carotene. Color depends on carotenoid structure; red color in tomato is

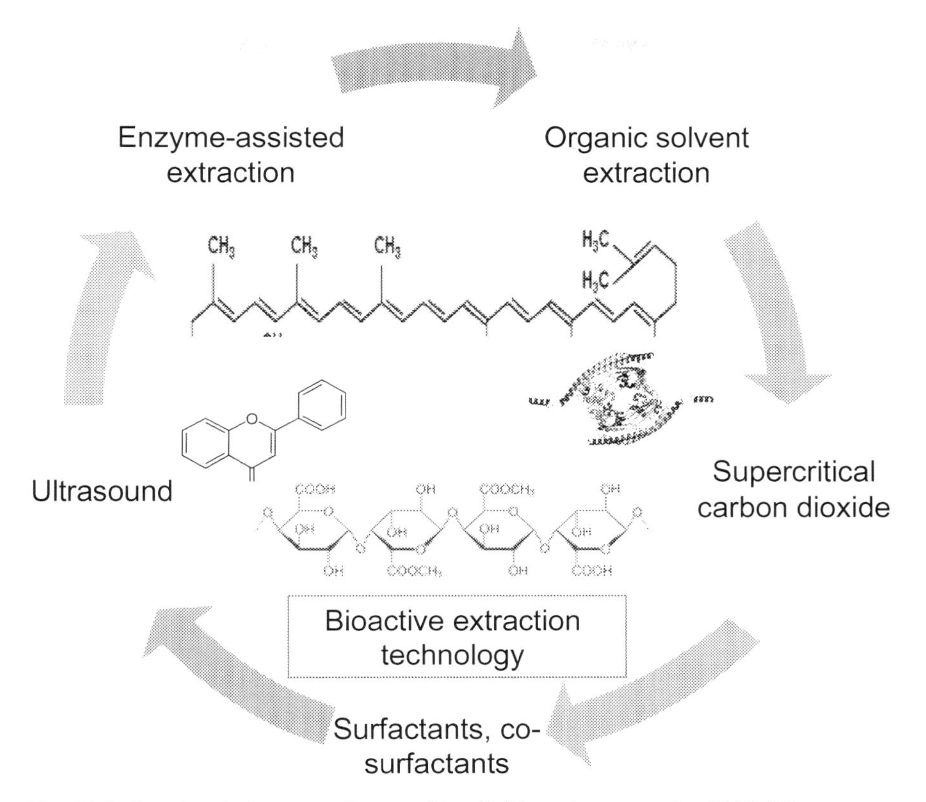


**Fig. 11.3** ***Set of techniques used to eco-friendly bioactive extraction*** [2,59,60].

associated to lycopene (main carotenoid in this case), which is an aliphatic polyunsaturated hydrocarbon, with 11 conjugated double bonds. In addition, lycopene has antioxidant properties and can reduce the single oxygen quencher (ten times higher than that of α-tocopherol) [61]. Lycopene is associated to redox regulation, in both agro-industrial activities and diseases, namely cardiovascular disorders or cancer (homeostasis redox plays an important role). β-Carotene is another carotenoid that may be found in tomato and is associated to yellow color. β-Carotene is synthetized from lycopene, which transforms its structure to two beta-ionone rings at the end. The main property of β-Carotene is its role as precursor of vitamin A. Also, β-Carotene has been described as enhancer of immunological function and has an important antitoxic activity [48].

A recurring aspect in the discussion of potential uses of carotenoids is the viability of their use as functional food ingredients and in pharmacology. In fact, only 20 from 750 carotenoids identified have been found in human blood and tissue. In the case of lycopene, is not synthesized by humans and must be ingested from food, but is absorption is associated to cis-isomers [62]. However, is mainly present in tomato trans-isomers, due to thermodynamic stability; isomer transformation needs an energy supplement and load to chemical degradation [63]. Extraction conditions (technique, storage conditions, etc.) must be implemented according to optimal purity, stability, and bioactivity. Lycopene is extracted in two steps. Followed by structural degradation, lycopene extraction uses chemical solvents, supercritical carbon dioxide (SC-$CO_2$), or vegetable oils [18,64].

#### 11.3.1.1 Conventional extraction

In tomato cells, carotenoids are stored in organelles, named chromoplasts, which are a type of plastids. The extraction methodology requires a degradation of the cell structure, followed by solubilization. Carotenoids are insoluble in polar solvent, as water, but have a high solubility in nonpolar solvents such as lipids, hexane, ethyl acetate. The solvent efficiency using conventional organic solvents was checked by Strati et al., [65]. They carried out a Box-Behnken experimental design, using solvents (ethanol, hexane, ethyl acetone, acetone) and solvent mixture (hexane-ethanol, hexane-acetone, and hexane-ethyl acetone), solvent-to-waste ratio and particle size as independent variables. The response variable was carotenoid yield (mg/kg of dry waste). Results shown that a solvent mixture (hexane/ethyl acetone, 45/55 v/v) with a solvent/waste ratio of 9.1:1 and 0.56 mm particle size provided the best solution. According to this, the highest solvent extraction yield was 36.5 mg/kg. Other solvent combinations provided lower yields. This result was also confirmed by Poojary et al. [50], which carried out an experimental design using four independent variables, namely, extraction temperature, time, percentage of acetone in n-hexane and solvent volume; the dependent variable was carotenoid yield (either mg/g or in percentage). In this study, the main objective was to produce an optimized high-purity trans-lycopene from pulp waste. Results exposed that, a solvent mix of 25% acetone in n-hexane (v/v), at 20°C, during 40 min of extraction, with 40 mL of solvent volume provided the highest lycopene yield (94.7%, 0.036 mg/g lycopene), from which 98% was trans-lycopene. It was found that low temperature reduces terminal degradation. Another study focused on the effect of solvent addition sequence on caustic tomato slurry. The experimental design was carried out in three steps, using different combination of solvents. The first step used a solvent combination, such as ethanol and acetone, to cellular structure degradation, while the second step used a combination of hexane, ethanol and acetone to extract lycopene. The last step used water as solvent, to separate polar from non-polar components. Previously, samples were pre-treated with a membrane filter to reduce the alkalinity, due to the complex chemical matrix of tomato slurry waste. Extraction step in samples, before and after filtration, was also checked and the results illustrated more than 90% of extraction efficiency. The best combination was carried out using the following sequence: 12.5% ethanol + 12.5% acetone (step 1), 25% hexane + 25% ethanol (step 2), and 25 % water (step 3), with 94.6% extraction efficiency. Results also showed that samples without pre-treatment achieved lower yield (80%), being step one more determinant in these samples [51].

Literature on carotenoid extraction for its use as a value-added product mainly focuses on required solvents. However, the discussion is being extended to other parameters derived from its physical-chemical properties. According to the revision of Strati et al. [8], extraction yield may be influenced by solvent-to-feed ratio, moisture content, physical parameters (i.e., temperature, extraction time, particle size) and methodology parameters (i.e., number of stages). Kaur et al. [66], studied the effect of solvent/feed ratio (20:1 to 60:1) particle size (0.05 to 0.43 mm), extraction number (1 to 5) and temperature (20 to 60°C), using central composite rotatable design. In all experiments, a mix of hexane: acetone: ethanol (2:1:1) was used as solvent. The maximum lycopene extraction yield (1.98 mg/100 g) was achieved at 30/1 solvent-to-TW ratio, 50°C, during 8 min, using 0.15 mm particle size and 4 extraction steps. Furthermore, the study focused on correlations between parameters like solvent ratio, number of extractions, temperature, and time. Results also revealed the existing dependence between different parameters. For the solvent/feed ratio parameter, a low impact in lycopene yield extraction was found, compared to other parameters. Similar results were found by Silva et al. [67], using as solvent a mix of ethyl lactate/ethyl acetate solvents. In this case, the experiment design revealed a positive effect of the solvent content. According to the literature, the importance of the solvent ratio is required to balance insufficient mixing or saturation effect and high cost. There is a tendency to prefer higher ratios (between 30% and 40%) when supercritical $CO_2$ and Soxhlet methods are used. However, the ratio is decreased when ultrasound, enzymatic reaction or pulsed electric fields (PEF) are used. Table 11.4 shows different solvent ratio values for each of the extraction technique.

Starti et al. [68], studied the effect of temperature in carotenoid extraction. They found that there is a balance between solvent boiling point and carotenoid thermal stability. Lycopene has a thermal stability in the range of 80–100°C; as temperature rises up to 120°C, (trans to cis) isomerization increases. When temperatures are above 140°C, degradation takes place [69]. Optimal carotenoid yield (243 mg/kg) was provided using ethyl lactate at 70°C. Finally, results shown that the higher the temperature, the higher the yield. Furthermore, there is a strong correlation with time. Manzo et al. [70] studied lycopene degradation kinetics at 100°C and found 90% lycopene content in tomato paste, degraded during 75 min. Kaur et al. [16] carried out a response surface methodology (RSM) with temperature and time intervals from 20 to 60°C and from 4 to 20 min, respectively. Results showed that, low temperatures required long treatment time, thus providing a yield increase. However, high temperatures required a reduced treatment time, thus reducing yield. Optimum lycopene extraction yield of 2.0 mg/100 g was achieved at 20°C after 20 min. Silva el al. [67], found a similar tendency with a temperature and time intervals from 40 to 70°C and from 20 to 50 min. In this case, optimum lycopene extraction yield of 1.2 mg/100 g was found at 63.4°C during 20 min of treatment. The main differences between both studies may be due the solvent.

Another parameter affecting carotenoid extraction is the effect of the extraction step. In this step, the solvent polarity affects solvent extraction efficiency. Nonpolar solvents (i.e., hexane and ethyl acetate) are able to extract a higher amount of carotenoids, in the first extraction step (regardless of temperature) in a range of 70%–75%, compared with polar solvents (i.e., acetone, ethanol, or ethyl lactate) in a range from 55%–68%. For both kind of solvents, the yield decreases in a range between 19%–23% for nonpolar solvents and between 24% and 34% for polar ones [69]. The effect of extraction step and time is also affected by polarity. Extraction step is correlated with diffusion phenomenon and extraction rate, which decreases with time; equilibrium is reached after 30 min [52].

Solvent extraction technologies are controversial due to questionable environmental sustainability, and even food safety concerns due to potential solvent resides. Conventional technologies, like Soxhlet, agitation or centrifugation are being displaced by other processes, which can be carried out in either one or multiple steps. These steps include a pretreatment, that may consist in physical processes (i.e., cooking, drying, milling, osmotic shocking, freeze-thawing, cryogenic griding, ultrasound), chemical (acid, base, surfactant), enzymatic or biological ones (fermentation). The objective of this pretreatment is to produce cell wall degradation, while either reducing solvent/feed ratio or directly using an eco-friendly solvent. Table 11.2 summarizes the new technologies implemented in this field.

#### 11.3.1.2 Ultrasound-assisted extraction

Ultrasound (US) is an economical, efficient, and simple widely known technology in food industry, applied to processes like emulsification, crystallization, filtration, separation, extrusion and microbial inhibition. From an environmental point of view, US may help reducing the use of toxic solvents, such as hexane, or enhancing the use of friendly ones. According to the literature, US reduces both carotenoid extraction time and solvent/feed ratio, while increases carotenoid yield around 10%. US leads to the disruption of cellular structure, enabling the diffusion of the solvent into the cellular material. US produces a mechanical effect (in contrast to that of hexane) and increases mass transference. This physical phenomenon is named cavitation, which is responsible of the formation and further implosion of bubbles inside a liquid. This phenomenon involves high pressure (up to 50 MPa) and temperature (5500°C) [71]. As a result, energy produces disruption in chemical structures (i.e. polymers, proteins and chemical bonds), turbulence and liquid circulation. The advantage of this technique is based on its non-thermal nature, which makes it compatible with heat labile compounds. Furthermore, US can be combined with other technologies (Table 11.4) like enzymatic, high pressure, and freeze drying.

Kumcuoglu et al. [17], compared lycopene extraction yield using both conventional (organic solvent) extraction (CE) and UAE. The solvents used were a mix of hexane – acetone – ethanol (2:1:1). CE experiments were carried out using temperature (20–60°C), solvent/feed ratio (20:1 to 50:1) and time (10–40 min) as independent variables. The results indicated that optimal extraction conditions were found at 60°C for 40 min reaction and 50:1 ratio, providing a yield of 93.9 mg/kg. UAE experiments were carried out using power (50–90 W), extraction time (1–30 min), and solvent/feed ratio (20:1 to 50:1) as independent variables. Maximum yield extraction (90.1 mg/kg) was lower than that provided by CE. However, both extraction time and solvent/feed ratio were reduced in 10 min and from 50.1 to 35.1, respectively. A similar study was carried out by Yilmaz et al. [72], where CE (solvent hexane-acetone-ethanol as solvent, 15°C and 35/1 ratio) and UAE (hexane-acetone-ethanol as solvent, 15°C, 35/1 ratio, and 90 W US power) were compared. As a conclusion of this the extraction if carotenoid were increased when US was used. The maximum amount of lycopene extracted with CE was 52 mg/kg, while UAE provided a yield of 70.1 mg/kg.

The use of eco-friendly solvents with UAE has been widely reported in the literature. Silva et al. [67] showed a 9.4% yield increase using 40 kHz of frequency and 100 W of power, which provided an optimal lycopene concentration of 1334.8 μg/g. Oils have also been used as solvents. In the previous case, olive oil was used to form a miscellaneous solvent with lecithin, 1-propanol and water. Previously, TW samples were pre-treated with pectinolytic enzymes (Table 11.4) [20]. US optimization treatment included 4 extraction cycles of 30 s, 50 and a microemulsion of 53.33% lecithin: 26.67% 1-propanol: 10% olive oil: 10% water. The extraction yield was 88.04% (the initial lycopene content was 371.26 μg/g, while the final one was 328.82 μg/g) [73]. A similar study was carried out by Rahami et al. [74], using sunflower oil as alternative solvent. The experimental design considered US intensity (30-70W/$m^2$), solid/oil ratio (3.18-36.82% w/v) and extraction time as independent variables. Results were compared with CE (using a solvent mixture of hexane: acetone: methanol at 2:1:1 v/v). UAE optimum extraction results provided 91.49 mg/kg of lycopene content (70 W, ratio 20:1 v/w at 10 min), above that provided by CE (74.89 mg/kg, after 1 h of treatment).

UAE is frequently assisted by auxiliary techniques, as shown in Table 11.4 Luengo et al. [71] optimized carotenoid extraction (mainly lycopene) by means of a two-step experiment. Firstly, the influence of pressure (0–100 kPa) and US amplitude (0–94 μm) were studied. Secondly, temperature (25–45°C) and hexane/ethanol ratio were considered. The increase of pressure showed a positive effect over yield extraction, with a 2% increase in concentration obtained without treatment. The effect of pressure in US is associated to an increase of the number of cavitating bubbles and, thus, its implosion energy [75]. Another accompanying methodology consists on previously freeze drying; this methodology provided an increase more than doubled that provided by CE (CE: 45.51 μg/g; UAE: 104.10 μg/g and freeze dried + UAE: 138.82 μg/g).

Overall, UAE results in a gentler treatment, with a linear correlation between US energy increase, cavitation phenomenon increase and, as a consequence, carotenoid yield increase. There is also a correlation between thermal effect and US intensity; for low intensity, heat is diffused through the medium and negative effects are reduced. However, as the intensity increases, the cavitation effect decreases and the thermal effects begin to prevail. Several works have fixed optimal working power for extraction at 100 W, but it also produces excessive energy that can be associated to other parallel chemical reactions, that is, carotenoid isomerization, thus producing radicals, like $H^+$, ROS, and $H_2O_2$ [76–79].

#### 11.3.1.3 Supercritical fluid extraction

SFE has been studied as an alternative to (toxic) organic solvents, in order to improve lycopene solubility and stability, as it reduces the impact of treatment temperature. SFE uses $CO_2$ as main solvent, due to its low cost, high purity, low critical temperature, low toxicity and safety. $CO_2$ changes its polarity in function of its density, which increases with pressure, thus improving lycopene extraction [80].

Nobre et al. [81] studied the correlation between moisture, feed initial composition, particle size, solvent flow ratio, pressure and temperature extraction. Results shown that, the concentration of moisture content plays an important role, and the optimum values were observed at 4.6% (w/w). Hence, higher or lower moisture content produced a negative effect over lycopene extraction. Low values are related to dry processes, which lead to lycopene degradation [82]. Also, the lower the particle size the higher the lycopene extraction. One effect derived from small particle size is the competition between carotenoids and triglycerides, due to their higher solubility in $CO_2$. Channeling effect is another problem related to size, due to extraction bed packing. The study revealed that, the higher the pressure and temperature, the better the extraction. However, high solvent flow-ratio provided low lycopene recovery [83]. The flow-ratio roof was fixed at 1.05 g/min (surface velocity, 2.58 cm/min). Machmudah et al. [84], checked previous parameters, besides the effect of both the rising method and tomato peel-to-seed ratio. The rising methodology influences efficiency extraction, due to the solvent (mainly tetrahydro furan, THF) used to remove the remains from tubing. Also, back pressure regulator may produce contamination in the next extraction. In this study, best extraction was achieved without THF, in presence of seed oil, which can replace the effect of solvent, while removing accumulated carotenoids. The effect of peels/seed ratio, with a high seed proportion, showed a lycopene and β-carotene increase, due to oil presence that leads to increased solubility. However, results showed that a content above 63% of seed declined extraction yield.

From an industrial point of view, scale-up methodologies have been developed. On one hand, Silva et al. [85] and Scaglia et al. [86] agreed to include, in the optimization process, an evaluation of manufacturing cost. On the other hand, Perretti et al. [59] developed a lycopene extraction methodology based on the concept of pilot plant. Silva et al. [58], used RSM for process optimization, through 36 experiments. Pressure (300–500 bar), temperature (70–90°C), and solvent ($CO_2$) were used as independent variables. Cost of manufacture analysis included investment, labor, waste treatment and raw material costs. The minimum cost was 1.8 k€/kg of lycopene, associated to optimal conditions (500 bars, 90°C, 25 kg $CO_2$/kg sample). The most significant finding from this study was the correlation between cost and optimal conditions, which showed that the higher the pressure (linear plus quadratic function) and temperature (linear function), the lower the cost; spent $CO_2$ solvent has a parabolic effect on cost of manufacturing and a crossed effect with pressure. Results shown that extraction time played a significant role in final cost of annual lycopene production (predicted to be above 2000 kg). Scaglia et al. [59] proposed a biorefinery model based on SFE-$CO_2$ extraction and AD. Lycopene extraction yield was found to be above 95% (w/w) and the final cost was 787.9 €/t of TW. The experiment, carried out in a pilot plant, by Perretti et al. [60] introduced a three-stage fractional column (2 l volume, 3 m length and 3 cm diameter), with three individual sections conditioned with different temperatures. The aim was to both produce a solvent density gradient and collect different fractions, with different triglyceride and lycopene composition. Time was also checked at 30, 60 and 120 min. RSM showed that the highest concentration of lycopene was achieved after 120 min, at 30 MPa, with a flow rate of 15 kg/h.

#### *11.3.1.4 Encapsulation technology*

This technology protects bioactive compound of hydrolytic reactions, oxidation reaction, polymerization of polyphenols, degradation temperature, photooxidation, and other variables such as pH, moisture or oxygen content [87]. Encapsulation mechanism is based on a protective membrane to enclose the bioactive compound. There are different encapsulation methodologies, namely physical (spray drying, freeze drying, supercritical fluid, ultrasound fluid, etc.), physicochemical (spray cooling, ionic gelatinating, solvent evaporation extraction), and chemical (polycondensation, *in-situ* polymerization) [88,89]. Encapsulation efficiency depends on parameters like feed flow rate, air inlet/outlet temperature, feed temperature, and wall material (also called packeting material, encapsulating agent, carrier, shell, etc.). The most conventional technologies are spray drying and freeze drying.

The wall material election plays a significant role in encapsulation yield, stability and properties of the final product. Among the materials used, polysaccharides, such as maltodextrins (using freeze drying, which help glass transition temperature) [90], which are low-cost starch and present good emulsifying properties, are found. However, starch may block the spray system (starch caramelizes when hot air is used). Other polysaccharides, such as gum arabic (which has good emulsifying properties and presents low viscosity in aqueous solution), chitosan (soluble in acid aqueous media), and inulin (low-cost oligosaccharide) also are also used. Proteins from soy, whey, pea, and gelatin have antioxidant and good binding properties.

Cyclodextrins (CD) are a good example of wall material for encapsulation, convened with SC-$CO_2$ extraction. CD are cyclic oligosaccharides from starch, which are transformed by the action of amylase (enzyme that degrades starch to glucose) and cyclomaltodextrin glucotransferase (responsible of the cyclic ring). CD can be classified in function on glucose residues (GR), α-Cyclodextrin(6GR), β-Cyclodextrin(7GR), and γ-Cyclodextrin(8GR). Internal surfaces form a hydrophobic (apolar) cavity with encapsulating capacity. These molecules have high thermal stability, are biocompatible and non-toxic, allowing their use in nutraceutical, cosmeceutical and pharmaceutical applications. Initially, the process starts with SC-$CO_2$ extraction, which produces an oil rich in carotenoids. It is followed by a maceration step of oil/water, using CD as surfactants [91,92]. According to Durante et al. [93], carotenoid entrapment efficiency (yield extraction) was tested for β-CD, providing 62.4% (mg/100 g oil). Carotenoid biostability and antioxidant activity were also studied. Both parameters shown an increase of 30% and 25%, respectively, compared to conventional extraction.

Spray-drying methodology is the most popular technology, as it has the ability to adapt to industrial processes, good reconstitution characteristic, low water activity and good behavior during transport and storage. Spray-drying process starts mixing and homogenizing both core and wall materials, followed by spraying the mixture to microparticles, using a nozzle. Hot air is used during the process, with the aim to remove water, thus providing a dryer powder. Freeze-drying or lyophilization is another popular technology used to encapsulation, which is based on sublimation. Firstly, TW is frozen and then, internal water is directly transformed into vapor through the action of a vacuum. Compared to spray-drying, lyophilization encapsulation shows better storage properties, due to lack of water activity in dry powder, which reduces microorganism degradation and provides resistance to chemical reactions that is, oxidation. Also, low temperature allows to preserve antioxidant activity and half-life of carotenoids [87,89].

#### *11.3.1.5 Pulsed electric fields*

PEF is a technique based on the disruption of the cellular membrane; short voltage pulses allow cell membrane permeabilization. The external electric field leads to membrane thickness reduction and pores formation. Considering intensity, the process may be reversible or irreversible. Carotenoid accessibility depends on process parameters, matrix origin, pre-treatments and solvent selection [94]. Table 11.4 summarizes few of the most prominent processes used by different authors. PEF values vary between 1-20 kV/cm, whereas carotenoid (mainly lycopene) content ranges from 17.53 to 58.81 μg/g.

Bot et al. [95] studied the effect of PEF in cell membrane permeability, using different tomato fractions, namely tissue, cell cluster, single cell and chromoplast, listed from the one with the highest to the one with the lowest structural barriers. To evaluate changes in microstructure and carotenoid bioaccessibility, heating and a combination of PEF plus heat treatment were also used. PEF effect was evaluated in a range from 1 kHz to 1 MHz. Results showed that low frequency was linked to low conductivity, as cell membrane prevents the flow of electrons, by acting as a capacitor. As the frequency increased, electrical flow and, therefore, conductivity increased. Higher conductivity values were achieved with a combination of FED plus heat treatment, due to the effect of pores formation (PEF effect) and cell wall pectin depolymerization (heat effect) [96]. The highest PEF yield (1,096.80 μg lycopene/g) was found in single cells, while lycopene concentration increased up to 2,191.70 μg/g in chloroplasts. Authors recommended a working temperature of 45°C. Also, Pataro et al. [19,97,98], studied the effect of steam-assisted peeling pretreatment and found the synergistic effect of PEF plus heat treatment. Steam blanching experiments were carried out at different temperatures (50, 60, and 70°C) during 1 min, at constant PEF treatment (0.5 kV/cm, 1 kJ/kg energy input). Results showed an increase of lycopene yield extraction. Optimal conditions for maximum thermal damage were fixed at 60°C, providing a carotene concentration of 360 μg/g (1.89-times increment). In further studies, authors moderately increased the energy input (5 kV/cm and 5 kJ/kg), which allowed reducing pre-treatment temperature, thus improving antioxidant activity. Their last study included acetone solvent as an ecofriendly solvent and compared its use with ethyl lactate. The results showed that both solvents are aprotic ones, as well as, ethyl lactate having a good permeability in dry matter. In presence of water (wet matter), permeability is enhanced due to hydrogen bonds, thus affecting extraction yield. These findings are opposite to those found from Strati and Oreopoulou [65].

Number of pulses is another parameter under study, due to its relation with energy input. The pulse produces an increase of the pressure and hydraulic effect, which produce plant tissue degradation. The hydraulic effect produces a wave, which expands at high speed and produces a temperature increase [99]. Gachovska et al. [100] increased lycopene extraction up to 68.80% (w/w) using 50 pulses (16 kV/cm). Andreu et al. [101], fixed an optimal lycopene extraction (142.90 μg/g) at 200 pulses (2.0 kV/cm) and 22.80 kJ/kg of energy input.

## 11.3.2 Antioxidant components

Oxidation is a chemical reaction involved in human diseases, i.e. cancer, heart disease, or stroke and, also, in agri-food industry, as is responsible of food spoilage. Oxidation is due to free-floating electrons, which come from free radicals, like hydroxyl ($OH^-$), superoxide ($O_2^-$), nitric oxide ($NO^-$) and lipid peroxyl ($LOO^-$) radicals. These radicals produce damage on cells and chemical substances. On the other hand, antioxidants are substances that reduce the effect of free radicals [102]. Tomato is rich in phytochemicals, like lycopene (followed by β-carotene), that is the main bioactive compound from TW, already discussed. However, TW antioxidants involve more compounds, like polyphenols, flavonoids or proteins.

Polyphenol is a secondary metabolite, with antioxidant, antimutagenic and cytotoxic activity. Flavonoid and polyphenol content depends on the tissue; skin presents the highest concentration in comparation with pulp [103]. Perea-Dominguez et al. [104], isolated phenolics in four fractions, namely soluble-free-phenolic (SFP), acid-hydrolysable phenolics (AHP), alkaline-hydrolysable phenolic (AKHP), and bound phenolic (BP) fractions, each one associated to different tomato fractions. Also, in this study, the phenolic compound profile, which can be divided in two fractions, was analyzed. Phenolic acid fraction is composed of caffeic, chlorogenic, ferulic, gallic, and ρ-coumaric acids. While flavonoid fraction is composed by catechin, kaempferol, naringenin, quercetin and rutin. The results showed that, BP was the main TW fraction (formed by peels and seeds) (51.30 mg GAE/g, 49.20%), with maximum antioxidant activity (83.56 DPPH μmol TE/g, 852.40 ORAC μmol TE/g). This fraction is associated to cell wall and its structural components, such as pectins or structural proteins [105].

Extraction of BP fraction is difficult due to the hindered wall structure [106]. Bao et al. [107], studied the effect of high voltage cold plasma as pretreatment to increase phenolic extractability. The methodology contributed to surface rupture, increasing hydrophilicity and breaking the covalent phenolic bond, thus increasing the diffusion of molecules in cell structure [108]. Highest total phenolic content, using nitrogen as gas, was 1 mg GAE/g in dry mater; extraction yield was increased above 10% compared to control sample. Best antioxidant capacity was also achieved with nitrogen (0.9 mgTE/g dry mater). Other study developed by Grassino et al. [43] focused on bioactive extraction and checked polyphenol extraction in depeptidized samples. Authors used UAE pretreatment (400W, 30 Hz, 95% continued amplitude) and ethanol as solvent. Pectin sample, before pectin extraction, showed a total polyphenol concentration (3266.28 mg/100 g) higher than samples without pectin (1625.65 mg/100 g).

Microwave-assisted antioxidant extraction was also evaluated. Pinela et al. [109] carried out an experimental design using extraction time (t), temperature (T), ethanol concentration (Et), and solid/liquid ratio (S/L) as independent variables. The proposed methodology exhibited an ABTS (2,2′-azino-bis(3-ethylbenzothiazoline-6-sulfonic acid)), activity increase, when Et and S/L parameters decreased. On the other hand, the higher the presence of Et and S/L, the higher the oxidative hemolysis inhibition assay antioxidant activity. Optimal conditions were fixed at 20 min, 180°C, 47.40% of Et concentration, 45g/L of S/L, and 200 W of microwave power.

### 11.3.3 Pectins

One important bioactive compound included in TW is formed by the structural heteropolysaccharide called pectin, usually used in food, cosmetics and pharmaceutical industry. Pectin is a structural polymer that appears in citrus peels (20%–40%, dry basis) and apple pomace (10%–20%). It is formed by three different poly-sugar chains, being poly-galacturonic the predominant one (around 65%) and may include between 300–1000 units of galacturonic acid. The second major block units are included in galacturonan chain, which is linked in its side chains to other sugars like apiose, fucose, arabinose, and xylulose. The last type of chain is the polysaccharide named rhamnagalacturonan. Rhamnose contains side chains formed by arabinan and galactan chain [110,111].

A characteristic feature of these poly-sugars is the carboxylic groups and the degree of esterification (DE). According to the methoxy content, they may be classified as high methoxy (more than 50% esterified) or low methoxy (less than 50% esterified) poly-sugars. The DE and its distribution, pH and $Ca^{2+}$ affect the gel-forming capacity of pectin [112].

Pectin extraction conditions determine pectin yield and quality, physicochemical properties (molecular weight), degree of methyl esterification and neutral sugar, which depends on the methodology used [113]. Traditionally, solvent extraction by stirring and heating, heat refluxing heating and new green techniques, including microwave heating extraction (MWE), UAE, and hydrostatic pressure extraction have been used

[55,114]. Also, extraction agents have usually included water, ethylenediaminetetraacetic acid, cyclohexanediaminetetraacetic acid and hot dilute acid or basic. However, the most popular solvent used in industry is the last one due to its competitive price, high yield and pectin quality [115–117]. Alancay et al. [15], developed a design of experiments to compare conventional pectin extraction with HCl (acid treatment pectin, ATP) and water treatment pectin extraction (WTP) (Table 11.5). ATP was carried out under different temperatures (68-85°C) at a fixed time of one hour, while WTP was carried out in a higher range of temperatures (75–95°C) and different exposure time (0.5–3 h). ATP depicted the best result (208 g/kg dry tomato) compared to WTP (87% g/kg dry tomato). One important factor affecting pectin extraction is the solvent penetration and its capacity to dissolve the pectin, followed by the diffusion of the pectin into the solid. The interaction with other components from TW matrix, i.e. proteins, other fibers, lipids etc. is also important [118]. Current research not only focuses on extraction yield, but also on polymer quality, anhydrogalacturonic acid (AGA) concentration (relative purity measure), neutral sugar, DE (that influences pectin behavior and functionality), viscosity, and molecular weight (behavior of polysaccharide dispersion). Color test also provides valuable information as it correlates color with pigments, i.e. lycopene, polyphenol, etc. The test has three levels "L" (lightness, where 0 is black and 100 is white), "a" (−a means greenness, +a means redness), and "b" (−b is blueness, +b is yellowness). The color tendency depends on extraction methodology. In this sense, UAE is associated to high L values [55]. Alancay et al. [113], studied pectin properties after ATP and WTP extraction methods and compared them with commercial pectin from citrus (CPC). Results showed that acid treatment produced a polysaccharide with similar properties to CPC. ATP extracted pectin purity degree was 70.9%, above that using WTP (57.5%). However, pectin from WTP had a higher degree methylation (87.8%) followed by that from ATP (76.3%) and that from CPC (60.2%). Also, intrinsic viscosity (η), which provides information of hydrodynamic volume, was calculated. η depends on molecular weight and polymer conformation. Obtained values showed a value of 4.2 dl/g for pectin extracted using ATP, similar to that from high-methoxyl pectin (HMC) (3.9 dl/g) and that from WTP (1.8 dl/g). The result of this analysis, together with the molecular weight values, showed that selected extraction treatment definitely affects the size of polymer molecule.

Table 11.5 shows the yield (10%–32%, w/w) achieved from different approaches and techniques, provided by best yield does not necessarily

Table 11.5 Pectin extraction technology from tomato waste.

| Pectin Tomato residue | Extraction technique | Design of experiments | Chemical conditions | Optimal conditions | Yield | Reference |
|---|---|---|---|---|---|---|
| Husk tomato | Extraction with acid | Design of experiments: Blanching time (min): 10/15; Thermal treatment (°C): 100; Thermal treatment (min): 15, 20, 25 | Acid hydrolysis: HCl 0.1N Precipitation step: ethanol 95% (v/v) (1:3 extract: alcohol) | Best treatment: Blanching/thermal treatment: 10/20 min | 19.8 ± 0.35% (w/w) | [139] |
| Tomato processing waste: skin, seed, pulp | Extraction with acid | Design of experiments: A) Acid treatment: HCl 6N, pH 2.0, 1h; Temperature (°C): 68, 75, 85 B) Water treatment: Temperature (°C): 75, 85, 95 Time (h): 0.5, 1, 3 | Acid hydrolysis: HCl 6N Precipitation EtOH: 96% | Acid treatment: Temperature (°C): 85 Time (min): 60 Water treatment: Temperature (°C): 95 Time (min): 180 | Acid extraction yield: 280.9± 12.8% (g/kg dry TW) Water extraction yield: 87 ± 0.35 % (w/w) | [15] |
| Dried tomato by-product | Extraction reflux and extraction ultrasound | Design of experiments: A) Reflux treatment: 24 h at 60, 80°C B) Ultrasound: 37 kHz at 60, 80°C Time (min): 15, 30, 45, 60, 90 | Solvent oxalate/ oxalic acid Precipitation EtOH: 96% | Two step extraction using optimal condition from previous experiment: Acid treatment: reflux 60°C, 48 h; Ultrasound treatment: 80°C. 15 min | Reflux treatment extraction yield: 31.2% (w/w) Ultrasound treatment extraction yield: 35% (w/w) | [43] |

| | | | | | | |
|---|---|---|---|---|---|---|
| / industrial tomato waste | Eco-friendly technologies | | Precipitation EtOH: 96% | Ultrasound assisted extraction: 600W, 8.61 min, 60°C<br>Microwave extraction: 900W, 3.34 min, 88.7°C<br>Ohmic heating assisted extraction: 60V, 5.0 min, 81°C<br>Ultrasound assisted microwave extraction: Ultrasound: 450 W, 8.0 min/ microwave: 540 W, 85.1°C<br>Ultrasound assisted Ohmic extraction: Ultrasound 450 W, 10 min/Ohmic 60V, 5 min, 68.9°C | Ultrasound assisted extraction yield: 15.21% (w/w)<br>Microwave extraction: 25.42%<br>Ohmic heating-assisted extraction: 10.65% (w/w)<br>Ultrasound-assisted microwave extraction: 18.00% (w/w)<br>Ultrasound-assisted ohmic extraction: 14.60 % (w/w) | [153] |

means best quality. Morales-Contreras et al. [113] used husk tomato discarded from local market for pectin extraction. The experimental methodology focused on treatment time. Initially, the extraction process consisted in a steam-blanched pretreatment, at two different times, followed by hydrolysis with HCl 0.1N, at 100°C, which was carried out at 15, 20 and 25 min. According to the findings, an increasing tendency with hydrolysis increasing time was indicated, while the optimum results (19.8%, w/w) was achieved after 10 min pre-treatment and 25 minutes of treatment. AUA content (relative to purity) increased with hydrolysis duration, due to temperature influence, which reduced the concentration of other fibers that is, hemicellulose and galactans [119]. Results also showed that, pretreatment and hydrolysis reduced DE; in this sense, a 10/15 ratio (pre-treatment time/hydrolysis time) provided a DE of 91.6%. Results showed that pectin is a good candidate as gelling compound in food industry, cosmetics, etc. Another study from Morales-Contreras et al. [120] focused on rheological behavior of pectin extracted from husk TW. Rheological properties provide knowledge about gel type and gelification mechanism, with the objective to introduce TW in industry [111]. The extraction methodology used HCl as acid agent for hydrolysis, while citric acid content, blanching time and hydrolysis time values were the same as in their previous work. Sample flow behavior revealed a shear thinning or pseudoplastic behavior, that was enhanced when citric acid was used. Similar results to other pectins such as orange peels, were also obtained from Hosseini et al. [121]. The frequency sweep test focuses on the mechanical response of gel. Material properties, i.e. storage modulus (elastic, G') and loss modulus (viscous, G") were calculated for different frequencies (rad/s). According to Clark et al [122], if G" > G', then polymer depicts a gel behavior, as it was the case of husk TW pectin, that formed a network structure.

Emulsion is another pectin property, which may reduce interfacial tension, stabilizing the oil and water phase. The emulsifying agent has a hydrophobic and hydrophilic domain. Pectin polysaccharides have lower emulsion power compared to proteins, due to less hydrophobic polarity. Nevertheless, pectin has a higher functional capacity due to its structural versatility [123] and the presence hydrophobic groups (i.e. protein groups) and neutral sugar side chain [124]. Zhang et al. [125] reported the use of black tomato pomace for its emulsifying properties. In this study, a sequential extraction methodology, including water, cyclohexane-trans-1,2-diamine tetraacetic acid and $Na_2CO_3$ as solvent was used. Different fractions of polysaccharides were obtained and their analysis shown a high-methylated degree, with values from 60% to 70% (w/w). The percentage of linear homogalacturonan was between 60% and 80% (w/w), while the branched rhamnogalacturonan was found between 10% and 30% (w/w). The emulsifying properties showed that polymers obtained from different fractions were able to reduce the surface tension of fluid with 50% oil phase.

Carbohydrate polymers may be considered natural inhibitors of corrosion. Pectin may reduce metallic degradation due to its chemical composition, which is rich in carbonyl, carboxylic and ionic carboxyl groups, neutral sugars, epidermal wax and phenols. There is a growing interest about the use of pectin in the canning industry, which use template for packing and food preservation. Template is composed by low carbon mild steel and tin, which serves as a coating. Corrosion of this material is due to interactions between reactive chemical species contained in food and the metallic substrate. Other factors that may also be involved are pH, oxidizing agents (nitrites, oxygen, etc.) and storage conditions. Grassino et al. [43] check corrosion activity of pectin from TW and compared it to that of apple pomace. Electrochemical measurements were carried out by three glass electrode cells in a test solution composed of 2% NaCl, 1% acetic acid and 0.5% citric acid. Pectin concentration was found to be 0.2–4.0 g/l. Optimum result for corrosion current density ($\mu A/cm^2$) was provided by 4 g/l of pectin. Pectin from apple pomace gave a density value of 12.6 $\mu A/cm^2$, while that of TM was 8.15 $\mu A/cm^2$. In conclusion of this work, authors found an inhibitor effect for tin corrosion from TW pectin, when low concentration of extract was used. Halambek el al. [126] studied tin corrosion inhibitor effect of pectin from tomato peel waste, using an experimental design, in a corrosion aggressive scenery (2%, w/v NaCl and 1%, v/v acetic acid, pH = 3). To study the corrosion effects of the test solution, gravimetric measurements were carried out. To study the inhibition effect, 20 g/l of pectin from TW and apple pomace was added. Samples were immersed in different solutions for 3, 6, 9, 12, and 24 h, at 25°C. Three parameters were defined to evaluate the efficiency of the corrosion inhibition. Firstly, corrosion ratio ($W_{corr}$), which measures the difference in weight as a function of surface and exposure time. Secondly, surface coverage ($\theta_w$), which provides the difference of $W_{corr}$ in presence and absence of inhibitor. Finally, corrosion index, which is defined as $\eta_w = 100\ \theta_w$. Current density was also measured by electrochemical analysis. Gravimetric results shown a corrosion index ($\eta_w$) of 74% for pectin from tomato waste (PTW), while apple pomace pectin shown a slightly higher result (76%). Electrochemical results exposed similar behavior to the previous study, viewing a reduction of corrosion current density with pectin content. Both studies revealed that PTW may be used as cathodic corrosion inhibitor.

To improve the environmental efficiency of pectin extraction, new extraction tools and solvents have been tested. Grassino et al. [55], carried out a study using two different reflux treatment conditions (60 and 80°C) and UAE, using the same temperatures but different times (15, 30, 45, 60, and 90 min). They used the same solvent (oxalate/oxalic acid) in all experiments and the process was developed in two stages. In the first stage, the extraction optimization was carried out. Conventional technique provided the best result (21.1% yield) at 60°C during 48 h, while UAE needed 80°C during 15 min to provide 15.2% yield. Results provided by UAE are similar to those achieved with other pectins from lemon (16.71%), orange (15.92%) and grape (15.70%) [110]. In the second stage, yields improved and UAE provided the best result (34%–36%), thus indicating that UAE affects degradation cell wax matrix and allows a better interaction with solvent. The DE, in both cases, was above 80%; authors categorized the pectin as set gelling pectin. Authors concluded that UAE reduces time by 192 times. Sengar et al. [127], conducted experiments with four different extraction techniques (UAE, MWE, Ohmic heating-assisted extraction or $O_H$AE and combination of them, UAE-MWE and UAE- $O_H$AE). The highest extraction yield (25.42%) was reached using MWE, with 900 W, 3.34 min and 88.7°C as optimal conditions. Results provided by UAE were similar to those found by Grassino et al., with a yield of 15.21%. Combination of UAE-MWE provided an intermediate result (18%) between both techniques. $O_H$AE and UAE-$O_H$AE yields were below 15%. Results also shown that yield was a function of energy. Moreover, increase of temperature was related to pectin degradation, which can reduce pectin in its monomers [128]. A drawback derived from the use of UAE is the formation of bubbles around the tip of the probe, which may reduce the energy available in the reaction. According to Xu et al. [129] this effect is considered as saturation effect. These authors found that all techniques provided similar pectin purity (between 678 to 913 g/kg pectin) and concluded that UAE-MWE provided the best result considering extraction yield and pectin quality (yield: 18%, gaA 913.3 g/kg, DE: 73.33, and L as color of pectin).

### 11.3.4 Oleoresins

Oleoresins result from solvent evaporation and consist on a semi-solid and complex matrix of chemical molecules. Fatty acids are their main component, followed by lycopene and other compounds, i.e. antioxidants, flavor compounds and vitamins. Its composition makes it suitable for food, nutraceutical and pharmaceutical industry. Oleoresin plays a thermal stability roll due to increase of lycopene stability. Hackett et al. [130] studied the thermal stability of lycopene in oleoresin from several tomato varieties (Roma, Roma peel waste, high lycopene and tangerine). Experiments were carried out at 25, 50, 75, and 100°C; storage conditions were also studied. Results shown that lycopene activation energy degradation in oleoresin (11.5–15 kcal/mol) was lower than that in pure lycopene (14.5 kcal/mol) and higher than that in oil-water emulsion (5.0 kcal/mol) as a consequence of the activation energy [131, 132]. Lycopene degradation increased with temperatures in the range of 25–50°C; oxidation degradation, but no isomerization (trans to cis) was noticed. However, isomerization appeared in the range of 75–100°C. Finally, it was concluded that the addition of antioxidants, like α-Thocopherol or butylated hydroxytoluene (BHT), increased the stability of oleoresin, with a half-life increase of 6 to 12 days.

Vallecilla-Yepez et al. [133] focused on techniques to increase cis-isomerization. As the trans-to-cis lycopene isomerization mechanism is not completely understood, authors designed experiments using SC-$CO_2$ technology, under heating and lighting conditions, using pure lycopene, different tomato matrix, etc. Results suggested that energy isomer interconversion is very similar, while activation energy is high [134–136]. Moreover, results showed that cis-lycopene is achieved under high pressure and temperature; in tomato peels, there is an increase from 0.5 g/kg oleoresin (30 MPa/40°C) to 1.4 g/kg oleoresin (50 MPa/80°C). The study also compared isomer distribution between oil and insoluble fraction, using two extraction techniques, namely SC-$CO_2$ extraction and hexane extraction. Results shown that cis-lycopene was present in higher concentration in the oil fraction at 80°C, than in insoluble fraction using hexane extraction. However, trans-lycopene depicted an inverse behavior.

In food industry, additives like butylated hydroxyanisole or BHT are usually used, despite their potential correlation with cancer disease. However, oleoresin can be used as natural antioxidant and may help reducing the negative effect of synthetic additives on human health. In fact, oleoresin has been checked in oil and meat (i.e. beef patties) industry. Lycopene-rich oleoresin from TW may reduce food chemical reactions like lipid oxidation. Food quality is altered in its texture, shelf-life and nutritional value; this alteration can be produced in any step of the production chain [18,137]. Kehili et al. [138] checked the efficiency of lycopene extraction, using four different oleoresin concentrations (250, 500, 1000 and 2000 μg/g) in refined olive and sunflower oils. The experiments included the self-life evaluation in function of the peroxide value, acidity, conjugated dienes and trienes. Samples were subjected to thermal stress at 50°C. Oleoresin extraction was carried out by maceration in hexane. Results shown a reduction of the oxidative effect in refined olive oil and protective effect in primary oxidation in both oils. An oleoresin concentration between 0.55 to 2 g/kg showed an improvement in the half live of the product, compared to that of tomato powder (50 kg/g) [139].

Due to complex chemical matrix from TW oleoresin, the lipid (polyhydroxylated fatty acid) and non-hydrolysable fractions have a significant potential to be employed as bio-based polymers, with hydrophovic, insoluble, infusible, and thermally stable properties. When tomato is processed to tomato pomace, the skin residue is mainly composed by plan cutin (around 80%), which a fatty acid composition rich in palmitic acid (C16) and stearic acid (C18) [140]. Esterification may lead to biopolymer production. Benítez et al. [13] used melt-polycondensation for biopolymer film production, from fatty acid TW. Results showed a hydrophobic, infusible, amorphous surface, and thermostable properties, being 300°C its decomposition temperature.

## 11.4 Anaerobic digestion of tomato waste

AD is a sequential process based on microorganism consortia. AD involves two to three stages, which are related to biomass degradation. First step is carried out by hydrolytic bacterial that degrade biomass structural fiber and protein to simple sugars and amino acids (simple blocks) that are used by acidogenic bacteria (second stage). In the second stage, simple blocks are transformed into $CO_2$, $H_2$, ammonia, and organic acids. In the last stage, organic acids are used to produce methane by methanogenesis [141, 142]. Due to connections between AD stages, a correct balance between microorganisms and nutrients is necessary. Currently, multiple methodologies based on co-digestion are being developed. The use of farm residues at ratio 3:2 (proportion calculated on volatile solids, VS) showed a high methane production 0.35–0.43 $m^3$ $CH_4$/kg VS [143,144]. Wardet et al. [145], Enzinar et al. [146] and Gonzalez-Gonzalez el al. [147] reported biomethane production from 199 to 384 Nml/gVS. Table 11.6 summarizes some of the

**Table 11.6** Anaerobic digestion applications using tomato waste.

| Technique | Tomato residue | Experimental conditions | Conclusions | Reference |
|---|---|---|---|---|
| Anaerobic digestion | Tomato processing waste (TPW) | NaOH dosage: 1% and 5%. Pretreatment contact time: 4 and 24 h; fermentation time: 30 days. | Yield:320 NmL/g VS. | [149] |
| Mesophilic anaerobic digestion (widely employed treatment for sewage sludge) | Sewage sludge (hazardous waste) | Mixture sewage sludge and TW evaluated was 95:5 (wet weight). | Methane yield: 159 L/kg VS Biodegradability: 95% (in VS). | [152] |
| Anaerobic digestion | Iberian pig slaughterhouse and tomato industry wastes | Codigestion operation conditions: ratio slaughterhouse wastes/tomato industry wastes: 60%:40% (optimal ratio) | Methane Yield: 6–19 $Nm^3$/ $m^3$ substrate methane production. | [155] |
| Anaerobic digestion/ bio-electrochemical system | Tomato plant residues (TPR) | Three methanogenic reactors: CSTR, CFT, BER + CFT | CSTR gas production: 133.3 mL/L day (68.3% methane), CFT gas production: 180.9 mL/L day (72.4% methane), BBER + CFT gas production 197.2 mL/L day (77.3% methane) | [153] |
| Solid state anaerobic codigestion (SS-AD) | Tomato residues with dairy manure and corn stover | Tomato waste fraction: 0%, 20%, 40%, 60%, 80% and 100%, based on VS weight codigested: with dairy manure and corn stover at 15% total solids. | Methane yield: 379.1 L/kg VS feed. | [151] |

*(continued)*

**Table 11.6** (Cont'd)

| Technique | Tomato residue | Experimental conditions | Conclusions | Reference |
|---|---|---|---|---|
| Anaerobic digestion (single-stage digester) | Tomato processing residues | To produce biogas, the digester was controlled under mesophilic conditions (35°C) with continuous mixing. | Yield biogas: 0.14 $m^3$/kg VS. (60.50% methane) | [151] |
| Ultrasound (US) pre-treatment + anaerobic digestion | Tomato pomace | The effect of ultrasound pre-treatment: Time: 5, 15, and 30 min Amplitude: 152 μm (0.9 W/ml) AD time: 4 and 22 days | Methane yield increase did not compensate US required electricity (between 3.3 and 19.5 MJ/kg VS). | [156] |
| Anaerobic digestion | Tomato canning waste | Substrate-to-inoculum ratios: 0.25, 0.5, 1, 2, 3, and 4 Temperature: 37 (mesophilic conditions) Time: 45 days. | Methane yield: 260.98 ± 7.24 mL/g VS. | [157] |

CFT, carbon fiber textile reactor; CFT+VER, carbon fiber textile reactor + bio-electrochemical reactor (BER with electrochemical regulation (BER + CFT); CSTR, continuously stirred tank reactor; VS, volatile solid.

applications derived from AD processes, as lycopene extraction processes; conventional and alternative processes can be distinguished depending on the techniques used for the disruption of the cell wall.

TW presents high humidity and nitrogen content, as illustrated in Tables 11.1 and 11.2. Another important aspect is pH; TW pH is close to 4.5 (acidic), so it may inhibit aerobic processes [148]. Calabrò et al. [149] studied neutralization pretreatments with alkaline reactives. Results shown no significant differences in methane yield, using an alkaline treatment. Methane production was close to 320 NmL/g VS in all experiments. The effect of NaOH concentrations above 5 % (w/w) showed a reduction of methane production due to ammonium generation (that is an inhibitor of AD) and salts. However, substrate biodegradability significantly improved.

Co-digestion is considered as an important approach to achieve nutrient balance. Trujillo et al. [150] studied the co-digestion of tomato-plant wastes (611g/kg of total solid (TS) and 538 g/kg of VS) and rabbit wastes (224 g/kg TS and 181g/kg VS). Four series of experiments with different ratios (1:1, 1:2, 1:3, and 2:3) were carried out and the results indicated an increase in methane yield, when waste from the plant is added in proportions above 40% (1:1 ratio). Similar result was published by Li et al. [151], adding 20%–40% TW to a diary mature and corn stover mixture; methane yield increased. Belhadj et al. [152] studied the impact of TW in codigestion with sewage sludge, which is rich in heavy metals that can produce AD inhibition [146]. Results showed an improved methane yield (159L/kg VS). In this case, the ratio used was 95:5 (sewage sludge: TW) and organic loading rate (OLR) interval increased to 400 L/m$^3$.

Another current approach is based on the electrochemical effect in methane fermentation. Hirano et al. [153] used a bio-electrochemical reactor (BER), which produce electrochemical regulation of microorganism redox homeostasis, affecting NADH/NAD$^+$ ratio and, thus, producing acetate production inhibition. Experiments were carried out using three different methanogenic reactors: a continuous stirred tank reactor (CSTR), carbon fiber textile (CFT) reactor, and CFT with electrochemical regulation (BER + CFT). The highest methane yield was achieved when a potential of −0.8 V (using electrodes Ag/AgCl) was applied.

Finally, an interesting experiment were carried out by Manfredini el al. [154], who evaluated the use of tomato pomace and buffalo slurry (12:1, 8:1 and 6:1 ratios) for biofuel production and fertilizer, through AD. The highest biofuel result (5.47 Nml $CH_4$/day and 19.04 NmL $H_2$/day) was achieved using the lowest ratio (6:1). A positive bio stimulating effect over soil microbiota was observed. The most successful metabolic experiment was achieved using a digestate dilution of low ratio (6:1), with 10 ppm of dissolved organic carbon (DOC). Experiments shown the potential application a TW-based biorefinery that allow to couple two processes, energy production and agronomic applications. Digestate is rich in nutrients, such as P, N, and trace elements, like Cu, Ni and Zn, which can improve soil organic matter. Some of the most relevant studies are summarized in Table 11.6.

## 11.5 Conclusions

As illustrated, TW could be integrated into a potential biorefinery containing multiple operations and products. The composition of these residues allows a first extraction operation of micronutrients, i.e. lycopene, carotenoids, antioxidants, essential oils and fibers, like pectin. These extractions will provide high added-value products, that can be integrated in agri-food and pharmaceutical industries. In this way, synthetic substances will be replaced by natural components. Throughout the chapter, techniques developed to reduce environmental impact and improving extraction processes have been demonstrated. An integrated approach is also a biorefinery process which is focused on the production of biofuels, like methane and hydrogen, *via* AD. The combination of both processes can be positive, since the extraction treatment, whose objective is cell disruption, may improve fermentation viability. In addition, extraction of volatile fatty acids may help reducing their negative effects on AD. One of the major problems associated with the generation of TW from industry, is its accumulation in landfills. The operations mentioned above may provide a residual fraction rich in N and P that may be used as fertilizer. In sum, feasibility of valorization of TWs through the concept of biorefinery has been demonstrated, thus contributing to circular economy and European Green Deal contributing the reduction of climate change.

## Acknowledgments

Authors are grateful to the Spanish Ministry of Science and Research, for grant no. PID2019-105936RB-C21.

## References

[1] M. Añibarro-Ortega, et al., Valorisation of table tomato crop by-products: phenolic profiles and in vitro antioxidant and antimicrobial activities, Food Bioprod. Process. 124 (2020) 307–319.
[2] E. Luengo, I. Álvarez, J. Raso, Improving carotenoid extraction from tomato waste by pulsed electric fields, Front. Nutr. 1 (2014) 12.

[3] FAO, Global food losses and food waste. 2011.
[4] C.C. Jenny Gustavsson, U. Sonesson, A. Emanuelsson, Global losses and food waste-Extent, causes and prevention, Food and Agriculture organization of the United Nations, Rome, 2011.
[5] K.-R. Bräutigam, J. Jörissen, C. Priefer, The extent of food waste generation across EU-27: different calculation methods and the reliability of their results, Waste Manag. Res. 32 (8) (2014) 683–694.
[6] A. Agapios, et al., Waste aroma profile in the framework of food waste management through household composting, J. Clean. Prod. 257 (2020) 120340.
[7] P. Loizia, N. Neofytou, A.A. Zorpas, The concept of circular economy strategy in food waste management for the optimization of energy production through anaerobic digestion, Environ. Sci. Pollut. Res. 26 (15) (2019) 14766–14773.
[8] A.A. Zorpas, I. Voukkali, P. Loizia, The impact of tourist sector in the waste management plans, Desalin. Water Treat. 56 (5) (2015) 1141–1149.
[9] A.A. Zorpas, et al., Waste prevention campaign regarding the waste framework directive, Fresenius Environ. Bull. 23 (11a) (2013) 2876–2883.
[10] N. Scialabba, Food Wastage Footprint and Climate Change, Food and Agriculture Organization of the United Nations, Rome, 2016.
[11] B. Buchner, C. Fischler, E. Gustafson, J. Reilly, G. Riccardi, C. Ricordi, U. Veronesi, Food waste: causes, impacts and proposal, Barilla center. Food nutrition (2012) 53–61.
[12] M. Del Valle, M. Cámara, M.-E. Torija, Chemical characterization of tomato pomace, J. Sci. Food Agric. 86 (8) (2006) 1232–1236.
[13] J.J. Benítez, et al., Valorization of tomato processing by-products: fatty acid extraction and production of bio-based materials, Materials 11 (11) (2018) 2211.
[14] M.Y. Ali, et al., Nutritional composition and bioactive compounds in tomatoes and their impact on human health and disease: a review, Foods 10 (1) (2020) 45.
[15] M.M. Alancay, et al., Extraction and physicochemical characterization of pectin from tomato processing waste, J. Food Meas. Charact. 11 (4) (2017) 2119–2130.
[16] Z. Lu, et al., Sustainable valorisation of tomato pomace: a comprehensive review, Trends Food Sci. Technol. 86 (2019) 172–187.
[17] S. Kumcuoglu, T. Yilmaz, S. Tavman, Ultrasound assisted extraction of lycopene from tomato processing wastes, J. Food Sci. Technol. 51 (12) (2014) 4102–4107.
[18] E. Bravi, et al., Antioxidant effects of supercritical fluid garlic extracts in sunflower oil, J. Sci. Food Agric. 97 (1) (2017) 102–107.
[19] G. Pataro, et al., Recovery of lycopene from industrially derived tomato processing by-products by pulsed electric fields-assisted extraction, Innov. Food Sci. Emerg. Technol. 63 (2020) 102369.
[20] A. Amiri-Rigi, S. Abbasi, Microemulsion-based lycopene extraction: effect of surfactants, co-surfactants and pretreatments, Food Chem. 197 (2016) 1002–1007.
[21] S. Dahiya, et al., Food waste biorefinery: sustainable strategy for circular bioeconomy, Bioresour. Technol. 248 (2018) 2–12.
[22] C. Du, et al., A wheat biorefining strategy based on solid-state fermentation for fermentative production of succinic acid, Bioresour. Technol. 99 (17) (2008) 8310–8315.
[23] S.S. Hassan, G.A. Williams, A.K. Jaiswal, Lignocellulosic biorefineries in Europe: current state and prospects, Trends Biotechnol. 37 (3) (2019) 231–234.
[24] D.A. Teigiserova, L. Hamelin, M. Thomsen, Review of high-value food waste and food residues biorefineries with focus on unavoidable wastes from processing, Resour. Conserv. Recycl. 149 (2019) 413–426.
[25] United Nations, Transforming Our World: The 2023 Agenda for Sustainable Development, New York (USA), 2015.
[26] European., C., A New Circular Economy Action Plan for a Cleaner and More Competitive Europe, European Parliament Commission, Brussels, 2020.

[27] P. Loizia, et al., Measuring the level of environmental performance in insular areas, through key performed indicators, in the framework of waste strategy development, Sci. Total Environ. 753 (2021) 141974.
[28] A.A. Zorpas, Strategy development in the framework of waste management, Sci. Total Environ. 716 (2020) 137088.
[29] D. Symeonides, P. Loizia, A.A. Zorpas, Tire waste management system in Cyprus in the framework of circular economy strategy, Environ. Sci. Pollut. Res. 26 ((35) (2019) 35445–35460.
[30] N.A. Antoniou, A.A. Zorpas, Quality protocol and procedure development to define end-of-waste criteria for tire pyrolysis oil in the framework of circular economy strategy, Waste Manag. 95 (2019) 161–170.
[31] I. Voukkali, et al., Urban strategies evaluation for waste management in coastal areas in the framework of area metabolism, Waste Manag. Res. 39 (3) (2021) 448-465.
[32] A.A. Zorpas, I. Voukkali, P. Loizia, A prevention strategy plan concerning the waste framework directive in Cyprus, Fresenius Environ. Bull. 26 (2) (2017) 1310–1317.
[33] A.A. Zorpas, et al., Household compost monitoring and evaluation in the framework of waste prevention strategy, J. Clean. Prod. 172 (2018) 3567–3577.
[34] N. Meza, J. Manzano, Z. Méndex, Characteristics of Tree Tomato (Cyphomandra betaceae [Cav] Sendtn) Fruits Based in Aril Coloration, at the Venezuelan Andes Zone, University de Oriente Press, Cumaná, Venezuela, 2009.
[35] D. Bilalis, et al., Energy inputs, output and productivity in organic and conventional maize and tomato production, under mediterranean conditions, Not. Bot. Horti Agrobot. Cluj Napoca 41 (1) (2013) 190.
[36] C. Rock, et al., Conventional and alternative methods for tomato peeling, Food Eng. Rev. 4 (1) (2012) 1–15.
[37] S. Azabou, et al., Towards sustainable management of tomato pomace through the recovery of valuable compounds and sequential production of low-cost biosorbent, Environ. Sci. Pollut. Res. 27 (31) (2020) 39402–39412.
[38] S. Cheikh-Rouhou, et al., Nigella sativa L.: chemical composition and physicochemical characteristics of lipid fraction, Food Chem. 101 (2) (2007) 673–681.
[39] H. Najafi, et al., The potential application of tomato seeds as low-cost industrial waste in the adsorption of organic dye molecules from colored effluents, Desalin. Water Treat. 57 (32) (2016) 15026–15036.
[40] M. Carmona-Cabello, et al., Valorization of food waste from restaurants by transesterification of the lipid fraction, Fuel 215 (2018) 492–498.
[41] S. Pinzi, et al., The ideal vegetable oil-based biodiesel composition: a review of social, economical and technical implications, Energy Fuels 23 (5) (2009) 2325–2341.
[42] J.H. Van Gerpen, B.B. He, Biodiesel and renewable diesel production methods, in: K. Waldron (Ed.), Advances in Biorefineries, Woodhead Publishing, Cambridge, UK, 2014, pp. 441–475.
[43] A.N. Grassino, et al., Utilization of tomato peel waste from canning factory as a potential source for pectin production and application as tin corrosion inhibitor, Food Hydrocoll. 52 (2016) 265–274.
[44] A. Sarkar, P. Kaul, Evaluation of tomato processing by-products: a comparative study in a pilot scale setup, J. Food Process Eng. 37 (3) (2014) 299–307.
[45] E. Santangelo, et al., Evaluation of tomato introgression lines diversified for peel color as a source of functional biocompounds and biomass for energy recovery, Biomass Bioenergy 141 (2020) 105735.
[46] V. Coman, et al., Bioactive potential of fruit and vegetable wastes, In: F. Toldrá (Ed.), Advances in Food and Nutrition Research, Academic Press, Cambridge (Ma), USA, 2020, pp. 157–225.
[47] M. Kehili, et al., Optimization of lycopene extraction from tomato peels industrial by-product using maceration in refined olive oil, Food Bioprod. Process. 117 (2019) 321–328.

[48] K. Szabo, A.-F. Cătoi, C. Vodnar, Bioactive compounds extracted from tomato processing by-products as a source of valuable nutrients, Plant Food Hum. Nutr. 73 (4) (2018) 268–277.
[49] D. Górecka, et al., Lycopene in tomatoes and tomato products, Open Chem. 18 (1) (2020) 752–756.
[50] M.M. Poojary, P. Passamonti, Optimization of extraction of high purity all-trans-lycopene from tomato pulp waste, Food Chem 188 (2015) 84–91.
[51] D.M. Phinney, et al., Effect of solvent addition sequence on lycopene extraction efficiency from membrane neutralized caustic peeled tomato waste, Food Chem. 215 (2017) 354–361.
[52] I.F. Strati, V. Oreopoulou, Recovery of carotenoids from tomato processing by-products – a review, Food Res. Int. 65 (2014) 311–321.
[53] S. Azabou, et al., Phytochemical profile and antioxidant properties of tomato by-products as affected by extraction solvents and potential application in refined olive oils, Food Biosci. 36 (2020) 100664.
[54] A.C. Fărcaş, et al., Tomato waste as a source of biologically active compounds, Bull. Univ. Agric. Sci. Vet. Med. Cluj Napoca. Food Sci. Technol. 76 (1) (2019) 85.
[55] A.N. Grassino, et al., Ultrasound assisted extraction and characterization of pectin from tomato waste, Food Chem 198 (2016) 93–100.
[56] M. Kehili, et al., Supercritical $CO_2$ extraction and antioxidant activity of lycopene and β-carotene-enriched oleoresin from tomato (Lycopersicum esculentum L.) peels by-product of a Tunisian industry, Food Bioprod. Process. 102 (2017) 340–349.
[57] L. Corrêa-Filho, et al., Microencapsulation of tomato (Solanum lycopersicum L.) pomace ethanolic extract by spray drying: optimization of process conditions, Appl. Sci. 9 (3) (2019) 612.
[58] D. Dussault, K.D. Vu, M. Lacroix, In vitro evaluation of antimicrobial activities of various commercial essential oils, oleoresin and pure compounds against food pathogens and application in ham, Meat Sci. 96 (1) (2014) 514–520.
[59] G. Perretti, et al., Production of a lycopene-enriched fraction from tomato pomace using supercritical carbon dioxide, Journal Supercrit. Fluid 82 (2013) 177–182.
[60] S. Rahimpour, S. Taghian Dinani, Lycopene extraction from tomato processing waste using ultrasound and cell-wall degrading enzymes, J. Food Meas. Charact. 12 (4) (2018) 2394–2403.
[61] P. Di Mascio, S. Kaiser, H. Sies, Lycopene as the most efficient biological carotenoid singlet oxygen quencher, Arch. Biochem. Biophys. 274 (2) (1989) 532–538.
[62] C. Gärtner, W. Stahl, H. Sies, Lycopene is more bioavailable from tomato paste than from fresh tomatoes, Am. J. Clin. Nutr. 66 (1) (1997) 116–122.
[63] J. Shi, et al., Lycopene degradation and isomerization in tomato dehydration, Food Res. Int. 32 (1) (1999) 15–21.
[64] S.K. Clinton, et al., cis-trans lycopene isomers, carotenoids, and retinol in the human prostate, Cancer Epidemiol Biomarkers Prev. 5 (10) (1996) 823–833.
[65] I.F. Strati, V. Oreopoulou, Process optimisation for recovery of carotenoids from tomato waste, Food Chem. 129 (3) (2011) 747–752.
[66] D. Kaur, et al., Effect of extraction conditions on lycopene extractions from tomato processing waste skin using response surface methodology, Food Chem. 108 (2) (2008) 711–718.
[67] Y.P.A. Silva, et al., Optimization of lycopene extraction from tomato processing waste using an eco-friendly ethyl lactate–ethyl acetate solvent: a green valorization approach, Waste Biomass Valor. 10 (10) (2019) 2851–2861.
[68] I.F. Strati, V. Oreopoulou, Effect of extraction parameters on the carotenoid recovery from tomato waste, Int. J. Food Sci. Technol. 46 (1) (2011) 23–29.
[69] J. Shi, et al., Effect of heating and exposure to light on the stability of lycopene in tomato purée, Food Control 19 (5) (2008) 514–520.

[70] N. Manzo, et al., Degradation kinetic (D100) of lycopene during the thermal treatment of concentrated tomato paste, Nat. Prod. Res. 33 (13) (2019) 1835–1841.
[71] E. Luengo, et al., Improving the extraction of carotenoids from tomato waste by application of ultrasound under pressure, Sep. Purif. Technol. 136 (2014) 130–136.
[72] T. Y1lmaz, et al., Ultrasound-assisted extraction of lycopene and beta-carotene from tomato-processing wastes, Yilmaz, T.; Kumcuoglu, S.; Tavman, S. Ital. J. Food Sci. 29 (1) (2016) 186-194.
[73] A. Amiri-Rigi, S. Abbasi, Extraction of lycopene using a lecithin-based olive oil microemulsion, Food Chem. 272 (2019) 568–573.
[74] S. Rahimi, M. Mikani, Lycopene green ultrasound-assisted extraction using edible oil accompany with response surface methodology (RSM) optimization performance: application in tomato processing wastes, Microchem. J. 146 (2019) 1033–1042.
[75] K.S. Suslick, Sonochemistry, Science 247 (4949) (1990) 1439–1445.
[76] V. Sivakumar, et al., Ultrasound assisted enhancement in natural dye extraction from beetroot for industrial applications and natural dyeing of leather, Ultrason. Sonochem. 16 (6) (2009) 782–789.
[77] Y.-Q. Ma, et al., Simultaneous extraction of phenolic compounds of citrus peel extracts: effect of ultrasound, Ultrason. Sonochem. 16 (1) (2009) 57–62.
[78] A.O. Adekunte, et al., Effect of sonication on colour, ascorbic acid and yeast inactivation in tomato juice, Food Chem. 122 (3) (2010) 500–507.
[79] J. Chen, et al., Comparison of lycopene stability in water- and oil-based food model systems under thermal- and light-irradiation treatments, LWT - Food Sci. Technol. 42 (3) (2009) 740–747.
[80] R.N. Lichtenthaler, Gerd Brunner, Gas extraction - an introduction to fundamentals of supercritical fluids and the application to separation processes, In: H. Baumgärtel, E.U Franck, W. Grünbein (Eds.), Topics in Physical Chemistry, Steinkopff, Darmstadt/Springer, New York, 1994.
[81] B.P. Nobre, et al., Supercritical CO2 extraction of trans-lycopene from Portuguese tomato industrial waste, Food Chem. 116 (3) (2009) 680–685.
[82] G. Vasapollo, et al., Innovative supercritical CO2 extraction of lycopene from tomato in the presence of vegetable oil as co-solvent, J. Supercrit. Fluid. 29 (1–2) (2004) 87–96.
[83] E. Sabio, et al., Lycopene and β-Carotene extraction from tomato processing waste using supercritical CO2, Ind. Eng. Chem. Res. 42 (25) (2003) 6641–6646.
[84] S. Machmudah, et al., Lycopene extraction from tomato peel by-product containing tomato seed using supercritical carbon dioxide, J. Food Eng. 108 (2) (2012) 290–296.
[85] A.F. Silva, M.M.R. De Melo, C.M. Silva, Supercritical solvent selection (CO2 versus ethane) and optimization of operating conditions of the extraction of lycopene from tomato residues: Innovative analysis of extraction curves by a response surface methodology and cost of manufacturing hybrid ap, J. Supercrit. Fluid. 95 (2014) 618–627.
[86] B. Scaglia, et al., Development of a tomato pomace biorefinery based on a CO2-supercritical extraction process for the production of a high value lycopene product, bioenergy and digestate, J. Clean. Prod. 243 (2020) 118650.
[87] P. Labuschagne, Impact of wall material physicochemical characteristics on the stability of encapsulated phytochemicals: a review, Food Res. Int. 107 (2018) 227–247.
[88] R.C. Ranveer, et al., Microencapsulation and storage stability of lycopene extracted from tomato processing waste, Braz. Arch. Biol. Technol. 58 (6) (2015) 953–960.
[89] J.-B. Eun, et al., A review of encapsulation of carotenoids using spray drying and freeze drying, Crit. Rev. Food Sci. Nutr. 60 (21) (2020) 3547–3572.
[90] S.Y. Quek, N.K. Chok, P. Swedlund, The physicochemical properties of spray-dried watermelon powders, Chem. Eng. Process. 46 (5) (2007) 386–392.
[91] S.M.T. Gharibzahedi, S.M. Jafari, Nanocapsule formation by cyclodextrins, in: S.M. Jafari (Ed.), Nanoencapsulation Technologies for the Food and Nutraceutical Industries, Academic Press, Cambridge (Ma), USA, 2017, pp. 187–261.

[92] G. Astray, et al., A review on the use of cyclodextrins in foods, Food Hydrocoll. 23 (7) (2009) 1631–1640.
[93] M. Durante, et al., Tomato oil encapsulation by α-, β-, and γ-cyclodextrins: a comparative study on the formation of supramolecular structures, antioxidant activity, and carotenoid stability, Foods 9 (11) (2020) 1553.
[94] U. Zimmermann, Electrical breakdown, electropermeabilization and electrofusion, Rev. Physiol. Biochem. Pharmacol. 105 (1986) 176–256.
[95] F. Bot, et al., The effect of pulsed electric fields on carotenoids bioaccessibility: the role of tomato matrix, Food Chem. 240 (2018) 415–421.
[96] K.R.N. Moelants, et al., A review on the relationships between processing, food structure, and rheological properties of plant-tissue-based food suspensions, Compr. Rev. Food Sci. Food Saf. 13 (2014) 241–260.
[97] G. Pataro, et al., Improved extractability of carotenoids from tomato peels as side benefits of PEF treatment of tomato fruit for more energy-efficient steam-assisted peeling, J. Food Eng. 233 (2018) 65–73.
[98] G. Pataro, D. Carullo, G. Ferrari, Effect of PEF pre-treatment and extraction temperature on the recovery of carotenoids from tomato wastes, Chem. Eng. Trans 75 (2019) 139-144.
[99] H. Akiyama, Streamer discharges in liquids and their applications, IEEE Trans. Dielectr. Electr. Insul. 7 (5) (2000) 646–653.
[100] Gachovska, T.K., et al. Enhancement of Lycopene Extraction from Tomatoes Using Pulsed Electric Field. Abstracts IEEE International Conference on Plasma Science (ICOPS), 2013, pp. 1-1, doi: 10.1109/PLASMA.2013.6633411.
[101] V. Andreou, et al., Application of pulsed electric fields to improve product yield and waste valorization in industrial tomato processing, J. Food Eng. 270 (2020) 109778.
[102] K. Bagchi, S. Puri, Free radicals and antioxidants in health and disease: a review, EMHJ - Eastern Mediterranean Health Journal, 4 (2) (1998) 350-360.
[103] G. Ćetković, et al., Valorisation of phenolic composition, antioxidant and cell growth activities of tomato waste, Food Chem. 133 (3) (2012) 938–945.
[104] X.P. Perea-Domínguez, et al., Phenolic composition of tomato varieties and an industrial tomato by-product: free, conjugated and bound phenolics and antioxidant activity, J. Food Sci. Technol. 55 (9) (2018) 3453–3461.
[105] D.W. Wong, Feruloyl esterase: a key enzyme in biomass degradation, Appl. Biochem. Biotechnol. 133 (2) (2006) 87–112.
[106] A. Khoddami, M. Wilkes, T. Roberts, Techniques for analysis of plant phenolic compounds, Molecules 18 (2) (2013) 2328–2375.
[107] Y. Bao, L. Reddivari, J.-Y. Huang, Development of cold plasma pretreatment for improving phenolics extractability from tomato pomace, Innov. Food Sci. Emerg. Technol. 65 (2020) 102445.
[108] Y. Gao, et al., Characterization of free, conjugated, and bound phenolic acids in seven commonly consumed vegetables, Molecules 22 (11) (2017).
[109] J. Pinela, et al., Valorisation of tomato wastes for development of nutrient-rich antioxidant ingredients: A sustainable approach towards the needs of the today's society, Innov. Food Sci. Emerg. Technol. 41 (2017) 160–171.
[110] V.O. Aina, et al., Extraction and characterization of pectin from peels of lemon (Citrus limon), grape fruit (Citrus paradisi) and sweet orange (Citrus sinensis), Brit. J. Pharmacol. Toxicol. 3 (6) (2012) 259–262.
[111] S.Y. Chan, et al., Pectin as a rheology modifier: origin, structure, commercial production and rheology, Carbohydr. Polym. 161 (2017) 118–139.
[112] R.Y. Lochhead, et al., The use of polymers in cosmetic products, in: K. Sakamoto et al (Ed.), Cosmetic Science and Technology, Elsevier, Amsterdam, 2017, pp. 171–221.
[113] B.E. Morales-Contreras, et al., Husk tomato (Physalis ixocarpa Brot.) waste as a promising source of pectin: extraction and physicochemical characterization, J. Food Sci. 82 (7) (2017) 1594–1601.

[114] X. Guo, et al., Extraction of pectin from navel orange peel assisted by ultra-high pressure, microwave or traditional heating: A comparison, Carbohydr. Polym. 88 (2) (2012) 441–448.
[115] C.D. May, Industrial pectins: sources, production and applications, Carbohydr. Polym. 12 (1) (1990) 79–99.
[116] H. Garna, et al., Effect of extraction conditions on the yield and purity of apple pomace pectin precipitated but not washed by alcohol, J. Food Sci. 72 (1) (2007) C001–C009.
[117] O. Yuliarti, et al., Characterization of gold kiwifruit pectin from fruit of different maturities and extraction methods, Food Chem. 166 (2015) 479–485.
[118] P. Methacanon, J. Krongsin, C. Gamonpilas, Pomelo (Citrus maxima) pectin: effects of extraction parameters and its properties, Food Hydrocoll. 35 (2014) 383–391.
[119] B.M. Yapo, Pectin quantity, composition and physicochemical behaviour as influenced by the purification process, Food Res. Int. 42 (8) (2009) 1197–1202.
[120] B.E. Morales-Contreras, et al., Pectin from husk tomato (Physalis ixocarpa Brot.): rheological behavior at different extraction conditions, Carbohydr. Polym. 179 (2018) 282–289.
[121] S.S. Hosseini, F. Khodaiyan, M.S. Yarmand, Aqueous extraction of pectin from sour orange peel and its preliminary physicochemical properties, Int. J. Biol. Macromol. 82 (2016) 920–926.
[122] A.H. Clark, S.B. Ross-Murphy, Structural and Mechanical Properties of Biopolymer Gels, Springer Berlin Heidelberg, Berlin, Heidelberg, 1987.
[123] T. Funami, et al., Structural modifications of sugar beet pectin and the relationship of structure to functionality, Food Hydrocoll. 25 (2) (2011) 221–229.
[124] J.-S. Yang, T.-H. Mu, M.-M. Ma, Extraction, structure, and emulsifying properties of pectin from potato pulp, Food Chem. 244 (2018) 197–205.
[125] W. Zhang, et al., Emulsifying properties of pectic polysaccharides obtained by sequential extraction from black tomato pomace, Food Hydrocoll. 100 (2020) 105454.
[126] J. Halambek, I. Cindrić, A. Ninčević Grassino, Evaluation of pectin isolated from tomato peel waste as natural tin corrosion inhibitor in sodium chloride/acetic acid solution, Carbohydr. Polym. 234 (2020) 115940.
[127] A.S. Sengar, et al., Comparison of different ultrasound assisted extraction techniques for pectin from tomato processing waste, Ultrason. Sonochem. 61 (2020) 104812.
[128] L. Zhang, et al., Ultrasound effects on the degradation kinetics, structure and rheological properties of apple pectin, Ultrason. Sonochem. 20 (1) (2013) 222–231.
[129] Y. Xu, et al., Effects of ultrasound and/or heating on the extraction of pectin from grapefruit peel, J. Food Eng. 126 (2014) 72–81.
[130] M.M. Hackett, et al., Thermal stability and isomerization of lycopene in tomato oleoresins from different varieties, J Food Sci. 69 (7) (2006) 536–541.
[131] M.T. Lee, B.H. Chen, Stability of lycopene during heating and illumination in a model system, Food Chem. 78 (4) (2002) 425–432.
[132] K. Ax, et al., Stability of lycopene in oil-in-water emulsions, Eng. Life Sci. 3 (4) (2003) 199–201.
[133] L. Vallecilla-Yepez, O.N. Ciftci, Increasing cis-lycopene content of the oleoresin from tomato processing byproducts using supercritical carbon dioxide, LWT 95 (2018) 354–360.
[134] J.F. Rinaldi de Alvarenga, et al., Home cooking and ingredient synergism improve lycopene isomer production in Sofrito, Food Res. Int. 99 (Pt2) (2017) 851–861.
[135] G.A. Chasse, et al., An ab initio computational study on selected lycopene isomers, J. Mol. Struc-THEOCHEM 571 (1) (2001) 27–37.
[136] Honda, M., et al., The E/Z isomer ratio of lycopene in foods and effect of heating with edible oils and fats on isomerization of (all-E)-lycopene. Eur. J Lipid Sci. Technol. 2017 119 8.

[137] Ben-Ali, M., et al., Stabilization of sunflower oil during accelerated storage: use of basil extract as a potential alternative to synthetic antioxidants. 2014. 17(7):1547–1559.
[138] M. Kehili, et al., Oxidative stability of refined olive and sunflower oils supplemented with lycopene-rich oleoresin from tomato peels industrial by-product, during accelerated shelf-life storage, Food Chem. 246 (2018) 295–304.
[139] A. Sánchez-Escalante, et al., Stabilisation of colour and odour of beef patties by using lycopene-rich tomato and peppers as a source of antioxidants, J. Sci. Food Agric. 83 (3) (2003) 187–194.
[140] E. Domínguez, J.A., Heredia-Guerrero, A. Heredia, The biophysical design of plant cuticles: an overview, New Phytol. 189 (4) (2011) 938–949.
[141] E. Puértolas, I. Martínez de Marañón, Olive oil pilot-production assisted by pulsed electric field: Impact on extraction yield, chemical parameters and sensory properties, Food Chem 167 (2015) 497–502.
[142] E. Tsouko, et al., Valorization of by-products from palm oil mills for the production of generic fermentation media for microbial oil synthesis, Appl. Biochem. Biotechnol. 181 (4) (2017) 1241–1256.
[143] A. Lehtomäki, S. Huttunen, J.A. Rintala, Laboratory investigations on co-digestion of energy crops and crop residues with cow manure for methane production: Effect of crop to manure ratio, Resour. Conserv. Recycl. 51 (3) (2007) 591–609.
[144] V.N. Gunaseelan, Biochemical methane potential of fruits and vegetable solid waste feedstocks, Biomass Bioenergy 26 (4) (2004) 389–399.
[145] A.J. Ward, et al., Optimisation of the anaerobic digestion of agricultural resources, Bioresour. Technol. 99 (17) (2008) 7928–7940.
[146] J.M. Encinar, J.F. González, G. Martínez, Energetic use of the tomato plant waste, Fuel Process Technol. 89 (11) (2008) 1193–1200.
[147] A. González-González, F. Cuadros, Continuous biomethanization of agrifood industry waste: a case study in Spain, Process Biochem. 48 (5) (2013) 920–925.
[148] H. Bouallagui, et al., Improvement of fruit and vegetable waste anaerobic digestion performance and stability with co-substrates addition, J. Environ. Manag. 90 (5) (2009) 1844–1849.
[149] P.S. Calabrò, et al., Anaerobic digestion of tomato processing waste: effect of alkaline pretreatment, J. Environ. Manag. 163 (2015) 49–52.
[150] D. Trujillo, J.F. Pérez, F.J. Cebreros, Energy recovery from wastes. Anaerobic digestion of tomato plant mixed with rabbit wastes, Bioresour. Technol. 45 (2) (1993) 81–83.
[151] Y. Li, et al., Reactor performance and energy analysis of solid state anaerobic co-digestion of dairy manure with corn stover and tomato residues, Waste Manag. 73 (2018) 130–139.
[152] S. Belhadj, et al., Evaluation of the anaerobic co-digestion of sewage sludge and tomato waste at mesophilic temperature, Appl. Biochem. Biotechnol. 172 (8) (2014) 3862–3874.
[153] S.-I. Hirano, N. Matsumoto, Analysis of a bio-electrochemical reactor containing carbon fiber textiles for the anaerobic digestion of tomato plant residues, Bioresour. Technol. 249 (2018) 809–817.
[154] A. Manfredini, et al., Assessing the biological value of soluble organic fractions from tomato pomace digestates, J. Soil Sci. Plant Nutrition 21 (1) (2020) 301-314.
[155] A. González-González, et al., Energy-environmental benefits and economic feasibility of anaerobic codigestion of Iberian pig slaughterhouse and tomato industry wastes in Extremadura (Spain), Bioresour. Technol. 136 (2013) 109-116.
[156] F. Girotto, et al., Bio-methane production from tomato pomace: preliminary evaluation of process intensification through ultrasound pre-treatment, J. Mater. Cycles Waste Manag. 23 (1) (2021) 416-422.
[157] K. Kaidi, K., et al., Valorization study of the organic waste resulting from the tomato canning by methanisation, University Politechnica of Bucharest Scientific Bulletin Series B-Chemistry and Materials Science 82 (2) (2020) 95-108.

# Index

Page numbers followed by "*f*" and "*t*" indicate, figures and tables respectively.

Printed in the United States
by Baker & Taylor Publisher Services